Lecture Notes in Physics

Springer
Berlin
Heidelberg
New York
Barcelona
Hong Kong
London
Milan
Paris
Singapore
Tokyo

Editorial Policy

The series *Lecture Notes in Physics* (LNP), founded in 1969, reports new developments in physics research and teaching -- quickly, informally but with a high quality. Manuscripts to be considered for publication are topical volumes consisting of a limited number of contributions, carefully edited and closely related to each other. Each contribution should contain at least partly original and previously unpublished material, be written in a clear, pedagogical style and aimed at a broader readership, especially graduate students and nonspecialist researchers wishing to familiarize themselves with the topic concerned. For this reason, traditional proceedings cannot be considered for this series though volumes to appear in this series are often based on material presented at conferences, workshops and schools (in exceptional cases the original papers and/or those not included in the printed book may be added on an accompanying CD ROM, together with the abstracts of posters and other material suitable for publication, e.g. large tables, colour pictures, program codes, etc.).

Acceptance

A project can only be accepted tentatively for publication, by both the editorial board and the publisher, following thorough examination of the material submitted. The book proposal sent to the publisher should consist at least of a preliminary table of contents outlining the structure of the book together with abstracts of all contributions to be included.
Final acceptance is issued by the series editor in charge, in consultation with the publisher, only after receiving the complete manuscript. Final acceptance, possibly requiring minor corrections, usually follows the tentative acceptance unless the final manuscript differs significantly from expectations (project outline). In particular, the series editors are entitled to reject individual contributions if they do not meet the high quality standards of this series. The final manuscript must be camera-ready, and should include both an informative introduction and a sufficiently detailed subject index.

Contractual Aspects

Publication in LNP is free of charge. There is no formal contract, no royalties are paid, and no bulk orders are required, although special discounts are offered in this case. The volume editors receive jointly 30 free copies for their personal use and are entitled, as are the contributing authors, to purchase Springer books at a reduced rate. The publisher secures the copyright for each volume. As a rule, no reprints of individual contributions can be supplied.

Manuscript Submission

The manuscript in its final and approved version must be submitted in camera-ready form. The corresponding electronic source files are also required for the production process, in particular the online version. Technical assistance in compiling the final manuscript can be provided by the publisher's production editor(s), especially with regard to the publisher's own Latex macro package which has been specially designed for this series.

Online Version/ LNP Homepage

LNP homepage (list of available titles, aims and scope, editorial contacts etc.):
http://www.springer.de/phys/books/lnpp/

LNP online (abstracts, full-texts, subscriptions etc.):
http://link.springer.de/series/lnpp/

T. Dauxois S. Ruffo
E. Arimondo M. Wilkens (Eds.)

Dynamics and Thermodynamics of Systems with Long-Range Interactions

Springer

Editors

Thierry Dauxois
ENS Lyon
Laboratoire de Physique
46, allée d'Italie
69007 Lyon, France

Ennio Arimondo
Universitá degli Studi di Pisa
Dipartimento di Fisica
Via F. Buonarroti, 2
56127 Pisa, Italy

Stefano Ruffo
Universitá di Firenze
Dipartimento di Energetica "S. Stecco"
Via S. Marta, 3
50139 Firenze, Italy

Martin Wilkens
Universität Potsdam
Institut für Physik
Am Neuen Palais 10
14469 Potsdam, Germany

Cover Picture: (see contribution "The Hamiltonian Mean Field Model: From Dynamics to Statistical Mechanics and Back" by T. Dauxois et al. in this volume)

Cataloging-in-Publication Data applied for

A catalog record for this book is available from the Library of Congress.

Bibliographic information published by Die Deutsche Bibliothek

Die Deutsche Bibliothek lists this publication in the Deutsche Nationalbibliografie; detailed bibliographic data is available in the Internet at http://dnb.ddb.de

ISSN 0075-8450
ISBN 3-540-44315-0 Springer-Verlag Berlin Heidelberg New York

Springer-Verlag Berlin Heidelberg New York
a member of BertelsmannSpringer Science+Business Media GmbH

http://www.springer.de

Typesetting: Camera-ready by the authors/editor
Camera-data conversion by Steingraeber Satztechnik GmbH Heidelberg
Cover design: *design & production*, Heidelberg

Printed on acid-free paper
SPIN: 10894435 54/3141/du - 5 4 3 2 1 0

Preface

This volume contains the lecture notes of the conference held in the Center of Physics of Les Houches, France, on February 18-22 2002. The lectures connected through the heading "Dynamics and Thermodynamics of Systems with Long-Range Interactions", explored a variety of topics involving long-range interactions.

This field is entering the 21st century in an interesting and exciting way. Properties of systems with long-range interactions are to a large extent only poorly understood although they concern a wide range of problems in physics (astrophysics, nuclear physics, plasmas physics, Bose–Einstein condensates, atomic clusters, hydrodynamics,...). However, recently, the disclosure of new methodologies to approach the study of these systems has revealed its importance also in a trans-disciplinary perspective. The main challenge of this workshop was represented by the construction of a thermodynamic treatment of systems with long-range forces and by the understanding of analogies and differences between the numerous domains of applications.

Some promising results in this direction have been recently obtained in the attempt of combining tools developed in the framework of usual statistical mechanics with concepts and methods of dynamical systems. Particularly arduous, but very exciting, is the understanding of phase transitions that shows several new features like negative specific heat, non-extensivity, inequivalence of ensembles, etc. Moreover, such systems are a perfect laboratory for studying aspects related to non-equilibrium phenomena and their description in terms of dynamical concepts (self-consistent chaos, slow relaxation, formation and role of structures,...).

This fundamental and methodological study should help us to detect the depth and the origin of the analogies found in the different domains mentioned above or, on the contrary, emphasize their specificities. In particular, we would like to put a special emphasis on Bose–Einstein Condensation (BEC), which could be an important field of applications, since experiments and theoretical ideas have reached an impressive quality in the last decade. In that domain, inequivalences between ensembles have been reported and should be clarified. Moreover, long range interactions in BEC have opened very exciting new perspectives to consider BEC as a model for other systems.

The main lecturers of the Les Houches school were Ofer Biham (University of Jerusalem), Pierre-Henri Chavanis (University of Toulouse), Philippe

Chomaz (CEA, Caen), Eddie Cohen (University of Rockefeller), Jean Dalibard (ENS Paris), Diego Del Castillo-Negrete (Oak Ridge National Laboratory), Yves Elskens (University of Marseille), Dieter Gross (HMI, Berlin), Martin Holthaus (University of Oldenburg), Gershon Kurizki (Weizmann Institute), Ulf Leonhardt (University of Saint Andrews), David Mukamel (Weizmann Institute), Thanu Padmanabhan (Pune, India), Andrea Rapisarda (University of Catania), Constantino Tsallis (University of Rio de Janeiro). Additional materials such as slides of conferences and written contributions by people who have attended the conference are available on the electronic site http://www.ens-lyon.fr/~tdauxois/procs02/.

Here we would like to express our sincere gratitude to the lecturers for all their efforts in preparing, presenting and finally writing up their lectures. Our thanks are also due to the main sponsors of this conference: the "Département Sciences Physiques et Mathématiques" of CNRS, the program "Quantum degenerate dilute systems Bose–Einstein condensation and beyond" of the European Science Foundation and the Conseil Régional Rhônes-Alpes. Support for this school comes also from the Groupement de Recherche CNRS "Phénomènes hors d'équilibre" and from the "Ecole Doctorale de Physique et d'Astrophysique de Lyon". Finally, we would like to thank the director of the Center of Physics of Les Houches, Martial Ducloy, for his continuous interest and assistance. We also acknowledge the staff of the secretary office of the school for the help in all the aspects of the organization.

Sainte Foy lès Lyon,
July, 2002

Thierry Dauxois, Stefano Ruffo
Ennio Arimondo, Martin Wilkens

Contents

List of Contributors

Ennio Arimondo
INFM
Dipartimento di Fisica
Università di Pisa
Via Buonarroti 2
I-56127 Pisa, Italy
arimondo@mail.df.unipi.it

Alexandre Artemiev
Department of Physics
Texas A&M University
College Station,
Texas, 77843-4242, USA
artemiev@chemphys.weizmann.ac.il

Fulvio Baldovin
Centro Brasileiro de Pesquisas Físicas
Rua Xavier Sigaud 150
22290-180 Rio de Janeiro, RJ, Brazil
baldovin@cbpf.br

Julien Barré
Laboratoire de Physique
ENS Lyon
46, allée d'Italie
F-69007 Lyon, France
Julien.Barre@ens-lyon.fr

Ofer Biham
Ofer Malcai
Racah Institute of Physics
The Hebrew University
Jerusalem 91904, Israel
biham@phys.huji.ac.il

Dave Boers
Fachbereich Physik
Carl von Ossietzky Universität
D-26111 Oldenburg, Germany
boers@marvin.physik.uni-oldenburg.de

Pierre-Henri Chavanis
Laboratoire de Physique Quantique
Université Paul Sabatier
118, route de Narbonne
31062 Toulouse, France
chavanis@irsamc2.ups-tlse.fr

Philippe Chomaz
GANIL (DSM-CEA/IN2P3-CNRS)
B.P.5027
F-14021 Caen cedex, France
chomaz@ganil.fr

Eddie G. D. Cohen
Center for Studies in Physics
and Biology
The Rockefeller University
1230 York Avenue
New York, NY 10021, USA
egdc@mail.rockefeller.edu

Jean Dalibard
Laboratoire Kastler Brossel
24 rue Lhomond
75005 Paris, France
Jean.Dalibard@lkb.ens.fr

Thierry Dauxois
Laboratoire de Physique

ENS Lyon
46, allée d'Italie
F-69007 Lyon, France
Thierry.Dauxois@ens-lyon.fr

Diego del-Castillo-Negrete
Oak Ridge National Laboratory
Oak Ridge
TN, 37831-8071, USA
delcastillod@ornl.gov

Yves Elskens
UMR 6633 CNRS
Université de Provence
Equipe turbulence plasma
case 321, campus Saint-Jérôme
13397 Marseille cedex 20, France
elskens@newsup.univ-mrs.fr

Stefano Giovanazzi
Department of Chemical Physics
Weizmann Institute of Science
76100 Rehovot, Israel
stevbolz@yahoo.it

Dieter H. E. Gross
Hahn-Meitner Institute
Freie Universität Berlin
Fachbereich Physik
Glienickerstr. 100
14109 Berlin, Germany
gross@hmi.de

Francesca Gulminelli
GANIL (DSM-CEA/IN2P3-CNRS)
B.P.5027
F-14021 Caen cedex, France
gulminelli@caelav.in2p3.fr

Martin Holthaus
Fachbereich Physik
Carl von Ossietzky Universität
D-26111 Oldenburg, Germany
martin.holthaus@uni-oldenburg.de

Iaroslav Ispolatov
Departamento de Física
Universidad de Santiago de Chile
Casilla 302, Correo 2
Santiago, Chile
iispolat@usach.cl

Gershon Kurizki
Department of Chemical Physics
Weizmann Institute of Science
76100 Rehovot, Israel
Gershon.Kurizki@weizmann.ac.il

Vito Latora
Dipartimento di Fisica e Astronomia
Universitá di Catania
Corso Italia 57
95129 Catania, Italy
latora@ct.infn.it

Ofer Malcai
Racah Institute of Physics
The Hebrew University
Jerusalem 91904, Israel
malcai@phys.huji.ac.il

Oliver Morsch
INFM
Dipartimento di Fisica
Università di Pisa
Via Buonarroti 2
I-56127 Pisa, Italy
morsch@df.unipi.it

David Mukamel
Department of Physics of Complex Systems
The Weizmann Institute of Science
Rehovot 76100, Israel
fnmukaml@wisemail.weizmann.ac.il

Duncan O'Dell
Sussex Centre for Optical
and Atomic Physics, C.P.E.S.
University of Sussex
Brighton BN1 9QH, UK
D.H.J.Odell@sussex.ac.uk

Thanu Padmanabhan
IUCAA
Pune University Campus
Ganeshkhind
Pune - 7, India
nabhan@iucaa.ernet.in

Andrea Rapisarda
Dipartimento di Fisica e Astronomia
Universitá di Catania
Corso Italia 57
95129 Catania, Italy
andrea.rapisarda@ct.infn.it

Stefano Ruffo
Dipartimento di Energetica
"S. Stecco"
Università di Firenze
via S. Marta, 3
I-50139 Firenze, Italy
ruffo@avanzi.de.unifi.it

Constantino Tsallis
Centro Brasileiro de Pesquisas Físicas
Rua Xavier Sigaud 150
22290 180 Rio de Janeiro, RJ, Brazil
tsallis@cbpf.br

Martin Wilkens
Universität Potsdam
Institut fuer Physik
Am Neuen Palais 10
14469 Potsdam, Germany
martinw@elektra.quantum.physik.uni-potsdam.de

Dynamics and Thermodynamics of Systems with Long-Range Interactions: An Introduction

Thierry Dauxois[1], Stefano Ruffo[2], Ennio Arimondo[3], and Martin Wilkens[4]

[1] Laboratoire de Physique, UMR CNRS 5672, ENS Lyon, 46, allée d'Italie, F-69007 Lyon, France
[2] Dipartimento di Energetica "S. Stecco", Università di Firenze, via S. Marta, 3, I-50139 Firenze, Italy
[3] Dipartimento di Fisica, Università degli Studi di Pisa, Via F. Buonarroti 2, I-56127, Pisa, Italy
[4] Universität Potsdam, Institut fuer Physik, Am Neuen Palais 10, 14469 Potsdam, Germany

Abstract. We review theoretical results obtained recently in the framework of statistical mechanics to study systems with long-range forces. This fundamental and methodological study leads us to consider the different domains of applications in a trans-disciplinary perspective (astrophysics, nuclear physics, plasmas physics, metallic clusters, hydrodynamics,...) with a special emphasis on Bose–Einstein condensates.

The main issues discussed in this context are: non additivity, ensemble inequivalence, thermodynamic anomalies at phase transitions (e.g. negative specific heat), "convex intruders" in the entropy, non-extensive statistics and new entropies, coherent structures and self-consistent chaos, laser induced long-range interactions in cold atomic systems.

1 Introduction

Properties of systems with long-range interactions are to a large extent only poorly understood although they concern a wide range of problems in physics. Recently, the disclosure of new methodologies to approach the study of these systems has revealed its importance also in a trans-disciplinary perspective (astrophysics, nuclear physics, plasmas physics, Bose–Einstein condensates, atomic clusters, hydrodynamics,...). The main challenge is represented by the construction of a thermodynamic treatment of systems with long-range forces and by the understanding of analogies and differences among the numerous domains of applications.

Some promising results in this direction have been recently obtained in the attempt of combining tools developed in the framework of standard statistical mechanics with concepts and methods of dynamical systems. Particularly arduous, but very exciting, is the understanding of phase transitions for such systems which must be treated separately in the different statistical ensembles and reveal anomalies like negative specific heat and temperature jumps in the microcanonical ensemble. Important are also those aspects of non-equilibrium phenomena that involve the formation of chaotic coherent structures of extraordinary stability.

This fundamental and methodological study should help us to detect the depth and the origin of the analogies found in the different domains mentioned above or on the contrary emphasize their specificities. In particular, we would like to put a special emphasis on Bose–Einstein Condensation (BEC), which could be the main field of applications, since experiments and theoretical ideas have reached an impressive quality in the last decade. In this domain, many inequivalences between ensembles have been reported and should be clarified. Moreover, long-range interactions in BEC have opened very exciting new perspectives to consider BEC as a model for other systems.

2 Why Systems with Long-Range Interactions Are Important?

2.1 The Problem of Additivity

The methods to describe a given system of N particles interacting via a gravitational potential in $1/r$ are dramatically dependent on the value N. If Newton showed the exact solution for $N = 2$, and one can expect to get a numerical solution in the range $N = 3 - 10^3$, the results are clearly out of reach for a larger number of particles. In addition, it is clear that the detailed knowledge of the evolution of the different trajectories is completely useless, since it is well known that these systems are chaotic as soon as N is greater than two. Therefore, one needs to get a *statistical* analysis, in order to get insights into the thermodynamical properties [1] of the system under study.

However, such statistical study leads immediately to unexpected behaviors for physicists used to neutral gases, plasmas or atomic lattices. The underlying reason is directly related to the long range of the interaction, and more precisely to the non additivity of the system.

To avoid misunderstandings, let us first clarify the definition of *extensivity* with respect to *additivity*. A thermodynamic variable, like the energy or the entropy, would be extensive, if it is proportional to the number of elements, once the intensive variables are kept constant. To be more precise, let us consider the mean-field Ising Hamiltonian,

$$H = -\frac{J}{N}\left(\sum_{i=1}^{N} S_i\right)^2 \quad , \tag{1}$$

where the spins $S_i = \pm 1$, $i = 1, \ldots, N$, are all coupled. Without the $1/N$ prefactor such a Hamiltonian would have an ill defined thermodynamic limit. This is correctly restored by applying the Kac prescription [2], within which the potential is rescaled by an appropriate volume dependent factor, here proportional to N: such a Hamiltonian is therefore extensive. Let us note in passing that this regularization is not always accepted. In cases with a kinetic energy term, such a regularization corresponds to a renormalization of the time scale. On the contrary, this Hamiltonian is not additive. Indeed, let us divide a system,

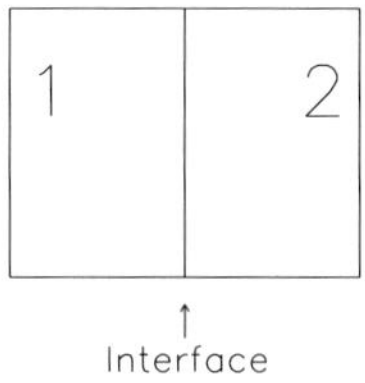

Fig. 1. Schematic picture of a system separated in two equal parts.

schematically pictured in Fig. 1, in two equal parts. In addition, one considers the particular case with all spins in the left part are equal to 1, whereas all spin in the right part are equal to -1. It is clear that the energy of the two different parts, will be $E_1 = E_2 = -\frac{J}{N}\left(\frac{N}{2}\right)^2 = -\frac{JN}{4}$. However, if one computes the total energy of the system, one gets $E = -\frac{J}{N}\left(\frac{N}{2} - \frac{N}{2}\right)^2 = 0$. It is therefore clear that such a system is not additive, since one cannot consider that $E_1 + E_2 = E$, even approximately. The energy of the interface, usually neglected, is clearly of the order of the energies of the two different parts: the system is not additive. The underlying reason is that Hamiltonian (1) is long (strictly speaking infinite) range, since every spin interact with all others: moreover, as the interaction is not dependent on the distance between spins, this is a mean-field model. This example is further elaborated in [4].

This non additivity has strong consequences in the construction of the canonical ensemble. Once the microcanonical ensemble has been defined, the usual construction of this ensemble is usually taught as follows. The probability that system 1 has an energy within $[E_1, E_1 + dE_1]$, given that the system 2 has an energy E_2, is proportional to $\Omega_1(E_1)\,\Omega_2(E_2)\,dE_1$, where the number of states of a system with a given energy E, $\Omega(E)$, is related to the entropy via the classical Boltzmann formula $S(E) = \ln E$ (we omit the k_B factor for the sake of simplicity). Using the additivity of the energy, and considering the case where system 1 is much smaller than system 2, one can expand the term $S_2(E - E_1)$, as shown in the following different steps

$$
\begin{aligned}
\Omega_1(E_1)\,\Omega_2(E_2)\,dE_1 &= \Omega_1(E_1)\,\Omega_2(E - E_1)\,dE_1 & (2)\\
&= \Omega_1(E_1)\,e^{S_2(E - E_1)}\,dE_1 & (3)\\
&= \Omega_1(E_1)\,e^{\left(S_2(E) - E_1 \frac{\partial S_2}{\partial E_2}_{|E} + \ldots\right)}\,dE_1 & (4)\\
&\propto \Omega_1(E_1)\,e^{-\beta E_1}\,dE_1 \quad , & (5)
\end{aligned}
$$

where $\beta = \frac{\partial S_2}{\partial E_2}_{|E}$. One ends up with the usual canonical distribution. It is clear that additivity is crucial to justify (2), which means that non additive systems will have a very peculiar behavior if there are in contact with a thermal reservoir. This is one of the topic discussed in this paper, and in numerous contributions in this book.

2.2 Definition of Long-Range Systems

To define now systems with long-range interactions, let us consider the potential energy for a given particle, situated in the center of a sphere of radius R, where mass or charge is homogenously distributed. We will omit at this stage the interaction of matter situated in a small neighborhood of radius $\varepsilon \ll R$ (see Fig. 2). The reason for excluding this neighborhood will be explained in the following subsection.

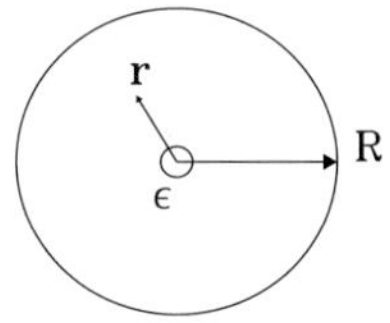

Fig. 2. Schematic picture of a particle interacting with all particles located in a homogeneous sphere of radius R, except the closest ones located in the sphere of radius ε.

If one considers that particles interact via a potential energy proportional to $1/r^\alpha$, where α is the key-parameter defining the range of interaction, we obtain in the three dimensional space

$$U = \int_\varepsilon^R 4\pi r^2 dr\, \rho\, \frac{1}{r^\alpha} = 4\pi\rho \int_\varepsilon^R r^{2-\alpha} dr \propto \left[r^{3-\alpha} \right]_\varepsilon^R \tag{6}$$

where ρ is the particle density. When increasing the radius R, the contribution due to the surface of the sphere, $R^{3-\alpha}$, could be neglected when $\alpha > 3$, but diverges if $\alpha < 3$. In the latter case, surface effects are important and therefore additivity is not fulfilled.

If one generalizes this definition to long-range systems in d dimensions, one easily shows that energy will not be additive if the potential energy behaves at long distance as

$$V(r) \sim \frac{1}{r^\alpha} \quad \text{with} \quad \frac{\alpha}{d} < 1 \quad . \tag{7}$$

Mean-field models, like Hamiltonian (1) correspond to the value $\alpha = 0$, since the interaction does not depend on the distance. They are therefore not additive as shown in Sect. 2.1. J. Barré et al consider [4] such a mean field model: the Blume-Emery-Griffith (BEG) model with infinite range interactions. The gravitational problem, which is at the origin of this study, and corresponds to $\alpha = 1$ in three dimensions, clearly belongs to this category, but presents also additional difficulties.

2.3 Difficulties with the Gravitational Problem

This problem is particularly tedious because, in addition to the non additivity due to long-range character, such a system needs a careful regularization at short distances to avoid collapse. To be more specific, let us consider the configurational partition function of a system of N particles

$$Z_U = \int_V d^{3N}\vec{r_i}\ e^{-\beta U(\vec{r_i})} \quad , \tag{8}$$

where one notes $U(\vec{r_i})$ the gravitational potential energy, β the inverse of the temperature and V the volume of the system. From the shape of the potential energy depicted in Fig. 3, one clearly see that Z_U will diverge if all particles collapse towards the same point. This difficulty arises because the potential energy is *not bounded from below* as for a Lennard-Jones or a Morse potential. This effect is of course physically forbidden by the Pauli principle. However, to avoid the use of Quantum Mechanics, the usual trick is to introduce an ad-hoc cut-off. The potential is therefore "regularized" by introducing the value $-C$, shown in Fig. 3. Thus, the inequality $U(\vec{r_i}) \geq -C$ allows easily to find a finite upper bound for the configurational partition function $Z_U \leq V\ e^{\beta C}$, where V is the volume of the system.

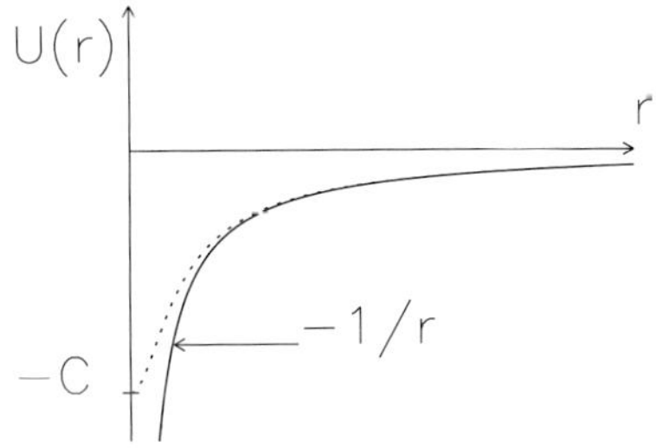

Fig. 3. The gravitational potential energy as a function of the distance r is represented by the solid curve, whereas the dotted one shows the regularized potential energy to avoid gravitational collapse.

However, there is a third difficulty in the case of gravitational interaction: the system is open, i.e. without boundary, strictly speaking. In the microcanonical ensemble, the number of states

$$\Omega(E) = \int \prod_i dp_i \int \prod_i dq_i\ \delta\left(E - H\left(q_i, p_i\right)\right) \tag{9}$$

will diverge if the system is not confined. This divergence is actually not restricted to the gravitational interaction but would also occur if one considers a perfect gas in an infinite volume. However, one considers of course always a gas in a finite domain, i.e. in a finite volume. This is not any more possible for the gravitational interaction where the system is clearly infinite.

Despite these additional difficulties, the astrophysics community has obtained an impressive quantity of results in this domain. Thanu Padmanabhan [5] describes several remarkable features, both for isolated gravitating systems as well as for systems undergoing nonlinear clustering in an expanding background cosmology. The emphasis is on general results and he brings out the inter-relationships of this subject with topics in fluid mechanics, condensed matter and renormalization group theory.

Similarly, Pierre-Henri Chavanis [6] presents how the structure and the organization of stellar systems (globular clusters, elliptical galaxies,...) in astrophysics can be understood in terms of a statistical mechanics for a system of particles in gravitational interaction. Finally, Eddie Cohen and Iaroslav Ispolatov [7] consider the related gravitational-like collapse of particles with an attractive $1/r^{\alpha}$ potential. Using mean field continuous integral equation, they determine the saddle-point density profile that extremizes the entropy functional. For all $0 < \alpha < d = 3$, a critical energy is determined below which the entropy of the system exhibits a discontinuous jump.

2.4 Applications to Large Systems

A growing scientific community has recently begun to tackle the problem of long-range interactions with different viewpoints. One of the fascinating aspects of this problem is that, in addition to gravitating systems, it concerns a large variety of systems that we would like to discuss briefly in the following section.

Plasmas. Rarefied plasmas share many properties with collisionless stellar systems, and in particular the property that the mean field of the system is more important than the fields of individual nearby particles. Here again, the Coulomb force is of long-range character. However, there is a fundamental difference between plasmas and gravitation. Plasmas have both positive and negative charges, so that they are neutral on large scales and can form static homogeneous equilibria; on the contrary, gravitating systems can never form static homogeneous equilibria. This so-called Debye screening explains why many techniques of plasma physics can not be transferred immediately to the gravitational problem. Yves Elskens [8] and Diego Del Castillo Negrete [9] present some of their results in the framework of plasma physics.

2D Hydrodynamics. Two-dimensional incompressible hydrodynamics is another important case where the interaction is long-range. Indeed, the stream-function ψ is related to the modulus of the vorticity ω, via the Poisson equation $\Delta\psi = \omega$. Using the Green's function technique, one easily finds that the solution is

$$\psi(\vec{r}) = -\frac{1}{2\pi}\int_D d^2\vec{r}\,'\ \omega(\vec{r}\,')\, G(\vec{r} - \vec{r}\,') \quad , \tag{10}$$

where $G(\vec{r} - \vec{r}\,')$ depends on D, but $G(\vec{r}) \sim |\ln \vec{r}|$, when $\vec{r} \to 0$. The kinetic energy being conserved by the Euler equation (dissipativeless), it is straightforward to compute it on the domain D, with boundary ∂D,

$$E = \int_D d^2\vec{r}\,\frac{1}{2}\,(\nabla\psi)^2 \tag{11}$$

$$= \oint_{\partial D} \vec{n}\,dl\;\psi\nabla\psi + \frac{1}{2}\int_D d^2\vec{r}\;\omega(\vec{r})\psi(\vec{r}) \tag{12}$$

$$= -\frac{1}{4\pi}\int\int_D d^2\vec{r}\,d^2\vec{r}'\;\omega(\vec{r}')\omega(\vec{r})\ln|\vec{r}-\vec{r}'| \tag{13}$$

since $\psi = 0$ on ∂D. This emphasizes that one gets a logarithmic interaction. The analogy is even more clear if one approximates the vorticity field by point vortices $\omega(\vec{r}) = \sum_i \Gamma_i\delta(\vec{r}-\vec{r}_i)$, located at $\vec{r}_i$, with a given circulation Γ_i. The energy of the system reads now

$$E = \frac{1}{2}\sum_{i\neq j}\Gamma_i\Gamma_j\ln|\vec{r}_i-\vec{r}'_j| \quad . \tag{14}$$

The interaction among vortices has a logarithmic character, which corresponds to $\alpha = 0$.

Pierre-Henri Chavanis [6] studies carefully the analogy between the statistics of large-scale vortices in two-dimensional turbulence and self-gravitating systems. This analogy concerns not only the equilibrium states, i.e. the formation of large-scale structures, but also the relaxation towards equilibrium and the statistics of fluctuations. Diego Del Castillo Negrete [9] discusses also his results in the framework of hydrodynamics.

Dipolar Interactions. Dielectrics and diamagnets in an external electric or magnetic field exhibit a shape dependent thermodynamic limit [11]. This is due to the marginal decay of the potential energy $\alpha = d = 3$ for systems of dipoles. There is some approach to the solution of this problem only in zero field and in the absence of spontaneous ferromagnetism [12]. This is a border case for the long-range interactions, but it deserves a special attention.

Fracture. Let us examine analytical solutions for the plane stress and displacement fields around the tip of a slit-like plane crack in an ideal Hookean continuum solid. The classic approach to any linear elasticity problem of this sort involves the search for a suitable "stress function" that satisfies the so-called biharmonic equation $\nabla^2(\nabla^2\psi) = 0$ where ψ is the Airy stress function, in accordance with appropriate boundary conditions. The deformation energy density is then defined as $U \propto \sigma\varepsilon$ where σ is the fracture stress field around the tip, whereas ε is the deformation field. Considering a crack-width a in a two-dimensional material and using the exact Muskhelishvili's solution [10], one obtains the elastic potential energy due to the crack

$$U \simeq \frac{\sigma_\infty^2(1-\nu)}{2E}\,\frac{a^2}{r^2} \quad , \tag{15}$$

where E is the Young modulus, σ_∞ the stress field at infinity, ν the Poisson coefficient and r the distance to the tip: the elasticity equation in the bulk of solids leads therefore, again, to a border case for the long-range interactions since $U \sim 1/r^2$ in two dimensions. It appears that, despite of its engineering applications, the dynamics of this non conservative system has been very little studied, presumably because of its long-range character. In addition, in such a two dimensional material, the presence of several fractures could exhibit very interesting screening effects.

Table 1. Table listing different applications where systems are governed by long-range interactions. *Large systems* where the interactions is truly long-range and *small systems* where the range of the interactions is of the order of the size of the system are separated.

Interactions	α	α/d	Comments
Large systems			
Gravity	1	1/3	long-range
Coulomb	1	1/3	long-range with Debye screening
Dipole	3	1	Limiting value
2D Hydrodynamics	0	0	Logarithmic interactions
Fracture	2	1	Stress field around the tip
Small systems			
atomic and molecular clusters			
nuclei			
BE Condensation			

2.5 Applications to Small Systems

In addition to large systems where the interactions are truly long range, one should consider small systems where the range of the interactions is of the order of the size. In these cases, the system would not be additive either, and many similarities with the long-range case will be encountered. Phase transitions are universal properties of interacting matter which have been widely studied in the thermodynamic limit of *infinite* systems. However, in many physical situations this limit is not attained and phase transitions should be considered from a more general point of view. This is for example the case of some microscopic or mesoscopic systems: atomic clusters can melt, small drops of quantum fluids may undergo a Bose–Einstein condensation or a super-fluid phase transition, dense hadronic matter is predicted to merge in a quark and gluon plasma phase while nuclei are expected to exhibit a liquid-gas phase transition. For all these systems the experimental issue is how to characterize a phase transition in a *finite* system.

Philippe Chomaz and Francesca Gulminelli [13] discuss results from nuclear physics as well as from clusters physics. In particular, they propose a definition of first order phase transitions in finite systems based on topology anomalies of the event distribution in the space of observations. This generalizes the definitions based on the curvature anomalies [14] of thermodynamical potentials and provides a natural definition of order parameters. The new definitions are constructed to be directly operational from the experimental point of view. Finally, they show why, without the thermodynamic limit or at phase-transitions, the systems do not have a single peaked distribution in phase space.

In a closely related contribution, Dieter Gross [15] makes the statement, that the microcanonical ensemble with Boltzmann's principle $S = k_B \ln \Omega$ is the only proper basis to describe the equilibrium of a closed "small" system. Phase-transitions are linked to convex (upwards bending) intruders of the entropy, where the canonical ensemble defined by the Laplace transform to the intensive variables becomes multi-modal, non-local, and violates the basic conservation laws. The one-to-one mapping of the Legendre transform being lost, Gross claims that it is all possible to define phase transitions without invoking the thermodynamic limit, extensivity, or concavity of the entropy.

3 Thermodynamics

3.1 Inequivalence of Statistical Ensembles

Following the example exhibited long time ago by Hertel and Thirring [16], it is striking that these systems could lead to inequivalences between microcanonical, canonical or grand canonical ensembles. In this book, the first example is given by Barré et al [4] who present the Blume-Emery-Griffiths (BEG) model which allows a deep understanding of the fundamental reason why this happens. They studied the spin-1 BEG model both in the canonical and in the microcanonical ensemble. The canonical phase diagram exhibits a first order and a continuous transition lines which join at a tricritical point. It is shown that in the region where the canonical transition is first order, the microcanonical ensemble yields a phase diagram which differs from the canonical one. In particular it is found that the microcanonical phase diagram exhibits energy ranges with negative specific heat and temperature jumps at the transition energies. The global phase diagrams in the two ensembles and their multicritical behavior are calculated and compared.

Pierre-Henri Chavanis [6] shows similar features in self-gravitating systems where canonical and microcanonical tricritical points do not coincide either, as shown in Fig. 4 in the framework of self-gravitating fermions. Let us emphasize that this property survives to the introduction of a finite cut-off instead of quantum degeneracy as discussed by Chavanis.

3.2 Negative Specific Heats

This fact produces striking phenomena in the microcanonical ensemble, since it may result in a negative specific heat, as was emphasized by Eddington in

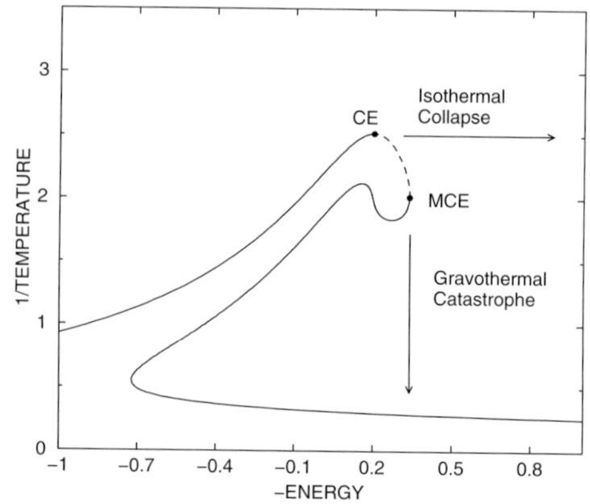

Fig. 4. Inverse temperature as a function of the energy for self-gravitating fermions without cut-off. CE (MCE) is the transition point in the canonical (microcanonica) ensemble. The dashed curve between CE and MCE has negative specific heat.

1926 [17] and then discussed by Lynden-Bell [18]. A first remark on the possibility of having negative specific heat in the microcanonical ensemble can even be found in the seminal paper on statistical mechanics by J.C. Maxwell [19]. Thirring [16] has finally clarified the point by showing that the paradox disappears if one realizes that only the microcanonical specific heat could be negative.

Indeed, in the canonical ensemble the mean value of the energy of a system with different energy levels E_i is

$$\langle E \rangle = \frac{\sum_i E_i \, e^{-\beta E_i}}{Z} = -\frac{\partial \ln Z}{\partial \beta} \tag{16}$$

where Z is the partition function. It is then straightforward to compute the specific heat

$$C_v = \frac{\partial \langle E \rangle}{\partial T} \propto \langle (E - \langle E \rangle)^2 \rangle > 0 \quad . \tag{17}$$

This clearly shows that the canonical specific heat is always positive. Notice also that this condition is true for systems of any size, regardless of whether a proper thermodynamic limit exists or not.

This is not the case if the energy is constant as shows the simplified following derivation for the example of interacting self-gravitating systems. Using the virial theorem for such particles

$$2\langle E_c \rangle + \langle E_{pot} \rangle = 0 \quad , \tag{18}$$

one gets that the total energy

$$E = \langle E_c \rangle + \langle E_{pot} \rangle = -\langle E_c \rangle \quad . \tag{19}$$

As the kinetic energy E_c is by definition proportional of the temperature one gets that

$$C_v = \frac{\partial E}{\partial T} \propto \frac{\partial E}{\partial \langle E_c \rangle} < 0 \tag{20}$$

Loosing its energy, the system is becoming hotter.

It is important at this stage to make a short comment on the Maxwell construction, usually taught in the framework of the Van der Waals liquid-gas transition. The existence of a negative specific heat region corresponds to a convex intruder in the entropy-energy curve, as shown in Fig. 5. When the interactions are short range, the system will phase separate in two parts, corresponding to the two phases 1 and 2 with a molar fraction x, so that the free energy $xF_1+(1-x)F_2$ is lower than the original free energy. This is clearly possible if the energy cost of the interface is proportional to the surface whereas the energy gain is proportional to the volume of the phase. However, this is not any more possible when the interactions are long-range since, on one hand, it is not straightforward to define a phase and, moreover, the system is not additive. The Maxwell construction has to be redefined in this new framework.

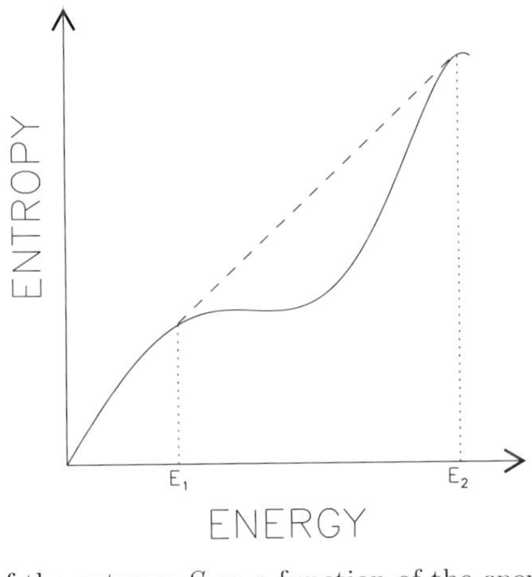

Fig. 5. Schematic shape of the entropy S as a function of the energy E with a convex intruder: the solid curve corresponds to the microcanonical result, whereas the dashed line to the canonical one.

Let us note that the microcanonical entropy as a function of the energy and of the order parameter generically leads to the landscape presented in Fig. 6. The projection for the critical points of the surface onto the entropy-energy plane produces the well known "swallowtail" catastrophe [20], depicted on the right of the figure. This corresponds to still another strange feature of the microcanonical ensemble, the presence of temperature jumps [4,6,7].

This concept of negative specific heat is now widely accepted in the astrophysical community, and was popularized in particular by Hawking [21] in 1974, with

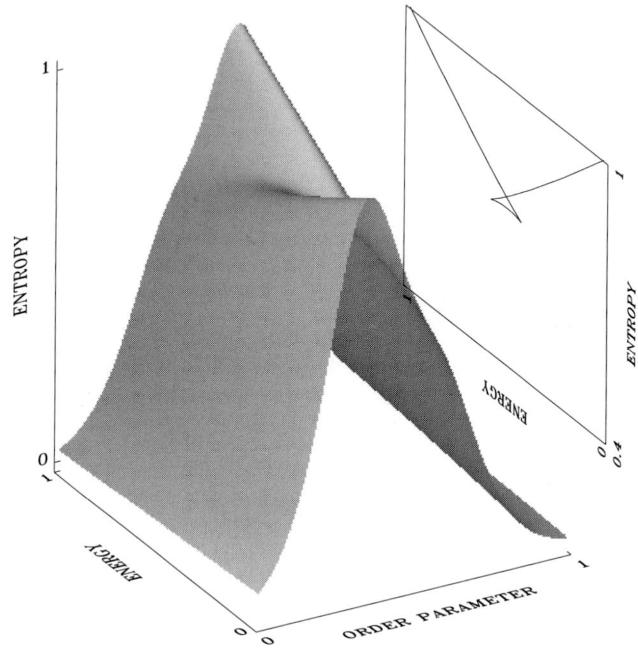

Fig. 6. A stylized microcanonical entropy as a function of the energy and of the order parameter mimicks an Alpine landscape where the workshop took place. The projection for the critical points of the surface onto the entropy-energy plane produces the well known "swallowtail" catastrophe.

some esoteric applications to black holes. The caloric curve of self-gravitating fermions derived by Chavanis and shown in Fig. 4 emphasizes such negative specific heat branch: the dotted branch is one example. Similarly one gets negative specific heat branch in the BEG model proposed by Barré et al [4]. In the canonical ensemble, they correspond to local maxima or saddle point of the corresponding free energy; it is the constraint of keeping the energy constant that stabilizes these canonical unstable states in the microcanonical ensemble.

Experimental groups have recently claimed signatures of negative specific heats in small systems. The first one corresponds to nuclear fragmentation [22], even if the authors use prudently the word "indication" of negative specific heat. The latter being inferred from the event by event study of energy fluctuations from Au + Au collisions. However, the signatures correspond to indirect measurements.

In the clusters community, two experimental groups have very recently reported negative specific heat. The first system [23] corresponds to atomic sodium clusters, namely Na_{147}^{+} and the negative microcanonical specific heat has been found near the solid to liquid transition. The cluster ion are produced in a gas aggregation source and then thermalized with Helium gas of controlled temperature. Accelerated thanks to the charge in a mass spectrometer, they are finally irradiated by a laser to determine the energy from the evaporation of several atoms after laser irradiation, also called photofragmentation. However, the control of equilibrium is as always the key point and therefore the evaluation of the temperature seems to be questionable, in particular since the temperature could not be constant during the motion of the ions.

In the Lyon's molecular cluster experiment [24], with H_{17}^+, the energy and the temperature are determined from the size distribution of fragments after collision of the cluster with a Helium projectile. To simplify the method, the larger the ratio of small fragments versus large ones, the larger is the temperature determined using the Bonasera et al procedure [25]. The reported caloric curve [24] shows a plateau. Work along this line is in progress and seems to show a negative specific heat region.

3.3 Non Extensive Statistics

Constantino Tsallis, Andrea Rapisarda, Vito Latora and Fulvio Baldovin [26] review the generalized non-extensive statistical mechanics formalism and its implications for different physical systems. The original very interesting idea is to generalize Boltzmann's entropy by defining

$$S_q = k_B \, \frac{1 - \sum_i {p_i}^q}{q-1} \tag{21}$$

where $\sum_i p_i = 1$. Using either the L'Hopital rule or a first order expansion of the term p_i^q in power of q, one immediately notices that

$$\lim_{q \to 1} S_q = -k_B \sum_i p_i \ln p_i \quad , \tag{22}$$

i.e. the well known Shannon entropy, known to be equivalent to the Boltzmann's one.

However, for q different from 1, this generalized entropy S_q is non additive, and one gets

$$S_q(A+B) = S_q(A) + S_q(B) + (1-q)\, S_q(A) S_q(B) \quad . \tag{23}$$

They illustrate in particular its application and the meaning of the entropic index q for conservative and dissipative low-dimensional maps. They also report on non Boltzmann-Gibbs behavior [26] and hindrance of relaxation for Hamiltonian systems with long-range interaction, where fingerprints of the generalized statistics have recently emerged.

This very interesting proposal [27] had however until now no strong foundations and many physicists were not ready to admit that the exponential Boltzmann distribution of states is at equilibrium only a particular case of a generalized distributions, with power tails. Dieter Gross [15] in particular makes different comments to this point. On the contrary, Tsallis et al emphasize also different situations were the Boltmann-Gibbs behavior is clearly not appropriate.

Recently, Beck and Cohen [28] showed that considering different statistics with large fluctuations, one can obtain generalized results, called superstatistics, with the Tsallis formalism being presumably so far the most relevant example. Moreover, Baldovin and Robledo worked out [29,26] exactly the q indices for the generalized largest Lyapunov exponent proposed by Tsallis for the logistic map. This an important step toward the derivation of a complete theory which, in particular, should help to understand the limits of its applications.

4 Dynamical Aspects

An essential peculiarity of these physical systems, and of some of their simplified models, is that a classical system of particles with long-range interactions will display strong non-equilibrium features. Dynamics is typically chaotic and self-consistent, since all particles give a contribution to the field acting on each of them: one calls this *self-consistent chaos*. Numerous physical systems fall in this category: galactic dynamics, dynamics of a plasma, vorticity dynamics,....

It is therefore essential to study the thermodynamic stability of these systems and in particular to understand the formation of structures trough *instabilities*. They should have logical similarities with the Jean's instability of self-gravitating systems, or with the modulational instability, leading to the formations of localized structures, as confirmed by preliminary results. Additional dynamical effects, like anomalous diffusion and Levy walks, which are reported in the negative specific heat regions, should be linked to these uncommon characteristics of thermodynamics [30].

In particular, Diego Del Castillo Negrete [9] discusses a mean-field single-wave model that describes the collective dynamics of marginally stable fluids and plasmas. He shows thus the role of self-consistent chaos in the formation and destruction of coherent structures, and presents a mechanism for violent relaxation of far from equilibrium initial conditions. The model bears many similarities with toy-models used in the study of systems with long range interactions in statistical mechanics, globally coupled oscillators, and gravitational systems.

One of these toy models is for example studied by Dauxois et al [31]. They consider the dynamics of the Hamiltonian Mean Field model which displays several interesting and new features. They show in particular the emergence of collective properties, i.e. the coherent (self-consistent) behavior of the particle motion. The space-time evolution of such coherent structures can sometimes be understood using the tools of statistical mechanics , otherwise can be a manifestation of the solutions of an associated Vlasov equation. Both cases in which the interaction among the particles is attractive and the one where it is repulsive are interesting to study: they offer different views to the process of cluster formation and to the development of the collective motion on different time-scales. The clustering transition can be first or second order, in the usual thermodynamical sense. In the former case, ensemble inequivalence naturally arises close to the transition. The behavior of the Lyapunov spectrum is also commented and the 'universal' features of the scaling laws that it involves.

Yves Elskens [8] shows that plasmas are a most common example of systems with long-range interactions, where the interplay between collective (wave) and individual (particle) degrees of freedom is well known to be central. This interplay being essentially non-dissipative, its prototype is described by a self-consistent Hamiltonian, which provides clear and intuitive pictures of fundamental processes such as the weak warm beam instability and Landau damping in their linear regimes. The description of the nonlinear regimes is more difficult. In the damping case, new insight is provided by a statistical mechanics approach, which identifies the distinction between a trapping behavior and linear Landau behavior in terms of a phase transition. In the unstable case, the model has shown that the commutation of long-time and large-N limits is not guaranteed.

Chavanis considers also dynamical aspects in the framework of stellar systems and two-dimensional vortices. He discusses in particular two possible relaxation scenarios: one due to collisions (or more precisely to discrete interactions) and the second one, called violent relaxation, really collisionless but due to the mean field effect and the long-range of the interaction.

Finally, the dynamical processes that give rise to power-law distributions and fractal structures have been studied extensively in the recent years. Ofer Biham and Ofer Malcai [32] describe recent studies of self-organized criticality in sandpile models as well as studies of multiplicative dynamics, giving rise to power-law distributions. Sandpile models turn out to exhibit universal behavior while in the multiplicative models the powers vary continuously as a function of the parameters. They consider the formation of a fractal object in the presence of a dynamical mechanism that generates a power-law distribution and present a model that demonstrates clustering when the probability of adding a particle decays with a power $\alpha > d$, so it has a short-range nature.

5 Bose–Einstein Condensation

Finally, we would like to put a special emphasis on Bose–Einstein Condensation (BEC), predicted by Bose and Einstein in 1924, which could be an important field of application. With the recent achievement [33] of Bose–Einstein condensation in atomic gases thanks to the evaporation cooling technique, it becomes possible to study these phenomena in an extremely diluted fluid, thus helping to bridge the gap between theoretical studies, only tractable in dilute systems, and experiments. In the BEC, atoms are trapped at such low temperatures that they tumble into the same quantum ground state creating an intriguing laboratory for testing our understanding of basic quantum phenomena.

First, Jean Dalibard [34] presents how coherence and superfluidity are hall mark properties of quantum fluids and encompass a whole class of fundamental phenomena. He reviews several experimental facts which reveal these two remarkable properties. Coherence appears in interference experiments, carried out either with a single condensate or with several condensates prepared independently. Superfluidity can be revealed by studying the response of the fluid to a rotating perturbation, which involves the nucleation of quantized vortices.

Second, Ennio Arimondo and Oliver Morsch [35] present the current investigations of Bose–Einstein condensates within optical lattices, where the long-range interactions are an essential part of the condensate stability. Previous work with laser cooled atomic gases is also briefly discussed.

On the theoretical side, the fluctuations of the number of particles in ideal Bose–Einstein condensates within the different statistical ensembles has shown interesting differences. Martin Holthaus explains [36] why the usually taught grand canonical ensemble is inappropriate for determining the fluctuation of the ground-state occupation number of a partially condensed ideal Bose gas: it predicts r.m.s.-fluctuations that are proportional to the total particle number at vanishing temperature. In contrast, both the canonical and the microcanonical ensemble yields fluctuations that vanish properly for the temperature going toward zero. It turns out that the difference between canonical and microcanonical fluctuations can be understood in close analogy to the familiar difference between the heat capacities at constant pressure and at constant volume. The detailed analysis of ideal Bose–Einstein condensates turns out to be very helpful for understanding the occupation number statistics of weakly interacting condensates.

Ulf Leohnardt [37] shows that Bose–Einstein condensates can serve as laboratory systems for tabletop astrophysics. In particular, artificial black holes can be made (sonic or optical black holes). A black hole represents a quantum catastrophe where an initial catastrophic event, for example the collapse of the hole, triggers a continuous emission of quantum radiation (Hawking radiation). The contribution summarizes three classes of quantum catastrophes, two known ones (black hole, Schwinger's pair creation) and a third new class that can be generated with slow light.

Finally, Gershon Kurizki presents [38] an exciting theoretical idea to induce long-range attractions between atoms that acts across the whole Bose–Einstein condensate. He shows that dipole-dipole interatomic forces induced by off-resonant lasers

$$V_{dd} = V_0 \left[\frac{2z^2 - x^2 - y^2}{r^3} (\cos qr + qr \sin qr) - \frac{2z^2 + x^2 + y^2}{r} q^2 \cos qr \right] \quad (24)$$

allow controllable drastic modifications of cold atomic media. "Sacrifying strength for beauty", Kurizki proposed [40] to average out the first term in $1/r^3$ of the dipole-dipole interaction by the different lasers, in order to keep only the last one with a $1/r$ interaction. The important point is that induced gravity-like force would be strong enough to see it acting among atoms in the BEC: i.e. that, having induced the gravity-like attraction in the BEC, one could switch off the trap used originally to create the BEC, and it will remain stable, holding itself together. Depending on the number of lasers, the resulting gravity-like force could be anisotropic for three lasers, or strictly identical to gravity with eighteen lasers ! If the last proposal is presumably too speculative and if the difficulties (the power of the laser required being really huge) facing the experimentalists are a real challenge, the ability to emulate gravitational interactions in the laboratory is of course fascinating. Indeed, these modifications may include the formation of self-gravitating "bosons stars" and their plasma-like oscillations, self-bound quasi-one-dimensional Bose condensates and their "supersolid" density modulation, giant Cooper pairs and quasibound molecules in optical lattices and anomalous scattering spectra in systems of interacting Bosons or Fermions. These novel regimes set the arena for the exploration of exotic astrophysical and condensed -matter objects, by studying their atomic analogs *in the laboratory.*

6 Conclusion

The dynamics and thermodynamics of long-range system is a rich and fascinating topic. We want to conclude with the following comments:

- long-range interactions are a rich laboratory for statistical physics. Let us only mention a few of the interesting phenomena and features: inequivalence of ensembles, negative specific heat, collisionless relaxation, role of coherent structures, nonadditivity, generalizations of entropy.
- This problem has also the nice property to be related to neighboring scientific disciplines. Let us mention mathematics, with the application of catastrophe theory [39] and large deviations theory [41], and computer science. In the latter, because of the long-range interactions, naive numerical codes are of order N^2, and the developments of efficient algorithms such as the heap based procedure [42] or local simulation algorithm for Coulomb interaction [43] is needed.
- This methodological and fundamental effort should provide a general approach to the problems arising in each specific domain which has motivated this study: astrophysical objects, plasmas, atomic and molecular clusters, fluid dynamics, fracture, Bose–Einstein condensation, ... in order to detect the depth and the origin of the observed analogies or, on the contrary, to emphasize their specificities.

Many of these different aspects are considered in this book but it is clear that, rather than closing the topic, it opens the pandora box.

Acknowledgements

This work has been partially supported by th EU contract No. HPRN-CT-1999-00163 (LOCNET network), the French Ministère de la Recherche grant ACI jeune chercheur-2001 N° 21-311. This work is also part of the contract COFIN00 on *Chaos and localization in classical and quantum mechanics.*

References

1. T. Padmanabhan: Physics Reports **188**, 285 (1990)
2. M. Kac, G.E. Uhlenbeck, P.C. Hemmer: J. Math. Phys. **4**, 216 (1963)
3. T. Dauxois, S. Ruffo, E. Arimondo, M. Wilkens: "Dynamics and Thermodynamics of Systems with Long-Range Interactions", Lecture Notes in Physics Vol. 602, Springer (2002).
4. J. Barré, D. Mukamel, S. Ruffo: *Inequivalence of ensembles in mean-field models of magnetism* in [3] (in this volume)
5. T. Padmanabhan: *Statistical mechanics of gravitating systems in static and expanding backgrounds* in [3] (in this volume)
6. P.-H. Chavanis: *Statistical mechanics of two-dimensional vortices and stellar systems* in [3] (in this volume)
7. E. G. D. Cohen, I. Ispolatov: *Phase transitions in systems with* $1/r^{\alpha}$ *attractive interactions* in [3] (in this volume)
8. Y. Elskens: *Kinetic theory for plasmas and wave-particle hamiltonian dynamics* in [3] (in this volume)
9. D. Del Castillo-Negrete: *Dynamics and self-consistent chaos in a mean field Hamiltonian model* in [3] (in this volume)
10. N. I. Muskhelishvili, in *Some Basic Problems of the Mathematical Theory of Elasticity*, P. Noordhoff, Groningen (1953)
11. L. D. Landau, E. M. Lifshitz: *Course of Theoretical Physics. T. 8: Electrodynamics of continuous media* (1984)
12. R. B. Griffiths: Physical Review **176**, 655 (1968). M.E. Fisher, Arch. Rat. Mech. Anal. **17**, 377 (1964)
13. Ph. Chomaz, F. Gulminelli: *Phase transitions in finite systems* in [3] (in this volume)
14. D. H. E. Gross: *Microcanonical Thermodynamics*, World Scientific, Singapore (2001)
15. D. H. E. Gross: *Thermo-Statistics or Topology of the Microcanonical Entropy Surface* in [3] (in this volume)
16. W. Thirring: Z. Phys. **235**, 339 (1970)
17. A. S. Eddington: *The internal constitution of stars*, Cambridge University Press (1926)
18. D. Lynden-Bell, R. Wood: Mont. Not. R. Astron. Soc. **138**, 495 (1968)
19. J.C. Maxwell: *On Boltzmann's theorem on the average distribution of energy in a system of material points*, Cambridge Philosophical Society's Trans., vol XII, p. 90 (1876)
20. T. Poston, J. Stewart, *Catastrophe Theory and its Application*, Pitman, London (1978)
21. S. W. Hawking: Nature **248**, 30 (1974)
22. D'Agostino et al: Physics Letters B **473**, 219 (2000)
23. M. Schmidt et al: Physical Review Letters **86**, 1191 (2001)
24. F. Gobet et al: Physical Review Letters **87**, 203401 (2001)
25. M. Belkacem, V. Latora, A. Bonasera: Physical Review C **52**, 271 (1995)
26. C. Tsallis, A. Rapisarda, V. Latora, F. Baldovin: *Nonextensivity: from low-dimensional maps to Hamiltonian systems* in [3] (in this volume)
27. C. Tsallis: Journal of Statistical Physics **52**, 479 (1988)
28. C. Beck, E.G.D. Cohen, [cond-mat/0205097]
29. F. Baldovin and A. Robledo, [cond-mat/0205356].

30. A. Torcini, M. Antoni: Physical Review E **59**, 2746 (1999)
31. T. Dauxois, V. Latora, A. Rapisarda, S. Ruffo, A. Torcini: *The Hamiltonian mean field model: from dynamics to statistical mechanics and back* in [3] (in this volume)
32. O. Biham, O. Malcai: *Fractals and Power-Laws* in [3] (in this volume)
33. M. H. Anderson et al: Science **269**, 198 (1995). C. C. Bradley et al, Physical Review Letters **75**, 1687 (1995). K. B. Davis et al, Physical Review Letters **75**, 3669 (1995)
34. J. Dalibard: *Coherence and superfluidity of gaseous Bose–Einstein condensates* in [3] (in this volume)
35. O. Morsch, E. Arimondo: *Ultracold atoms and Bose–Einstein condensates in optical lattices* in [3] (in this volume)
36. D. Boers, M. Holthaus: *Canonical statistics of occupation numbers for ideal and weakly interacting Bose–Einstein condensates* in [3] (in this volume)
37. U. Leonhardt: *Quantum catastrophes* in Proceedings of the Conference "Dynamics and thermodynamics of systems with long-range interactions", Les Houches, France, February 18-22 2002, T. Dauxois, E. Arimondo, S. Ruffo, M. Wilkens Eds., published on http://www.ens-lyon.fr/~tdauxois/procs02/
38. G. Kurizki, D. O'Dell, S. Giovanazzi, A. I. Artemiev: *New regimes in cold gases via laser-induced long-range interactions* in [3] (in this volume)
39. J. Barré, F. Bouchet, in preparation (2002)
40. D. O'Dell, S. Giovanazzi, G. Kurizki, V. M. Akulin: Physical Review Letters **84**, 5687 (2000)
41. R. S. Ellis, K. Haven, B. Turkington: Journal of Statistical Physics **101**, 999 (2000)
42. A. Noullez, D. Fanelli, E. Aurell: "Heap base algorithm", cond-mat/0101336 (2001)
43. A. C. Maggs, V. Rossetto, Physical Review Letters **88**, 196402 (2002)

Part I

Statistical Mechanics

Thermo-statistics or Topology of the Microcanonical Entropy Surface

Dieter H.E. Gross

[1] Hahn-Meitner Institute
[2] Freie Universität Berlin Fachbereich Physik,
Glienickerstr. 100, 14109 Berlin, Germany

Abstract. Boltzmann's principle $S(E, N, V \cdots) = \ln W(E, N, V \cdots)$ allows the interpretation of Statistical Mechanics of a *closed* system as Pseudo-Riemannian geometry in the space of the conserved parameters $E, N, V \cdots$ (the conserved mechanical parameters in the language of Ruppeiner [1]) without invoking the thermodynamic limit. The topology is controlled by the curvature of $S(E, N, V \cdots)$. The most interesting region is the region of (wrong) positive maximum curvature, the region of phase-separation. This is demonstrated among others for the *equilibrium* of a typical *non-extensive* system, a self-gravitating and rotating cloud in a spherical container at various energies and angular-momenta. A rich variety of realistic configurations, as single stars, multi-star systems, rings and finally gas, are obtained as *equilibrium* microcanonical phases. The global phase diagram, the topology of the curvature, as function of energy and angular-momentum is presented. No exotic form of thermodynamics like Tsallis [2,3] non-extensive one is necessary. It is further shown that a finite (even mesoscopic) system approaches equilibrium with a change of its entropy $\Delta S \geq 0$ (Second Law) even when its Poincaré recurrence time is not large.

1 Introduction

Why this paper?

Since more than 100 years does thermo-statistics emphasize the canonical- or the grand-canonical ensemble in the thermodynamic limit as the appropriate microscopic description of an equilibrized system. Here a *homogeneous*, practically infinite, system is controlled by intensive parameters like the temperature. Though in textbooks the microcanonical ensemble is considered as the fundamental ensemble, due to mathematical difficulties it is quickly left in favor of the canonical ones. Intensive variables like temperature, pressure and chemical potential are used instead of the mechanical defined extensive energy, volume and particle number.

The intensive variables are even canonized to found the axiomatic definition of an orthode [4] from where Statistical Mechanics (at least its intensive or canonical form) is deduced. Lebowitz [5,6] considers the thermodynamic limit and Elliott Lieb [7] extensivity, which also needs the thermodynamic limit to ignore surface effects, as the condition sine-qua-non.

In the thermodynamic limit $\lim_{V,N\to\infty,\ N/V=\rho}$ all surface effects may be ignored. – Usually – It is clear that this approach cannot do justice to phase separations. In fact, the gain in entropy when a system splits into different phases by interphase surfaces is the essence of phase transitions of first order. A liquid-gas transition is experimentally detected just by the interface between the liquid and the gas. Consequently, in the (grand)-canonical approach, phase-transitions are indicated by the Yang-Lee singularities [8] where the grand-canonical potentials are non-analytic in $z = e^{\beta\mu}$ or singular. These indicate the break-down of the (grand)-canonical formalism. In remarkable contrast, the microcanonical ensemble has no problems at phase-separations and the microcanonical density of states remains single-valued and multiply differentiable in all conserved control parameters also here, see below. This is certainly the strongest argument in favor of the fundamental significance of the micro-ensemble.

At phase-separation the entropy $S(E, N, \cdots)$ has a positive curvature. Ruppeiner's Riemannian geometry of fluctuations [1] must be reformulated there as Pseudo-Riemannian. This leads to a negative heat capacity and a violation of Clausius' formulation of the Second Law ("heat flows always from the hot to the cold system"). Phase-separations demand an essential, fundamental, change of conventional classical thermo-statistics. Thermodynamics, however, was invented in the 19.century to explain the working of steam engines. I.e. one of its primary issues were just phase-separations.

What was said applies to large systems with short-range forces. The largest systems in nature, self-gravitating astro-physical systems, are subjected to forces (gravitation) with a range comparable to the linear extension of the system. These systems are naturally inhomogeneous and non-extensive. A description of their equilibrium by intensive variables with a homogeneous spatial distribution misses these most interesting situations. Ironically, the thermodynamic limit does not apply to these really large systems. Tsallis [2] on the other hand addresses non-extensive systems explicitely but keeps the description in terms of intensive variables which fix the relevant conserved parameters only on average. However, non-extensive systems are usually not self-averaging. He believes the equilibrium statistics of Hamiltonians systems demands a new definition of entropy. This, is not so [9], see Sect. 3.2.

It is well known that self-gravitating systems collapse to a star in a more or less void background at low energies, the "gravothermal catastrophe" [10,11]. This is of course quite interesting but outside of any homogeneous canonical thermodynamics. There is nothing peculiar with this in microcanonical thermodynamics. Certainly, angular-momentum can change this essentially. However, the microcanonical equilibrium configuration of a self-gravitating system under larger angular-momentum has not been investigated yet (exception [12]). I will show in Sect. 3.2 how the competition between gravitational collapse and centrifugal disruption leads in a natural manner to a breaking of rotational symmetry and to multi-star configurations with a large variety of different but quite realistic configurations. In this section the global phase diagram of self-gravitating and rotating many-body systems as function of energy and angular-momentum is presented, c.f. [12]. It is for the first time that these various realistic stellar configurations are interpreted as global equilibrium configurations.

The thermodynamic limit is also invoked since Boltzmann to deduce the Second Law of Thermodynamics [6,13–15] from microscopic reversible dynamics. Then Zermelo's [16] paradox becomes blunted as the Poincaré recurrence time t_{rec} is much longer than any physically relevant observation time. This is different for a finite, eventually small system.

By all these reasons a reinvestigation of the microscopic foundation of Statistical Mechanics starting from Newtonian *reversible* mechanics of the *finite* many-body system using a minimum of assumptions but avoiding the thermodynamic limit is highly needed. The "Geometric Foundation of Thermo-Statistics" proposed in [17] and further developed here offers a new, deeper, and much simpler understanding of the microscopic foundation of Thermodynamics.

2 The Few Essentials of Statistical Mechanics

2.1 Why Does Conventional Statistics Like the Thermodynamic Limit?

The relative fluctuations of a macroscopic observable A in pure phases of an extensive system vanish in the large N limit:

$$\frac{\langle A^2\rangle - \langle A\rangle^2}{\langle A\rangle^2} \propto \frac{1}{N}$$

We call this behavior *self-averaging*. Then the probability aspect of statistics becomes unimportant.

Here we want to study also non-extensive situations, therefore, we are not allowed to go to the thermodynamic limit. Fluctuations must be taken seriously.

2.2 Obsolete Gospels of Conventional Thermo-statistics

Then, most axioms which are mistaken to be fundamental for Statistical Mechanics even turn out to be violated:

- Phase transitions (*do not*) exist only in the thermodynamic limit
- Specific heat is (*not*) $c \sim \left\langle (\delta E)^2 \right\rangle > 0$ or $dT/dE > 0$
- Heat does (*not*) always flow from hot to cold
- Thermodynamic stability does (*not*) necessarily imply the concavity of $S(E, N, \cdots)$
- (*No*) extensivity of S, (*no*) scaling with N
- (*No*) unique Legendre mapping, $T \longrightarrow E$, etc.
- Rise of entropy is (*not*) necessarily connected to trend towards uniformization
- Second Law (*not*) only in infinite systems, i.e. second law (*not*) only when the recurrence time is much larger than the observation-time $t_{rec} \gg t_{obs}$

- A system at equilibrium is (*not*) necessarily an orthode in the sense of Gallavotti [4]. "Boltzmann's heat theorem", i.e.

$$\frac{dE(T,P)}{T} + \frac{PdV(T,P)}{T} \tag{1}$$

is (*not*) necessarily a total differential $dS(T,P)$, because $S(T,P)$ is not always a smooth, one valued, function of (T,P), see Sect. 3.1, order parameter.

Violations of these gospels seem shocking statements:
Lebowitz [5] and Lieb[18,14] believe these make thermo-statistics impossible to exist. Nevertheless these violations are valid building stones of statistics. They are even necessary for thermo-statistics to apply to the original goal of Thermodynamics, the description of phase-separation, the scenario in which steam engines do work. At closer inspection these violations are not so strange. Recapitulating the main roots of statistical mechanics we will see that it makes a lot of sense to formulate it without invoking the above axioms and without using the thermodynamic limit. The only essential axiom needed to define equilibrium statistics is Boltzmann's principle c.f. Sect. 2.4, (3), once we agree not to use the thermodynamic limit. To concentrate on this is a great advantage as this principle has a very simple *geometrical* meaning.

It is a benefit of the new, extended theory which I offer here, that by reducing its axiomatic basis to this single principle it applies also to Hamiltonian non-extensive systems and among them to the really large systems as astrophysical ones, which are far larger than the thermodynamic "limit" of any homogeneous system allows. A whole new world for applications of thermo-statistics opens. Of course, then one cannot separate volume from surface effects. This is anyhow dubious for non-extensive systems or at phase-separation.

Here however, I must make it very clear that in any cases where the thermodynamic limit of a homogeneous phase does exist, the geometrical theory is in complete agreement with conventional Thermodynamics and conventional extensive Statistical Mechanics.

2.3 Thermodynamics, a Probabilistic Theory; Control Parameters

Thermodynamics is a macroscopic description of a many-body system within a few ($M \sim 3$) macroscopic control parameters and where ($6N - M \gg M$) dof's remain uncontrolled. Therefore, Thermodynamics describes all systems with the same M simultaneously. All systems under the same macroscopic constraints are simultaneously addressed by the theory. Statistical mechanics describes the whole $6N - M$ dimensional manifold, i.e. all points in then N-body phase-space with same energy E , the microcanonical ensemble $\mathcal{E}$[1]. Consequently, it gives only *probabilistic* predictions how the *average* of the systems in the ensemble behave [19].

[1] We denote manifolds in phase-space by calligraphic letters like $\mathcal{E}$.

A large extensive system in a pure phase is self-averaging. In the thermodynamic limit a sharp peak of the probability distribution guarantees the identity of the average with the most likely configuration.

But what if the thermodynamic limit does not exist like for a non-extensive system? For a small system like a nucleus or an atomic cluster the same measurement must be performed very often and the average be taken before its thermodynamic behavior can be seen.

2.4 Boltzmann's Principle, the Microcanonical Ensemble

The key quantity of statistics and thermodynamics is the *entropy* S. Its most fundamental definition is as the logarithm of the *area* $W(E)$ of the manifold $\mathcal{E}$ in the N-body phase-space by Boltzmann's principle [20]:

$$\begin{aligned} W(E,N,V) &= \epsilon_0 tr\delta(E-H_N) \\ tr\delta(E-H_N) &= \int \frac{d^{3N}p\, d^{3N}q}{N!(2\pi\hbar)^{3N}}\delta(E-H_N). \end{aligned} \tag{2}$$

$$\boxed{\boxed{\mathrm{S{=}k{\cdot}lnW}}} \tag{3}$$

Boltzmann's principle is the only axiom necessary for thermo-statistics. With it Statistical Mechanics and also Thermodynamics become *geometric* theories. For instance all kinds of phase-transitions are entirely determined by *topological* peculiarities of $\mathcal{E}(E,N,\cdots)$ and thus of $S(E,N,\cdots)$ see below.

Local or Global Constraints?

In microcanonical statistics the "extensive", better conserved, control parameters as energy, volume, particle number etc. can be determined macroscopically sharp. There may well be small, microscopic violations of some microscopic conservation laws due to the non-ideal nature of the container. Therefore, we allow small fluctuations in these microscopically conserved quantities. Evidently, it does not matter whether the entropy $S(E,N,V)$ has an extremum or not. Its *local* value is significant. It is uniquely defined by (2,3) as a high-dimensional integral. It is thus everywhere multiply differentiable, – certainly the most important difference to canonical statistics. This is especially important at phase-separations where the curvature of $S(E,N,V)$ is positive c.f. Sect. 3.1.

Why Not Canonical?

In the canonical statistics, also in Tsallis "non-extensive statistics" e.g. the energy is fixed in the mean by Lagrange parameters like $\beta = 1/T$. This works only if the microcanonical ensemble is self-averaging. Now for non-extensive situations like at phase-separations e.g. the energy $E(T)$ as function of the Lagrange parameter like T is multi-valued c.f. Sect. 3.1(ambiguity $\cdots$). This leads to the (in view of standard Thermodynamics) surprising negative heat capacity c.f. Sect. 3.1(negative heat capacity $\cdots$), which is well documented even experimentally c.f. [11,21–33].[2]

[2] In fact the paper [24] pointed explicitely to the fundamental failure of the canonical ensembles near first order phase transitions in general and its non-equivalence to

At Points of Negative Curvature of $S(E, N, \cdots)$ the Canonical Ensemble is Not an Orthode
In Chap. 1.5 of his book[4] Gallavotti presented an axiomatic deduction of thermo-statistics starting from the definition of an orthode.

Following Boltzmann's heat theorem [34] he defines an ensemble to be an orthode when an infinitesimal change of the energy dE and volume dV makes

$$\frac{dE}{T} + \frac{PdV}{T} \tag{4}$$

an exact differential, at least in the thermodynamic limit. Here T the "temperature" is the average kinetic energy per particle and P the "pressure" is defined as the average momentum transfer per unit time and unit surface area of the container.

Clearly this definition is of little help for a non-extensive system and/or when the thermodynamic limit does not exist, where a given T or P does not fix the energy or volume. It fails in situations where the canonical ensemble is not equivalent to the microcanonical one, i.e. also at phase-transitions.

3 Equilibrium Thermodynamics

3.1 Phase Transitions

Relation to Yang-Lee Theory
In conventional extensive thermodynamics phase transitions are indicated by the Yang-Lee zeros of the grand-canonical partition sum Z *in the thermodynamic limit*. In order to generalize the definition of phase transitions also to non-extensive systems I start for the moment with the Laplace transform from the microcanonical density of states e^S to the grand-canonical one (here the discreteness of N does not matter):

$$\begin{aligned} Z(T, \mu, V) &= \iint_0^\infty \frac{dE}{\epsilon_0} dN\, e^{-[E-\mu N-TS(E)]/T} && (5) \\ &=: \frac{V^2}{\epsilon_0} \iint_0^\infty de\, dn\, e^{-V[e-\mu n-Ts(e,n)]/T} && (6) \\ &\approx \qquad e^{\,\text{cons.+lin.+quadr.}} \end{aligned}$$

the fundamental micro-ensemble which shows a negative heat capacity there. It was rejected by Gary S. Grest, the Divisional Associate Editor of statistical mechanics of PRL, June 3 1997, with the argument: "I am not convinced that the microcanonical ensemble is more physical than the canonical ensemble. After all, phase transitions in the experimental world are at constant temperature and not at constant energy." If this would be true, a ship would not have been able to sail on the surface of the ocean, America would never have been discovered and PRL would not even exist.

The linear term is explicitely put to 0 by solving

$$\frac{1}{T} = \left.\frac{\partial S}{\partial E}\right|_{e_s,n_s} \tag{7}$$

$$\frac{\mu}{T} = -\left.\frac{\partial S}{\partial N}\right|_{e_s,n_s} \tag{8}$$

$$\left(\frac{P}{T} = \left.\frac{\partial S}{\partial V}\right|_{e_s,n_s}\right). \tag{9}$$

If $s(e,n)$ is concave (has negative principal curvatures), and there is a single solution e_s,n_s of (7) and (8), the stationary point, where integral (6) is a double Gaussian integral along the two principal curvatures $\boldsymbol{v}_1, \boldsymbol{v}_2$ and the free-energy density is:

$$\frac{F(T,\mu,V)}{V} = \frac{-T\ln(Z)+\mu N_s}{V} \to e_s - Ts_s + \frac{T\ln\left(\sqrt{(-\lambda_1)}\sqrt{(-\lambda_2)}\right)}{V} + o\left(\frac{\ln V}{V}\right)$$

The curvature matrix $c(e,n)$ of $s(e,n)$

$$c(e,n) = \begin{pmatrix} \frac{\partial^2 s}{\partial e^2} & \frac{\partial^2 s}{\partial n \partial e} \\ \frac{\partial^2 s}{\partial e \partial n} & \frac{\partial^2 s}{\partial n^2} \end{pmatrix} \tag{10}$$

has the eigenvalues :$\lambda_1 \geq \lambda_2$ $\longrightarrow$eigenvectors : $\boldsymbol{v}_1, \boldsymbol{v}_2$ (11)

$$\text{Hessian } d(e,n) = \|c(e,n)\| = \lambda_1\lambda_2 \tag{12}$$

In general λ_1 can have either sign. This leads to a new, deeper definition of thermodynamic phases:

Classification by the Local Topology of Curvature

- A single stable phase is defined by $\lambda_1 < 0$. Here $s(e,n)$ is concave (downwards bended) in both directions, Fig. 1. There is only a single solution of:

$$\frac{1}{T} = \left.\frac{\partial S}{\partial E}\right|_{e_s,n_s} \tag{13}$$

$$\frac{\mu}{T} = -\left.\frac{\partial S}{\partial N}\right|_{e_s,n_s}. \tag{14}$$

  Here is a one to one mapping of the (grand)-canonical $\leftrightarrow$the micro-ensemble. The (grand)-canonical has in the thermodynamic limit the same analytical properties as the micro-ensemble. It is everywhere smooth, multiply differentiable. There are no discontinuities neither in $\ln Z$ nor its derivatives.
- A transition of first order with phase-separation and surface tension is indicated by the maximum curvature $\lambda_1(e,n) > 0$. $s(e,n)$ has a convex intruder (upwards bended with Pseudo-Riemannian geometry) in the direction $\mathbf{v}_1$ of

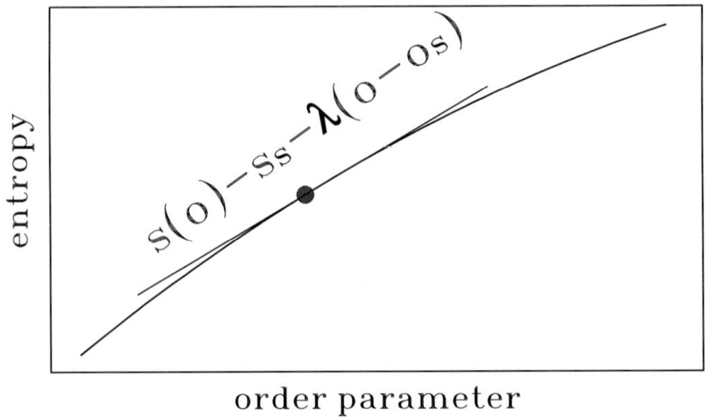

Fig. 1. Mono-phase. The order parameter o is defined in subsection "order parameter".

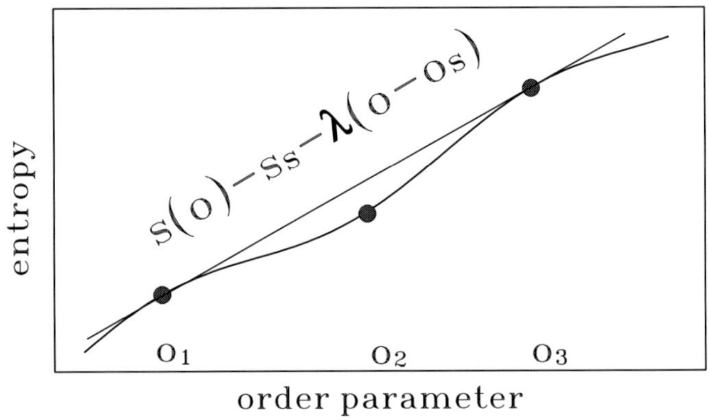

Fig. 2. Phase-separation, Gibbs double tangent.

the largest curvature (order parameter). There are at least three solutions (e_s, n_s): o_1, o_2, o_3 see Fig. 2 of

$$\beta = \frac{1}{T} \quad = \left.\frac{\partial S}{\partial E}\right|_{e_s,n_s} \tag{15}$$

$$\nu = -\frac{\mu}{T} = \left.\frac{\partial S}{\partial N}\right|_{e_s,n_s} \tag{16}$$

The whole region $\{o_1, o_3\}$ is mapped by (6) into a single point (T_{tr}, ν_{tr}) in the canonical ensemble which is consequently non-local in o and *bimodal*. This is the origin of the Yang-Lee singularities. I.e. if the largest curvature of $s(e,n)$ is $\lambda_1 \geq 0$ both ensembles are not equivalent as already pointed out by us in 1993[35,36,24,37] see also Barré et al [38]. The possibility of positive curvatures (Pseudo Riemannian geometry) is the main difference to the Riemannian geometry proposed by G. Ruppeiner[1,39]. Ruppeiner discusses shortly phase transitions of first order and points out that there is no divergence of the correlation length ξ like at second-order transitions and ξ remains of the order of the interface thickness, i.e. finite. He does not mention (or is not interested in) the positivity of the curvature of $s(e,n)$, which encodes important information about the surface tension, see below and which is also the essential reason why the canonical formalism does not apply to non-extensive systems.

- A continuous ("second order") transition with vanishing surface tension, where two neighboring phases become indistinguishable. I.e. where the three stationary solutions o_1, o_2, o_3 move into one-another. This is at the extremum of λ_1 in the direction of order parameter $\mathbf{v}_{\lambda=0} \cdot \nabla\lambda_1 = 0$. These are the catastrophes of the Laplace transform (6) $E \to T$ and the critical points of the transition. If it is also on the border line ($\lambda_1 = 0$) of the first order transition (where $\lambda_1 < 0$), it is the critical end-point of the transition. It is an open question whether a line of second-order transition is also the locus of the critical end-point of a first-order transition in a hidden parameter [37].

Physical Origin of Positive Curvature, the Surface Tension

For short range forces it is linked to the interphase surface tension. This is demonstrated for a system of 1000 Na-clusters by Fig. 3.

Negative Heat Capacity as Signal for a Phase Transition of First Order

As explained in Fig. 4 for the example of the $q = 10$ Potts-model, a positive curvature (convex intruder) of $S(E)$ in the energy direction corresponds to a characteristic *backbending* of the caloric curve $T(E)$ or $\beta(E)$, and to a negative

Table 1. Parameters of the liquid–gas transition of small sodium clusters (MMMC-calculation [37]) in comparison with the bulk for a rising number N_0 of atoms, N_{surf} is the average number of surface atoms (estimated here as $\sum N_{cluster}^{2/3}$) of all clusters with $N_i \geq 2$ together. $\sigma/T_{tr} = \Delta s_{surf} * N_0/N_{surf}$ corresponds to the surface tension. Its bulk value is adjusted to agree with the experimental values of the a_s parameter which we used in the liquid-drop formula for the binding energies of small clusters, c.f. Brechignac et al. [40], and which are used in this calculation [37] for the individual clusters.

	N_0	200	1000	3000	**bulk**
	T_{tr} $[K]$	940	990	1095	**1156**
	q_{lat} $[eV]$	0.82	0.91	0.94	**0.923**
Na	s_{boil}	10.1	10.7	9.9	**9.267**
	Δs_{surf}	0.55	0.56	0.44	
	N_{surf}	39.94	98.53	186.6	∞
	σ/T_{tr}	2.75	5.68	7.07	**7.41**

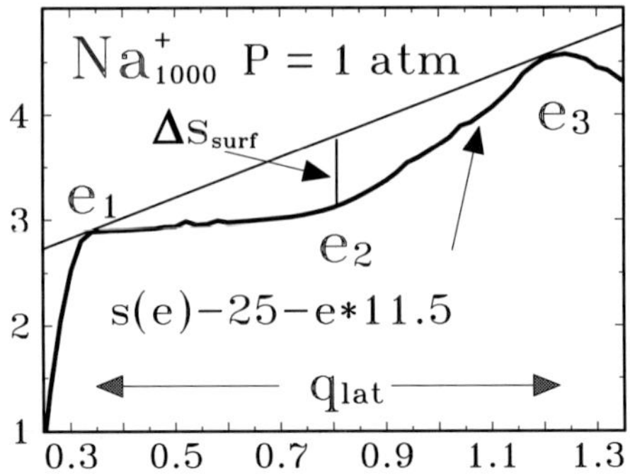

Fig. 3. MMMC [37] simulation of the entropy $s(e)$ per atom (e in eV per atom) of a system of $N_0 = 1000$ sodium atoms at an external pressure of 1 atm. At the energy $e \leq e_1$ the system is in the pure liquid phase and at $e \geq e_3$ in the pure gas phase, of course with fluctuations. The latent heat per atom is $q_{lat} = e_3 - e_1$. Attention: the curve $s(e)$ is artifically sheared by subtracting a linear function $25 + e * 11.5$ in order to make the convex intruder visible. *$s(e)$ is always a steeply monotonic rising function.* We clearly see the global concave (downwards bending) nature of $s(e)$ and its convex intruder. Its depth is the entropy loss due to additional correlations by the interfaces. It scales $\propto N^{-1/3}$. From this one can calculate the surface tension per surface atom $\sigma_{surf}/T_{tr} = \Delta s_{surf} * N_0/N_{surf}$. The double tangent (Gibbs construction) is the concave hull of $s(e)$. Its derivative gives the Maxwell line in the caloric curve $T(e)$ at T_{tr}, Fig. 4. In the thermodynamic limit the intruder would disappear and $s(e)$ would approach the double tangent from below. Nevertheless, even there, the probability of configurations with phase-separations are suppressed by the (infinitesimal small) factor $e^{-N^{2/3}}$ relative to the pure phases and the distribution remains *strictly bimodal in the canonical ensemble*. The region $e_1 < e < e_3$ of phase separation gets lost.

heat-capacity c, the general signal for a phase transition of first order as proposed by us more than 15 years ago [41,21,35,25,37,42]:

$$\begin{aligned} c &= \frac{\partial E}{\partial T} \\ &= -\left(\frac{\partial S}{\partial E}\right)^2 / \frac{\partial^2 S}{\partial^2 E} \end{aligned} \tag{17}$$

This was later-on further elaborated by Chomaz and Gulminelly [28,43–46] and experimentally verified [30,31].

Connecting such a system at $e_1 + \delta E, T_1$ with another one at $e_3 - \delta E > e_1 + \delta E, T_3$ and $T_1 > T_3$ then the latter one heats up to $T_{combined} = T_3 + \Delta T$ by loosing energy, whereas the former one cools down to $T_{combined} = T_1 - \Delta T$ by gaining energy. Here one of the Clausius formulation of the Second Law gets invalidated: *"Heat flows always from the hotter to the cooler body."* Or with other words in the region of negative heat capacity a system acts *in equilibrium* as a refrigerator. This is well within ordinary classical thermodynamics! However, here Ruppeiner's Riemannian geometry [1] must be extended to a Pseudo-Riemannian geometry.

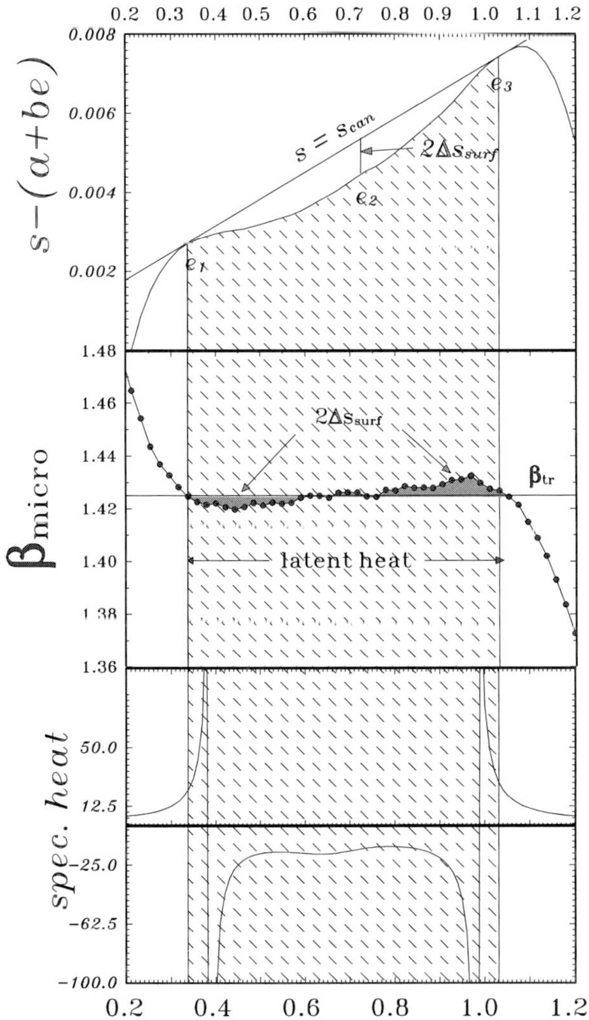

Fig. 4. a) Specific entropy $s(e) = \int_0^e \beta_{micro}(\bar{e})d\bar{e}$ vs. the specific energy e for the 2-dim. Potts model with $q = 10$ spin orientations per lattice point on a $100 * 100$ lattice. In order to visualize the anomaly of the entropy the linear function $a + be$ ($a = s(0.2119)$, $b = 1.4185$) was subtracted. Because we use periodic boundary conditions one needs two cuts to separate the phases and the depth of the convex intruder is twice the surface-entropy.
b) Inverse temperature $\beta_{micro}(e) = 1/T(e)$ as directly calculated by *MMMC*
c) Specific heat $c(e) = -\beta^2/(\partial\beta/\partial e)$. The canonical ensemble of the bulk jumps over the shaded region between the vertical lines at e_1 and e_3. This is the region of the coexistence of two phases one with ordered spins, the other with disordered spins. Here $c(e)$ has two poles and in between it becomes negative. Canonical thermodynamics is blind to this region. Observe that the poles are *inside* $e_1 \le e \le e_3$, i.e the canonical specific heat (non-dashed region) remains finite and positive as it should, from [25].

The Topology of the Curvature $c(e, n)$ De-covers the Global Phase-Diagram

Figure 5 shows the example of a micocanonical global phase diagram of the Potts ($q = 3$) lattice-gas as function of the energy $e = E/L^2$ per lattice-point and the relative occupancy $n = N/L^2$ [47]. $L \times L$ is the size of the lattice taken to be $L = 50$, and $0 \le N \le L^2$ is the number of occupied sites.

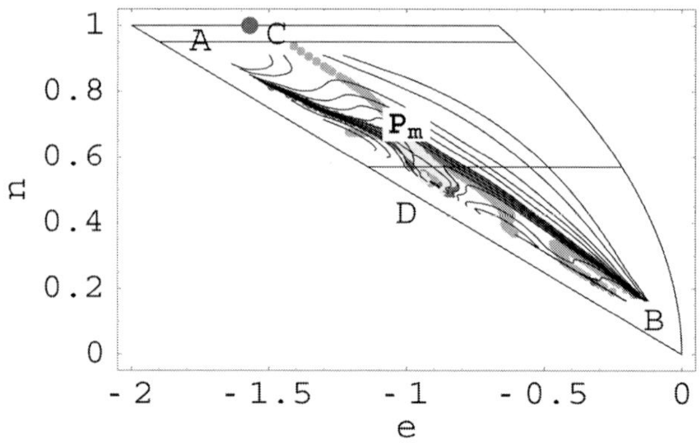

Fig. 5. Global phase diagram or conture plot of the curvature determinant (Hessian), (12), of the 2-dim Potts-3 lattice gas with $50 * 50$ lattice points, n is the number of particles per lattice point, e is the total energy per lattice point; the line (-2,1) to (0,0) is the ground-state of the lattice-gas (18) as function of n. The most right curve is the locus of configurations with completely random spin-orientations (maximum entropy). The whole physics of the model is between these two boundaries. At the dark-gray lines the Hessian is det $= 0$, this is the boundary of the region of phase separation (the triangle AP_mB) with a negative Hessian. The region of Pseudo-Riemannian geometry; At the light-gray lines is a minimum of $\det(e,n)$ in the direction of the largest curvature ($\boldsymbol{v}_{\lambda_{max}} \cdot \boldsymbol{\nabla} \det = 0$), these are lines of second order transition; In the triangle AP_mC is the pure ordered (solid) phase (det > 0); Above and right of the line CP_mB is the pure disordered (gas) phase (det > 0); The crossing P_m of the boundary lines is a multi-critical point. It is also the critical end-point of the region of phase separation (det < 0). The light-gray region around the multi-critical point P_m corresponds to a flat (cylindric) region of $\det(e,n) \sim 0$ and $\boldsymbol{\nabla}\ \lambda_1 \sim \mathbf{0}$, details see [47]; C is the analytically known position of the critical point which the ordinary $q = 3$ Potts model (without vacancies)*would have in the thermodynamic limit* $N \to \infty$, from [47].

The Hamiltonian of the lattice gas is:

$$\begin{aligned} H &= -\sum_{i,j}^{n.n.pairs} o_i o_j \delta_{\sigma_i,\sigma_j} \qquad (18) \\ n &= L^{-2} N = L^{-2} \sum_i o_i . \end{aligned}$$

$$\text{with the occupancy } o_i = \begin{cases} 1 \text{ , spin particle in site } i \\ 0 \text{ , vacancy in site } i \end{cases} .$$

Order Parameters

Definition: In the geometric theory, the order parameter o of a phase-transition is defined as the length of the trajectory along the direction of maximum curvature in the global phase-diagram see Fig. 6. Progressing in that direction one experiences the transition from one phase to the other. In the thermodynamic limit (when it exists) the region of positive maximum curvature is jumped over

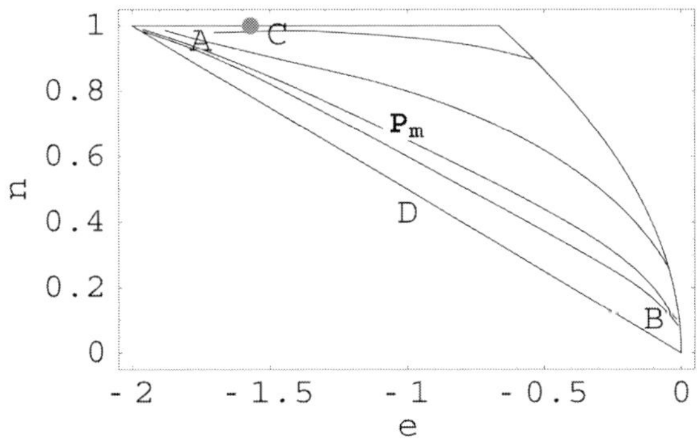

Fig. 6. Various trajectories of maximum curvature λ_1, $\boldsymbol{v_1}$ =order parameter, for the global phase-diagram of Fig. 5. Left and below the multi-critical point P_m, in the region of phase-separation (positive maximum curvature λ_1) we see an approximately linear behavior.Here $\beta(e,n)$ =const. is approximately paralell to $\nu(e,n)$ =const. and paralell to $\boldsymbol{v_1}(e,n)$.

by the canonical ensemble and the order parameter jumps here also (traditional definition of the order parameter). It is important to notice that the order parameter is *not* a simple linear function of the control parameters like E, N c.f. Fig. 6.

If there are more control parameters $(e, n, \cdots)$ there might be a situation where because of some underlying symmetry the eigenvalue of largest curvature λ_1 is degenerate. In those cases the order parameter is multi-dimensional. All these features, convex regions, curvatures of $s(e, n, \cdots)$ are *topological* properties of the entropy surface $s(e, n, \cdots)$ determined by the area of the manifold of constant energy etc. within the N-body phase space.

Ambiguity of Intensive Parameters and the Canonical Ensemble
Figure 7 explains what happens if one plots the entropy $s(e, n)$ vs. the "intensive" quantities $\beta = \partial S/\partial E$ and $\nu = \partial S/\partial N$ as one would do for the grand-canonical ensemble: As there are several points e_i, n_i with identical β, ν, $s_{micro}(\beta, \nu)$ is a *multi-valued function of* β, ν. The entropy surface $s_{micro}(e, n)$ is folded onto itself. In the projection in Fig. 7 on the β, ν-plane, these points show up as a black critical line (dense region). This black line continues over the multi-critical point P_m towards C indicating the direction towards the critical point of the ordinary $q = 3$ Potts model at $n = 1$ (zero vacancies). Between P_m and C the slopes

$$\left.\frac{\partial s}{\partial \beta}\right|_{\nu} = \frac{1}{d}[\beta s_{nn} - \nu s_{ne}] \tag{19}$$

or (20)

$$-\left.\frac{\partial s}{\partial \nu}\right|_{\beta} = \frac{1}{d}[\beta s_{en} - \nu s_{ee}] \tag{21}$$

are negative large but finite.

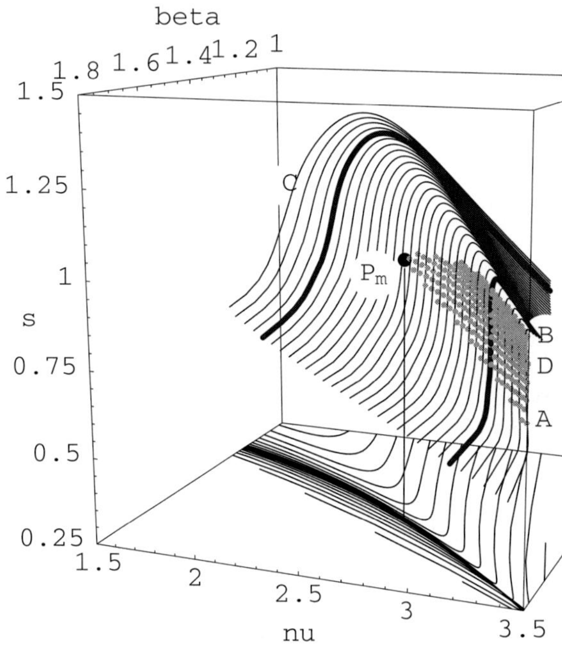

Fig. 7. Plot of the entropy $s(e, n)$ as function $s_{micro}(\beta, \nu)$ of the "intensive" variables ($\nu = \partial s/\partial n = -\beta\mu$) in the figure labeled as "nu" and $\beta = \partial s/\partial e$ is called "beta". The lines which build the surface are lines for β =cons. The positions of the points A,D,B,C defined in Fig. 5 are only roughly indicated. The convex intruder between the lines $\widehat{AP_mB}$ and the ground-state line $\widehat{ADB}$ of Fig. 5, where $s_{micro}(\beta, \nu)$ becomes multi-valued as function of $\nu > \nu_{P_m}$ and $\beta > \beta_{P_m}$, here seen from the side, is indicated by shadowing. This corresponds to phase-separation at a first order transition. At the bottom the projection of the entropy surface onto the $\{\beta, \nu\}$ plane is shown as contour plot (lines of equal $S(\beta, \nu)$). The convex part (region of phase-separation) is hidden behind the dark "critical" line in the (β, ν)-plane.

The information given by the projection would be all information which can be obtained from the conventional grand-canonical entropy $s(T, \mu, V)$, if we would have calculated it from the Laplace transform, (6).

The back folded branches, the convex intruder of $s(e, n)$, the region of phase-separation, is jumped over in the Laplace transform (6) and gets consequently lost in $Z(T, \mu)$. Here $s(T, \mu)$ becomes non-analytical $\longrightarrow$Yang-Lee singularity. This demonstrates the far more detailed insight into phase transitions and critical phenomena obtainable by the geometrical interpretation of microcanonical thermo-statistics [17] but which is not accessible by the canonical treatment, c.f. the similar arguments of Gibbs [48].

3.2 Rotating Self-gravitating Systems

Stars and Multi-star Clusters

The most interesting and important non-extensive systems are self-gravitating ones. I.e. systems with the Hamiltonian

$$H_N \equiv H_N(\{\boldsymbol{r}_i\},\{\boldsymbol{p}_i\}) = \frac{1}{2m}\sum_{i=1}^{N} p_i^2 + \Phi(\{\boldsymbol{r}_i\}) \tag{22}$$

with the gravitation $\Phi(\{\boldsymbol{r}_i\}) = -Gm^2\sum_{i<j}|\boldsymbol{r}_i - \boldsymbol{r}_j|^{-1}$. $\boldsymbol{r}_i$, $\boldsymbol{p}_i$ and m denote, respectively, the position, momentum and mass of the i-th particle. Because of the long-range of the gravitation the total potential energy is $\propto N^2$ and consequently non-extensive.

The statistical equilibrium of self-gravitating systems without angular-momentum was first considered by Thirring [11] who pointed out that microcanonically these systems have a negative heat capacity and therefore the microcanonical and canonical ensembles are not equivalent.

Here, we overcome the simple Thirring model and investigate the equilibrium of rotating, self-gravitating systems in a spherical symmetric box under various energies E and angular-momenta L. These calculations are done by E.V.Votyakov [12].

The following approximations are used:

a) Mean-field approximation: We approximate any N-body spatial density $\rho_N = \rho(\boldsymbol{r}_1,\dots,\boldsymbol{r}_N)$ by the non-correlated product of single-particle densities [12] and work henceforth in the single-particle μ-space.

$$\rho_N(\boldsymbol{r}_1,\dots,\boldsymbol{r}_N) \approx \prod_{i=1}^{N}\rho(\boldsymbol{r}_i) \tag{23}$$

and the gravitation interaction

$$\Phi[\rho_N] \approx -Gm^2\sum_{i<k}\int\frac{\rho(\boldsymbol{r}_i)\rho(\boldsymbol{r}_k)}{|\boldsymbol{r}_i-\boldsymbol{r}_k|}d\boldsymbol{r}_i d\boldsymbol{r}_k \tag{24}$$

The aim is now to find those density profiles $\rho(\boldsymbol{r})$ that maximize the (mean-field) entropy ($k=1$)

$$S_N^m(E,\boldsymbol{L}) = \ln W_N^m(E,\boldsymbol{L}) \tag{25}$$

W_N^m being the mean-field approximation to the microcanonical "partition sum", i.e. the sum of all uncorrelated many-body states ($W_N^m \le W_N$). *After integrating over the N-momenta*:

$$W_N^m(E,\boldsymbol{L}) = \frac{A}{N!}\int\left[E - \frac{1}{2}\boldsymbol{L}^T\mathbb{I}^{-1}\boldsymbol{L} - \Phi[\rho]\right]^{\frac{3N-5}{2}} P[\rho]d\rho(\boldsymbol{r}) \tag{26}$$

where $P[\rho]$ is the probability to observe a density profile $\rho \equiv \rho(\boldsymbol{r})$.

We use the same trick as Lynden-Bell [49] and avoid configurations with high densities where other physical processes like nuclear reactions become more important than gravity. To achieve this, we subdivide the spherical volume V into K identical cells labeled by the positions of their centers. The idea is to replace the integral over V with a sum over the cells. In order to avoid overlapping and to cure the short-distance singularity of the Newtonian potential we assume that each cell may host up to n_0 particles ($1 \ll n_0 \ll N$ but $Kn_0 > N$). This condition is essentially equivalent to consider hard spheres instead of point particles. The probability $P[\rho]$ to find a given density profile $\rho(\boldsymbol{r})$ is now proportional to the number of ways in which our N particles can be distributed inside the K cells with maximal capacity n_0 and individual occupancies $n(\boldsymbol{r}_k)$ or $\rho(\boldsymbol{r}_k) = n(\boldsymbol{r}_k)K/V$. A simple combinatorial reasoning leads to

$$P(\rho) = \prod_{\text{cells } k, \sum n(r_k)=N} \binom{n_0}{n(\boldsymbol{r}_k)}. \tag{27}$$

This looks analogous to a Fermi-Dirac statistics, however, here *only in coordinate space*. This guarantees the strict non-overlapping condition. ρ can nowhere be larger than $\rho_0 = n_0 K/V$. In contrast, in a recent paper Chavanis and Ispolatov [50] use Fermi-Dirac statistics in *phase-space*.

We express S_N^m as a functional of the density profile $\rho(\boldsymbol{r})$, such that

$$\int_V \rho(\boldsymbol{r}) d\boldsymbol{r} = N \tag{28}$$

and subsequently find the $\rho(\boldsymbol{r})$'s that maximize S_N^m. This leads to the self-consistent integral equation:

$$\log \frac{\rho(\boldsymbol{x})}{\rho_0 - \rho(\boldsymbol{x})} = -\frac{\beta}{\Theta} U(\boldsymbol{x}) + \frac{1}{2}\beta(\boldsymbol{\omega} \times \boldsymbol{x})^2 - \mu \tag{29}$$

or, equivalently,

$$\rho(\boldsymbol{x}) = \frac{\rho_0}{1 + e^{\frac{\beta}{\Theta}U(\boldsymbol{x}) - \frac{1}{2}\beta(\boldsymbol{\omega}\times\boldsymbol{x})^2 + \mu}} \tag{30}$$

where $\boldsymbol{\omega}[\rho]$ is the angular velocity (related to the total angular momentum by the relation $\boldsymbol{L} = \boldsymbol{I}[\rho]\boldsymbol{\omega}[\rho]$), and $\beta[\rho]$ and $U(\boldsymbol{x})$ are respectively defined as

$$\beta = \frac{3/2}{E - \frac{1}{2}\boldsymbol{L}^T(\boldsymbol{I}[\rho])^{-1}\boldsymbol{L} - \Phi[\rho]} \tag{31}$$

$$U(\boldsymbol{x}) = -\int \frac{\rho(\boldsymbol{x}')}{|\boldsymbol{x} - \boldsymbol{x}'|}\, d\boldsymbol{x}' \tag{32}$$

$$\Phi[\rho] = \frac{1}{2}\int U(\boldsymbol{x})\rho(\boldsymbol{x}) d\boldsymbol{x} \tag{33}$$

b) To solve (30), Votyakov [12] expands $\rho(\boldsymbol{r}_i)$ into spherical harmonics:

$$\rho(\boldsymbol{r}_i) = \sum_{l,m} b_{l,m}(|r_i|)\, Y_{lm}(\Theta_i, \phi_i) \tag{34}$$

Fig. 8. Global phase diagram of a self-gravitating system as function of energy E and angular-momentum L in dimension-less units. Systematic calculations were done only in the dotted region. There are 3 different mono-phases (regions of negative maximum curvature $\lambda_1 < 0$): A single star plus some gas at small E and L; at high energy and moderate L, a gas phase; and at large L configurations of broken spherical symmetries with double-star systems. The mixed phase is the region with positive maximum curvature $\lambda_1 > 0$ where at least two different phases (single star and gas (low L), single star and double star systems, eventually rings, at intermediate L are in competition, finally double-stars and gas compete with one-another at large L and E.

and ignores for simplicity reasons all odd l, m as well also all $l > 16$. I.e. he allows only for parity even, upside – down symmetric configurations (this is later overcome). The expansion into spherical harmonics has the advantage that the original non-linear self-consistent three-dimensional integral equation (30) becomes now a finite self-consistent set of coupled one-dimensional and two-dimensional integral equations. As function of energy and total angular-momentum the microcanonical global phase-diagram defined by the topology of the curvature (Hessian) of $S(E, N)$ shows an astonishing rich picture see Fig. 8.

This example proves the superiority of the geometrical, topological, interpretation of thermo-statistics. It reproduces realistically many different configurations of even gravitating systems. No canonical description nor Tsallis non-extensive statistics had achieved this.

4 Approach to Equilibrium, Second Law

4.1 Zermelo's Paradox

When Zermelo [16] argued against Boltzmann, that following Poincaré any many-body system must return after the Poincaré recurrence time t_{rec} and consequently its entropy cannot grow indefinitely, Boltzmann [34] answered that for

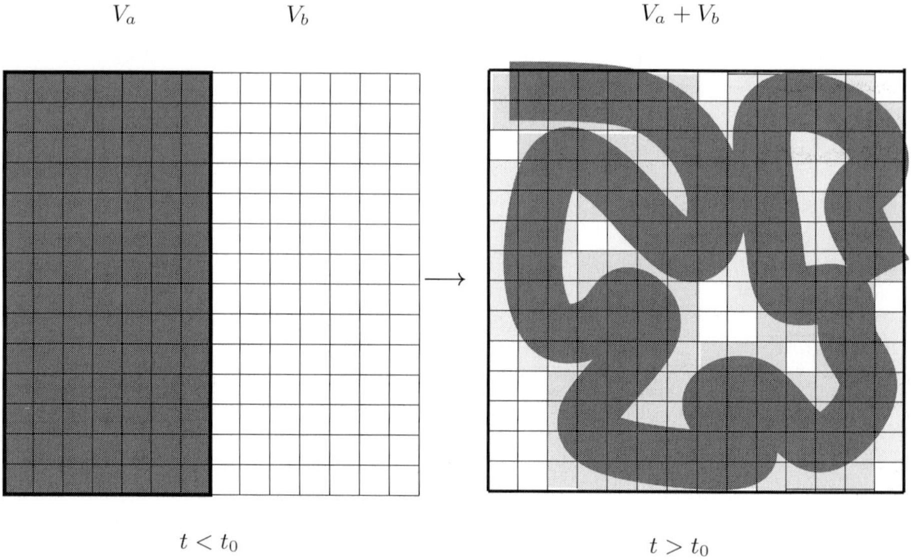

Fig. 9. The compact set $\mathcal{E}(t_0)$, left side, develops into an increasingly folded but non-crossing "spaghetti"-like distribution $\mathcal{E}(t, t_0)$ in the phase-space with rising time t after opening the volume V_b. The right figure shows only the early form of the distribution. At much later times it will become more and more fractal and finally dense in the new larger phase space. The grid illustrates the boxes of the box-counting method. All boxes which overlap with $\mathcal{M}(t, t_0)$ contribute to the box-counting volume and are shaded gray. Their number is N_δ.

any macroscopic system t_{rec} is of several orders of magnitude larger than the age of the universe, c.f. Gallavotti [4]. Still today, this is the answer given when the Second Law is to be proven microscopically, c.f. [51]. Then, Zermelo's paradox becomes blunted.

Here, I argue, even a *small* system approaches equilibrium with a rise of its entropy $\Delta S \geq 0$ under quite general conditions. Thus, Zermelo's objection must be considered much more seriously.

However, care must be taken, Boltzmann's definition of entropy (3) is only for systems at equilibrium. To be precise: in the following I will consider the equilibrium manifold $\mathcal{E}(E, V_a)$ at $t \leq t_0$. At t_0 the macroscopic constraint V_a is quickly removed e.g. a piston pulled quickly out to $V_a + V_b$, and the ensemble is followed in phase-space how it approaches the new equilibrium manifold $\mathcal{E}(E, V_a + V_b)$ sce Fig. 9.

4.2 The Solution

Entropy does not refer to a single point in N-body phase space but to the whole ensemble $\mathcal{E}$ of points. It is the $\ln(W)$ of the geometrical size W of the ensemble. Every trajectory starting at different points in the initial manifold $\mathcal{M}(t = t_0) = \mathcal{E}(E, V_a) \in \mathcal{E}(E, V_a + V_b)$ spreads in a non-crossing manner over the available phase-space $\mathcal{E}(E, V_a + V_b)$ but returns after t_{rec}. Different points of the manifold, or trajectories, have different t_{rec} which are normally incommensurable. I.e. the manifold $\mathcal{M}(t)$ spreads irreversibly over $\mathcal{E}(E, V_a + V_b)$.

Mixing

When the system is dynamically mixing then the manifold $\mathcal{M}(t)$ will "fill" the new ensemble $\mathcal{E}(E, V_a + V_b)$. Though at finite times the manifold remains compact due to Liouville and keeps the volume $W(E, V_a)$, but as already argued by Gibbs [52,53] $\mathcal{M}(t)$ will be filamented like ink in water and will approach any point of $\mathcal{E}(E, V_a + V_b)$ arbitrarily close. Then, $\lim_{t\to\infty} \mathcal{M}(t)$ becomes dense in the new, larger $\mathcal{E}(E, V_a + V_b)$. The *closure* $\overline{\mathcal{M}(t = \infty)}$ becomes equal to $\mathcal{E}(E, V_a + V_b)$. I.e. the entropy $S(t = \infty) > S(t_0)$. This is the Second Law for a *finite* system.

Macroscopic Resolution, Fractal Distributions and Closure [19]

We calculate the closure of the ensemble by box counting [54]. Here the phase-space is divided in N_δ equal boxes of volume δ^{6N}. The number of boxes which overlap with $\mathcal{M}(t)$ is N_δ and the box-counting volume is then:

$$\Omega_d(\delta) = N_\delta \delta^d, \qquad \text{here with } d = 6N - 1 \tag{35}$$

$$\Omega_d = \lim_{\delta\to 0} \Omega_d(\delta). \tag{36}$$

The box-counting method is illustrated in Fig. 9. The important aspect of the box-counting volume of a manifold is that it is equal to the volume of its closure.

At finite times $\mathcal{M}(t)$ is compact. Its volume $W(t)$ equals that of its closure $\equiv W(t_0)$. However, calculated with finite resolution $\delta > 0$, $W_\delta(t)$ becomes $\geq W(t)$ for t larger than some t_δ, where

$$W_\delta(t) = ⨌_d^\delta \frac{d^{3N}p\, d^{3N}q}{(2\pi\hbar)^{3N}}\, \delta\left(E - H_N\{q(t), p(t); [q(t_0), p(t_0) \in \mathcal{E}(V_a)]\}\right) \tag{37}$$

with

$$⨌_d^\delta f(q) d^d q = \Omega_d(\delta)\overline{f(q)}.$$

A natural finite resolution would be

$$\delta = \sqrt{2\pi\hbar}. \tag{38}$$

Of course the actual problem will often allow a much coarser resolution because of the insensitivity of the usual macroscopic observables. Then the equilibration time t_δ will also be much shorter.

Thus the new definition of Boltzmann's principle equation (2) is:

$$S = \ln(W_\delta), \tag{39}$$

or mathematically correct, though unphysical, at infinite times:

$$S = \lim_{\delta\to 0}\lim_{t\to\infty} \ln(W_\delta(t)) = N \ln[(V_a + V_b)/V_a] + S(t_0). \tag{40}$$

5 Conclusion

The geometric interpretation of classical equilibrium Statistical Mechanics by Boltzmann's principle offers an extension also to the equilibrium of non-extensive systems.

Because microcanonical Thermodynamics as a macroscopic theory controls the system by a few, usually conserved, macroscopic parameters like energy, particle number, etc. it is an intrinsically probabilistic theory. It describes all systems with the same control-parameters simultaneously. If we take this seriously and avoid the so called thermodynamic limit ($\lim_{V\to\infty,N/V=\rho}$), the theory can be applied to the really large, usually *inhomogeneous*, self-gravitating systems. In Chap. 3.2 it is shown how this new approach enables to view many realistic astro-physical configurations as *equilibrium configurations under the control of total energy and angular-momentum*, c.f.[12].

Within the new, extended, formalism several principles of traditional Statistical Mechanics turn out to be violated and obsolete. E.g. we saw that at phase-separation heat (energy) can flow from cold to hot. Or phase-transitions can be classified unambiguously in astonishingly small systems. These are by no way exotic and wrong conclusions. On the contrary, many experiments have shown their validity. I believe this approach gives a much deeper insight into the way how many-body systems organize themselves than any canonical statistics is able to. The thermodynamic limit clouds the most interesting region of Thermodynamics, the region of inhomogeneous phase-separation.

Because of the only *one* underlying axiom, Boltzmann's principle equation (3), the geometric interpretation keeps statistics most close to Mechanics and, therefore, is most transparent. The Second Law ($\Delta S \geq 0$) is shown to be valid in *closed, small* systems under quite general dynamical conditions.

Acknowledgements

I have to thank my various collaborators over the last 10 years. Their numerical work produced much of the material presented here. However, first of all I must mention E.V.Votyakov who developed the sophisticated program for the Potts lattice-gas and the self-gravitating and rotating system discussed in Sect. 3.2.

References

1. G. Ruppeiner: Rev. Mod. Phys. **67**, 605 (1995)
2. C. Tsallis: Braz. J. Phys. **29**, 1 (1999)
3. C. Tsallis: J. Stat. Phys **52**, 479 (1988)
4. G. Gallavotti, *Statistical Mechanics*, Springer, Berlin (1999)
5. J.L. Lebowitz: Rev. Mod. Phys. **71**, S346 (1999)
6. J.L. Lebowitz: Physica A **263**, 516 (1999)
7. E.H. Lieb, J. Yngvason: Notices Amer. Math. Soc. **45**, 571 (1998)
8. C.N. Yang, T.D. Lee: Phys. Rev. **87**, 404 (1952)
9. D.H.E. Gross: Chaos, Solitons and Fractals **13**, 417 (2002)

10. D. Lynden-Bell, R. Wood: Mon. Not. R. Astr. Soc. **138**, 495 (1968)
11. W. Thirring: Z. Phys. **235**, 339 (1970)
12. E.V. Votyakov, H.I. Hidmi, A. De Martino, D.H.E. Gross: Phys. Rev. Lett. **89**, 031101 (2002)
13. P. Gaspard: J. Stat. Phys. **88**, 1215 (1997)
14. B.H. Lavanda, J. Dunning-Davies: Foundations of Physics Letters **5**, 435 (1990)
15. E. H. Lieb, J. Yngvason: The mathematical structure of the second law of thermodynamics. math-ph/0204007v1 (2002)
16. E. Zermelo: Wied. Ann. **60**, 392 (1897)
17. D.H.E. Gross: PCCP **4**, 863 (2002)
18. E. H. Lieb, J. Yngvason: Physics Report **310**, 1 (1999)
19. D.H.E. Gross. Ensemble probabilistic equilibrium and non-equilibrium thermodynamics without the thermodynamic limit. In Andrei Khrennikov, editor, Foundations of Probability and Physics, number XIII in PQ-QP: Quantum Probability, White Noise Analysis, pages 131, Boston, October 2001. ACM, World Scientific.
20. A. Einstein: Annalen der Physik **17**, 132 (1905)
21. D.H.E. Gross: Rep. Progr. Phys. **53**, 605 (1990)
22. A. Chbihi, O. Schapiro, S. Salou, L.G. Sobotka: Experimental evidence for a phase transition in nuclear evaporation process. preprint (1995)
23. R.M. Lynden-Bell: Negative specific heat in clusters of atoms. to be published in Galactic Dynamics (1995)
24. D.H.E. Gross, M.E. Madjet: Microcanonical vs. canonical thermodynamics. cond-mat/9611192 (1996)
25. D.H.E. Gross, A. Ecker, X.Z. Zhang: Ann. Physik **5**, 446 (1996)
26. D. Lynden-Bell: Physica A **263**, 293 (1999)
27. A. Chbihi, O. Schapiro, S. Salou, D.H.E. Gross: Eur. Phys. J. A **5**, 251 (1999)
28. F. Gulminelli, Ph. Chomaz: Phys. Rev. Lett. **82**, 1402 (1999)
29. L. Casetti, M. Pettini, E.G.D. Cohen, Phys. Rep. **337**, 238 (2000)
30. M. Schmidt et al: Negative heat capacity for a cluster of 147 sodium stoms. submitted to Nature (2000)
31. M. D'Agostino et al: Phys. Lett. B **473**, 219 (2000)
32. M. Schmidt et al: Phys. Rev. Lett. **86**, 1191 (2001)
33. I. Ispolatov, E.G.D. Cohen: Physica A **295**, 475 (2001)
34. L. Boltzmann: Entgegnung auf die wärmetheoretischen Betrachtungen des Hrn.E. Zermelo, In *Kinetic Theory*, S. Brush ed., Pergamon Press, Oxford, 1965-1972
35. D.H.E. Gross, R. Heck: Phys. Lett. B **318**, 405 (1993)
36. P.A. Hervieux, D.H.E. Gross: Z. Phys. D **33**, 295 (1995)
37. D.H.E. Gross: *Microcanonical thermodynamics: Phase transitions in "Small" systems*, volume 66 of *Lecture Notes in Physics*. World Scientific, Singapore (2001)
38. Julien Barré, David Mukamel, Stefano Ruffo: Phys. Rev. Lett. **87**, 030601 (2001)
39. B. Andresen: Rev. Gen. Therm. **35**, 647 (1996)
40. C. Bréchignac et al: J. Chem. Phys. **102**, 1 (1995)
41. D.H.E. Gross, H. Massmann: Nucl. Phys A **471**, 339c (1987)
42. D.H.E. Gross: Nucl. Phys. A **681**, 366c (2001)
43. F. Gulminelli, Ph. Chomaz, V. Duflot: Europhys. Lett. **50**, 434 (2000)
44. Ph. Chomaz, F. Gulminelli: Nucl. Phys. A **647**, 153 (1999)
45. Ph. Chomaz, F. Gulminelli, V. Duflot: Phys. Rev. E **64**, 046114 (2001)
46. Ph. Chomaz, V.Duflot, F. Gulminelli: Caloric curves and energy fluctuations in the microcanonical liquid–gas phase transition. preprint (2000)
47. D.H.E. Gross, E. Votyakov: Eur. Phys. J. B **15**, 115 (2000)

48. J.W. Gibbs: Graphical methods in the thermodynamics of fluids. volume I of *The scientific papers of J. Willard Gibbs*, Longmans, Green, NY (1906)
49. D. Lynden-Bell: Mon. Not. R. Astr. Soc. **136**, 101 (1967)
50. P.H. Chavanis, I. Ispolatov: Phys. Rev. E **66**, 036109 (2002)
51. T. Gilbert, J.R.Dorfman, P.Gaspard: Phys. Rev. Lett. **85**, 1606 (2000)
52. J.W. Gibbs: *Elementary Principles in Statistical Physics*, vol. II of *The Collected Works of J. Willard Gibbs*. Yale Univ. Press (1902)
53. J.W. Gibbs: Collected works and commentary, vol. I and II. Yale Univ. Press (1936)
54. Kenneth Falconer: *Fractal Geometry - Mathematical Foundations and Applications.* John Wiley & Sons, Chichester, New York, Brisbane, Toronto, Singapore (1990)

Ensemble Inequivalence in Mean-Field Models of Magnetism

Julien Barré[1,3], David Mukamel[2], and Stefano Ruffo[1,3]

[1] Ecole Normale Supérieure de Lyon, Laboratoire de Physique, 46 Allée d' Italie, 69364 Lyon Cedex 07, France
[2] Department of Physics of Complex Systems, The Weizmann Institute of Science, Rehovot 76100, Israel
[3] Dipartimento di Energetica "Sergio Stecco", Università di Firenze, via s. Marta 3, 50139 Firenze, Italy, INFM and INFN

Abstract. Mean-field models, while they can be cast into an *extensive* thermodynamic formalism, are inherently *non additive*. This is the basic feature which leads to *ensemble inequivalence* in these models. In this paper we study the global phase diagram of the infinite range Blume-Emery-Griffiths model both in the *canonical* and in the *microcanonical* ensembles. The microcanonical solution is obtained both by direct state counting and by the application of large deviation theory. The canonical phase diagram has first order and continuous transition lines separated by a tricritical point. We find that below the tricritical point, when the canonical transition is first order, the phase diagrams of the two ensembles disagree. In this region the microcanonical ensemble exhibits energy ranges with negative specific heat and temperature jumps at transition energies. These two features are discussed in a general context and the appropriate Maxwell constructions are introduced. Some preliminary extensions of these results to weakly decaying nonintegrable interactions are presented.

1 Introduction

Thermodynamics of systems with long-range interactions is quite distinct from that of systems where the interactions are short ranged. For pairwise interactions, long range potentials decay at large distances as $V(r) \sim 1/r^\alpha$ with $0 \leq \alpha \leq d$ in d dimensions. Such systems are *not extensive* , although models used to describe them can be made extensive by an appropriate volume dependent rescaling (the Kac prescription) [1] . However, even with such rescaling, these models remain inherently *non-additive*. Namely, when divided into two or more macroscopic subsystems, the total energy (or other extensive quantity) of the system, is not necessarily equal to the sum of the energies of the subsystems. Since additivity is an essential ingredient in the derivation of thermodynamics and statistical mechanics of systems with short range interactions, its violation may lead to interesting and unusual effects.

To illustrate the lack of additivity in these systems consider an extremely simple example of an Ising model with infinite range, mean-field like, interactions,

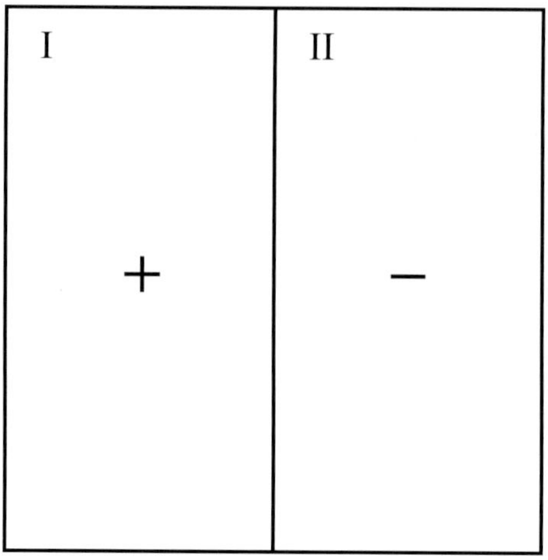

Fig. 1. A two phase configuration of the infinite range interaction Ising model for which the energy is non-additive

corresponding to $\alpha = 0$. The Hamiltonian takes the form

$$H = -\frac{J}{2N}\left(\sum_{i=1}^{N}\sigma_i\right)^2 \tag{1}$$

where $\sigma_i = \pm 1$ is the spin 1/2 variable on site i and N is the number of spins in the system. This Kac type pre-factor is taken to ensure *extensivity*. Let us consider for simplicity a system with vanishing magnetization, $M = \sum_{i=1}^{N}\sigma_i = 0$. It is clear that the energy, $E = H$, of the system satisfies $E = 0$. Now let us divide the system into two subsystems, I and II, each composed of $N/2$ sites, where all spins in subsystem I are up while those in subsystem II are down (see Fig. 1). The energies of the two subsystems, E_I and E_{II} satisfy $E_I = E_{II} = -JN/8$. Since the sum of the two energies $E_I + E_{II}$ is not equal to the total energy E the system is clearly not additive.

The lack of additivity may result in many unusual properties of systems with long-range interactions, like inequivalence of various ensembles, negative specific heat in the microcanonical ensemble and possible temperature discontinuity at first order transitions. For example, the usual argument for the non-negativity of the specific heat, or equivalently, for the concavity of the entropy-energy curve $s(\epsilon)$, makes use of the fact that systems with short range interactions are additive. Here $s = S/N$ and $e = E/N$ are the entropy and energy per particle, respectively. An entropy curve which is not concave in the energy interval $e_1 < e < e_2$ (see Fig. 2) is unstable for systems with short range interactions since entropy may be gained by phase separating the system into two subsystems with energy densities e_1 and e_2 while keeping the total energy fixed. In particular the average energy and entropy densities in the coexistence region is given by the weighted average of the corresponding densities of the two coexisting systems, and thus the correct entropy curve in that region is given by the common tangent line (see Fig. 2), resulting in an overall concave curve. On the other hand for systems with long-range interactions, additivity does not hold, and thus it is not always the case that entropy may be gained by phase separation. In such cases the non-concave entropy curve may in fact be the correct entropy of the system, resulting, for example, in negative specific heat.

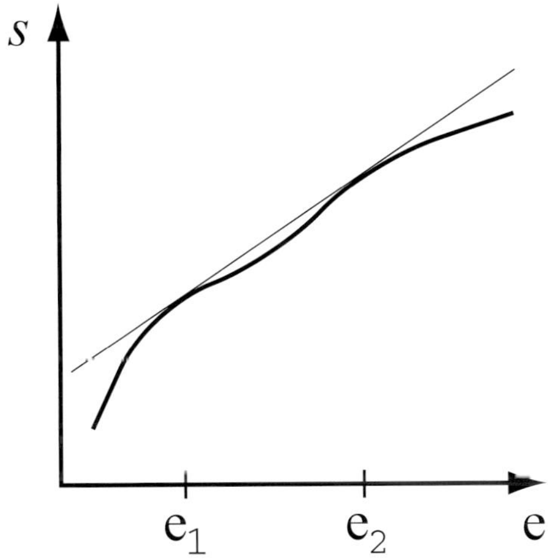

Fig. 2. A non-concave entropy curve, which for additive systems is made concave by considering two phase coexistence configurations in the region where the entropy of the homogeneous system is non-concave. This procedure is not applicable for non-additive systems like those with long-range interactions.

Self gravitating particles are perhaps the most extensively studied systems with long-range forces (corresponding to $\alpha = 1$ in $d = 3$ dimensions). The fact that the entropy of these systems is not necessarily a concave function of the energy has first been pointed out by Antonov [2]. The thermodynamical consequences of this observation have been elaborated using simple models by Lynden-Bell [3], Thirring [4] and others (see also the papers by Chavanis [5] and Padhmanaban [6] in this book for a review).

In fact, non-additivity is not limited to systems with long-range interactions. Finite systems with short range interactions, such as atomic clusters, are non-additive as long as they are small enough so that surface energy may not be neglected as compared with the bulk energy [7]. Negative specific heat in clusters of atoms has been discussed by R. M. Lynden-Bell [8] and it has been recently observed experimentally, as discussed in this book in Refs. [7,9].

In the present paper we consider a simple spin 1 mean-field model ($\alpha = 0$) and study its phase diagram both within the canonical and the microcanonical ensembles. This model, known as the Blume-Emery-Griffiths (BEG) model [10], may serve as a useful tool for studying thermodynamic effects which characterize systems with long-range interactions. The canonical phase diagram of this model has been studied in the past, and it exhibits a line of phase transitions separating a magnetically ordered phase from a paramagnetic one. The line is composed of a first order transition and a second order transition segments separated by a tricritical point. The microcanonical phase diagram on the other hand exhibits some rather distinct features [11]. While the second order segment of the canonical phase diagram is present also in the microcanonical ensemble, the nature of the first order segment, its location and the location of the tricritical point are rather different. In this ensemble, a region with negative specific heat is found, as well as a first order transition at which the temperatures of the two coexisting phases are different.

The paper is organized as follows: in Sect. 2 the BEG model is introduced and its canonical phase diagram is studied. The microcanonical phase diagram and the relationship between the two ensembles are studied in Sect. 3. Some features of the Maxwell constructions leading to temperature jumps are presented in Sect. 4. An alternative more general derivation of the expression for microcanonical entropy is presented in Sect. 5. The results presented in this paper are generalized to models with slowly decaying interactions (with $0 < \alpha < d$) in Sect. 6. Finally, concluding remarks are given in Sect. 7

2 The Blume-Emery-Griffiths Model in the Canonical Ensemble

We consider a spin model with infinite range, mean-field like, interactions whose phase diagram can be analyzed analytically both within the canonical and the microcanonical ensembles. This study enables one to compare the two resulting diagrams and get a better understanding of the effect of the non-additivity on the thermodynamic behavior of the system. The model we consider is a simple version of the Blume-Emery-Griffiths (BEG) model [10], known as the Blume-Capel model, with infinite range interactions. The model is defined on a lattice, where each lattice point i is occupied by a spin-1 variable $S_i = 0, \pm 1$. The Hamiltonian is given by

$$H = \Delta \sum_{i=1}^{N} S_i^2 - \frac{J}{2N} \left(\sum_{i=1}^{N} S_i \right)^2 \tag{2}$$

where $J > 0$ is a ferromagnetic coupling constant and Δ controls the energy difference between the magnetic ($S_i = \pm 1$) and the non-magnetic ($S_i = 0$) states. The canonical phase diagram of this model in the (T, Δ) plane has been studied in the past [10]. The free energy

$$f(\beta) = -\frac{1}{\beta} \lim_{N \to \infty} \frac{\ln Z}{N}, \tag{3}$$

where Z is the partition function

$$Z = \sum_{[S_i]} \exp -\beta H, \tag{4}$$

with $\beta = 1/k_B T$ and k_B the Boltzmann constant, can be exactly derived in the $N \to \infty$ limit. One uses the Hubbard-Stratonovich transformation, which in this case amounts to the simple Gaussian identity

$$\exp(ba^2) = \sqrt{\frac{b}{\pi}} \int_{-\infty}^{\infty} dx \exp(-bx^2 + 2abx), \tag{5}$$

with $a = \sum_i S_i/N$ and $b = \beta JN/2$. One then easily gets

$$Z = \sqrt{\frac{\beta JN}{2\pi}} \int_{-\infty}^{\infty} dx \exp(-N\beta\tilde{f}(\beta, x)) \tag{6}$$

where

$$\beta\tilde{f}(\beta, x) = \frac{1}{2}\beta Jx^2 - \ln[1 + e^{-\beta\Delta}(e^{\beta Jx} + e^{-\beta Jx})]. \tag{7}$$

Using Laplace's method to perform the integral in formula (6) in the $N \to \infty$ limit, one finally gets

$$f(\beta) = \min_x \tilde{f}(\beta, x) \tag{8}$$

At $T = 0$ the model has a ferromagnetic phase for $2\Delta/J < 1$ and a non-magnetic phase otherwise separated by a first order phase transition. Indeed, the paramagnetic zero temperature state $S_i = 0, \forall i$ is degenerate with the ferromagnetic state $S_i = 1, \forall i$ (or $S_i = -1, \forall i$) at $2\Delta/J = 1$ (this latter state being the ground state for $2\Delta/J < 1$), a typical scenario for first order phase transitions. At $\Delta = -\infty$ the model reduces to the mean-field Ising model and hence it has a second order phase transition at $k_B T/J = 1$. The (T, Δ) phase diagram displays a transition line separating the low temperature ferromagnetic phase from the high temperature paramagnetic phase. The transition line is found to be first order at high Δ values, while it is second order at low Δ. The second order line is found by locating the instability of the paramagnetic phase $x = 0$, which means finding the condition for the x^2 coefficient in formula (7) to be zero (note that $\tilde{f}(\beta, x)$ is even in x). One gets for the second order term

$$\beta J = \frac{1}{2}e^{\beta\Delta} + 1\ . \tag{9}$$

The two segments of the transition line (high and low Δ) are separated by a tricritical point located at $\Delta/J = \ln(4)/3 \simeq 0.4621$, $\beta J = 3$. This is obtained by requiring that both the x^2 coefficient of $\tilde{f}(\beta, x)$ in (7) and the x^4 coefficient

$$\frac{1}{3} - \frac{2\exp(-\beta\Delta)}{(1 + 2\exp(-\beta\Delta))} \tag{10}$$

vanish. The first order segment of the transition line is obtained numerically by equating the free energies of the ferromagnetic and the paramagnetic states. The behavior of the function $\beta f(\beta, x)$ as β varies is shown in Fig. 3 at a second order phase transition ($\Delta = 0$) and in Fig. 4 at a first order phase transition ($\Delta = 0.476190$).

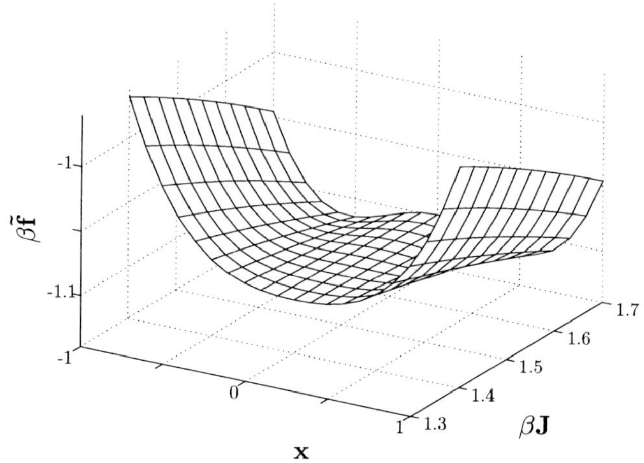

Fig. 3. $\beta\tilde{f}(\beta, x)$ as a function of $\beta J \in [1.4, 1.7]$ and $x \in [-1, 1]$ for $\Delta = 0$. The second order transition is at $\beta J = 3/2$ where the minimum at $x = 0$ becomes a maximum and two side minima develop

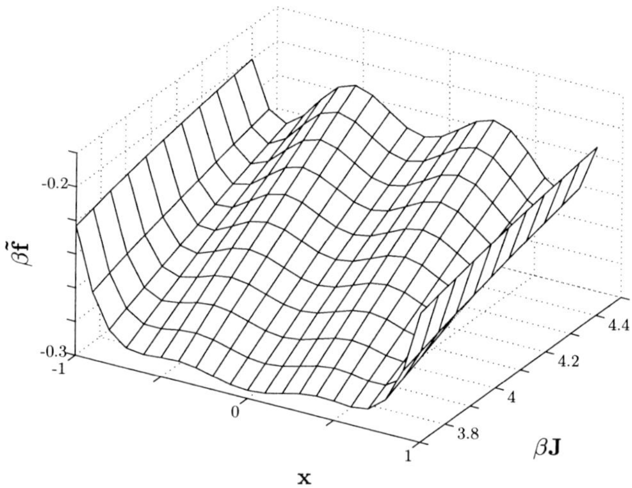

Fig. 4. $\beta\tilde{f}(\beta, x)$ as a function of $\beta J \in [3.7, 4.4]$ and $x \in [-1, 1]$ for $\Delta = 0.476190$. The first order transition is at $\beta J = 4$. Two side minima to $x = 0$ first develop and then reach the same hight at the transition point.

3 Microcanonical Solution of the Blume–Emery–Griffiths Model

The derivation of the phase diagram of the BEG model (2) in the *microcanonical ensemble* reduces to a simple counting problem, since all spins interact with equal strength, independently of their mutual distance (there is no space in the model). A given microscopic configuration is characterized by the numbers N_+, N_-, N_0

of up, down and zero spins, with $N_+ + N_- + N_0 = N$. The energy E of this configuration is only a function of N_+, N_- and N_0 and is given by

$$E = \Delta Q - \frac{J}{2N} M^2 \ , \tag{11}$$

where $Q = \sum_{i=1}^{N} S_i^2 = N_+ + N_-$ (the quadrupole moment) and $M = \sum_{i=1}^{N} S_i = N_+ - N_-$ (the magnetization) are the two order parameters. The number of microscopic configurations Ω compatible with the macroscopic occupation numbers N_+, N_- and N_0 is

$$\Omega = \frac{N!}{N_+!N_-!N_0!} \ . \tag{12}$$

Using Stirling's approximation in the large N limit, the entropy, $S = k_B \ln \Omega$, is given by

$$\begin{aligned} S = -k_B N[(1-q)\ln(1-q) + \frac{1}{2}(q+m)\ln(q+m) \\ + \frac{1}{2}(q-m)\ln(q-m) - q\ln 2] \ , \end{aligned} \tag{13}$$

where $q = Q/N$ and $m = M/N$ are the quadrupole moment and the magnetization per site, respectively. Let $\epsilon = E/\Delta N$ be the dimensionless energy per site, normalized by Δ. Equation (11) may be written as

$$q = \epsilon + K m^2 \ , \tag{14}$$

where $K = J/2\Delta$. Using this relation, the entropy per site $s = S/(k_B N)$ can be expressed in terms of m and ϵ. At fixed ϵ, the value of m which maximizes the entropy corresponds to the equilibrium magnetization. The corresponding equilibrium entropy $s(\epsilon) = \max_m s(\epsilon, m)$ contains all the information about the thermodynamics of the system in the microcanonical ensemble. For instance, temperature can be obtained from the relation

$$\frac{\Delta}{k_B T} = \frac{\partial s}{\partial \epsilon} \ . \tag{15}$$

As usual in systems where the energy per particle is bounded from above, the model has both a positive and a negative temperature region: entropy is a one humped function of the energy. The interesting features take place in the positive temperature range. In order to locate the continuous transition line, one develops $s(\epsilon, m)$ in powers of m, in analogy with what has been done above for the canonical free energy

$$s = s_0 + A m^2 + B m^4 + O(m^6) \ , \tag{16}$$

where

$$s_0 = s(\epsilon, m=0) = -(1-\epsilon)\ln(1-\epsilon) - \epsilon\ln\epsilon + \epsilon\ln 2 \ , \tag{17}$$

and

$$
\begin{aligned}
A &= -K \ln \frac{\epsilon}{2(1-\epsilon)} - \frac{1}{2\epsilon} \,, \\
B &= -\frac{K^2}{2\epsilon(1-\epsilon)} + \frac{K}{2\epsilon^2} - \frac{1}{12\epsilon^3} \,.
\end{aligned}
\tag{18}
$$

In the paramagnetic phase both A and B are negative, and the entropy is maximized by $m = 0$. The continuous transition to the ferromagnetic phase takes place at $A = 0$ for $B < 0$. In order to obtain the critical line in the (T, Δ) plane we first observe that temperature is calculable on the critical line ($m = 0$) using (15) and (17)

$$
\frac{\Delta}{k_B T} = \ln \frac{2(1-\epsilon)}{\epsilon} \,. \tag{19}
$$

Requiring now that $A = 0$, one gets the following expression for the critical line

$$
2\bar{\beta} K = \frac{1}{2} e^{\bar{\beta}} + 1 \,, \tag{20}
$$

where $\bar{\beta} \equiv \beta \Delta$. Equivalently, this expression may be written as $\bar{\beta} K = 1/2\epsilon$. The microcanonical critical line thus coincides with the critical line (9) obtained for the canonical ensemble. The tricritical point of the microcanonical ensemble is obtained at $A = B = 0$. Combining these equations with (19) one finds that at the tricritical point $\bar{\beta}$ satisfies

$$
\frac{1}{8\bar{\beta}^2} \frac{e^{\bar{\beta}} + 2}{e^{\bar{\beta}}} - \frac{1}{4\bar{\beta}} + \frac{1}{12} = 0 \,. \tag{21}
$$

Equations (20, 21) yield a tricritical point at $K \simeq 1.0813$, $\bar{\beta} \simeq 1.3998$. This has to be compared with the canonical tricritical point located at $K = 3/\ln(16) \simeq 1.0820$, $\bar{\beta} = \ln(4) \simeq 1.3995$. The two points, although very close to each other, do not coincide. The microcanonical critical line extends beyond the canonical one. In the region between the two tricritical points, the canonical ensemble yields a first order transition at a higher temperature, while in the microcanonical ensemble the transition is still continuous. It is in this region, as discussed below, that negative heat capacity appears. A schematic phase diagram near the canonical tricritical point (CTP) and the microcanonical one (MTP) is given in Fig. 5. Beyond the microcanonical tricritical point the temperature has a jump at the transition energy in the microcanonical ensemble. The two lines emerging on the right side from the MTP correspond to the two limiting temperatures which are reached when approaching the transition energy from below and from above (see Fig. 6d). The two microcanonical temperature lines and the canonical first-order transition line all merge on the $T = 0$ line at $2\Delta/J = 1$.

To get a better undertanding of the microcanonical phase diagram and also to compare our results with those obtained for self-gravitating and for finite systems we consider the temperature-energy relation $T(\epsilon)$. This curve has two branches: a high energy branch (19) corresponding to $m = 0$, and a low energy

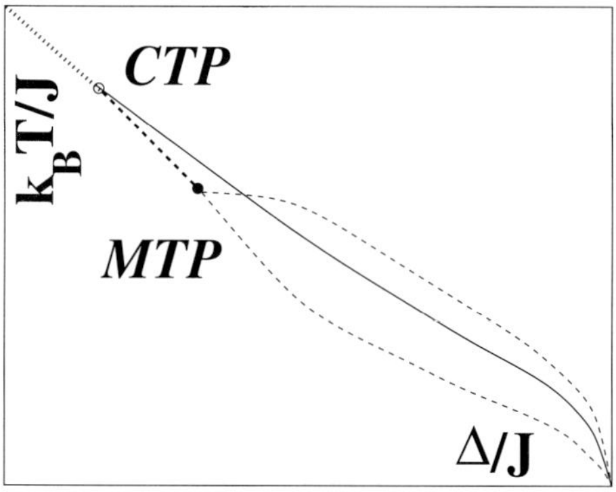

Fig. 5. A schematic representation of the phase diagram, where we expand the region around the canonical (CTP) and the microcanonical (MTP) tricritical points. The second order line, common to both ensembles, is dotted, the first order canonical transition line is solid and the microcanonical transition lines are dashed (with the bold dashed line representing a continuous transition)

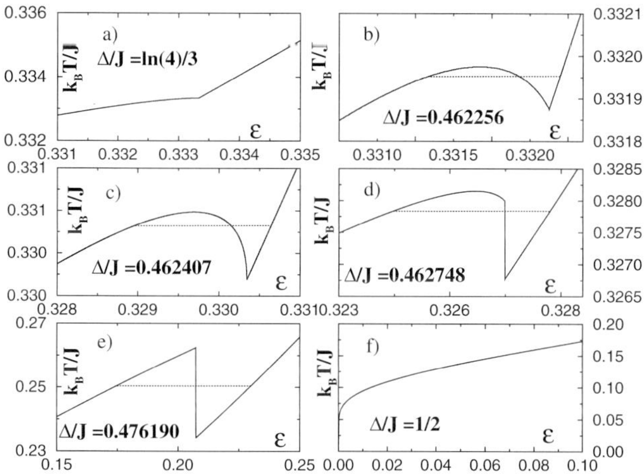

Fig. 6. The temperature-energy relation in the microcanonical ensemble for different values of Δ. The dotted horizontal line in some of the plots is the Maxwell construction in the canonical ensemble and identifies the canonical first order transition temperature at the point where two minima of the free energy coexist. The vertical line in panels d) and e) marks the presence of a temperature jump: for this energy two entropy maxima of the same height coexist

branch obtained from (15) using the spontaneous magnetization $m_s(\epsilon)$. At the intersection point of the two branches the two entropies become equal. Their first derivatives at the crossing point can be different, resulting in a jump in the temperature, i.e. *a microcanonical first order transition.* When the transition is continuous in the microcanonical ensemble, i.e. the first derivative of the entropy branches at the crossing point are equal, our model always displays, at variance with what happens for gravitational systems, a discontinuity in the second derivative of the entropy. This is due to the fact that here we have a true symmetry breaking transition. Figure 6 displays the $T(\epsilon)$ curve for increasing values of Δ. For $\Delta/J = \ln(4)/3)$, corresponding to the canonical tricritical point, the lower branch of the curve has a zero slope at the intersection point (Fig. 6a). Thus, the specific heat of the ordered phase diverges at this point. This effect signals the canonical tricritical point as it appears in the microcanonical ensemble. Increasing Δ to the region between the two tricritical points a *negative specific heat* in the microcanonical ensemble first arises ($\partial T/\partial\epsilon < 0$), see Fig. 6b. At the microcanonical tricritical point Δ the derivative $\partial T/\partial\epsilon$ of the lower branch diverges at the transition point, yielding a vanishing specific heat (Fig. 6c). For larger values of Δ a jump in the temperature appears at the transition energy (Fig. 6d). The lower temperature corresponds to the $m = 0$ solution (19) and the upper one is given by $\exp(\bar{\beta}) = 2(1 - q^*)/\sqrt{(q^*)^2 - (m^*)^2}$, where m^*, q^* are the values of the order parameters of the ferromagnetic state at the transition energy. The negative specific heat branch disappears at larger values of Δ, leaving just a temperature jump (see Fig. 6e). In the $\Delta/J \to 1/2$ limit the low temperature branch, corresponding to $q = m = 1$ in the limit, shrinks to zero and the $m = 0$ branch (19) occupies the full energy range (Fig. 6f).

We now present some general considerations concerning the intricate relation between canonical and microcanonical ensembles. The following discussion is quite general, and not specific to the BEG model. Let us first calculate the canonical partition sum using the density of states $\Omega = e^{Ns(e,m)}$, where $e = E/N$ is the energy density and m the order parameter (for the BEG model $e = \epsilon\Delta$, m is the magnetization and q is obtained from e and m by relation (14)).

$$\begin{aligned} Z &= \sum_{[S_i]} \exp\left(-\beta H(S_i)\right) \\ &= \sum_{[e,m]} e^{Ns(e,m)} e^{-N\beta e} , \end{aligned} \tag{22}$$

where we have replaced the sum over the configurations by the sum over the values of the energy per spin and the order parameter, which for finite N take discrete values. In the thermodynamic limit $N \to \infty$, the free energy in the canonical ensemble is thus given by

$$f(\beta) = \min_{e,m} \left(e - \frac{s(e,m)}{\beta} \right) . \tag{23}$$

The point where the minimum is reached yields the mean energy and magnetization per spin in the canonical ensemble. As explained above, the microcanonical entropy is

$$s(e) = \max_{m} s(e,m) \tag{24}$$

and the magnetization in the microcanonical ensemble is the point where the maximum is reached.

All questions are now reduced to the solution of the two variational problems (23) and (24). The first order conditions are the same for the two problems

$$\begin{aligned}\frac{\partial s}{\partial m} &= 0\\ \frac{\partial s}{\partial e} &= \beta,\end{aligned} \tag{25}$$

where the second condition is indeed given in the microcanonical case by the definition of temperature. We will denote by $e^*(\beta), m^*(\beta)$ the solution of the variational problem (25). Using (25), it is straightforward to verify that

$$\frac{d(\beta f)}{d\beta} = e^*(\beta), \tag{26}$$

meaning that the canonical mean energy coincides with the minimizing energy. Hence, the extrema in (23) and in (24) are the same. However this does not imply ensemble equivalence: we have to study the stability of these extrema.

In order to discuss the stability of the canonical solution one has to determine the sign of the eigenvalues of the Hessian of the function to be minimized in (23). The Hessian is

$$\mathcal{H} = \frac{-1}{\beta}\begin{pmatrix} s_{mm} & s_{me} \\ s_{em} & s_{ee}\end{pmatrix} \tag{27}$$

where, for example, s_{mm} is the second derivative of s with respect to m. The extremum is a minimum if and only if the determinant and the trace of the Hessian are positive

$$-s_{ee} - s_{mm} > 0 \tag{28}$$

$$s_{ee}s_{mm} - s_{me}^2 > 0, \tag{29}$$

which implies that s_{ee} and s_{mm} must be negative, and moreover $s_{ee}<-s_{me}^2/|s_{mm}|$. This has strong implications on the canonical specific heat, which must be positive. Let us prove it using the variational approach, instead of the usual Thirring argument [4], that uses the expression of the canonical partition sum. Indeed, taking the derivatives of (25) with respect to β, after having substituted into them $e^*(\beta), m^*(\beta)$, one gets

$$s_{ee}\frac{de^*}{d\beta} + s_{em}\frac{dm^*}{d\beta} = 1 \tag{30}$$

$$s_{me}\frac{de^*}{d\beta} + s_{mm}\frac{dm^*}{d\beta} = 0, \tag{31}$$

where it is understood that all second derivatives are computed at $e^*(\beta), m^*(\beta)$. Recalling now that the specific heat per particle at constant volume is,

$$\frac{C_V}{Nk_B} = \frac{de^*}{dT} = -\beta^2\frac{de^*}{d\beta} \tag{32}$$

one gets

$$\frac{C_V}{Nk_B} = \beta^2 \frac{s_{mm}}{s_{em}^2 - s_{ee}s_{mm}}, \tag{33}$$

which is always positive if the stability conditions (29) are satisfied. However, since the stability condition in the microcanonical ensemble only requires that s_{mm} is negative, in order to have an entropy maximum, this indirectly proves that a canonically stable solution is also microcanonically stable. The converse is not true: one may well have an entropy maximum, $s_{mm} < 0$, which is a free energy saddle point, with $s_{ee} > 0$. This implies that the specific heat (33) can be negative .

The above results are actually quite general, provided the canonical and microcanonical solutions are expressed through variational problems of the type (23) and (24). The extrema, and thus the caloric curves $T(e)$, are the same in the two ensembles, but the stability of the different branches is different (see also the important paper by Katz [12] and the contribution by Chavanis to this book [5] for a discussion of this point in connection to self-gravitating systems). Another example of this behavior is discussed for the HMF model in this book [16].

4 Maxwell Constructions

In this Section we briefly comment on the Maxwell construction leading to temperature jumps in the microcanonical ensemble. This is quite a new feature, a sort of analogue of the latent heat phenomenon in the canonical ensemble.

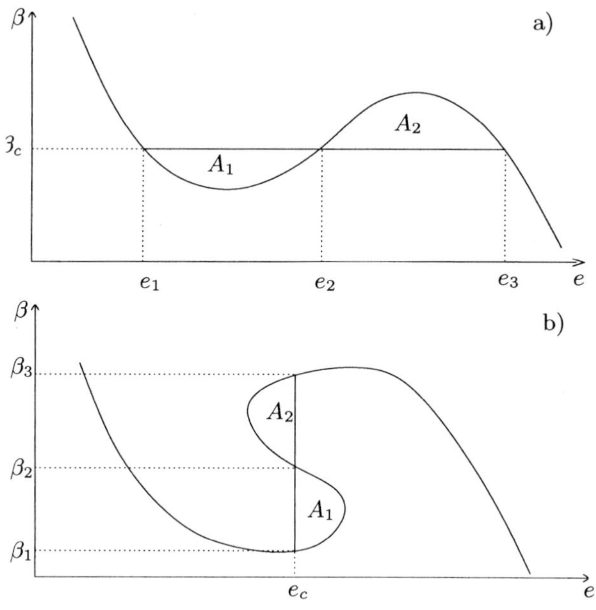

Fig. 7. Canonical (a) and microcanonical (b) Maxwell constructions.

Let us begin with the ordinary Maxwell construction in the canonical ensemble. We consider the curve $\beta(e)$ as in Fig. 7a. It is the situation of Fig. 6c, if for simplicity we disregard the discontinuity in the temperature derivative, which is irrelevant for the present reasoning. The equal area Maxwell condition $A_1 = A_2$ simply reads

$$\int_{e_1}^{e_3} [\beta(e) - \beta_c] \, de = 0 \ , \tag{34}$$

where β_c is the first order transition inverse temperature. Since $\beta = ds/de$, this implies

$$s(e_3) - s(e_1) - \beta_c(e_3 - e_1) = 0. \tag{35}$$

Using now the definition of free energy, $f = e - s/\beta$, we obtain the usual equal free energy condition at a first order phase transition,

$$f(e_1) = f(e_3) \ . \tag{36}$$

Near a discontinuous microcanonical transition, the $\beta(e)$ curve has the typical shape given in Fig. 7b. This is in fact the case of Fig. 6, if one continues the low energy branch of $\beta(e)$ above the microcanonical transition energy, drawing the metastable and the unstable lines as well. We invert this relation to obtain $e(\beta)$ in the vicinity of e_c. Then, the equal areas condition reads

$$\int_{\beta_1}^{\beta_3} [e(\beta) - e_c] \, d\beta = 0 \ . \tag{37}$$

using $e = d(\beta f)/d\beta$, we get

$$\beta_3 f(\beta_3) - \beta_1 f(\beta_1) - e_c(\beta_3 - \beta_1) = 0 \ , \tag{38}$$

which implies

$$s(\beta_3) = s(\beta_1) \ . \tag{39}$$

This is the equal entropy condition at a microcanonical discontinuous transition. Equation (37) is valid only for the simple configuration of Fig. 7b; for more complex curves one has to evaluate the areas more carefully (see also the discussion by Chavanis [5]).

5 The Microcanonical Solution by Large Deviation Techniques

In Sect. 3 we derived an expression for the logarithm of the density of states at fixed energy and magnetization, i.e. the entropy $s(e, m)$, using combinatorial techniques. We now present an alternative derivation of the entropy using Cramér's theorem [13]. This theorem, derived in the context of large deviation theory , states the following: consider a set of N independent and identically distributed random variables in R^d, $(X_i)_{i=1 \ldots N}$. One would like to know the probability distribution of the average $M_N = 1/N \sum X_i$. For this purpose, one defines, for $\lambda \in R^d$,

$$\psi(\lambda) = \ln \langle e^{\lambda \cdot X} \rangle, \tag{40}$$

where the average $\langle\rangle$ is taken with respect to the common Probability distribution Function (PdF) of all X_i. Now let $s(x)/k_B$, with $x \in R^d$, be the inverse Legendre transform of ψ

$$\frac{s(x)}{k_B} = - \sup_{\lambda \in R^d} \left(\lambda \cdot x - \psi(\lambda)\right) . \tag{41}$$

Then, under quite general conditions, Cramér's theorem states that the PdF of the average is

$$P(M_N = x) \sim e^{N \frac{s(x)}{k_B}} . \tag{42}$$

This yields an expression for the logarithm of the density of states $s(x)/k_B$. The great advantage of this method is, of course, its generality, since it applies also when combinatorial tools are not available.

Let us give a brief heuristic argument for this result in the $d = 1$ case, $\lambda, x \in R$. The probability of having $M_N = x$ is given by the volume in phase space compatible with $M_N = x$. Let $d\mu$ be the common measure of all X_i. One gets,

$$P(M_N = x) = \int d\mu(X_1) \dots d\mu(X_N) \delta(M_N - x) \tag{43}$$

$$= \frac{1}{2\pi i} \int_\Gamma d\lambda e^{-N\lambda x} \int d\mu(X_1) \dots d\mu(X_N) e^{\lambda(X_1 + \dots + X_N)} \tag{44}$$

$$= \frac{1}{2\pi i} \int_\Gamma d\lambda e^{-N\lambda x} \left[\langle e^{\lambda X} \rangle\right]^N , \tag{45}$$

where the Dirac δ function has been represented by a Laplace integral over a path Γ transverse to the real axis in the complex λ plane. Evaluating the last integral using the saddle point method one is lead to look for the critical points of $\lambda x - \psi(\lambda)$, with $\psi(\lambda) = \ln[\langle e^{\lambda X} \rangle]$, which justifies (41).

As an example, let us apply this method to the BEG model[1]. In this case the X_i variables are bidimensional ($d = 2$), $X_i = (S_i^2, S_i)$, so that

$$\begin{aligned} \psi(\lambda, \rho) &= \ln \langle e^{\lambda S_i^2 + \rho S_i} \rangle \\ &= \ln(1 + 2e^\lambda \cosh \rho) - \ln 3 \end{aligned} \tag{46}$$

To calculate $s(q, m)$ we now have to solve equations $\partial\psi/\partial\lambda = q$ and $\partial\psi/\partial\rho = m$ for λ and ρ. One gets

$$\lambda = \ln \frac{q\sqrt{1 - r^2}}{2(1 - q)} \tag{47}$$

$$\rho = \ln \sqrt{\frac{1 + r}{1 - r}} \tag{48}$$

[1] An application to the Ising model and to other systems can be found in [14]

where r is the ratio m/q. Substituting in (41), we obtain the expected result

$$\frac{s(q,m)}{k_B} = -\frac{q+m}{2}\ln\frac{q+m}{2} - \frac{q-m}{2}\ln\frac{q-m}{2} - (1-q)\ln(1-q) - \ln 3\ , \quad (49)$$

where the $-\ln 3$ constant is a normalization factor related to the fact that the total number of configurations is 3^N.

6 Slowly Decaying Interactions

So far, we have restricted our study to infinite-range models. Space is in this case irrelevant, which simplifies very much the study. However, the physically interesting interactions (gravity for instance) are not infinite range, and one may wonder whether the properties of the previous Sections still hold in a more general case. To answer this question we discuss in this Section spin systems on d-dimensional lattices with pairwise interaction potential between two sites i and j decaying with the distance r_{ij} like $1/r_{ij}^{\alpha}$ $(0 \le \alpha < d)$, so that the interaction is long-range. An Ising model with short range algebraically decaying interactions $(\alpha > d)$ has been studied in the past by Dyson [15]. More recently, versions of the Hamiltonian Mean Field model [16], with slowly decaying long-range interactions among rotors $(\alpha < d)$ have been considered by several authors [17,18] and the problem has been discussed in a more general framework in Ref. [19].

Our main result is that all peculiar features of the microcanonical ensemble of infinite range models extend to these slowly decaying interactions. We proceed by considering a continuum, or coarse grained version of magnetic systems. We then explain that in fact these coarse grained models can be derived from the microscopic models, like the Ising or the BEG models considered in this paper, using the large deviations tools introduced in Sect. 5 (see [23] for a more thorough discussion of this point).

We begin by analyzing the phase diagram of the continuum version of the Ising model with algebraically decaying interactions $(0 < \alpha < d)$. We show that as in the mean field model $(\alpha = 0)$ this model does not exhibit a phase transition in the microcanonical ensemble. We then consider the continuum version of the BEG model.

The Ising Hamiltonian functional with algebraically decaying interactions in $d = 1$ dimensions on the segment $[0,1]$ with periodic or free boundary conditions (to be discussed below) takes the form

$$h_{Ising} = -\frac{\bar{J}}{2}\int_0^1 dx \int_0^1 dy \frac{m(x)m(y)}{r(x,y)^{\alpha}}, \quad (50)$$

where $m(x)$ is the local magnetization, $r(x,y)$ is the distance between x and y on the segment (the shortest distance for periodic boundary conditions), and $\bar{J} > 0$ is a ferromagnetic coupling constant . The entropy density corresponding to a magnetization profile $m(x)$ of an Ising variable may be written as

$$s_{Ising} = -k_B \int_0^1 dx\left[\frac{(1+m)}{2}\ln\frac{(1+m)}{2} + \frac{(1-m)}{2}\ln\frac{(1-m)}{2}\right]\ , \quad (51)$$

Once the energy and entropy functionals are given, the microcanonical (canonical) solution is obtained by a maximization (minimization) of the entropy (free-energy) functional under the constraint of constant energy (temperature). We begin below with the discussion of these two variational problems for the Ising model

$$s_{Ising}(e) = \max_{m(x)} \left(s_{Ising}[(m(x)] \mid \quad h_{Ising} = e \right), \tag{52}$$

in the microcanonical ensemble , and

$$f_{Ising}(\beta) = \min_{m(x)} \left(h_{Ising}[m(x)] - \frac{1}{\beta} s_{Ising}[m(x)] \right) \tag{53}$$

in the canonical ensemble .

So far, although we have stated the problem for the Ising model, we could have followed the same path, whatever the lattice model and the boundary conditions are. We can reach a general conclusion at this point by writing down the extremality conditions of the two variational problems (52) and (53). The first order reads

$$\frac{\delta s_{Ising}}{\delta m(x)} = \beta \frac{\delta h_{Ising}}{\delta m(x)} , \tag{54}$$

in the microcanonical ensemble, where β is a Lagrange multiplier. In the canonical ensemble

$$\frac{\delta h_{Ising}}{\delta m(x)} - \frac{1}{\beta} \frac{\delta s_{Ising}}{\delta m(x)} = 0 . \tag{55}$$

Hence, the first order extremality conditions lead to the same equations for the two ensembles. However, this *does not* imply ensemble equivalence, since the stability of the solutions may differ in the two ensembles, as has been noted in Sect. 4.

Before proceeding, let us briefly recall the results of Ref. [17]. These authors, discussing the Hamiltonian Mean Field model with slowly decaying interactions [16,22], remarked that, choosing appropriately the renormalization factor in the Hamiltonian, the thermodynamic behavior of the system turned out to be independent of α; i.e. the system behaves at $0 < \alpha < 1$ (slowly decaying case) exactly as if $\alpha = 0$ (mean-field case). An explanation of this result was provided by Campa et al. [18] through a canonical analysis of long-range interacting systems on lattices with periodic boundary conditions, and independently by Vollmayr-Lee and Luijten [19] in a more general context. Unfortunately, neither of the two latter methods provide the microcanonical solution. The method we present here is, instead, fully general and allows to discuss in detail how to extend the results for $\alpha = 0$ to $\alpha > 0$.

Let us solve now the variational problem of quations (54) and (55) for the Ising model [21]. The magnetization profile that extremizes entropy and free energy satisfies the following self-consistency condition

$$m(x) = \tanh \left(\beta \bar{J} \int_0^1 \frac{m(y)}{r(x,y)^\alpha} \, dy \right) \tag{56}$$

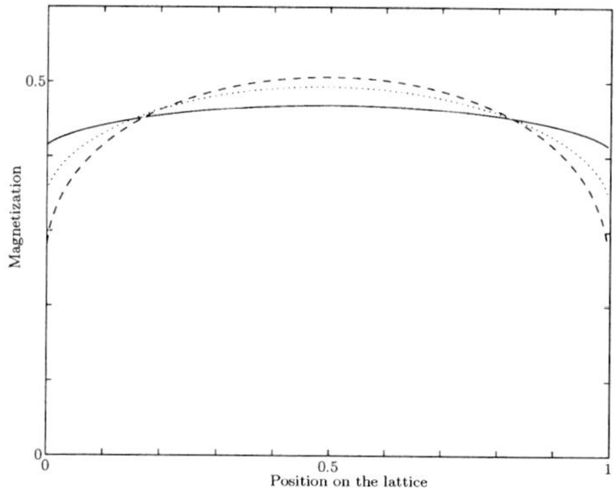

Fig. 8. Magnetization profile for the Ising model with slowly decaying interactions in the microcanonical ensemble. The energy density is fixed at $e = -0.1$ while α varies: $\alpha = 0.2$ (full line), $\alpha = 0.5$ (dotted line), $\alpha = 0.8$ (dashed line). The boundary conditions are free.

In the canonical ensemble, β is fixed, whereas in the microcanonical ensemble it has to be tuned in order to get the required energy. For periodic boundary conditions, the integral on the r.h.s. of (56) does not depend on x, and (56) always has a homogeneous solution with $m(x)$ independent of x. Equation (56) then reduces to the usual consistency equation of the mean-field Ising model, and the system behaves exactly as in the $\alpha = 0$ case, showing a phase-transition in the canonical ensemble (no transition is present in the microcanonical ensemble).

On the contrary, for free boundary conditions, there is no non-zero constant solution, and the system develops a non trivial magnetization profile that can be only numerically computed from (56). Moreover, in this case a phase transition is present in the canonical ensemble, while the microcanonical ensemble is always in a magnetically ordered phase. The magnetization profile in the microcanonical ensemble for free boundary conditions is shown in Fig. 8 for three different values of α.

Let us now discuss the behavior of the BEG model for slowly decaying interactions. The energy functional is

$$h_{BEG} = \Delta \int_0^1 dx q(x) - \frac{\bar{J}}{2} \int_0^1 dx \int_0^1 dy \frac{m(x)m(y)}{r(x,y)^\alpha}, \tag{57}$$

where all is defined as before and, moreover, $q(x)$ is the quadrupolar field in the continuum on the segment $[0,1]$ (obtained in an analogous way as for the magnetization $m(x)$). The entropy functional of the BEG model is obtained from expression (13) by replacing m, q by $m(x), q(x)$.

The BEG model is much richer, as has already been demonstrated for $\alpha = 0$. The analysis of the microcanonical phase diagram for periodic boundary conditions follows the same path as that presented above for the Ising model. It is straightforward to show that the constant profiles $m(x) = m^*$, $q(x) = q^*$, with m^* and q^* given by the solution of the $\alpha = 0$ model, are solutions of the equations (54) and (55) (where we replace the Ising subscript by the BEG one) also for $\alpha > 0$. Moreover, detailed calculations [24] show that these solutions are local maxima of the entropy and local minima of the free energy. Thus one concludes that for periodic systems the entropy and free energy are independent of α and the conclusions reached for $\alpha = 0$ are valid for $0 < \alpha < d$. In particular, ensemble inequivalence can be extended to the case of long-range algebraically decaying interactions.

Let us briefly sketch these calculations. To verify that uniform profiles are not destabilized by modulations, it is convenient to develop $q(x)$ and $m(x)$ in Fourier series: $q(x) = \sum_k \tilde{q}_k u_k(x)$, $m(x) = \sum_k \tilde{m}_k u_k(x)$, where k ranges from $-\infty$ to $+\infty$ and $u_k(x) = \cos kx$ if $k \geq 0$, $u_k(x) = \sin kx$ if $k < 0$.

The Fourier representation, due to the periodicity, diagonalizes the Hamiltonian. The energy may thus be written as

$$\epsilon = \tilde{q}_0 - \bar{K} \sum_{k=-\infty}^{+\infty} \lambda_k \tilde{m}_k^2 \ . \tag{58}$$

Here λ_k are the energy eigenvalues and $\bar{K} = 2^\alpha \bar{J}/2\Delta(1-\alpha)$ is chosen such that the largest eigenvalue is unity $\lambda_0 = 1$ (this amounts to choose the prefactor $\tilde{N}$ in the Kac prescription, see below).

The free energy functional is

$$f[q,m] = \epsilon - 1/\beta \int s[q(x), m(x)]\, dx \ , \tag{59}$$

so that we get the first order conditions

$$\delta_{k0}\beta - \int s_q[q,m] u_k(x)\, dx = 0 \tag{60}$$

$$-2\beta\bar{K}\lambda_k\tilde{m}_k - \int s_m[q,m] u_k(x)\, dx = 0 \tag{61}$$

We can easily check that the uniform profile $q = \tilde{q}_0$, $m = \tilde{m}_0$ (with $\tilde{q}_0$ and $\tilde{m}_0$ given by the $\alpha = 0$ case) is indeed a solution of these conditions. To prove that the uniform profile is a free energy minimum, we evaluate now the Hessian matrix at this point

$$\frac{\partial f}{\partial \tilde{q}_k \partial \tilde{q}_l} = -\frac{s_{qq}}{\beta}\delta_{kl} \tag{62}$$

$$\frac{\partial f}{\partial \tilde{q}_k \partial \tilde{m}_l} = -\frac{s_{qm}}{\beta}\delta_{kl} \tag{63}$$

$$\frac{\partial f}{\partial \tilde{m}_k \partial \tilde{m}_l} = -\frac{s_{mm}}{\beta}\delta_{kl} - 2\bar{K}\lambda_k\delta_{kl} \ . \tag{64}$$

This infinite Hessian matrix turns out to be positive definite. To prove this, it is sufficient to verify that each matrix H_k

$$H_k = -\begin{pmatrix} s_{qq} & s_{qm} \\ s_{qm} & s_{mm} + 2\bar{K}\beta\lambda_k \end{pmatrix} \tag{65}$$

is positive definite, when taken at the solution point. This follows easily from the fact that H_0

$$H_0 = -\begin{pmatrix} s_{qq} & s_{qm} \\ s_{qm} & s_{mm} + 2\bar{K}\beta \end{pmatrix} \tag{66}$$

is positive definite (because the point considered is the solution of the $\alpha = 0$ case), and recalling that $\lambda_k \leq 1$ for whatever k.

We have thus proved that the local stability of the uniform profiles at $\alpha = 0$ are not modified in the canonical ensemble when $\alpha > 0$. All canonical $T(e)$ curves shown in Fig. 6 thus extend to the $\alpha > 0$ case. We now make use of the general discussion at the end of Sect. 3. We recall that results there derived state, in particular, that a stable canonical solution is always a stable microcanonical one, and that if one follows a $T(e)$ curve, like in Fig. 6, a change of stability in the microcanonical ensemble may occur only where this curve has a vertical tangent (see [12] and the contribution by Chavanis [5] to this book). We thus conclude that the microcanonical part of Fig. 6 does not change either. This reasoning is however not rigorous, since we have considered only local stability; it may happen that when $\alpha > 0$, a heterogeneous solution appears and it could yield the true free energy minimum or entropy maximum. Some numerical investigations we have carried out did not detect any such heterogeneous solution.

We now turn to a different issue, namely the correspondence between the microscopic models and the continuum ones. In the following we show that the Ising energy functional (50) can be derived by coarse graining from the microscopic Ising Hamiltonian with slowly decaying interactions

$$H_{Ising} = -\frac{J}{2\tilde{N}} \sum_{i,j} \frac{\sigma_i \sigma_j}{r_{ij}^{\alpha}} , \tag{67}$$

where r_{ij} is the distance on a 1D lattice between spins at sites i and j. The interaction is *non integrable* for $\alpha \leq 1$. The normalization $\tilde{N} = 2^{\alpha} N^{1-\alpha}/(1-\alpha)$ ensures that the energy is extensive and the prefactor provides a convenient normalization for the eigenvalues of the Hamiltonian in the case of periodic boundary conditions. To make the correspondence between continuous and microscopic models explicit, one has to set $\bar{J} = (1-\alpha)J/2^{\alpha}$ in (50).

Analogously, the BEG microscopic model we refer to is

$$H_{BEG} = \Delta \sum_{i=1}^{N} S_i^2 - \frac{J}{2\tilde{N}} \sum_{i,j} \frac{S_i S_j}{r_{ij}^{\alpha}} . \tag{68}$$

The coarse-graining procedure we present, closely follows Ref. [20]. We will keep here a heuristic level of description, although all the following calculations can be made rigorous using the language and techniques of large deviation theory [13]. A more detailed account of this derivation is given in Ref. [23].

The first step involves a coarse-graining procedure. We divide the lattice in N/n boxes, each one containing n sites. We then describe macroscopically the system in the Ising case by an average magnetization in each box, or, in the BEG case, by the average m and q. In the limit $N \to \infty$, $n \to \infty$ with $N/n \to \infty$, the system is then described by continuous functions $m(x), q(x)$, with $x \in [0,1]$ if the system length is normalized. We have to show in the following that all the information lost in the coarse graining procedure is unessential.

In the mean-field case ($\alpha = 0$), the Hamiltonian had a natural and exact expression as a function of the macroscopic parameters m (Ising) or m and q (BEG). The long-range Ising case has already been discussed [21] and it has been shown that the intra-box couplings can be neglected in the thermodynamic limit, leading to the expression for the Hamiltonian functional (50) and for the entropy (51) as a function of the macroscopic parameter $m(x)$. The generalization to the BEG model of the method applied in Ref. [21] is straightforward and allows one to obtain the Hamiltonian functional (57). The estimation is uniform over all the microscopic configurations [23], and $r(x,y)$ is the distance between the two boxes located in x and y. On the contrary, if $\alpha > d$, there is no way to approximate the Hamiltonian in the continuum limit described above as a functional of $m(x)$ and $q(x)$.

We show now how to construct the entropy functional for the Ising model. Its extension to the BEG model is straightforward. We have to estimate the probability to obtain a certain given macroscopic configuration $m(x)$, assuming the equiprobability of microscopic configurations. For a finite number of boxes, N/n, we can evaluate the probability to get a given average magnetization in each box

$$\begin{aligned} P(m_1, m_2, \ldots, m_n) &= P(m_1)P(m_2)\cdots P(m_n) \\ &\simeq e^{ns_1(m_1)/k_B} \ldots e^{ns_n(m_n)/k_B} \end{aligned} \tag{69}$$

where s_i is the entropy associated to the i-th box. Letting n and N/n go to infinity, we obtain

$$P[m(x)] \simeq e^{\frac{N}{k_B}\int_0^1 s(m(x))\,dx} , \tag{70}$$

which defines the entropy functional s_{Ising} of the Ising model. Let us remark again that large deviation techniques [13] give a precise and rigorous meaning to these calculations, and allow one to extend this type of derivation to all sort of lattice models. The field $q(x)$ of the BEG model can be treated in a similar way as $m(x)$ and one gets the BEG entropy functional similarly.

Chavanis [5] and, Cohen and Ispolatov [25] analyze in this book more realistic off-lattice systems, self-gravitating or interacting through a $1/r^\alpha$ potential ($\alpha < 3$) . They study the thermodynamic properties of these models through a mean-field approximation, and the phenomenology is pretty much the same as the one decribed here. Unfortunately, to see the analogies, one must cope with discrepancies in the "vocabulary". For instance, the *microcanonical first order* transition described here is called *gravitational first order* transition, whereas the *canonical first order* one is called *normal first order* transition. Within the conventions of this paper one can see that, varying a control parameter (the interaction K for the BEG model, and a parameter controlling the short range cut-off in gravitational models) the system crosses over from a microcanonical first order transition to a microcanonical second order one (or no transition at all in the gravitational systems) associated with a canonical first order, and then recovers full ensemble equivalence, crossing the critical point.

All these off-lattices studies raise the question of the validity of the mean-field approximation, which is intimately related to the scaling with N of the thermodynamic variables and potentials. In our case, the mean field treatment is fully valid in the thermodynamic limit, but the problem was avoided using the physically unjustified Kac prescription [1] in which the coupling constants are rescaled by $\tilde{N}$.

7 Conclusions

In this paper a detailed comparison is made between the canonical and the microcanonical phase diagrams of the spin-1 BEG model with mean field long-range interactions. Since systems with long-range interactions, and particularly mean field models, are non-additive the two ensembles need not be equivalent. The BEG model, for which both phase diagrams can be evaluated analytically, provides a convenient and interesting ground for studying the distinctions between the two ensembles. Although the model is rather simple, it exhibits a non-trivial phase diagram, which in the canonical ensemble consists of both a first order and a second order transition lines separated by a tricritical point. It is shown that the microcanonical phase diagram, while it yields the same thermodynamic behavior in the region where the canonical transition is second order, it differs considerably from the canonical one in the region where the transition is first order. In particular it exhibits a region with a negative specific heat, a tricritical point whose location differs from that of the canonical ensemble, and a first order line at which the two coexisting phases have different temperatures. The mechanisms which lead to these features are rather general, and thus they are expected to take place in other models with long-range interactions within the microcanonical ensemble (see also [26,27]).

A straightforward generalization of these results to models with algebraically decaying interactions are considered. It is shown that the mean field phase diagram remains valid even for these interactions, as long as they are long-range.

It would be of great interest to extend this study in the future and to analyze and compare canonical and microcanonical phase diagrams in more complex cases, where higher order multicritical points are present. Such studies may eventually lead to general rules which should be obeyed by microcanonical phase diagrams of systems with long-range interactions, in analogy to the rules existing for systems with short range interactions, like the Gibbs phase rules, the Landau and Lifshitz symmetry rules for continuous transitions and others.

Acknowledgements

We would like to warmly thank our collaborators Freddy Bouchet and François Leyvraz for fruitful interactions. We thank Thierry Dauxois and Dieter Gross for very useful discussions. This work has been partially supported by the EU contract No. HPRN-CT-1999-00163 (LOCNET network) and the Région Rhône-Alpes for the fellowship N° 01-009261-01. This work is also part of the contract COFIN00 on *Chaos and localization in classical and quantum mechanics.*

References

1. M. Kac, G.E. Uhlenbeck and P.C. Hemmer, J. Math. Phys. **4**, 216 (1963).
2. V.A. Antonov Vest. Leningrad Univ. **7**, 135 (1962); Translation in IAU Symposium **113**, 525 (1995).
3. D. Lynden-Bell and R. Wood, Mon. Not. R. Astron. Soc. **138**, 495 (1968); D. Lynden-Bell, Physica A **263**, 293 (1999).
4. W. Thirring, *Z. Phys.* **235**, 339 (1970); P. Hertel and W. Thirring, *Ann. of Phys.* **63**, 520 (1971).
5. P.H. Chavanis, *Statistical mechanics of two-dimensional vortices and three-dimensional stellar systems*, in "Dynamics and Thermodynamics of Systems with Long-Range Interactions", T. Dauxois, S. Ruffo, E. Arimondo, M. Wilkens Eds., Lecture Notes in Physics Vol. 602, Springer (2002), (in this volume)
6. T. Padhmanaban, *Statistical mechanics of gravitating systems in static and cosmological backgrounds*, in "Dynamics and Thermodynamics of Systems with Long-Range Interactions", T. Dauxois, S. Ruffo, E. Arimondo, M. Wilkens Eds., Lecture Notes in Physics Vol. 602, Springer (2002), (in this volume)
7. D.H.E. Gross, *Microcanonical thermodynamics: Phase transitions in "small" systems*, (World Scientific, Singapore, 2000) and *Thermo-Statistics or Topology of the Microcanonical Entropy Surface*, in "Dynamics and Thermodynamics of Systems with Long-Range Interactions", T. Dauxois, S. Ruffo, E. Arimondo, M. Wilkens Eds., Lecture Notes in Physics Vol. 602, Springer (2002), (in this volume)
8. R.M. Lynden-Bell, in *Gravitational dynamics*, O. Lahav, E. Terlevich and R.J. Terlevich (eds.), Cambridge Univ. Press (1996); R.M. Lynden-Bell, Mol. Phys. **86**, 1353 (1995).
9. P. Chomaz and F. Gulminelli *Phase transitions in finite systems*, in "Dynamics and Thermodynamics of Systems with Long-Range Interactions", T. Dauxois, S. Ruffo, E. Arimondo, M. Wilkens Eds., Lecture Notes in Physics Vol. 602, Springer (2002), (in this volume)
10. M. Blume, V.J. Emery and R.B. Griffiths, Phys. Rev. A **4**, 1071 (1971).
11. J. Barré, D. Mukamel and S. Ruffo, Phys. Rev. Lett. **87**, 030601 (2001).
12. J. Katz, Mon. Not. R. Astr. Soc. **183**, 765 (1978).
13. A. Dembo, O. Zeitouni, *Large Deviations Techniques and Applications*, (Springer, Berlin, 1998).
14. R. S. Ellis, Physica D **133**, 106, (1999).
15. F.J. Dyson, Comm. Math. Phys., **12**, 91 (1969).
16. T. Dauxois, V. Latora, A. Rapisarda, S. Ruffo and A. Torcini, *The Hamiltonian Mean Field Model: from Dynamics to Statistical Mechanics and back*, in "Dynamics and Thermodynamics of Systems with Long-Range Interactions", T. Dauxois, S. Ruffo, E. Arimondo, M. Wilkens Eds., Lecture Notes in Physics Vol. 602, Springer (2002), (in this volume)
17. C. Anteneodo and C. Tsallis, Phys. Rev. Lett. **80**, 5313 (1998); F. Tamarit and C. Anteneodo, Phys. Rev. Lett. **84**, 208 (2000);

18. A. Campa, A. Giansanti and D. Moroni, Phys. Rev. E **62**, 303 (2000) and Chaos Solitons and Fractals, **13**, 407 (2002).
19. B.P. Vollmayr-Lee and E. Luijten, Phys. Rev. E **63**, 031108 (2001) and Phys. Rev. Lett. **85**, 470 (2000).
20. R. S. Ellis, K. Haven, B. Turkington, J. Stat. Phys. **101**, 999 (2000).
21. J. Barré, Physica A, **305**, 172 (2002).
22. C. Tsallis, A. Rapisarda, V. Latora and F. Baldovin, *Nonextensivity: from low-dimensional maps to Hamiltonian systems*, in "Dynamics and Thermodynamics of Systems with Long Range Interactions", T. Dauxois, S. Ruffo, E. Arimondo, M. Wilkens Eds., Lecture Notes in Physics Vol. 602, Springer (2002), (in this volume)
23. J. Barré and F. Bouchet, "Mean-Field justified by large deviations results in long-range interacting systems". Proceedings of the Confererence "Dynamics and thermodynamics of systems with long-range interactions", Les Houches, France, February 18-22 2002, Eds. T. Dauxois, E. Arimondo, S. Ruffo, M. Wilkens, published on http://www.ens-lyon.fr/~tdauxois/procs02/
24. J. Barré, D. Mukamel and S. Ruffo, to be published.
25. E.G.D. Cohen, I. Ispolatov, *Phase transitions in systems with* $1/r^{\alpha}$ *attractive interactions*, in "Dynamics and Thermodynamics of Systems with Long-Range Interactions", T. Dauxois, S. Ruffo, E. Arimondo, M. Wilkens Eds., Lecture Notes in Physics Vol. 602, Springer (2002), (in this volume)
26. F. Leyvraz and S. Ruffo, J. Phys. A, **35**, 285 (2002) and Physica A, **305**, 58 (2002).
27. I. Ispolatov and E.G.D. Cohen, Physica A, **295**, 475 (2001).

Phase Transitions in Finite Systems

Philippe Chomaz[1] and Francesca Gulminelli[2]

[1] GANIL (DSM-CEA/IN2P3-CNRS), B.P. 5027, F-14021 Caen cedex, France
[2] LPC Caen, (IN2P3-CNRS/ISMRA et Université), F-14050 Caen cedex, France

Abstract. In this series of lectures we will first review the general theory of phase transition in the framework of information theory and briefly address some of the well known mean field solutions of three dimensional problems. The theory of phase transitions in finite systems will then be discussed, with a special emphasis to the conceptual problems linked to a thermodynamical description for small, short-lived, open systems as metal clusters and data samples coming from nuclear collisions. The concept of negative heat capacity developed in the early seventies in the context of self-gravitating systems will be reinterpreted in the general framework of convexity anomalies of thermostatistical potentials. The connection with the distribution of the order parameter will lead us to a definition of first order phase transitions in finite systems based on topology anomalies of the event distribution in the space of observations. Finally a careful study of the thermodynamical limit will provide a bridge with the standard theory of phase transitions and show that in a wide class of physical situations the different statistical ensembles are irreducibly inequivalent.

1 Equilibrium and Information

1.1 States and Observables [1,2]

Modern physics associates to every physical system two different types of objects: *observables* that characterize the measurable physical quantities and *states* whose knowledge allows to predict the result of experiments. From the microscopic point of view, single realizations of systems with N degrees of freedom are characterized by a pure state (or microstate), that is a wave function, $|\Psi_N\rangle$, in quantum mechanics or a point in the 2N-dimensional phase space, $\mathbf{s} = (q_1, q_2, ..., q_N; p_1, p_2, ..., p_N)$, with q_i and p_i the position and momentum of each degree of freedom, in classical mechanics. If systems are sufficiently complex, the exact state is in general impossible to define and each actual realization corresponds to a microstate (n) with the probability $p^{(n)}$. In such a realistic case, one rather speaks of mixed states (or macrostates) described the density

$$\widehat{D} = \sum_n \left|\psi^{(n)}\right\rangle p^{(n)} \left\langle\psi^{(n)}\right| \quad \text{or} \quad \widehat{D}\left(\mathbf{s}\right) = \sum_n \delta\left(\mathbf{s} - \mathbf{s}^{(n)}\right)$$

Observables are operators defined on the Hilbert space or classically real functions of 2N real variables. The *information* that can be associated to the system is the ensemble of expectation values of the observables $\widehat{A}_l$ on the state, i.e. the ensemble of *observations* $\left\langle \widehat{A}_l \right\rangle = \sum_n p^{(n)} A_l^{(n)}$ where $A_l^{(n)}$ is the actual result of a measurement on the realization (n). In the quantum case $\left\langle \widehat{A}_l \right\rangle = \sum_n p^{(n)} \left\langle \psi^{(n)} \right| \widehat{A}_l \left| \psi^{(n)} \right\rangle = \mathrm{Tr}\left(\widehat{D} \widehat{A}_l \right)$. Both for pure and mixed states, if the information on the system is complete at the initial time, this stays true at any time because the dynamical evolution of states is governed by the deterministic Liouville Von Neumann equation $\partial_t \widehat{D} = \left\{ \widehat{H}, \widehat{D} \right\}$ where $\widehat{H}$ is the Hamiltonian of the system and where $\{.,.\}$ is the commutator divided by $i\hbar$ in quantum mechanics which reduces to the usual Poisson bracket at the classical limit. However in the case of complex systems, the initial conditions are in general incompletely known and an exact solution of the Liouville Von Neumann equation is out of reach. In general only a small set of pertinent observables is known at any time which is sufficient to determine the state (i.e. the totality of the $p^{(n)}$) because of the complexity of the density operator.

1.2 The Shannon Entropy [1,2]

The incompleteness of the available information can be measured through the lack of information or *statistical entropy*

$$S = -\sum_n p^{(n)} \ln p^{(n)} = -\mathrm{Tr}\left(\widehat{D} \ln \widehat{D} \right) \tag{1}$$

Let us show within a simple example that the statistical entropy (or Shannon entropy) indeed measures the lack of information.

Let us consider a system constituted of N identical boxes and an experiment consisting in putting randomly a ball in a box. The missing information S to know where the ball is depends first on the occupation probability of each box $S = S(p^{(1)}, ..., p^{(N)})$. Let us first consider equiprobable boxes $p^{(n)} = 1/N, \forall n$. In this case S depends only on the total number of boxes, $S = S(N)$.

Let us enumerate some fundamental properties of S :

- The lack of information must grow with the number of possible results $S(N_1) > S(N_2) \vee N_1 > N_2$.

Let us divide the N boxes into N_1 groups of N_2 boxes each, $N = N_1 N_2$. The experiment now consists in two successive steps, first find out in which group out of the N_1 equiprobable ones the ball is (which is associated to a lack of information $S(N_1)$) and then determine which of the N_2 equiprobable boxes belonging to the group the ball is (associated to a lack of information $S(N_2)$). The missing information of the two steps experiment is then $S(N_1) + S(N_2)$.

- The information cannot depend on the number of steps through which it is collected $S\,(N) = S(N_1 \cdot N_2) = S(N_1) + S(N_2)$.

The ensemble of these properties is fulfilled by a logarithmic function $S(N) = k \ln N$ where k is a constant.

We have just shown that the Shannon entropy coincides for equiprobable states with the Boltzmann entropy (or microcanonical entropy). Let us now turn to the more general case in which boxes are not equiprobable. To derive the associated information let us consider a big number W (eventually going to infinity) of experiments identical to the one described above. Among these W experiences, a number $N_i \neq W/N$ will lead to the observation of the ball in the i-th box. This experimental result defines *a posteriori* a probability $p_i = N_i/W$ for the i-th box.

Within this result $\{N_1, ..., N_k, ..., N_N\}$, the number of possible configurations for the box is given by the combinatorial

$$\Omega = \frac{W!}{N_1!\,(W-N_1)!} \cdot \frac{(W-N_1)!}{N_2!\,(W-N_1-N_2)!} \cdots = \frac{W!}{\prod_i N_i!}$$

where the first term represents the number of ways of choosing N_1 indistinguishable objects out of W, and so on. All the events Ω are equiprobable. The entropy is then

$$\begin{aligned} S(\Omega) &= k \ln \Omega = k(\ln W! - \sum_i \ln N_i!) \\ &= k(W \ln W - W - \sum_i (N_i \ln N_i - N_i)) \\ &= k(W \ln W - \sum_i (W p_i \ln W + W p_i \ln p_i)) \\ &= -kW \sum_i p_i \ln p_i \end{aligned}$$

where we have used the Stirling formula $\ln N! \approx N \ln N - N$. The additivity property introduced above allows to conclude that for a single experiment the missing information is given by the Shannon entropy $S(p^{(1)}, ..., p^{(N)}) = -k \sum_n p^{(n)} \ln p^{(n)}$.

It may be interesting to know that if the additivity property of the information is relaxed, it is possible to construct a non-extensive extension of the Shannon theory based on the so called q-statistics which has interesting applications in out of equilibrium situations as in the case of turbulent flows [3]. In the following of these lectures we will limit ourselves to the standard information kernel introduced above.

1.3 The Fundamental Postulate of Statistical Mechanics

The fundamental postulate of statistical mechanics can be expressed as follows

"The statistical distribution of microstates usually called the equilibrium is the one which maximizes the statistical entropy within the external constraints (i.e. the pertinent information) imposed to the system".

Indeed any other distribution would introduce an extra piece of information, in contrast with the statement that all the available information is given by the constraint.

It is important to remark that this postulate, though certainly intuitive and elegant, does not necessarily imply that the theory has any predictive power: the fact that we have only a limited amount of information on a system does not necessarily mean that the information contained in the system is objectively limited. In this series of lectures we shall anyway keep the fundamental postulate as the only reasonable working hypothesis in a complex system.

The fundamental postulate of statistical mechanics allows to determine the equilibrium values of the state probabilities $p^{(n)}$. This task is easily accomplished with the help of the method of Lagrange multipliers.

1.4 The Method of Lagrange Multipliers [4]

Let us consider the problem of finding an extremum of a two variables real function $f(x, y)$ along a curve defined by the relation $\omega(x, y) = \omega_0$. To this aim the standard way is to calculate the total differential

$$df = \frac{\partial f}{\partial x} dx + \frac{\partial f}{\partial y} dy$$

where dx and dy are linked by the relation

$$d\omega = \frac{\partial \omega}{\partial x} dx + \frac{\partial \omega}{\partial y} dy = 0$$

Expressing dy as a function of dx the differential reads

$$df = (\frac{\partial f}{\partial x} - \frac{\partial \omega / \partial x}{\partial \omega / \partial y} \frac{\partial f}{\partial y}) dx$$

Putting df to zero at the point (x_0, y_0) which fulfills the constraint $\omega(x_0, y_0) = \omega_0$, leads to

$$\frac{\partial \omega}{\partial y} \frac{\partial f}{\partial x}\rfloor_{x_0,y_0} = \frac{\partial \omega}{\partial x} \frac{\partial f}{\partial y}\rfloor_{x_0,y_0}$$

which defines the coordinates (x_0, y_0) of the extremum.

This same result can be obtained in a simpler way if we introduce a Lagrange multiplier λ and we define the auxiliary function $F = f - \lambda(\omega - \omega_0)$ that coincides with the function f on the curve we are interested in. Differentiating F respect to its two *independent* variables x and y

$$dF = (\frac{\partial f}{\partial x} - \lambda \frac{\partial \omega}{\partial x}) dx + (\frac{\partial f}{\partial y} - \lambda \frac{\partial \omega}{\partial y}) dy$$

the two partial derivatives have to go to zero separately at the extremum leading to a solution $(x_0(\lambda), y_0(\lambda))$. This extremum fulfills the condition

$$\frac{\partial \omega}{\partial y} \frac{\partial f}{\partial x}\rfloor_{x_0,y_0} = \frac{\partial \omega}{\partial x} \frac{\partial f}{\partial y}\rfloor_{x_0,y_0}$$

which exactly corresponds to the condition above if λ is such that $\omega(x_0(\lambda), y_0(\lambda)) = \omega_0$. The extension to a bigger number of variables and constraints is straightforward.

To summarize, this method allows to replace the study of a function of non independent variables to the study of an auxiliary function for which all variables are independent and the constraints are absorbed by real numbers (Lagrange multipliers).

1.5 The Equilibrium [4]

Let us use this method to maximize the statistical entropy $S = -\mathrm{Tr}\hat{D}\ln\hat{D}$ under the constraint of a given set of L observations $\left\langle \hat{A}_l \right\rangle$.

This situation corresponds to the L constraints $\mathrm{Tr}\hat{D}\hat{A}_l = \left\langle \hat{A}_l \right\rangle$ that has to be augmented with the extra constraint of the normalization of probability $\mathrm{Tr}\hat{D} = 1$ which can be incorporated as an additional observable $\hat{A}_0 = 1$. The auxiliary function is defined as

$$Y = -\mathrm{Tr}\hat{D}\ln\hat{D} - \sum_{l=0}^{L} \lambda_l \mathrm{Tr}\hat{D}\hat{A}_l$$

The variation of Y induced by a variation $\delta\hat{D}$ of the density matrix $\hat{D}$ reads

$$\delta Y = -\mathrm{Tr}\delta\hat{D}\left(\ln\hat{D} + 1 + \sum_{l=0}^{L} \lambda_l \hat{A}_l\right)$$

The extremum correspond to $\delta Y = 0$, with no restrictions on $\delta\hat{D}$ leading to the condition $\ln\hat{D} + 1 + \sum_{l=0}^{L} \lambda_l \hat{A}_l = 0$. The solution is the density matrix at equilibrium which is a function of the Lagrange multipliers λ_l

$$\hat{D}^0 = \frac{1}{Z}\exp - \sum_{l=1}^{L} \lambda_l \hat{A}_l \tag{2}$$

where we have already taken care of the normalization constraint by introducing the partition sum

$$Z = \mathrm{Tr}\exp - \sum_{l=1}^{L} \lambda_l \hat{A}_l \tag{3}$$

The link between a constraint $\left\langle \hat{A}_l \right\rangle$ (or observation, or *extensive* variable) and the associated Lagrange multiplier λ_l(or thermodynamically conjugated *intensive* variable) is given by an *equation of state*

$$\left\langle \hat{A}_l \right\rangle = -\frac{\partial \ln Z}{\partial \lambda_l} \tag{4}$$

It is also possible to express λ_l as a function of $\langle A \rangle$ by inverting the equation of state. Indeed the equilibrium corresponding to the considered constraints is associated to a value for the statistical entropy

$$S = \mathrm{Tr}\hat{D}^0 \ln \hat{D}^0 = \sum_l \lambda_l \left\langle \hat{A}_l \right\rangle + \ln Z \tag{5}$$

This last equation known as a *Legendre transform* gives the relation between the entropy and the partition sum and implies for the Lagrange multipliers

$$\lambda_l = \frac{\partial S}{\partial \left\langle \hat{A}_l \right\rangle} \tag{6}$$

It should be noticed that while D^0 and Z are functions of the intensive variables (λ_l), the Legendre transform S is a function of the associated extensive variables $\left\langle \hat{A}_l \right\rangle$.

Using (2,3,4) the whole thermodynamics of the system can be calculated if the constraints $\left\langle \hat{A}_l \right\rangle$ are known. It is important to remark that this formalism is completely general in the sense that it can be applied for an arbitrary number of bodies with no need of a thermodynamical limit (infinite systems), and that all observables (and not only variables conserved by the dynamics) can play the role of constraints. Moreover the maximization of entropy as a tool to deal with the general problem of missing information can be extended in dynamical situations and has shown to be a fruitful approach in the field of stochastic quantum transport [5].

1.6 The Usual Thermodynamics [4]

The usual ensembles of standard thermodynamics can also be obtained as applications of this general theory. Let us consider for example the case where the only constraint is the energy $\langle E \rangle = Tr\left(\hat{D}^0 \hat{H}\right) = \sum_n p_n E^{(n)}$ associated with the Lagrange multiplier β. The probability of the n-th energy eigenstate is then $p_n^0 = \frac{1}{Z_\beta} \exp(-\beta E^{(n)})$ while the energy probability distribution reads $p_\beta(E) = \frac{W(E)}{Z_\beta} \exp(-\beta E)$ where $W(E)$ is the number of states corresponding to an energy E. The Lagrange multiplier β has the physical meaning of the inverse of the temperature $T = \beta^{-1}$. The relation between the average energy and the temperature is given by the equation of state $\langle E \rangle = -\partial_\beta \ln Z_\beta$ and the Legendre transform $S(\langle E \rangle) = \ln Z_\beta + \beta \langle E \rangle$ represents the well known relation between the canonical entropy and the free energy $F_T = -\beta^{-1} \ln Z_\beta$.

The microcanonical ensemble can also be obtained from this general theory considering that in the absence of any constraint (except the normalization of probabilities) all states must be equiprobable. The microcanonical entropy is then obtained as the expression of the Shannon entropy within the equilibrium distribution $p_i^0 = 1/W(E)$, $S(E) = -\sum_{i=1}^{W} W^{-1} \ln W^{-1} = \ln W$.

2 Generalities About Phase Transitions

Generally speaking, for a given value of the *control parameters* (or *intensive variables*) λ_l, the properties of a substance are univocally defined, i.e. the conjugated *extensive variables* $\left\langle \hat{A}_l \right\rangle$ have a unique value unambiguously defined by the corresponding equation of state. For instance the volume occupied by n moles of an ideal gas at a given pressure P and temperature T is given by $V = nRT/P$. In reality we have seen in the previous chapter that extensive variables, being by definition expectation values of operators, are associated with a probability distribution unless the system is described by a pure state. The intuitive expectation that extensive variables at equilibrium have a unique value therefore means that the probability distribution is narrow and normal, such that a good approximation can be obtained by replacing the distribution with its most probable value.

In this case, as we will see in Sect. 2.1, the Legendre transform gives an exact mapping between the standard *intensive ensembles* in which the control parameter is intensive or equivalently only the average of the extensive variable is known and the more exotic *extensive ensembles* where an extensive variable is controlled event by event, demonstrating the equivalence between the different statistical ensembles. In the following we will often take as an paradigm of intensive ensembles the *canonical ensemble* for which the inverse of the temperature β^{-1} (or equivalently the average energy $\langle E \rangle$) is controlled while the archetype of the extensive ensemble will be the microcanonical one for which the energy is strictly controlled.

The normality of probability distributions is usually assumed on the basis of the central limit theorem that we will briefly review in Sect. 2.2. However some situations exist in which the probability distributions of extensive variables are abnormal and for example bimodal: in this case two different properties (*phases*) coexist for the same value of the intensive control variable. A first elementary description of phase coexistence using this intuitive bimodality argument will be given at the end of Sect. 2.2.

The topological anomalies of probability distributions and the failure of the central limit theorem in phase coexistence imply that in a first order phase transition the different statistical ensembles are in general *not* equivalent and different phenomena can be observed depending on the fact that the controlled variable is extensive or intensive. This general statement will be developed in great detail in Chap. 4 and its far reaching consequences will be analyzed in Chap. 6.

2.1 The Difference Between Laplace and Legendre

We have seen in the last chapter that the relation between the different thermostatistical potentials is given by the Legendre transform. It is important to distinguish between transformations within the same statistical ensemble as the Legendre transform (which gives for instance the link between the canonical partition sum and the canonical entropy) and transformations between different ensembles which are instead given by non linear integral transforms. Let us consider energy as the extensive observable and temperature as the conjugated intensive one. The definition of the canonical partition sum is

$$Z_\beta = \sum_n \exp(-\beta E^{(n)})$$

where the sum runs over the available eigenstates n of the Hamiltonian. If energy can be treated as a continuum variable this equation can be written as

$$Z_\beta = \int_0^\infty dE\, W(E) \exp(-\beta E) \tag{7}$$

which is nothing but a Laplace transform between the canonical partition sum and the microcanonical entropy $S_E = \ln W(E)$. If the integrand $f(E) = W(E) \cdot \exp(-\beta E)$ is a strongly peaked function the integral can be replaced by the maximum $f(\bar{E})$

$$Z_\beta \approx W(\bar{E}) \exp(-\beta \bar{E}) \tag{8}$$

which can be rewritten as

$$\ln Z_\beta \approx S_{\bar{E}} - \beta \bar{E} \tag{9}$$

or introducing the free energy $F_T = -\beta^{-1} \ln Z_\beta$

$$F_T \approx \bar{E} - T S_{\bar{E}}$$

Equation (9) has the structure of an approximate Legendre transform and shows that in the saddle point approximation (8) the ensembles differing at the level of constraints acting on a specific observable (here energy) differ only by a simple linear transformation. We will see in the next section and in more details in Chap. 6 that however the saddle point approximation (8) can be highly incorrect close to a phase transition. In particular, when the canonical distribution of energy is bimodal a unique saddle point approximation becomes inadequate. In this case (9) cannot be applied and (7) is the only possible transformation between the different ensembles.

To summarize one should not confuse

- the link between the thermodynamical potential of the intensive (e.g. log of canonical partition sum) and of the extensive ensemble (e.g. the microcanonical entropy) which are always related with a Laplace transform. This Laplace transform may lead to an approximate Legendre transformation for normal distributions but we know that this Legendre transformation is wrong if the distribution is abnormal.
- with the exact Legendre transform between the entropy of the intensive ensemble and the corresponding thermodynamical potential.

This simply corresponds to the fact that the microcanonical and canonical entropies can be very different.

2.2 The Central Limit Theorem and Phase Coexistence

The typical representation of the probability distribution of any generic random variable depending on a not too small number of degrees of freedom is a Gaussian distribution. The very general validity of the Gaussian is due to one of the most important theorems of statistics, the Laplace central limit theorem. Let us consider an extensive observable E (i.e., energy) that can be written as the sum of I independent contributions (i.e. the energy of the different particles constituting the system) $E = \sum_i^I e_i$, where the e_i follow an arbitrary probability distribution with the unique requirement that the global variance $\sigma_E^2 = \sum_i^I (\langle e_i^2 \rangle - \langle e_i \rangle^2)/I$ is finite. Then the central limit theorem states that the distribution of E tends to a Gaussian distribution with a width decreasing with the number of degrees of freedom

$$\lim_{I \to \infty} p(E) = \frac{1}{\sqrt{2\pi\sigma^2/I}} \exp(-\frac{(E - \langle E \rangle)^2}{2\sigma^2/I}) \tag{10}$$

According to the central limit theorem at the thermodynamical limit the distribution of an extensive variable $p(E)$ tends to a δ-function, implying as we have mentioned at the beginning of the chapter that the properties of the system are univocally defined by the value of the intensive parameter that controls the asymptotic value of $\langle E \rangle$ through the appropriate equation of state. Moreover in most cases a few tens of particles are enough for the Gaussian approximation to be correct, meaning that the limit appearing in (10) can be neglected in practical applications. Another consequence of the central limit theorem is that the Laplace transform becomes equivalent to a Legendre transform as we have discussed in the preceding section, leading to the equivalence of statistical ensembles.

However a situation can occur in which the probability distribution is bimodal and never tends to a Gaussian. Such a situation is called a first order phase transition. This patent violation of the central limit theorem is due to the fact that phase transitions are associated to long-range correlations and the independence hypothesis between the different degrees of freedom breaks down.

Let us illustrate the standard picture of phase coexistence within a simple example. Consider a molecular system in the canonical ensemble characterized by the free energy $F = -T \ln Z = \langle E \rangle - TS$. As we have demonstrated in Sect. 1.3 the maximization of the statistical entropy with the energy constraint is equivalent to the minimization of the free energy. At low temperature a minimization of F is approximately equivalent to a minimization of $\langle E \rangle$: the equilibrium state of the system will be given by a compact configuration (a crystal or a liquid) with free energy $F_L(A, V)$. On the other side at high temperature the minimization of F corresponds to a maximization of the canonical entropy, which will be achieved by a disordered rarefied state (a gas phase) with free energy $F_G(A, V)$. Phase coexistence means that at an intermediate temperature the two free energy solutions are allowed giving for the total free energy

$$F(A, V) = F_L(A_L, V_L) + F_G(A_G, V_G) \tag{11}$$

where A_L, V_L (A_G, V_G) are the fractions of total number of molecules A and volume V belonging to the ordered (disordered) phase

$$A = A_L + A_G \quad ; \quad V = V_L + V_G$$

The equilibrium sharing of A and V is given by the minimization of the free energy

$$\frac{\partial F}{\partial A_L} = \frac{\partial F_L}{\partial A_L} - \frac{\partial F_G}{\partial A_G} = 0 \quad ; \quad \frac{\partial F}{\partial V_L} = \frac{\partial F_L}{\partial V_L} - \frac{\partial F_G}{\partial V_G} = 0$$

implying the equality for the intensive variables conjugated to the mass number and the volume, namely the chemical potential and the pressure

$$p_L = p_G \quad ; \quad \mu_L = \mu_G$$

This procedure can be generalized to any statistical ensemble. If we consider for example the microcanonical ensemble, the absence of constraints means that the thermostatistical potential is directly the microcanonical entropy

$$S(A, E, V) = \ln W(A, E, V) = S_L(A_L, E_L, V_L) + S_G(A_G, E_G, V_G) \tag{12}$$

with the extra conservation law $E = E_L + E_G$. The extremization of S respect to V and A gives again the equality of the chemical potential and pressure for the two coexisting phases, while the derivative respect to the energy variable gives

$$T_L = T_G$$

where we have defined the microcanonical temperature as $T^{-1} = \partial_E S$ in analogy with the canonical Legendre transform $\beta = \partial_{\langle E \rangle} S_\beta$ (the justification of the physical meaning of $\partial_E S$ as an inverse temperature is postponed to Chap. 4).

Equilibrium between the two phases is characterized by the equality of the temperatures. On the other hand, the conjugated extensive variables are different in the two phases $E_L < E_G$. This means that at the transition temperature $T_{tr} = T_L = T_G$ the energy is discontinuous at the phase transition (latent heat).

To summarize, in this standard view first order phase transitions are characterized by

- the presence of two phases in contact
- a discontinuity in (one ore more) first order derivatives of the thermostatistical potential (energy, volume, mass number...).

To obtain this result we have written the thermostatistical potential as a simple sum of the contributions of the two phases (see (11,12)). This is true only if the free energy (or entropy) of the interface between the two phases is negligible, i.e. for large systems interacting through short range forces.

In the next sections we will illustrate this standard view of first order phase transitions within an exactly solvable model in one and two dimensions (Sects. 2.3, 2.4) and in three dimensions with the help of the mean field approximation (Chap. 3).

The additivity hypothesis of the thermostatistical potential breaks down for finite systems and even in the thermodynamical limit if the forces are long ranged. The far reaching consequences of dropping this approximation will be developed in Chap. 4.

2.3 Isomorphism Between Ising and Lattice Gas

Let us consider an ensemble of N classical spins which can take one of the two values $s_k = \pm 1$ on a lattice under the influence of an external magnetic field h and a constant coupling J between neighboring sites according to the Hamiltonian

$$H_{IS} = -h \sum_{k=1}^{N} s_k - \frac{J}{2} \sum_{k \neq j}^{N} s_k s_j \tag{13}$$

where the second sum extends over closest neighbors.

The Ising model (13) has been originally introduced to give a simple description of ferromagnetism (i.e. a spontaneous magnetization that some substances present in the absence of a field at low temperature). In reality the phenomenon of ferromagnetism is far too complicated to be treated in a satisfactory way by this oversimplified Hamiltonian; however the fact that the Ising model is exactly solvable in 1D and 2D and that very accurate numerical solutions exist for the three dimensional case makes this model a paradigm of first and second order phase transitions. The other appealing side of the Ising model is its versatility: introduced to explain magnetic phase transitions, it is also well adapted to describe fluid phase transitions. Indeed we can show that a close link exists between the Ising Hamiltonian (13) and the Lattice Gas Hamiltonian which is the simplest modelization of the liquid-gas phase transition

$$H_{LG} = \frac{1}{2m} \sum_{k=1}^{N} \boldsymbol{p}_k^2 n_k - \frac{\varepsilon}{2} \sum_{k \neq j}^{N} n_k n_j \tag{14}$$

In the Lattice Gas model, the same N lattice sites in D dimensions are characterized by an occupation $n_k = 0, 1$ and by a D components momentum vector $\boldsymbol{p}_k$. Occupied sites (particles) interact with a constant closest neighbor coupling ε.

Because of the transformation $n_k = (s_k + 1)/2$ the Ising Hamiltonian H_{IS} can be mapped into the interaction part H_{LG}^{int} of the Lattice Gas Hamiltonian H_{LG}. Indeed let us consider the interaction part of the Lattice Gas partition sum in the grancanonical ensemble

$$Z_{LG}^{int} = \sum_{n_1=0,1} \cdots\cdots \sum_{n_N=0,1} \exp(-\beta(H_{LG}^{int} - \mu A))$$

where $A = \sum_k^N n_k$ is the total number of particles and β, μ are Lagrange multipliers. The factor multiplied by $-\beta$ in the exponential can be written as

$$H_{LG}^{int} - \mu A = -\frac{\varepsilon}{8} \sum_{k \neq j}^{N} s_k s_j - \left(\frac{\varepsilon z + 2\mu}{4} \right) \sum_{i=1}^{N} s_i - N \left(\frac{\varepsilon z + 4\mu}{8} \right)$$

where $z = 2D$ is the number of closest neighbors. With the identification $J = \varepsilon/4$ and $h = (z\varepsilon + 2\mu)/4$, this equation shows that the grancanonical partition sum of the Lattice Gas interaction hamiltonian is isomorphous to the canonical partition sum of the Ising model in an external field

$$\beta\left(H_{LG}^{int} - \mu A\right) = \beta H_{IS} + K$$

where K is a constant. This result implies that all results obtained within the Ising model concerning magnetic transitions can be translated in terms of fluid transitions and vice-versa. In particular the magnetization $m = \left\langle \sum_k s_k \right\rangle / N$ is related to the matter density $\rho = \left\langle \sum_k n_k \right\rangle / N$ by $m = 2\rho - 1$.

2.4 Exact Solution of the Ising Model in 1D and 2D

The Ising model was proposed by Lenz to his student Ising in 1925. The exact solution of the model in one dimension is given in Ising's thesis.

Let us consider a one dimensional spin chain with periodic boundary conditions (spin ring). The Ising hamiltonian can be written as

$$H_{IS} = -\sum_{k=1}^{N} \left(h s_k + J s_k s_{k+1}\right)$$

and the partition sum results

$$Z_{IS} = \sum_{s_1=-1}^{1} \ldots\ldots \sum_{s_N=-1}^{1} \exp\left(\beta \sum_{k=1}^{N} \left(h s_k + J s_k s_{k+1}\right)\right) = \sum_{s_1=-1}^{1} \ldots\ldots \sum_{s_N=-1}^{1} \tau_{12}\tau_{23}\ldots\tau_{N1}$$

where we have introduced

$$\tau_{ij} = \exp\left(\frac{h}{2}\left(s_i + s_j\right) + J s_i s_j\right)$$

If we consider the τ_{ij} as the elements of a 2x2 matrix depending upon the two spins $s_i = \pm$ and $s_j = \pm$

$$T = \begin{pmatrix} \tau_{+\,+} & \tau_{+\,-} \\ \tau_{-\,+} & \tau_{-\,-} \end{pmatrix}$$

where the definition of the $\tau_{i\,j}$ implies

$$\begin{aligned} \tau_{+\,+} &= \exp \beta(J + h) \\ \tau_{-\,-} &= \exp \beta(J - h) \\ \tau_{+\,-} &= \tau_{-\,+} = \exp(-\beta J), \end{aligned}$$

then we can write $\sum_{s_j=-1}^{1} \tau_{ij}\tau_{jk} = T_{ik}^2$ and the partition sum becomes

$$Z_{IS} = \sum_{s_1=-1}^{1} T_{11}^N = Tr\left(T^N\right) = \lambda_1^N + \lambda_2^N = \lambda_1^N \left(1 + \left(\frac{\lambda_2}{\lambda_1}\right)^N\right) \underset{N\to\infty}{\longrightarrow} \lambda_1^N$$

where λ_1, λ_1 ($\lambda_1 > \lambda_2$) are the eigenvalues of the T matrix. The problem is then reduced to an eigenvalue problem

$$\det(T - \lambda I) = 0 \quad ; \quad \lambda^2 - (\tau_{+\,+} + \tau_{-\,-})\,\lambda + (\tau_{+\,+}\tau_{-\,-} - \tau_{+\,-}\tau_{-\,+}) = 0$$

After a little algebra we obtain the eigenvalues

$$\lambda = \exp(\beta J)\left(ch(\beta h) \pm \left(\exp(-4\beta J) + sh^2(\beta h)\right)^{\frac{1}{2}}\right)$$

and the partition sum

$$\ln Z_{IS} = N\left(\beta J + \ln\left(ch(\beta h) + \left(\exp(-4\beta J) + sh^2(\beta h)\right)^{\frac{1}{2}}\right)\right)$$

It is easy to verify that $\ln Z_{IS}$ is a continuous function with continuous derivatives for all orders: the Ising model in one dimension does not present a phase transition. In particular the magnetization

$$m = \frac{1}{N\beta}\frac{\partial \ln Z_{IS}}{\partial h} = \frac{1}{2} sh(\beta h)\left(\exp(-4\beta J) + sh^2(\beta h)\right)^{\frac{1}{2}}$$

is a continuous monotonic function which is zero at zero field: no spontaneous magnetization is observed.

The solution of the Ising model in two dimensions [6] is far too complicated to be developed here. Let us simply give the asymptotic result $N \to \infty$ in the zero field case

$$\begin{aligned} Z_{IS} &= \left(2ch\,(2\beta J)\exp I\right)^N \\ I &= \frac{1}{2\pi}\int_0^{\pi} d\phi \ln \frac{1 + \left(1 - x^2 \sin^2\phi\right)^{\frac{1}{2}}}{2} \\ x &= 2\frac{sh\,(2\beta J)}{ch^2\,(2\beta J)} \end{aligned} \tag{15}$$

With the partition sum of (15) the magnetization equation of state can be computed. It is easy to verify that for temperatures lower than the critical temperature $T_c = \beta_c^{-1}$ given by $sh\,(2J\beta_c) = 1$ the system presents a spontaneous magnetization at zero field [7]

$$m_0 = m(h = 0, \beta > \beta_c) = \frac{ch^2\,(2\beta J)}{sh^4\,(2\beta J)}\left(sh^2\,(2\beta J) - 1\right)^{1/8} \underset{\beta \to \beta_c}{\longrightarrow} (T_c - T)^{1/8}$$

The equation of state of the Ising model in more than one dimension is schematically shown in Fig. 1.

At subcritical temperatures a discontinuity in magnetization is seen at zero field, showing that a first order phase transition is taking place in agreement with the intuitive arguments of the preceding section. For $T = T_c$ the magnetization goes to zero as a power law (second order phase transition) while the equation of state is monotonous in the supercritical regime.

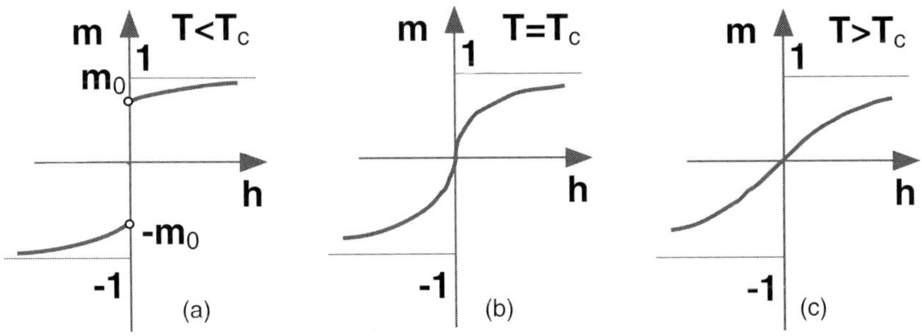

Fig. 1. Schematic representation of the average magnetization as a function of the applied external field for the Ising model in more than one dimension at subcritical (left), critical (center) and supercritical (right) temperature.

3 The Mean Field Approximation

Even for simplified models such as Ising no analytical solution exists for a number of dimension $D > 2$. This is the reason why mean field solutions have been developed. The idea of the mean field approximation is to replace the intractable N-body problem with an approximately equivalent analytical one body problem. Let us illustrate this method on the Ising case. If the Hamiltonian is composed of one body terms solely

$$H_{1b} = -\sum_{k=1}^{N} h_k s_k \tag{16}$$

with h_k a generic one body operator, the thermodynamics of the system is solved in one line. Indeed the partition sum in the canonical ensemble reads

$$Z_{1b} = \sum_{s_1=-1}^{+1} \cdots \sum_{s_N=-1}^{+1} \exp\left(-\beta H_{1b}\right) = \prod_{k=1}^{N} z_k = \left(\exp\left(-\beta h\right) + \exp\left(\beta h\right)\right)^N \tag{17}$$

where the last equality holds if $h_k = h \; \forall k$, and is promptly generalized to the non-local case.

To reduce the Hamiltonian to a one body interaction the correlations among the different sites have to be neglected such that the interaction on a given site depends only of the coordinates of the site. This chapter is devoted to the applications of this approximation to the Ising model (Sect. 3.1) and its general consequences for the problem of first order phase transitions (Sects. 3.2–3.3). We will see that an equivalent one body problem can be formulated and the two body character of the force results in a self-consistency problem for the equations of state which have to be solved iteratively.

It is important to stress that all mean field approaches are approximations which, because of the intrinsic lack of correlations, are especially bad in phase co-existence. In the recent years the enormous progress of computing machines has allowed the numerical solution of three dimensional models without any approximation with Monte-Carlo based methods. These exact solutions clearly show the inherent limitations of mean field approaches and will be discussed in Chap. 4.

3.1 Mean Field Approximation for the Ising Model

The interaction acting on the k-th site in the Ising model (13) is $h_k = h + J\sum_j s_j$, where the sum extends over the first neighbors of site k. A one body term is obtained if the spin of the neighboring sites s_j is assumed constant all over the lattice and equal to the average magnetization $s_j \approx \langle s \rangle = m$. In other words the exact interaction is approximated by the interaction the site would experience if the spin distribution was uniform. The Ising Hamiltonian can then be written as a one body Hamiltonian

$$H_{MF} = -\sum_{k=1}^{N} h_k s_k + K = -\sum_{k=1}^{N} (h + Jzm)\, s_k + K \tag{18}$$

within a constant K which has to be determined by imposing that the expectation value of H_{MF} is equal to the mean field energy

$$E_{MF} = -hNm + E_{MF}^{int} = -N\left(hm + \frac{Jz}{2}m^2\right) \tag{19}$$

where the last equality is obtained by writing the interaction energy as

$$E^{int} = -\frac{J}{2}\sum_{k=1}^{N}\sum_{j\neq k}\langle s_k s_j\rangle \approx -\frac{J}{2}\sum_{k=1}^{N}\sum_{j\neq k}\langle s_k\rangle\langle s_j\rangle = -\frac{J}{2}Nzm^2$$

which shows once again that the effect of the mean field approximation is the neglect of two body correlations. The comparison of (19) with the expectation value of (18) leads to the definition of the constant K as $K = JNzm^2/2$. In fact this energy correction exactly compensates the double counting of the two-body interaction due to the introduction of the average interaction of each spin with all its neighbors. The mean field partition sum as for (17) is factorized in the product of the individual partition sums of the different sites

$$Z_{MF} = \sum_{s_1=\pm 1} \dots \sum_{s_N=\pm 1} \exp\left(-\beta(H_{MF})\right) = z^N \tag{20}$$

where

$$\begin{aligned} z &= \sum_{s=\pm 1} \exp\left(-\beta\left(-(h + Jzm)\, s + \frac{Jz}{2}m^2\right)\right) \\ &= 2\exp\left(-\beta\frac{Jz}{2}m^2\right) ch\left(\beta\left(h + Jzm\right)\right) \end{aligned}$$

which leads to a self-consistent equation for the magnetization

$$m = \tanh\left(\beta\left(h + Jzm\right)\right) \tag{21}$$

Equation (21) is represented in Fig. 2 in the subcritical, critical and supercritical regime. If the behavior of the equation of state for $T \geq T_c = Jz$ is

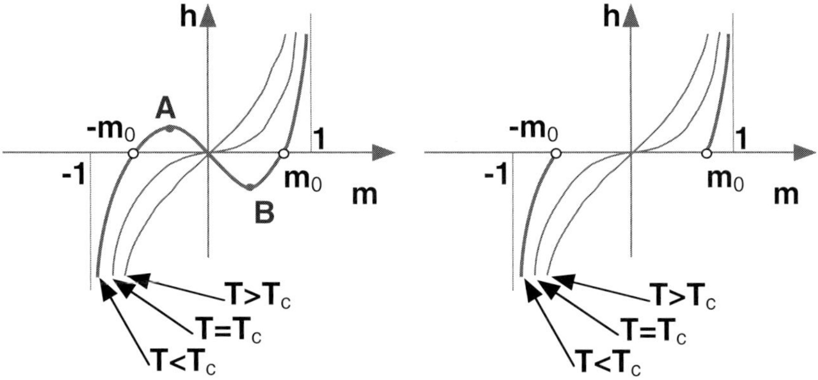

Fig. 2. Left side: relation between the average magnetization and the magnetic field at subcritical, critical and supercritical temperature for the three dimensional Ising model in the mean field approximation. Right side: Maxwell construction modifying the subcritical magnetization curve.

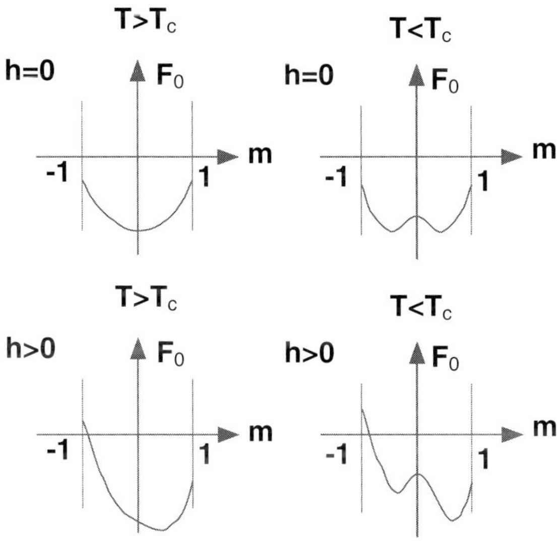

Fig. 3. Mean field free energy as a function of magnetization at zero (upper part) and positive (lower part) magnetic field, for a supercritical (left) and a subcritical (right) temperature.

qualitatively similar to the exact Onsager solution of Sect. 2.4, in the first order phase transition regime the mean field solution shows a backbending behavior with a negative susceptibility $\chi^{-1} = \partial_h m$ region. To understand the physical meaning of the backbending, the free energy $F = -\beta^{-1} \ln Z_{MF}$ is shown as a function of magnetization in Fig. 3 in the $h = 0$ and $h > 0$ case. From this figure one can see that the backbending corresponds to a maximum of the free energy, i.e. an instability. Indeed the coexistence between the two phases at different magnetization cannot be obtained in a mean field calculation because of the intrinsic homogeneity hypothesis $m = \langle s \rangle = const$. The backbending therefore reflects the instability of the homogeneous mean field solution with zero magnetization respect to the separation into two distinct phases at $m = \pm m_0$. At non zero field the magnetization oriented in the direction of the field has the minimum free energy, therefore will correspond to the unique equilibrium solution. In the zero field case the two solutions have the same energy. This implies that every linear combination of these solutions

$$m(h = 0, T < T_c) = \alpha m_0 + (1 - \alpha)(-m_0) \quad ; \quad 0 < \alpha < 1 \tag{22}$$

will have the same free energy; such a linear combination represents the coexistence between the two solutions as we have discussed in Chap. 2 and corresponds to an horizontal straight line in the $F - m$ and in the $h - m$ plane (tangent construction) as shown in the right part of Fig. 2.

If the lack of correlations of the mean field is cured by allowing a mixed phase according to (22), the usual shape of the phase transition is recovered (discontinuity in the first derivative of the thermodynamical potential).

To conclude this section we would like to comment the difference between a self consistent approach as the mean field approximation and a genuine one body Hamiltonian as in (16),(17). We have shown in Chap. 1 that the thermodynamics of a system is completely determined once the partition sum is known, since all thermodynamical quantities can be calculated as successive derivatives of $\ln Z$. The Hamiltonian entering the mean field approximation of the partition sum (20) differs from the mean field approximation of the Ising Hamiltonian because of the constant K which we have been forced to add for the Hamiltonian to have the correct expectation value. The constant K in the partition sum represents more than a trivial shift in the energy scale since K depends on m which in turn is calculated from $\ln Z$ showing the self-consistent character of the approach. Following (16),(17) one could be tempted to define from the mean field approximation to the Ising Hamiltonian a one body partition sum as

$$Z_{1b} = \left(\sum_{s= \pm 1} \exp\left(-\beta(h + Jzm)s\right) \right)^N$$

and the question arises weather thermostatistical observables can be obtained from the successive derivatives of Z_{1b}.

To answer to this question one has to use the formalism of Chap. 1 and explicitly calculate the statistical entropy

$$\begin{aligned} S_{MF} &= -\sum_i p_i \ln p_i \\ &= -\sum_i \left(p_i \left(-\beta H_{1b}^{(i)} - \ln Z_{1b} \right) \right) \\ &= \beta \langle H_{1b} \rangle + \ln Z_{1b} \end{aligned}$$

since the probability distribution for the mean field problem reads $p_i = \exp\left(-\beta H_{1b}^{(i)}\right)/Z_{1b}$. The expectation of H_{1b} is readily calculated as

$$\langle H_{1b}\rangle = -N\left(hm + Jzm^2\right) = -hNm + 2E_{MF}^{int} \tag{23}$$

The general relation between entropy and free energy $-\beta F \equiv \ln Z_{MF} = S_{MF} - \beta E_{MF}$ finally leads to

$$\ln Z_{MF} = \ln Z_{1b} + \beta E_{MF}^{int} \tag{24}$$

Equation (24) shows that because of the two body interaction the partition sum is different from the one body partition sum even in the mean field approximation. In fact the difference comes from the double counting of the two-body interaction if the energy is calculated as $\langle H_{1b}\rangle$.

The best way to understand the mean field approach is to consider mean-field solutions as a trial state to maximize the entropy completed by the constraints $(S - \beta\langle E\rangle)$ i.e. to variationally estimate the free energy $F = -\beta^{-1}(S - \beta\langle E\rangle)$. Then only the mean-field free energy can be considered as a good approximation of the exact free energy leading to $\beta F = -\ln Z_{1b} - \beta E_{MF}^{int}$ which is nothing but (24).

3.2 Implications for the Liquid-Gas Transition

We have seen in Sect. 2.3 that the isomorphism between the Ising model and the Lattice Gas model implies that all physical results concerning magnetic transitions can be easily translated in the fluid language and applied within minor modifications to the liquid-gas transition. To this aim the Ising canonical partition sum Z^{can}, or free energy $F = -\beta^{-1}\ln Z^{can}$, has to be transformed into the Lattice Gas grand-canonical partition sum Z^{GC} , or grand potential $\Omega = -\beta^{-1}\ln Z^{GC}$. If we only focus on the interaction part of the Lattice Gas model this leads to

$$\begin{aligned}\exp\left(-\beta\Omega_{LG}\right) &= \sum_n \exp\left(-\beta\left(H_{LG}^{(n)} - \mu A^{(n)}\right)\right)\\ &= \sum_n \exp\left(-\beta\left(H_{IS}^{(n)} - N\frac{zJ}{2} - N\frac{\mu}{2}\right)\right)\\ &= \exp\left(-\beta F_{IS}\right)\exp\left(\beta N\left(\frac{zJ}{2} + \frac{\mu}{2}\right)\right).\end{aligned}$$

In the mean field approximation $F_{IS} = -\beta^{-1}\ln Z_{MF}$ is given by (20) giving for the Lattice Gas grand potential

$$\frac{\Omega}{N} = \frac{1}{2}Jzm^2 - \beta^{-1}\ln\left(2ch\left(\beta\left(h + Jzm\right)\right)\right) - \frac{zJ}{2} - \frac{\mu}{2}$$

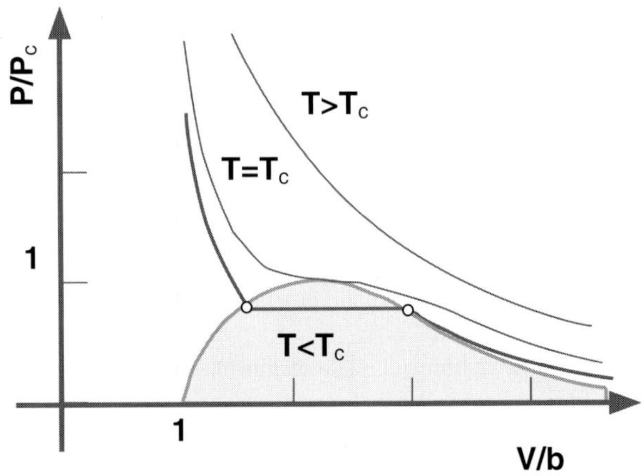

Fig. 4. Isotherms in the pressure versus volume (in cell units b) plane for the three dimensional Lattice Gas model in the mean field approximation with a Mawell construction of the mixed phase. The coexistence zone is also indicated.

The total number of lattice sites in the Lattice Gas framework represents the volume of the fluid $N = V$. The equation of state $p = \beta^{-1}\partial_V \ln Z$ allows to access the pressure

$$p = -\frac{\Omega}{V} = \frac{1}{2}\left(zJ + \mu - Jzm^2\right) + \beta^{-1}\ln\left(2ch\left(\beta\left(h + Jzm\right)\right)\right) \qquad (25)$$
$$= -\varepsilon z\rho^2 + \beta^{-1}\ln\frac{1}{1-\rho}$$

where the last equality is obtained using the magnetization equation of state (21) and the substitutions $m = 2\rho - 1$, $J = \varepsilon/2$. Fig. 4 shows some selected isotherms of the fluid equation of state. At subcritical temperatures $T < T_c$ a clear backbending is seen reflecting the instability of the homogeneous mean field solution respect to the separation into distinct phases as in Fig. 2 above. Once again, if a linear interpolation of the liquid and gas volume solutions is imposed, the usual plateau of the Maxwell construction is recovered. The critical point is defined as the ending point of the coexistence zone, i.e. the point at which the first as well as the second derivative of the equation of state are zero. Substituting in (25) we get $\rho_c = 1/2$, $T_c = Jz$, $p_c = T_c\left(\ln 2 - 1/2\right)$.

3.3 The Van der Waals Equation of State

The Ising model in the mean field approximation reflects the same physics as the Van der Waals equation of state which describes a classical canonical gas of N identical molecules in an external pressure field p_0 and volume V interacting via an attractive two body force. The free enthalpy connected to such a physical scenario is

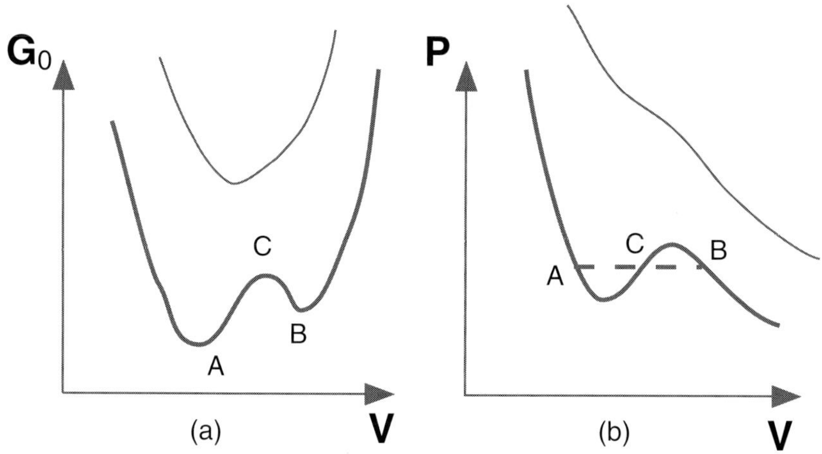

Fig. 5. Free energy and isotherms of the Van der Waals equation of state at a subcritical and supercritical temperature.

$$H = F + p_0 V = \frac{3}{2} NT - b\rho N - TS + p_0 V$$

Here $\rho = N/V$ is the density of matter and $b\rho N$ is the two body interaction energy in the mean field approximation. If we make an explicit use of the equivalence between ensembles at the macroscopic limit, the entropy S can be calculated in the mean field approximation as an effective one body problem

$$S = \ln W = \ln(W_r W_p) = \ln\left(w_r^N w_p^N\right) = N \ln \frac{V - Nv_0}{N} + \frac{3}{2} N \ln T$$

where the integral over the configuration space is

$$w_r^N = \left(\int_V d^3r \; \theta(r^3 - V) \right)^N /N! = (V - Nv_0)^N /N!$$

where v_0 is the volume occupied by each particle, the momentum space integral gives

$$w_p^N = \left(\int d^3p \; \exp(-p^2/2mT) \right)^N = (2\pi m T)^{3N/2}$$

and we have used the Stirling approximation of the factorials. Using as above the equation of state $p = \beta^{-1} \partial_V \ln Z = -\partial_V F$ or the extremum condition $\partial_V H = 0$ we get

$$p = \frac{NT}{V - Nv_0} - \frac{bN^2}{V} \tag{26}$$

The free energy together with the isotherms are represented in Fig. 5. The similarity with the microscopic results from the Ising model is evident. Once again the volume interval $V_A < V < V_B$ is unstable in the sense that if we mix up in linear proportions solutions of type A and of type B, the free enthalpy

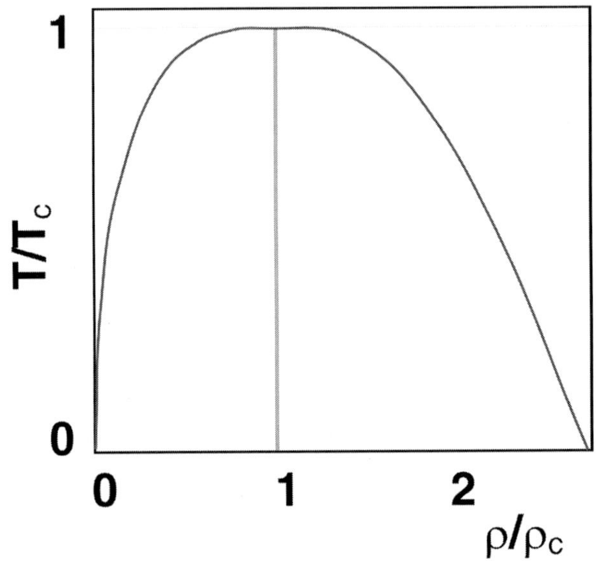

Fig. 6. Guggenheim's coexistence line in scaled variables obtained from many different substances.

(H) does not change. This tangent construction is the well known equal area Maxwell construction since

$$H(V_B) - H(V_A) = \int_{V_A}^{V_B} dV \; dH/dV = p_0 \, (V_B - V_A) - \int_{V_A}^{V_B} dV \; p(V) = 0$$

The coordinates of the critical point are found from $d_V \; p|_c = d_V^2 \; p|_c = 0$ as $V_c = 3Nv_0$, $T_c = 8b/27v_0$, $p_c = b/27v_0$. If we introduce the scaled variables $\upsilon = V/V_c$, $\tau = T/T_c$, $\pi = p/p_c$ one can readily verify that the Van der Waals equation (26) becomes

$$\pi = \frac{8\tau}{3\nu - 1} - \frac{3}{\nu^2}$$

with no dependence on b or v_0 i.e. on the quantities specific of the structure of the gas.

This feature is preserved in realistic gases for which all thermodynamical variables can be rescaled according to the critical values leading to the famous Guggenheim phase diagram [8] (shown in Fig. 6), namely a unique coexistence curve for many different substances which shows the universality character of phase transitions; this universality feature gives an a posteriori justification of the use of a schematic oversimplified model as the Ising model to describe complex and widely different physical phenomena.

3.4 The Landau Model

The simplest functional form of the thermodynamical potential as a function of the order parameter that contains all the physical situations discussed in the previous sections is given by

$$F(m) = C + N\left(a(T)m^2 + bm^4 - hm\right) \quad ; \quad a(T) = a_0\left(T - T_c\right) \tag{27}$$

where h is the intensive variable conjugated to the order parameter m. Note $b > 0$ in order to have a free energy bound from below in order to ensure that an absolute minimum (i.e. an equilibrium) does exist. Equation 27 is known in the literature as the result of Landau theory of phase transitions [9]. It is immediate to verify that in the proximity of the critical point $T \to T_c$, $h \to 0$ the order parameter and its derivatives follow a power law behavior $m \to |T_c - T|^\beta$, $dm/dh \to |T_c - T|^\gamma$ where $\beta = 1/2$, $\gamma = -1$ are typical mean-field critical exponents.

Summarizing the present chapter, the mean field approximation leads to the definition of phase transitions as universal phenomena with the following characteristics

- presence of two different phases (i.e. minima of the thermodynamical potential) that coexist in contact (via a non analyticity or tangent construction that mixes the two solutions in linear proportions)
- existence of critical points (or second order phase transitions) that correspond to the limit of the coexistence line
- definition of an order parameter (i.e. the extensive observable that allows to distinguish the two phases) that presents a discontinuity at the (first order) phase transition.

4 Finite Systems: Getting More from Pushing Less

In the preceding sections we have defined a first order phase transition as a discontinuity in the first derivative (or order parameter) m of the thermodynamical potential F as a function of the control parameter. Such a discontinuity can exist only in the thermodynamical limit since

- this discontinuity corresponds to phase coexistence according to the equation (see Sect. 2.2)

$$F(\alpha m_1 + (1 - \alpha)\, m_2) = \alpha F(m_1) + (1 - \alpha)\, F(m_2)$$

  which holds if the free energy per particle is independent of the number of particles, i.e. if surface can be neglected respect to volume which is only possible if the volume goes to infinity.
- if the system is finite the partition sum is a sum over a finite number of configurations, i.e. an analytic function. As a consequence, its first order derivative $\langle m \rangle$ according to the corresponding equation of state cannot present discontinuities.

For these reasons, it is often stated that phase transitions are only defined for infinite systems. Following this viewpoint, finite systems can present only smooth phase changes (cross-over); to demonstrate the asymptotic existence of a phase transition a careful study of the behavior of thermostatistical variables with the size of the system (finite size scaling [10]) has to be performed; in this context, finite size scaling is also the only way to determine the order of a transition if one believes that in finite systems all transitions are smooth.

These considerations are based on analyses where the intensive variable associated with the order parameter is controlled (i.e. the order parameter is only measured in average). In such a case, indeed the equation of state are always smooth in finite system. But what happens in other statistical ensembles?

A fundamental theorem in statistical mechanics, the Van Hove theorem, demonstrated in the next session, guarantees the equivalence between different statistical ensembles at the thermodynamical limit. (However, it tells nothing about finite systems except that the theorem cannot be demonstrated). However, if this equivalence between statistical ensembles is a good approximation even for finite systems, an experiment where the order parameter is controlled (e.g. for the Ising model all events in the statistical sample share the same value of the magnetization) will lead to the same equation of state as the one obtained when the conjugated field is imposed to the system but the magnetization remains free to fluctuate. Then the sudden jump of the magnetization observed in infinite systems will be replaced by a smooth variation. If this would be true, the existence of the transition in the finite system could only be proved through finite size scaling techniques.

This is in fact not the case.

In the next sections we will show that the Van Hove theorem is violated in first order phase transitions if the system is finite, and this violation can persist up to the thermodynamical limit in the case of long-range forces. A consequence of that is that it will be possible to give a rigorous definition of phase transitions even in finite systems, with the prediction (and in some case the experimental evidence, see Chaps. 7,8) of fancy phenomena as negative heat capacities, negative compressibilities and negative susceptibilities.

The non-equivalence of statistical ensembles has also important conceptual consequences. It implies that the value of thermodynamical variables for the very same system depends on the type of experiment which is performed (i.e. on the ensemble of constraints which are put on the system), contrary to the standard thermodynamical viewpoint that water heated in a kettle is the same as water put in an oven at the same temperature. This point will be discussed in Chap. 6.

4.1 The Van Hove Theorem

Let us consider a system in a volume V for which only the average value of energy and number of particles is defined (grancanonical ensemble). Let us divide $V = mV_0 + V_1$ in m equal boxes separated by a "corridor" of width b larger than the range of the force such that the interactions among particles in different boxes can be neglected, (see Fig. 7). Let us calculate the grand potential $\Omega = -T \ln Z$

$$Z_{\beta\mu}(V) = \sum_n \exp\left(-\beta\left(H^{(n)} - \mu N^{(n)}\right)\right)$$

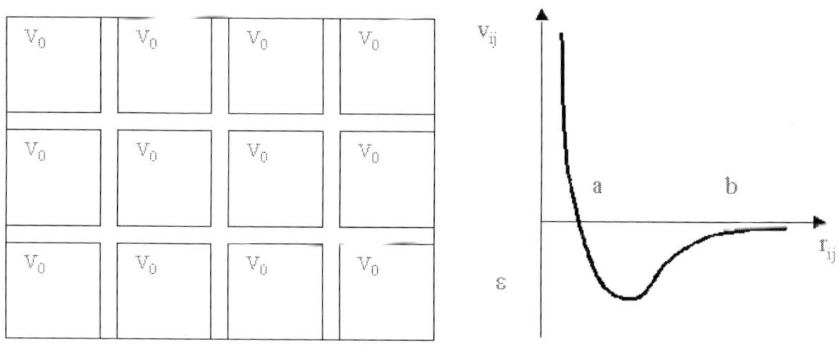

Fig. 7. Schematic representation of the Van Hove theorem demonstration (left) and the corresponding interparticle interaction

where the sum extends over all the possible configurations of the system, $H^{(n)} = K^{(n)} + U^{(n)}$ ($N^{(n)}$) represents the energy (number of particles) of the system in the configuration (n), and β, μ are the associated Lagrange multipliers, the inverse temperature and the chemical potential respectively. The partition sum results

$$Z_{\beta\mu}(V) = \sum_{N=0}^{\infty} \frac{1}{N!} \frac{1}{h^{3N}} \int_V d^{3N}r \int d^{3N}p \exp\left(-\beta\left(H - \mu N\right)\right) \tag{28}$$

$$Z_{\beta\mu}(V) = \sum_{N=0}^{\infty} z_k^N \frac{1}{N!} \int_V d^{3N}r \exp\left(-\beta U\right) = \sum_{N=0}^{\infty} z_k^N Z_\beta(N,V)$$

with $z_k = \exp\left(\beta\mu\right)\left(\frac{2m\pi}{h^2\beta}\right)^{3/2}$ and $Z_\beta(N,V) = \frac{1}{N!}\int_V d^{3N}r \exp\left(-\beta U\right)$. This represents the well known relation between the grancanonical partition sum $Z_{\beta\mu}(V)$ and the interaction part of the canonical one $Z_\beta(N,V)$. To calculate $Z_\beta(N,V)$ let us consider introduce the number of particles in the corridor N_1

$$Z_\beta(N,V) = \sum_{N_1=0}^{N} \frac{1}{N_1!} \frac{1}{(N-N_1)!} \int_{V_1} d^{3N_1}r \int_{V-V_1} d^{3(N-N_1)}r \exp\left(-\beta U\right)$$

Let us note ϵ the minimum of the two body interaction (see Fig. 7); the potential energy in the corridor satisfies then the inequality $U_{V_1} \geq \varepsilon\xi N_1$, where $\xi = (b/a)^3$ represents the maximum number of particles interacting with a given particle. For the total potential energy we can then write

$$U \geq \varepsilon\xi N_1 + \frac{1}{2} \sum_{i=N_1+1}^{N} \sum_{j=N_1+1}^{N} v_{ij}$$

which implies for the partition sum

$$Z_{\beta\mu}(V) \leq \sum_{N_1=0}^{\infty} \frac{1}{N_1!} V_1^{N_1} z_k^{N_1} \exp\left(-N_1\beta\varepsilon\xi\right) \sum_{N_2=0}^{\infty} \frac{1}{N_2!} z_k^{N_2} \int_{nV_0} d^{3N_2}r \exp\left(-\beta U\right)$$
$$= \exp\left(z_k V_1 e^{-\beta\varepsilon\xi}\right) Z_{\beta\mu}^m(V_0)$$

where the last equality stems from the fact that particles interact only within the same box again because of the short range of the force. Finally we get using $V_1 \propto mV_0^{2/3}$

$$\ln Z_{\beta\mu}(V) \leq kmV_0^{2/3} + m \ln Z_{\beta\mu}(V_0)$$
$$\frac{\ln Z_{\beta\mu}(V)}{V} \leq kV_0^{-1/3} + \frac{\ln Z_{\beta\mu}(V_0)}{V_0}$$

which gives in the thermodynamical limit (keeping m constant) $V \to \infty$, $V_0 \to \infty$, $V \to mV_0$

$$\frac{\ln Z_{\beta\mu}(V)}{V} \leq \frac{\ln Z_{\beta\mu}(V_0)}{V_0}$$

On the other side the opposite inequality is trivially true

$$Z_{\beta\mu}(V) \geq Z_{\beta\mu}^m(V_0)$$

since by neglecting the corridor in the integral (28) a positive term in the partition sum is neglected. In conclusion we have demonstrated that

$$\frac{\ln Z_{\beta\mu}(V)}{V} \underset{V\to\infty}{\longrightarrow} \frac{\ln Z_{\beta\mu}(V_0)}{V_0} \tag{29}$$

It is very important to stress that this result is true only in the thermodynamical limit and for short range interactions. For these specific systems the implications of (29) can be summarized as follow

- A thermodynamical limit exists for these systems. Indeed if the thermodynamical potential per unit volume tends to a constant independent of the volume (or the number of particles) $\ln Z_{\beta\mu}(V)/V \to \omega$ the observables distributions $P_N(A)$ will also tend to an asymptotic function

$$\lim_{N\to\infty} P_N(A) = \widetilde{p}\left(\frac{A}{N}\right)^N$$

- In the thermodynamical limit ensembles are equivalent. Indeed if $\omega = Z_{\beta\mu}(V_i)/V_i$ for an arbitrary subsystem V_i this implies that reduced extensive variables are intensive, i.e. that the asymptotic distribution $\widetilde{p}(\frac{A}{N})^N$ has a vanishing width in agreement with the central limit theorem Sect. 2.2. Since ensembles differ at the level of fluctuations, this demonstrates the equivalence between ensembles.

In the next section we will show that the violation of the Van Hove theorem in finite systems leads to the emergence of new thermodynamical phenomena in first order phase transitions of mesoscopic systems.

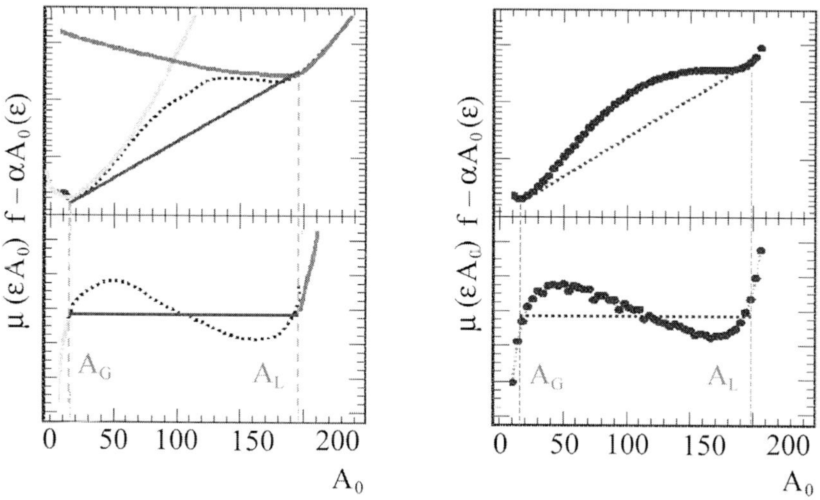

Fig. 8. Schematic representation of the liquid-gas phase transition in the canonical ensemble for an infinite system (left side) and a finite system (right side).

4.2 Convexity Anomalies and Phase Transitions

Let us consider the thermodynamics of a first order phase transition in an extensive ensemble, i.e. in the physical case where the order parameter is a controlled variable. A schematic representation of the liquid gas phase transition in the canonical ensemble is given in the left part of Fig. 8. The mean field solution for the free energy (dotted line) $F(A_0, V)$ at a given temperature shows two minima in the spirit of Landau theory, corresponding to a gas-like solution at low density (light grey) and a liquid-like solution at high density (dark grey). A tangent construction (straight line) corresponds to the inclusion in the partition sum of mixed partitions given by linear combinations of the two solutions; this linear interpolation is only possible in infinite systems for which we can neglect the role of the interface (the "corridor" between the two phases); this leads to a plateau in the conjugated intensive variable $\mu = \partial F/\partial A_0$. If the system is finite (right part) the free energy per particle of the liquid fraction is higher than in the case of the pure liquid solution because of the increased surface tension; as a result the free energy of the mixed configuration is higher than the tangent construction, i.e. is a concave function of A_0, giving rise to a backbending in the chemical potential and a negative susceptibility $\chi^{-1} = \partial_{A_0}\mu$.

This is illustrated in Figs. 9,10 that show the free energy, chemical potential and pressure obtained from a canonical Monte Carlo simulation of the lattice gas model in a cubic box of linear dimension $L = 6$ for different values of the temperature [11], [12]. The convexity anomaly of the partition sum is in the particle density $\rho = A_0/L^3$ direction which represents the order parameter of the transition; ρ can be seen both as a number of particles at fixed volume or as an inverse number of lattice sites at fixed number of particles. For this reason the same anomalous backbending presented by the chemical potential is observed in

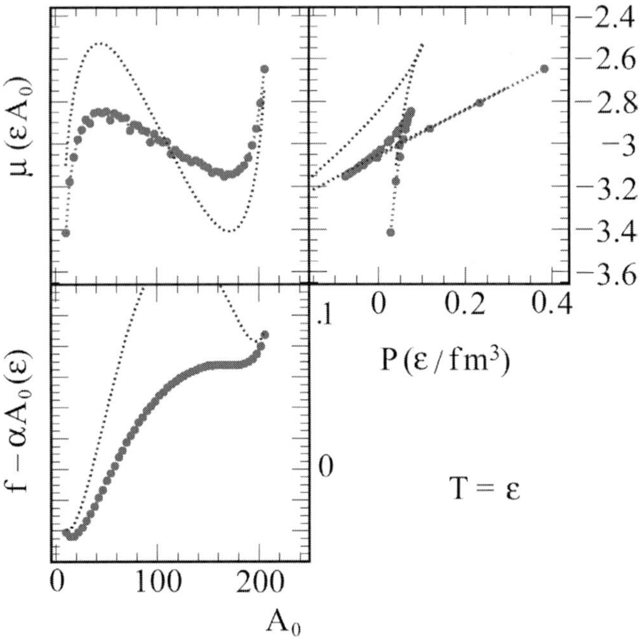

Fig. 9. Free energy, chemical potential and pressure for a 6x6x6 Lattice Gas at a sub-critical temperature. Dashed lines: mean field approximation; symbols: exact results.

the $p(V)$ equation of state as shown in Fig. 11. All isotherms up to the critical temperature backbend, i.e. show a region of negative compressibility.

In Figs. 9, 10, 11 the dashed lines represent the mean field approximation discussed in Chap. 3. Not surprisingly, the mean field badly fails in the phase transition region while it is close to the exact result for supercritical temperatures. In particular the Van der Waals loop is still present (see Fig. 10) at temperatures at which the two phases have merged into one in the exact calculation. This means that the critical temperature is overestimated by the mean field approximation because of its intrinsic lack of fluctuations.

It is important to stress that the physical origin of the backbendings and loops is completely different in the exact calculation and in the mean field. In the former the system inside coexistence presents inhomogeneous partitions that are stable equilibrium solutions with negative compressibility and susceptibility, while in the latter the backbending reflects the instability of the homogeneous mean field solutions respect to phase separation (tangent constructions).

The same reasoning as in Fig. 8 can be done for the microcanonical ensemble as schematically shown in Fig. 12 [13], [14]. If energy is controlled, the appropriate thermostatistical potential is the microcanonical entropy. A first order phase transition can be viewed as the sudden opening of a new disordered phase at a certain threshold energy. The number of states of the disordered phase grows much faster with energy than the one associated to the ordered phase, and this creates a convex intruder in the total microcanonical entropy. This convexity

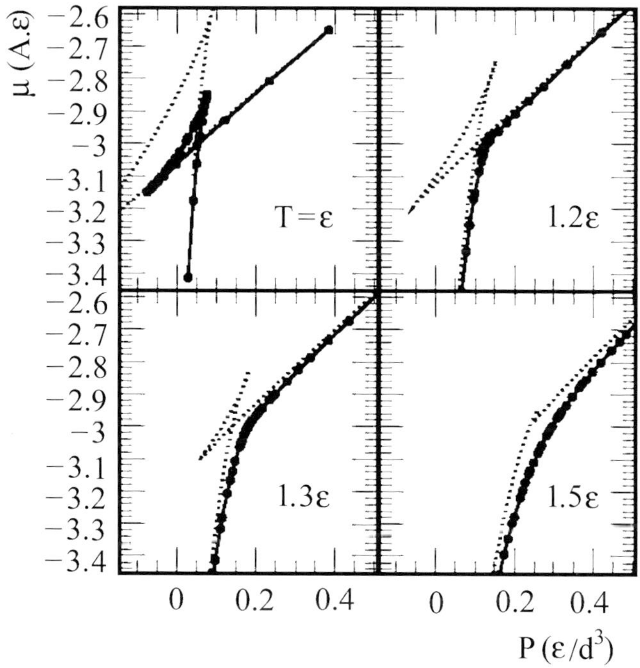

Fig. 10. Chemical potential as a function of pressure at different temperatures for a 6x6x6 Lattice Gas. Dashed lines: mean field approximation; symbols: exact results.

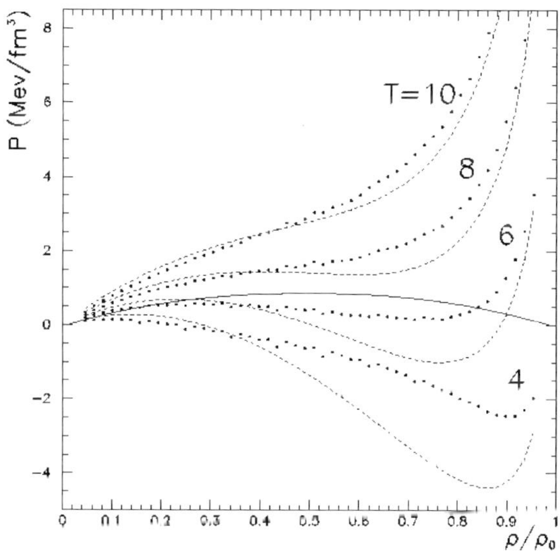

Fig. 11. Pressure versus density equations of state for a 6x6x6 Lattice gas exactly (symbols) and in the mean field approximation (dashed lines). Solid line: coexistence curve.

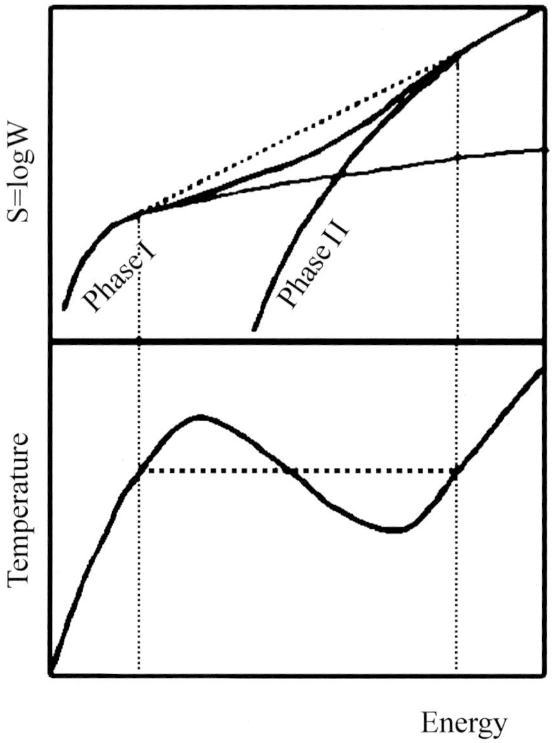

Fig. 12. Schematic representation of a first order phase transition in the microcanonical ensemble.

anomaly cannot be cured by a conventional tangent construction because of the non negligible surface entropy at the interface between the two phases. This phenomenon is in fact observed in a Monte Carlo simulation of the Lattice Gas model in the isobar microcanonical ensemble as shown in Fig. 13 [15], [16]. The convex intruder implies a backbending in the temperature $T^{-1} = \partial_E S$ and a negative branch for the heat capacity $C^{-1} = \partial_E T$ between two divergences. Density being the order parameter of the liquid gas phase transition, since the number of particles is fixed in this calculation ($A_0 = 216$) the volume has to increase with energy to allow the system explore the partitions belonging to the disordered phase. To this aim, the calculation shown in Fig. 13 are performed at constant pressure, where pressure is here defined as the Lagrange multiplier conjugated to the system volume. In this calculation the system volume is assumed to be the average cubic radius. The corresponding caloric curves at constant pressure and constant volume are shown in Fig. 14 . The isobaric curves show a backbending up to the critical pressure giving a clear definition of the coexistence zone.

To summarize, we have shown through some selected examples that a first order phase transition in a finite system is associated to a convexity anomaly in the appropriate thermostatistical potential; the direction of the anomaly in the space of observables can be defined as the order parameter of the transition

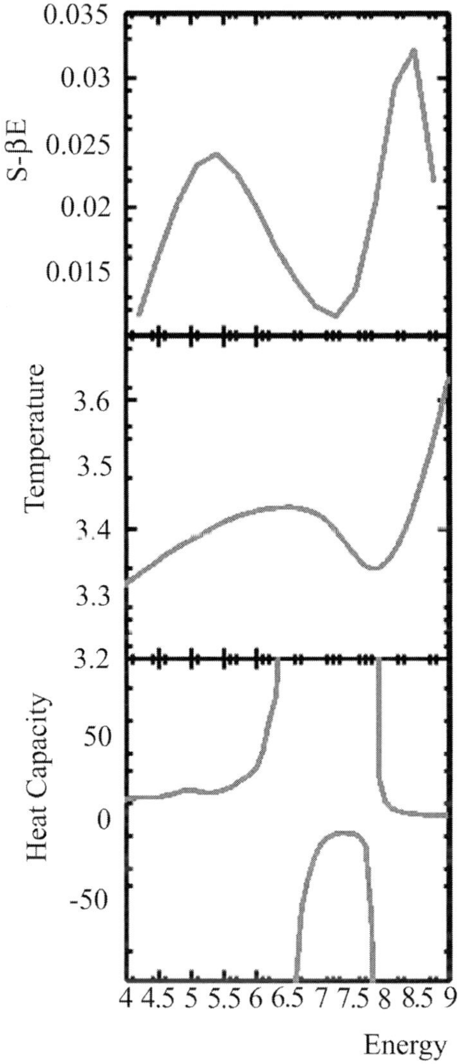

Fig. 13. Entropy, temperature and heat capacity for 216 particles in the microcanonical Lattice Gas model at a subcritical pressure.

and the conjugated intensive variable shows a backbending in the coexistence zone; energetic considerations suggest that this backbending transforms into a plateau (standard first order phase transition) if surfaces are negligible in the global energetic of the infinite system (c.f.. Van Hove theorem).

In particular in the microcanonical ensemble adding energy to the system can cause its temperature to decrease and we have intuitively associated the observation of negative heat capacity with the sudden opening of a disordered collective channel [17]. To clarify the link between channel openings, phase tran-

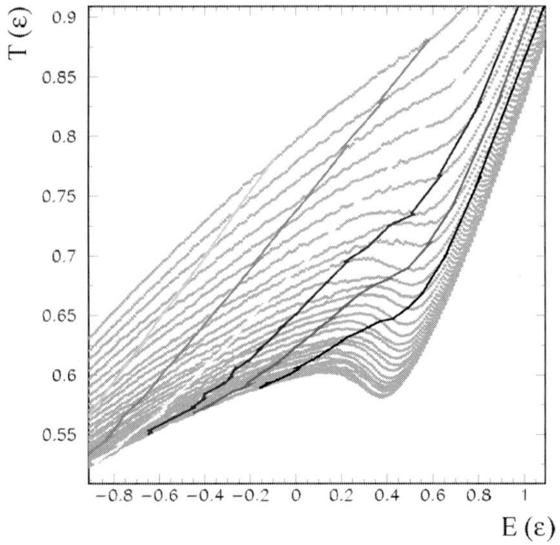

Fig. 14. Caloric curves at constant pressure (showing a backbending up to the critical pressure) and at constant volume (monotonically increasing) for 216 particles in the microcanonical Lattice Gas model.

sitions and negative heat capacities in the next sections we will consider some simple analytical equilibrium models that exhibit negative heat capacity.

4.3 Interacting Particles in Harmonic Potentials: The Classical Case

To see the link between the opening of a channel and negative heat capacity let us consider the simple example of A classical particles in two harmonic oscillators of different frequency ω_1, ω_2 such that the particles in the first oscillator interact all with a constant coupling ε, while they are free in the second one. This model can be seen as a schematic representation of the liquid-gas phase transition at constant pressure. The Hamiltonian reads

$$H = \sum_{i=1,2} \sum_{k=1}^{A_i} \frac{p_k^2}{2m} + \frac{1}{2} k_i x_k^2 + \varepsilon A_1^2 \tag{30}$$

where A_i is the number of particles in the i-th oscillator. The entropy of such a system is exactly calculable. Indeed the number of states with energy $e \leq E$ accessible to a particle in an harmonic oscillator with frequency $\omega = \sqrt{k/m}$ is

$$\Omega(E) = \int_{\sigma(E)} \frac{dx dp}{\hbar} = \frac{2\pi E}{\hbar\omega}$$

where $\sigma(E)$ is the surface in phase space defined by the iso-energy curve $p^2/2m + kx^2/2 = E$. Similarly the number of states with energy $e \leq E$ accessible to A

particles will be given by

$$\Omega_A(E) = \left(\frac{2}{\hbar\omega}\right)^A V_{2A}(\sqrt{E})$$

where $V_n(x)$ is the volume of an hypersphere in n dimensions with radius x. The state density $W(E) = d\Omega/dE$ results

$$W_A^{\omega}(E) = \left(\frac{2\pi}{\hbar\omega}\right)^A \frac{E^{A-1}}{(A-1)!}$$

This formula can be generalized to the case of A_1 (A_2) particles in the oscillator at frequency ω_1 (ω_2) such that the total sum gives A, i.e. the case (30) with no interaction, $\varepsilon = 0$:

$$\begin{aligned} \Omega_A(E) &= \sum_{A_1=0}^{A} \frac{A!}{A_1!(A-A_1)!} \left(\frac{2}{\hbar\omega_1}\right)^{A_1} \left(\frac{2}{\hbar\omega_2}\right)^{A-A_1} V_{2A}(\sqrt{E}) \qquad (31) \\ W_A(E) &= \left(\frac{2\pi}{\hbar}\right)^A \frac{E^{A-1}}{(A-1)!} \left(\frac{1}{\omega_1} + \frac{1}{\omega_2}\right)^A \end{aligned}$$

which is equivalent to a unique oscillator at intermediate frequency, $\omega^{-1} = \omega_1^{-1} + \omega_2^{-1}$. The entropy of this model $S(E) \propto \ln(E)$ is a regular function of positive concavity. Equation (31) shows that the possible existence of the system in two configurations of different density does not imply a phase transition. However if we introduce an interaction in the first oscillator ($\varepsilon \neq 0$) the situation drastically changes. Indeed in this case to have a total energy E, the energy of the $A_1 + A_2$ independent particles that enters the hypersphere volume has to be $E - \varepsilon A_1^2$ and the sum in (31) becomes

$$W_A^{\varepsilon}(E) = \sum_{A_1=0}^{A} \frac{A!}{A_1!(A-A_1)!} \left(\frac{\omega_1}{\omega_2}\right)^{A-A_1} W_A^{\omega_1}(E - \varepsilon A_1^2) \qquad (32)$$

If the second oscillator is sufficiently soft the microcanonical entropy presents a convex intruder leading to a backbending of the caloric curve and negative heat capacity as shown in Fig. 15. The same anomaly observed in the density of states is present in the average kinetic energy (dashed line in Fig. 15) showing that the convex intruder corresponds indeed to a physical cooling of the system (see Sect. 4.6 for a deeper discussion on this subject).

4.4 Interacting Particles in Harmonic Potentials: The Quantum Case

In the classical model we have studied a back bending in the caloric curve appears evident. However, one may worry about the generality of such a statement. Do such anomalies also exist in quantum systems or is this definition of phase

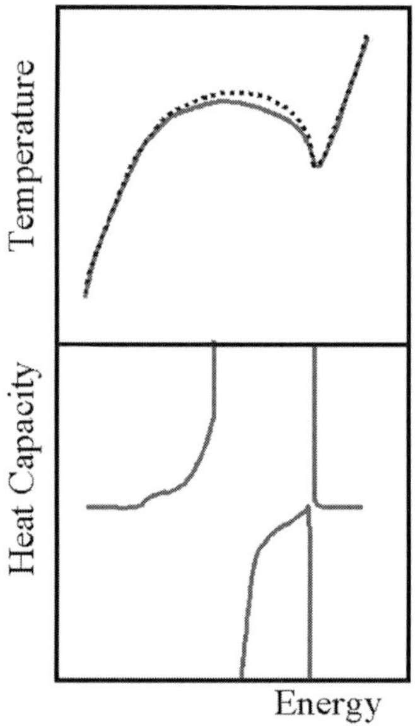

Fig. 15. Temperature and heat capacity for the classical two oscillator model in the subcritical regime. Full curves: derivatives of the entropy. Dashed curve: average kinetic energy per degree of freedom.

transition restricted to classical systems? In order to address this question we can solve the quantum analog of the model of Sect. 4.3. Let us consider again A particles which can jump from one harmonic oscillator to another. In the first one all particles strongly interact while in the second one they are free. The curvature of the second well plays the role of a confining potential i. e. of a pressure. The corresponding Hamiltonian reads

$$\hat{\mathbf{H}} = \omega_1 \left(\hat{\mathbf{N}}_1 - \frac{A_1}{2} \right) - \varepsilon \left(\hat{\mathbf{A}}_1^2 - A_{tot}^2 \right) + \omega_2 \left(\hat{\mathbf{N}}_2 - \frac{A_2}{2} \right) \tag{33}$$

with the operators

$$\hat{\mathbf{N}}_i = \sum_{n=1}^{A} \delta_{i_n}^i \left(a_n a_n^+ \right) \tag{34}$$

$$\hat{\mathbf{A}}_i = \sum_{n=1}^{A} \delta_{i_n}^i \tag{35}$$

where i_n is the harmonic well occupied by the particle i. Using this Hamiltonian we can compute the level density and its associated entropy. To simplify the calculation we have chosen ω_1, ω_2 and ε to be commensurable, $\omega_2 = \omega_1/\lambda$ and $\varepsilon = \omega_1/\mu$. Then, we get the energy

$$E/\varepsilon_1 = N_1 - \frac{A_1}{2} + \left(A_{tot}^2 - A_1^2 \right) / \mu + N_2/\lambda - \frac{A_{tot} - A_1}{2} / \lambda \tag{36}$$

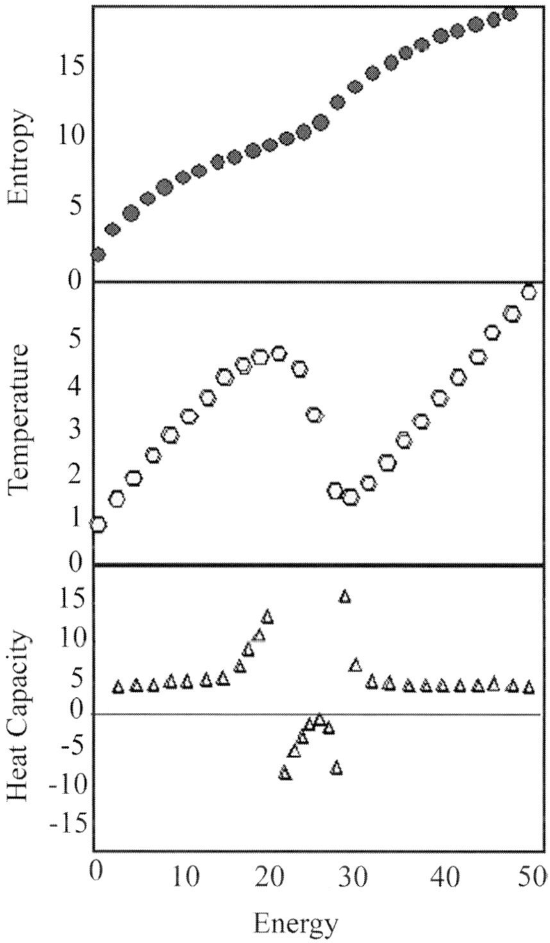

Fig. 16. Microcanonical entropy, temperature and heat capacity for the two quantum oscillators model in the subcritical regime.

In a harmonic oscillator the number of states associated with N_1 quanta carried by A_1 particles is $\binom{N_1 - 1}{A_1 - 1}$ so that for the double oscillator system this corresponds to

$$W^{A_{tot}}(E) = \binom{A_{tot}}{A_1}\binom{N_1 - 1}{A_1 - 1}\binom{N_2 - 1}{A_{tot} - A_1 - 1}$$

Then, we can compute the temperature and the associated heat capacity (see Fig. 16). We observe that the system indeed presents an anomaly in the curvature of the entropy. Back-bending and negative heat capacities automatically follow.

From this analytical example of the occurrence of a negative heat capacity, we can come to the following conclusions:

1. negative heat capacity is not an artifact of imperfect numerical simulations: a $C < 0$ system can be thermodynamically stable.
2. the existence of two different kinds of states for a system appears as a phase coexistence only if their respective level densities are sufficiently different (for instance particles interacting only in the low energy phase ($\varepsilon > 0$) and a much higher volume (ω_1/ω_2 large) available for the high energy phase) so that their addition presents a concavity anomaly.
3. if the average energy of the two phases is not the same (i.e. energy is an order parameter) a finite isolated system in equilibrium at phase coexistence presents a negative heat capacity.

4.5 Surface Tension and Negative Heat Capacity

We have seen that $C < 0$ is a generic feature of finite systems at coexistence and is expected to disappear at the thermodynamical limit (at least if the interactions are short-ranged, see Chap. 5). One can ask how much this behavior is confined to the microscopic world: how small a system has to be for the convex intruder to be sizeable? To answer to this question let us consider the macroscopic analytical example of a liquid drop in equilibrium with its vapor [18].

The bulk free energy of an incompressible liquid can be parametrized in the spirit of the Landau theory as $F = F_0 + Nk(v - v_0)^2/2$ where k is the compressibility, v the reduced volume $v = V/N = \rho^{-1}$ and v_0 the saturation point. For a finite liquid drop one needs to introduce an additional term coming from the surface tension σ leading to the free energy per particle of the drop $f_L = f_0 + k(v - v_0)^2/2 + \sigma s N^{-1/3}$ where the surface coefficient $s = SN^{-2/3}$. The free energy of the vapor can be analytically calculated under the approximation that the vapor is an ideal classical monatomic gas. For N indistinguishable non interacting particles in a volume V we have $Z = z^N/N!$ with the single particle partition sum

$$z = V\frac{1}{h^3}\int d^3p \exp(-\beta\frac{p^2}{2m}) = V\left(\frac{m\pi}{2\beta\hbar^2}\right)^{3/2}$$

which finally gives for the free energy per particle

$$f_G = -T\ln v - \frac{3}{2}T\ln T + K$$

The free energies of the two phases are schematically shown in Fig. 17. Phase coexistence implies the equality between the two partial pressures $\partial_v f_L = \partial_v f_G$ which gives the usual tangent construction (dashed line). The finiteness of the system appears in the constraint of mass conservation. As we increase v we dive inside coexistence with an increased proportion of the vapor fraction respect to the liquid fraction; this leads to an overall increase of the free energy of the drop and a consequent increase of the slope of the tangent construction as a function of v. The net result is a convexity anomaly of the free energy, i.e. a negative compressibility. The backbending in the $p(v)$ equation of state can be analytically calculated using the Clapeyron equation

$$\frac{dp}{dT} = \frac{\Delta h}{(v_G - v_L)\,T}$$

where the enthalpy of vaporization per particle is given by $\Delta h = \Delta h_0 - 3c_s v_L/r$, $c_s$is the surface energy coefficient and r is the drop radius. Assuming the bulk vaporization enthalpy Δh_0 as well as the specific volume of the drop v_L as constant, the Clapeyron equation can be directly integrated giving

$$p = p_{bulk} \exp\left(\frac{3c_s v_L}{rT}\right)$$

which grows as the drop radius decreases (and consequently the specific volume increases) showing that the compressibility is negative at coexistence even for macroscopic droplets, while the plateau is recovered in the bulk limit.

This schematic example shows that the convexity anomalies associated to first order phase transitions can be relevant even on a mesoscopic scale. An interesting consequence of that is that the value of physical observables can be drastically different in the different ensembles still at a mesoscopic scale. As an example, a caloric curve is always by definition monotonic in the canonical ensembles while we have seen that temperature can decrease for increasing excitation if the system is isolated. Of course the quantity called temperature is not defined in the same way in the two ensembles; as long as ensembles are not equivalent and $\beta \neq \partial_E \ln W$ we can wonder weather $\partial_E \ln W$ still represents the physical temperature of the system: does the anomaly that we have analytically recognized in the density of states really imply that pumping energy out of a system heats it up, or is it rather a mathematical curiosity? This question, addressed in the next section, can of course be generalized to any intensive variable conjugated to an order parameter in a generic first order phase transition.

4.6 What Is Temperature?

We all know that the second law of thermodynamics states that temperature measures the increase rate of entropy, $dS = dQ/T$ where Q is the (disordered) thermal energy we have called E in these lectures. Let us show that the rate of entropy is indeed the response of a thermometer loosely coupled to the system under study. In such a case loosely coupled means that the states of the total system are the independent tensorial products of the states of the thermometer times the one of the system. Moreover, the total energy is also simply the sum of the two partial energies.

From a macroscopic point of view, the equilibrium between the system and the thermometer requires the total entropy $S_{tot} = S_{sys} + S_{therm}$ to be a maximum under the constraint of the total energy $E_{tot} = E_{sys} + E_{therm}$. The definition $T^{-1} = dS/dE$ leads then to the equality of temperatures for the system

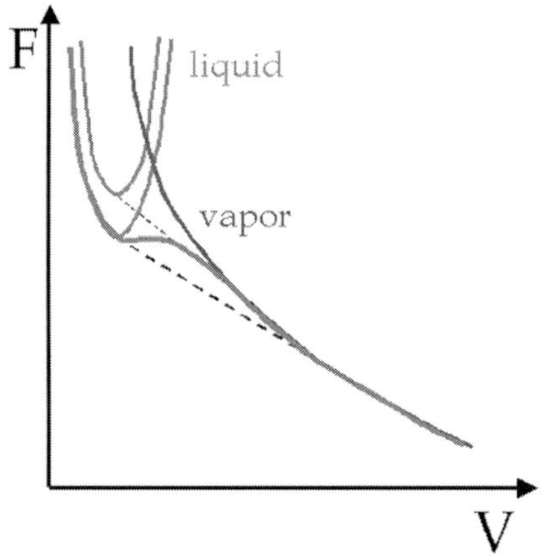

Fig. 17. Schematic representation of the free energy of a liquid drop of different sizes (parabolic curves) in equilibrium with its vapor.

and the thermometer , $T_{sys} = T_{therm}$ in agreement with the zeroth's law of thermodynamics.

However we have also seen in the first chapter that starting from the same Shannon information kernel many different entropies exist according to the different constraints that define the equilibrium under study. The various entropies only converge toward a unique quantity at the thermodynamical limit if this latter exists. On the other side the quantity that backbends is only the microcanonical temperature $T^{-1} = d \ln W/dE$. We therefore ask the question weather a physical thermometer applied to an isolated system measures the microcanonical temperature, i.e. weather the negative heat capacity discussed above is a real measurable physical phenomenon. A thermometer by definition loosely interacts with the system. This means that if a thermometer (of energy E_{therm}) is put into a system (of energy E_{sys}), E_{therm} and E_{sys}can be considered as independent variables. The ensemble of system plus thermometer is isolated with a total energy E_{tot}, therefore the equiprobability of microstates and the factorization of the Hilbert space into the system and thermometer parts, leads to the equilibrium probability distribution for the thermometer energy

$$p_{E_{tot}}(E_{th}) = \frac{W_{th}(E_{th})W_{sys}(E_{tot} - E_{th})}{W_{tot}(E_{tot})} \tag{37}$$

where W_{th} (W_{sys}) is the number of states of the thermometer (system) and where $W_{tot}(E_{tot}) = \int dE_{th}W_{th}(E_{th})W_{sys}(E_{tot} - E_{th})$ is the number of states of the total system (thermometer + system). Temperature is then defined as the response of the thermometer in the most probable state; if we maximize the probability (37) we get

$$\frac{d \ln W_{th}}{dE_{th}} = \frac{d \ln W_{sys}}{dE_{sys}} \tag{38}$$

which shows that the quantity shared at the most probable energy partition is indeed the microcanonical temperature. This result is not in contradiction with the standard idea that for a thermostat, the physical temperature is the intensive variable conjugated to the energy, i.e. the (inverse of the) β Lagrange multiplier. Indeed if we consider a thermometer inside a thermal bath its most probable energetic state will be given by the maximum of the canonical distribution of energies

$$p(E_{th}) = W_{th}(E_{th}) \exp(-\beta E_{th})/Z_\beta$$

which gives an energy $\overline{E}_{th}$ such that $\partial_E \ln W_{th}\left(\overline{E}_{th}\right) = \beta$.

Equation (37) shows that from a microscopic point of view the temperature is indeed an ensemble property: the properties of a thermometer fluctuate from event to event. Only at the thermodynamical limit (for both the system and the thermometer) the fluctuations will be reduced to zero and the distribution of thermometer responses to a unique value.

An example of thermometer is given by the kinetic energy: indeed for a classical system (37) holds with $E_{sys} = E_{pot}$ and $E_{th} = E_{kin}$. Equation (38) then implies that the most probable kinetic energy measures the microcanonical temperature.

5 A Typical Finite System: The Explosion of a Supernova

In the previous chapter we have shown that negative heat capacities systematically occur in first order phase transitions of finite systems when energy is an order parameter. From an historical point of view however, the problematics linked to negative heat capacity has started in the early seventies in a very different context, namely as a specificity of self-gravitating systems. Indeed a star that has exhausted its nuclear fuel radiates (i.e. loses energy) and heats up because of the gravitational contraction, i.e. behaves as a negative heat capacity system. Such a thermodynamical interpretation of the last steps of stellar evolution was considered as a triviality by the astronomers and as an absurdity by thermodynamicists. Let us summarize the two opposite arguments here.

- The astronomers' point of view. If we consider an isolated system with a potential energy $\propto r^{-n}$, the virial theorem states
$$2E_{kin} + nE_{pot} = 0 \tag{39}$$
In the case of the gravitational potential $n = 1$ and we get for N particles $E_{tot} = -E_{kin} = -3/2NT$ which implies $C = -3/2N < 0$ showing that a self-gravitating object has always $C < 0$.
- The thermodynamicists' point of view. For a generic equilibrated system in the canonical ensemble
$$C = -\beta^2 \frac{d\langle E\rangle}{d\beta} = \beta^2 \left\langle (E - \langle E\rangle)^2 \right\rangle > 0 \tag{40}$$
showing that the heat capacity corresponds to the energy fluctuation which cannot be negative.

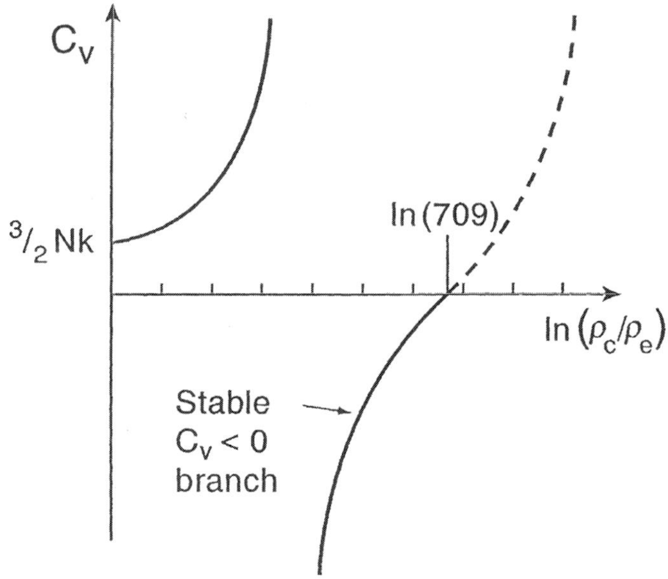

Fig. 18. Microcanonical heat capacity as a function of the density contrast for hard spheres with gravitational interactions from [20].

In the following sections we will show that both statements are incorrect, and that negative heat capacity in macroscopic self-gravitating systems is a physical equilibrium phenomenon exactly equivalent to the first order phase transitions in finite systems studied in the last chapter. Indeed a mesoscopic boiling droplet and an collapsing (or exploding) star are very similar in the sense that in both cases the size of the system is comparable to the range of the force.

5.1 $C < 0$ in Self-gravitating Systems and the Gravothermal Catastrophe

The confusion arising from the two contradictory arguments about the sign of the heat capacity was increased by the famous gravothermal catastrophe predicted first by Antonov [19] from the study of the extrema of the microcanonical entropy of a self-gravitating object in the mean field approximation as a function of the one body matter density. The one body Shannon entropy reads

$$S = -\int d^3r d^3p \ f(\vec{r}, \vec{p}) \ln f(\vec{r}, \vec{p})$$

The microcanonical equilibrium can be found as usual by maximizing the entropy in a fixed volume V under the particle number and energy constraint (see Chap. 1)

$$
\begin{aligned}
0 &= dS - \alpha dN - \beta dE \\
N &= \int d^3r d^3p \; f(\vec{r}, \vec{p}) \\
E &= \int d^3r d^3p \frac{p^2}{2m} \; f(\vec{r}, \vec{p}) - \frac{Gm^2}{2} \int d^3r d^3p d^3r' d^3p' \frac{f \; f'}{|\vec{r} - \vec{r}'|}
\end{aligned}
$$

The result is a self-consistent equation for the equilibrium one body matter density $\rho(\vec{r}) = \int d^3p \; f(\vec{r}, \vec{p})$ that can be solved to give the density contrast ρ_c/ρ_e between the center and the surface of the box. The result is that the density contrast is an increasing function of the volume. Fig. 18 shows an exact calculation of this problem for the case of identical hard spheres [20]. For extremely large volumes $\rho_c/\rho_e > 709$ the extremum of the entropy is a minimum meaning that no equilibrium solution exists in this case (Antonov gravothermal catastrophe). This result looks particularly strange since the instability provoked by the gravitational potential does not appear when the box is too small (and the interaction energy is strong) but when the box is too large (and the interaction energy is loose). Moreover for intermediate boxes $32.2 < \rho_c/\rho_e < 709$ the extremum is a maximum , i.e. the solution is stable, but the heat capacity contains a pole and a negative branch, which traditionally was associated to an instability (see Chap. 3).

5.2 Solution of the Antonov Paradox

The disagreement between the gravothermal catastrophe and (40) clearly implies that the Van Hove theorem (see Sect. 4.1) is violated. This can only be due to the long-range of the force. The incoherence with the virial theorem (39) can be explained by considering that if we are dealing with N particles the density of states (see Sect. 4.3)

$$
W \propto \int d^{3N}r \left(E - \sum_{ij} \frac{m_i m_j}{r_{ij}} \right)^{3N/2}
$$

diverges unless a short range cut-off and a constraining potential (or boundary condition) are introduced. This means that the virial theorem has to be corrected for the short range repulsion and for the boundary condition. A schematic modelization of this problem has been proposed by W. Thirring as a constant interaction among all particles inside a volume $V_0 < V$ and no interaction outside [21]. This model is equivalent to the two coupled harmonic oscillator problem proposed in Sect. 4.3. The finite liquid drop of section 4.3 is now an infinite star, while the saturated vapor corresponds to the star atmosphere. As for the model 4.3, for the two boxes model a $C < 0$ is found [21] whenever the difference between the two volumes is important, i.e. for big total volumes in agreement with the Lynden-Bell result [20]. Following the reasoning of Chap. 4 we can then associate the $C < 0$ regime to a phase coexistence between the star and

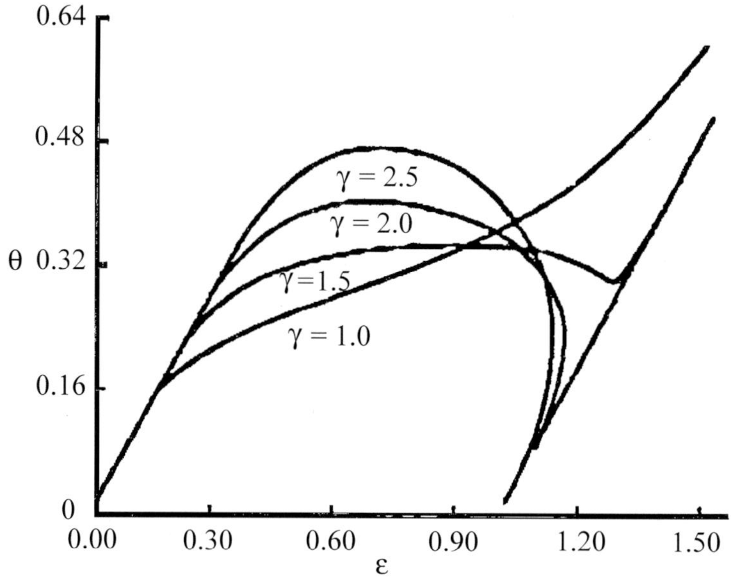

Fig. 19. Caloric curves in the subcritical regime for different ranges of the interaction from [21].

its atmosphere that stabilizes the stellar nuclear plant, and the gravothermal catastrophe to a first order phase transition when the nuclear fuel is exhausted.

If we consider the generic case for the interaction in the dense phase $E = \varepsilon N^\gamma$ we can study the effect of the range of the force on the heat capacity in the thermodynamical limit. The resulting caloric curves are given in Fig. 19 [21]. In agreement with Chap. 4, we can see that the negative heat capacity disappears at the thermodynamical limit for short range interactions ($\gamma = 1$) but all ranges $\gamma > 1$ lead to a backbending that is preserved when $N \to \infty$, i.e. this phenomenon is not specific of the gravitational interaction. We can also understand why in the short range case the same phenomenology appears in self-gravitating macroscopic system, as long as the range of the force is comparable with the linear dimension of the system.

5.3 Thermal Contact Between $C < 0$ Systems

Phase coexistence in standard macroscopic thermodynamics is a trivial phenomenon: the tangent construction implies that the thermodynamical properties of phase coexistence are completely determined by the properties of the corresponding pure phases. On the other hand if surfaces cannot be neglected respect to bulk properties (i.e. if the system is finite or the interaction is long-range) new unexpected phenomena peculiar to the coexistence phase appear, as negative compressibility or negative heat capacity. The peculiar thermodynamical properties of the coexistence phase can be better understood if we consider in some detail the problem of thermal contact between $C < 0$ systems.

Let us consider a system with $C_1 < 0$ in thermal contact with a second system with a heat capacity C_2. Let us distinguish some different cases.

1. If $C_2 < 0$ no equilibrium is possible between the two systems. Indeed if $T_2 > T_1$ energy will be transferred from system (2) to system (1) and system (2) will get even hotter while (1) becomes even cooler. This implies that it is not possible to divide a $C < 0$ system into two parts each with $C < 0$; in other words $C < 0$ systems are not extensive.
2. A thermal equilibrium is possible if the $C_1 < 0$ system is in contact with a $C_2 > 0$ system small enough such that $C_2 < |C_1|$. In this case the temperature increase of the initially slightly hotter system (1) is compensated by system (2) which has positive heat capacity and receives energy, and therefore increases its temperature more rapidly; a final equilibrium state will then be achieved with $T_{eq} > \max(T_1, T_2)$. In particular if the two systems are independent we can write for the microcanonical distribution of energy $E_1 : \ln P_E(E_1) = S_1(E_1) + S_2(E_2) - S_{tot}(E)$ where $E = E_1 + E_2$. The stability condition (i.e. the curvature of $\ln P$ around the extrema) can then be written as
$$\frac{d^2 S_{tot}}{dE_1^2} - -\frac{1}{2T^2}\left(\frac{1}{C_1} + \frac{1}{C_2}\right) < 0 \tag{41}$$
where $T_1 = T_2 = T$ is the equilibrium temperature (the microcanonical temperature associated with the most probable energy partition). This is the case of a living star: the central part has $C_1 < 0$, the surface has $C_2 > 0$, the global system (or coexistence phase) has $C = C_1 + C_2 < 0$ and is stable since the core transfers energy to the atmosphere in the form of radiation. The gravothermal catastrophe occurs when $C = 0$ (see Fig. 18) when the gravitational contraction cannot be stopped by nuclear reactions leading to the whole well known phenomenology of stellar collapse [22].
3. A $C < 0$ system cannot reach thermal equilibrium with a thermostat. Indeed if (1) is much smaller than (2) (41) is violated and the total entropy is a minimum. This implies that a negative heat capacity is impossible in the canonical ensemble coherently with (40): the $C < 0$ system in contact with a thermal bath stops to hesitate between the two coexisting phases and jumps to its stable equilibrium state, i.e. makes a phase transition. These considerations allow to gain an intuitive understanding of the fact that the $C < 0$ of a boiling liquid drop (see Chap. 4) transforms into a plateau when a water kettle is put on a gas cooker in the macroscopic world. Indeed any microscopic portion of the boiling portion once isolated would appear as a microcanonical $C < 0$ system; the thermal contact with the rest of the system (which can be considered a thermostat because of the short range of the force) forces the stable $C < 0$ system to choose between the liquid and the gas solution giving rise to the Maxwell construction.

All we have discussed in this chapter has been presented in the early seventies; it may be surprising then that negative heat capacity has not been recognized at this time as a general paradigm of phase coexistence in non-extensive (finite or infinite) systems. Indeed it is only in the recent years, following the success of the experimental application of these ideas to mesoscopic systems (see Chaps. 7 and 8) that these concepts have been developed further. The conceptual difficulty of accepting a stable equilibrium with negative heat capacity is due to different reasons.

First, the inhomogeneity of the star makes it difficult to consider it as a single thermodynamical object; following the reasoning of Chap. 4 we however understand that on the contrary this inhomogeneity is an essential feature of the thermodynamics of coexisting phases.

Moreover the fact that the energy exchange between the star and the atmosphere consists in radiation (taking their energy from nuclear reactions) makes the problem an out of equilibrium transport problem.

Most important, the fact that it is not possible to define a canonical equilibrium for a $C < 0$ system means that the validity itself of such an equilibrium is not easy to understand: how can one speak of temperature in the absence of thermal contact, and what is in this case the meaning of a temperature decrease with increasing energy? We have already discussed (and solved!) the problem of the physical meaning of the microcanonical temperature in Sect. 4.5; let us just recall here that if a $C < 0$ cannot be put in contact with a thermostat, on the other hand a thermometer put in a $C < 0$ system is perfectly well defined, and measures the microcanonical temperature. This is the case 2 of the discussion above (with $C_1 < 0$ and $C_2 < |C_1|$)

6 Abnormal Topology of Event Distributions

In most textbooks the equivalence between the different statistical ensembles is either postulated or demonstrated at the thermodynamical limit through the Van Hove theorem (see Sect. 4.1).

In the previous chapters we have shown that ensembles may not be equivalent. For finite systems, two ensembles which put different constraints on the fluctuations of the order parameter lead to very different equations of states close to a first order phase transition. As an example the microcanonical heat capacity may diverge to become negative while the canonical one remains always positive and finite (see Sects. 4.2, 4.3). In Chap. 5 we have moreover seen that such inequivalences may survive at the thermodynamical limit for systems involving long-range forces.

In this chapter we will try to formalize all these findings looking at the general properties of the order parameter distribution. This will allow us to propose a definition of phase transitions in finite systems also for intensive ensembles (i.e. ensembles where the order parameter is not controlled on an event by event basis) based on topology anomalies of the event distribution in the space of observations. We will show in the next sections that this generalizes the definitions based on the curvature anomalies of thermostatistical potentials presented in Chap. 4. Such a definition gives a clear understanding of the physical meaning of an order parameter as the best variable to separate the two maxima of the distribution, and can be directly used experimentally (see Chaps. 7,8).

In Sect. 6.6 we will come back to the problem of the thermodynamical limit. If the order parameter is sufficiently collective the anomaly represented by the phase transition may survive until the infinite volume and infinite number limit. According to the specific properties of the Hamiltonian we will then work out a sufficient condition for the finite size phase transition to become the one known in the bulk.

6.1 Negative Heat Capacity and Bimodal Energy Distributions

Let us first concentrate on finite systems. To begin we will consider the specific example of the microcanonical and the canonical ensemble characterized by the energy E and the temperature β^{-1} respectively.

In Sect. 4.5 we have shown that the extrema of the probability distribution in the canonical ensemble

$$P_\beta(E) = \exp(S(E) - \beta E - \log Z_\beta)$$

are given by the equation

$$T^{-1} \equiv \partial_E S\left(\overline{E}_\beta\right) = \beta \tag{42}$$

If this extremum is unique, we can perform a saddle point approximation around the most probable energy $\overline{E}_\beta$ leading to the average energy

$$\langle E\rangle_\beta = \int dE E e^{-\frac{(E-\overline{E}_\beta)^2}{2C}} g_\beta(E-\overline{E}) \tag{43}$$

with $g_\beta(x) = c_0 + c_3x^3 + c_4x^4 + \ldots$. If P_β is symmetric, $\langle E\rangle_\beta = \overline{E}_\beta$ and $T^{-1}(\langle E\rangle) =.\beta$ meaning that the microcanonical caloric curve $T(\overline{E})$ exactly coincides with the canonical one $\beta^{-1}(\langle E\rangle)$. If the distribution is not symmetric some corrective terms arise : $\langle E\rangle_\beta = \overline{E}_\beta + \delta_\beta$, where $\delta_\beta = \int dx\, x \exp(-x^2/2C)\tilde{g}_\beta(x) = 3c_3\sqrt{2\pi C^5} + \ldots$ with $\tilde{g}_\beta$ the series of the odd terms of g_β. However, the shift δ is in most cases small so that when P_β has a unique maximum the ensembles are almost equivalent even for a finite system. The same approximation can be applied to the partition sum, which is linked to the entropy by an exact Laplace transform

$$Z_\beta = \int dE\, W(E)e^{-\beta E}$$

leading to

$$Z_\beta \approx W(\overline{E})e^{-\beta\overline{E}} \approx W(\langle E\rangle)e^{-\beta\langle E\rangle}$$

which corresponds to a simple linear transformation between the thermodynamical potentials, i.e. an approximate Legendre transform

$$\ln Z_\beta \approx \ln W(\langle E\rangle) - \beta\langle E\rangle$$

As we have already mentioned in Sect. 2.1, this expression has not to be mixed up with the true (and exact) Legendre transform $\ln Z_\beta = S_\beta(\langle E\rangle) - \beta\langle E\rangle$ which gives the relation between the partition sum and the Shannon entropy within the canonical ensemble.

However in first order phase transitions P_β has a characteristic bimodal shape (see Sect. 2.2) [23–25] with two maxima $\overline{E}_\beta^{(1)}$, $\overline{E}_\beta^{(2)}$ that can be associated with the two phases and a minimum $\overline{E}^{(0)}$. These three solutions of (42) imply a backbending for the microcanonical caloric curve. Indeed a minimum of P_β corresponds to a convexity of the entropy according to $d_E^2 \ln P_\beta = d_E^2 \ln W$. A single saddle point approximation is not valid in this case; however it is always possible to write $P_\beta = m_\beta^{(1)} P_\beta^{(1)} + m_\beta^{(2)} P_\beta^{(2)}$ with $P_\beta^{(i)}$ mono-modal normalized probability distribution peaked at $\overline{E}_\beta^{(i)}$. The canonical mean energy is then the weighted average of the two energies

$$\langle E\rangle_\beta = \tilde{m}_\beta^{(1)}\overline{E}_\beta^{(1)} + \tilde{m}_\beta^{(2)}\overline{E}_\beta^{(2)} \tag{44}$$

with $\tilde{m}_\beta^{(i)} = m_\beta^{(i)} \int dE P_\beta^{(i)}(E) E / \overline{E}_\beta^{(i)} \simeq m_\beta^{(i)}$, the last equality holding for symmetric distributions $P_\beta^{(i)}$. Since only one mean energy is associated with a given temperature β^{-1}, the canonical caloric curve is monotonous. In particular in the backbending region the mean energy is an interpolation between the two energies $\overline{E}_\beta^{(1)}$, $\overline{E}_\beta^{(2)}$ associated with the considered β, the weighting factor being the probability of each phase $\tilde{m}_\beta^{(i)}$.

If instead of looking at the average $\langle E\rangle_\beta$ we look at the most probable energy $\overline{E}_\beta$, this (unusual) canonical caloric curve is almost identical to the microcanonical one (see (42)) up to the transition temperature β_t^{-1} for which the two components of $P_\beta(E)$ have the same height. At this point the most probable energy jumps from the low to the high energy branch of the microcanonical caloric curve. The most probable canonical energy is still a monotonic curve but it presents a plateau at β_t^{-1} which is equivalent to the Maxwell construction since

$$S(\overline{E}_\beta^{(2)}) - S(\overline{E}_\beta^{(1)}) = \int_{\overline{E}_\beta^{(1)}}^{\overline{E}_\beta^{(2)}} \frac{dE}{T} = \beta\left(\overline{E}_\beta^{(2)} - \overline{E}_\beta^{(1)}\right) \tag{45}$$

Therefore, the difference between the canonical and microcanonical caloric curves remains when one is looking at the most probable energy instead of the average. The connection between the canonical energy distribution and the microcanonical equation of state is presented for the three dimensional lattice gas model in Fig. 20. The bimodality of the canonical energy distribution as well as the discontinuity in the most probable energy value are definitions of the phase transition exactly equivalent to the convexity anomaly of the entropy discussed in Chap. 4 and clearly defined even for very small system (216 particles are considered for the calculation of Fig. 20).

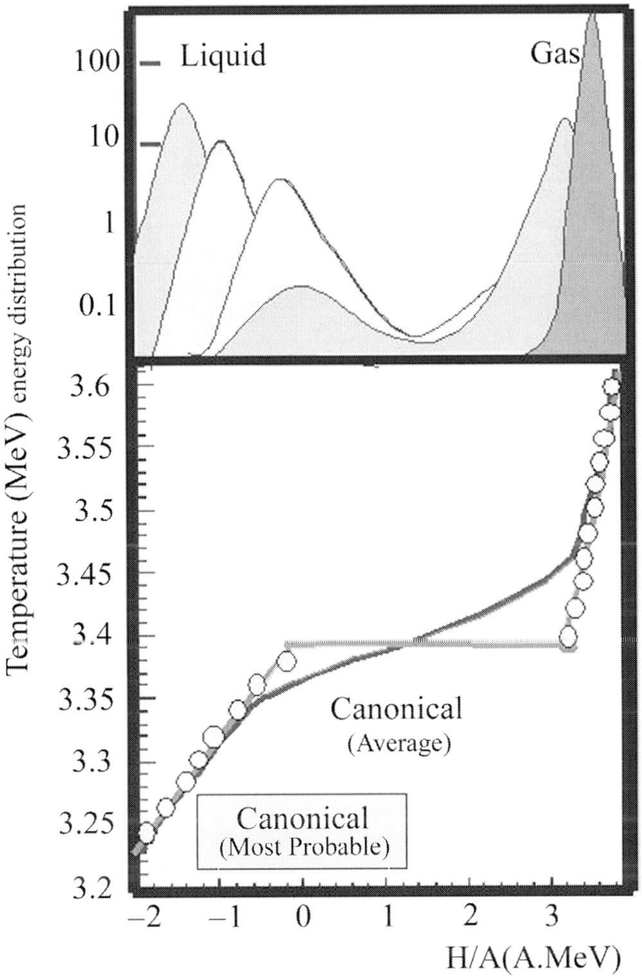

Fig. 20. Canonic thermodynamics of 216 particles in the isobar Lattice Gas model. Upper part: energy distributions for different temperatures. Lower part: caloric curve from the average and the most probable energy.

6.2 Convexity Anomalies and Bimodal Probability Distributions

This discussion can be generalized to any couple of extensive/intensive ensemble. Fig. 21 shows the example of the Ising model at zero field. The bimodal structure in the m direction corresponds to a negative susceptibility in a constant magnetization ensemble. In this case the projection on the energy axis does not show anomalies : the microcanonical heat capacity remains positive and the energy is not an order parameter. At T_c the distribution presents a curvature anomaly only on the low energy side respect to the maximum. Therefore at this point the curvature passes through zero signalling a second order phase transition. Since in finite systems the canonical distribution for any β, h (μ) allows a

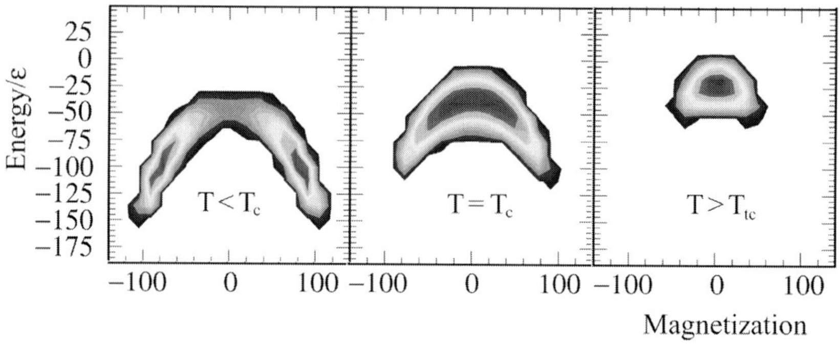

Fig. 21. Event distribution in the energy versus magnetization plane for a 6x6x6 Ising model with zero field at a subcritical, critical and supercritical temperature.

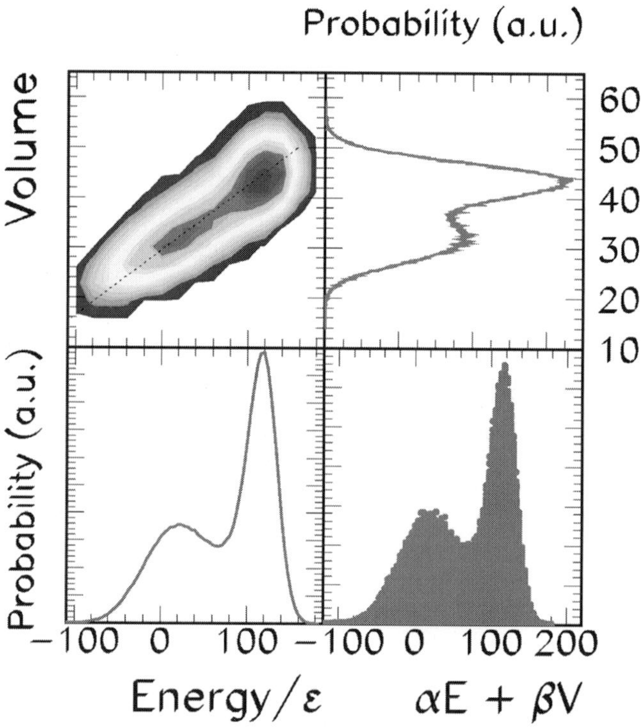

Fig. 22. Event distribution in the volume versus energy plane for 216 particles in the isobar canonical Lattice Gas model and projection over the two axes. Bottom right: projection of the distribution over the direction separating at best the two distributions.

complete exploration of the microcanonical entropy surface (in the limit of the total number of events analyzed), the whole microcanonical phase diagram can in principle be drawn from any single canonical temperature using the fact that the distribution is in fact the entropy

$$S(E) = \log P_{\beta,h=0}(E,M) + \ln Z_{\beta,h=0} + \beta E.$$

As an example the croissant shape of the distribution at T_c not only defines the critical energy e_c and magnetization m_c of the second order phase transition but also allows to infer the coexistence line where the first order phase transition takes place. Indeed a constant energy cut of the distribution below e_c directly represents the entropy as a function of magnetization and has a bimodal shape.

6.3 Liquid Gas Phase Transition and Volume Fluctuations

Let us now take the example of the liquid-gas phase transition in a system of n particles for which the volume is not controlled on an event by event basis but is at best known in average. In such a case, we can define an observable $\hat{B}_1$ as a measure of the size of the system; for example the cubic radius $\hat{B}_1 = \frac{4\pi}{3n}\sum_i \hat{r}_i^3 \equiv \hat{V}$ where the sum runs over all the particles . Then a Lagrange multiplier λ_V has to be introduced which has the dimension of a pressure divided by a temperature. In a canonical ensemble with an inverse temperature β we can define different distributions which are illustrated in Fig. 22. A complete information is contained in the distribution $P_{\beta\lambda_v}(e,v) = \bar{W}(e,v)\, Z^{-1}_{\beta\lambda_v} \exp -(\beta e + \lambda_v v)$ since events are sorted according to the two thermodynamical variables, e and v. This leads to the density of states $\bar{W}(e,v)$ with a volume v and an energy e. One can see that in the first order phase transition region the probability distribution is bimodal. In the spirit of the principal component analysis we can look for an order parameter $\hat{Q} = x\hat{H} + y\hat{V}$ which provides the best separation of the two phases. A projection of the event on this order parameter axis is also shown in Fig. 22. One can see a clear separation of the two phases. On the other hand if we cannot measure both the volume v and the energy e we are left either with

$$P_{\beta\lambda_v}(e) = \bar{W}_{\lambda_v}(e)\, Z^{-1}_{\beta\lambda_v} \exp(-\beta e)$$

giving access to the microcanonical partition sum $\bar{W}_{\lambda_v}(e)$ at constant λ_v or with the probability

$$P_{\beta\lambda_v}(v) = \bar{Z}_\beta(v)\, Z^{-1}_{\beta\lambda_v} \exp(-\lambda_v v)$$

leading to the isochore canonical partition sum $\bar{Z}_\beta(v)$. Since both probability distribution $P_{\beta\lambda_v}(e)$ and $P_{\beta\lambda_v}(v)$ are bimodal the associated partition sum do have anomalous concavity intruders ,*i.e.* negative heat capacity as well as negative compressibility.

As a general statement, we can define a first order phase transition for any number of particles as a bimodality in the probability distribution of an arbitrary observable; any observable that allows to separate the two maxima of the event distribution can then be considered as an order parameter.

6.4 A Mesoscopic Example: Negative Magnetic Susceptibility

To better explore the connection between the distribution of the order parameter and the equation of state, let us consider the Ising model (see Sect. 2.3) at fixed magnetization. The partition sum can be written as

$$Z_\beta(m) = \int dE\ W(E,m)\exp(-\beta E)$$

where $m = \sum_i^N s_i/N$ is the magnetization. As shown in Fig. 23 the phase transition is signalled by a backbending of the equation of state $\widetilde{h}(m) = \beta^{-1}\partial_m \ln Z_\beta$ (m) that replaces the discontinuity in the equation of state $\langle m\rangle(h)$ of the standard Ising model with partition sum

$$Z_{\beta h} = \int dmdE\ W(E,m)\exp(-\beta(E+mNh))$$

Note that in the case of this specific model finite size correction are particularly small and an almost perfect jump is observed already for a $3D$ lattice size of linear dimension $L = 6$.

This same information concerning a negative magnetic susceptibility can be obtained by looking at the magnetization distribution of the standard Ising model with zero field

$$P_{\beta h}(m) = \frac{1}{Z_{\beta h}}\int dE\ W(E,m)\exp(-\beta(E+mNh)) = \frac{Z_\beta}{Z_{\beta h}}\exp(-\beta mNh) \quad (46)$$

From (46) we can see that $d_m^2 \ln N_{\beta h} = d_m^2 \ln Z_\beta$. This shows that a minimum in the magnetization distribution (left part of Fig. 23) implies a convexity anomaly in the constant magnetization partition sum (right part of Fig. 23) i.e. a negative magnetic susceptibility for mesoscopic ferromagnets sorted in constant magnetization bins.

6.5 First Order Phase Transitions and Bifurcations

Another way to understand a first order phase transition is to relate it with a bifurcation. Let us for example look at the grancanonical lattice gas model at the critical chemical potential and let us look at the mass distribution as a function of the temperature (see Fig. 24). Above the critical temperature the mass distribution is normal but at the critical temperature it bifurcates into a bimodal distribution with two peaks:

- the gas one at low mass, i.e. low density,
- and the liquid phase at high mass.

This sudden bifurcation is typical of a phase transition.

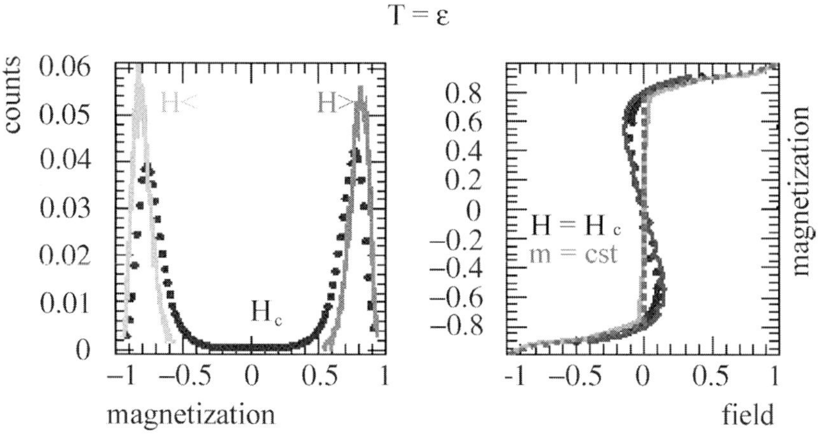

Fig. 23. Left side: magnetization distribution in a 6x6x6 Ising model at a subcritical temperature for three different values of the external field. Right side: equation of state for Ising and for the constant magnetization ensemble (backbending curve).

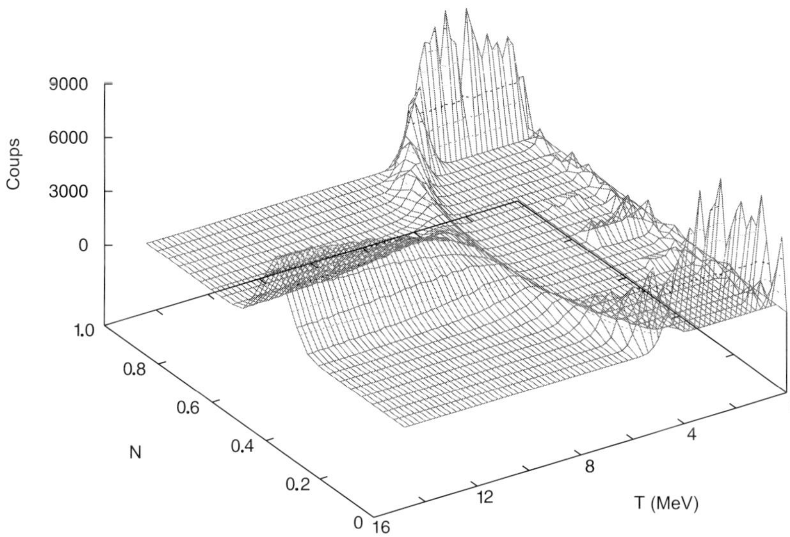

Fig. 24. Mass distribution as a function of the system temperature for the grancanonical lattice gas model.

6.6 The Thermodynamical Limit and the Yang Lee Theorem

The definition of phase transition proposed in Sect. 6.1 can be applied to a wide range of situations even out of equilibrium [26] and can be directly implemented on experimental data (see Chaps. 7,8). However it is clear that not all topological anomalies will survive up to the bulk limit and give rise to a conventional thermodynamical phase transition. The transition between two isomeric states, the breaking of a Cooper pair, ionization are all examples of state changes that do not lead to discontinuities in the bulk limit, i.e. do not converge to a thermodynamical first order phase transition. It is therefore of extreme interest to study the thermodynamical limit of the order parameter distributions. Since we have seen that the different statistical ensembles are in general not equivalent, the thermodynamical limit has to be considered separately for the intensive (say, canonical) and extensive (say, microcanonical) ensemble.

The thermodynamical limit can be expressed as the fact that the thermodynamical potentials per particle converge when the number of particles N goes to infinity : $f_{N,\beta} = \beta^{-1} \log Z_\beta / N \to \bar{f}_\beta$ and $s_N(e) = S(E)/N \to \bar{s}(e)$ where $e = E/N$. Let us also introduce the reduced probability $p_{N,\beta}(e) = (P_\beta(N,E))^{1/N}$ which then converges towards an asymptotic distribution $p_{N,\beta}(e) \to \bar{p}_\beta(e)$ where $\bar{p}_\beta(e) = \exp\left(\bar{s}(e) - \beta e + \bar{f}_\beta\right)$. Since $P_\beta(N,E) \approx (\bar{p}_\beta(e))^N$ one can see that when $\bar{p}_\beta(e)$ is normal the relative energy fluctuation in $P_\beta(N,E)$ is suppressed by a factor $1/\sqrt{N}$. At the thermodynamical limit P_β reduces to a δ-function and the ensemble equivalence is recovered (see Sect. 2.2). To analyze the thermodynamical limit of a first order phase transition (bimodal $p_{N,\beta}(e)$), let us introduce as before $\beta_{N,t}^{-1}$, the temperature for which the two maxima of $p_{N,\beta}(e)$ have the same height. For a first order phase transition $\beta_{N,t}^{-1}$ converges to a fixed point $\bar{\beta}_t^{-1}$ as well as the two maximum energies $e_{N,\beta}^{(i)} \to \bar{e}_\beta^{(i)}$. For all temperature lower (higher) than $\bar{\beta}_t^{-1}$only the low (high) energy peak will survive at the thermodynamical limit since the difference of the two maximum probabilities will be raised to the power N. Therefore, below $\bar{e}_\beta^{(1)}$ and above $\bar{e}_\beta^{(2)}$ the canonical caloric curve coincides with the microcanonical one in the thermodynamical limit. In the canonical ensemble the temperature $\bar{\beta}_t^{-1}$ corresponds to a discontinuity in the state energy irrespective of the behavior of the entropy between $\bar{e}_\beta^{(1)}$ and $\bar{e}_\beta^{(2)}$. A more rigorous demonstration can be done with the help of the Yang Lee unit circle theorem [25].

The Yang Lee theorem [27] considers the distribution of the zeros of the partition sum Z_β in the complex β plane. Under very general conditions it is possible to demonstrate [27] that the zeros form a line that cuts across the real axis with a density increasing with the number of particles of the system, leading to a vanishing imaginary part for $N \to \infty$, i.e. a first order phase transition at a definite (real) temperature. The partition sum for a complex parameter $\beta = \beta + i\eta$ is nothing but the Laplace transform of the probability distribution $P_{\beta_0}(e)$ for a temperature parameter β_0 [28,29]

$$Z_\beta = \int de Z_{\beta_0} \, P_{\beta_0}(e) \, e^{-(\beta - \beta_0)e}. \equiv \int de \; p_\beta(e) \, e^{-i\eta e}$$

In order to study the thermodynamical limit (when it exists), if $p_\beta(e)$ is monomodal we can use a saddle point approximation around the maximum $\bar{e}_\beta$ giving $Z_{\beta}. = e^{\phi_\beta(\bar{e}_\beta)}$, with

$$\phi_\beta(e) = \log p_\beta(e) - i\eta e + \eta^2 C(e)/2 + \log\left(\frac{2\pi C(e)}{2}\right)$$

where $C^{-1} = \partial_e^2 \log p_{\beta_0}(e)$. However, if $\bar{W}_{\beta_0}(e)$ has a curvature anomaly it exists a range of β for which the equation $\partial_e \log(W_{\beta_0}(e)) - (\beta - \beta_0) = 0$ has three solutions e_1, e_2 and e_3. Two of these extrema are maxima so that we can use a double saddle point approximation which will be valid close to thermodynamical limit [28]

$$z_\beta = e^{\phi_\beta(e_1)} + e^{\phi_\beta(e_3)} = 2e^{\phi_\beta^+} \cosh\left(\phi_\beta^-\right)$$

where $2\phi_\beta^+ = \phi_\beta(e_1) + \phi_\beta(e_3)$ and $2\phi_\beta^- = \phi_\beta(e_1) - \phi_\beta(e_3)$. The zeros of Z_β then correspond to $\phi_\beta^- = i(2n+1)\pi$. The imaginary part is given by $\eta = 2(2n+1)\pi/(e_3 - e_1)$ while for the real part we should solve the equation$\Re\left(\phi_\gamma^-\right) = 0$. In particular, close to the real axis this equation defines a β which can be taken as β_0. If the bimodal structure persists when the number of particles goes to infinity, the loci of zeros corresponds to a line perpendicular to the real axis with a uniform distribution as expected for a first order phase transition.

This demonstrates the link of the curvature anomalies of thermodynamical potentials which are equivalent to bimodalities of probability distribution with the usual definition of first order phase transitions when the thermodynamical limit is reached.

6.7 The Death of Thermodynamics

Let us now investigate the more general situation when the system size goes to infinity even if the Van Hove theorem do not apply. The microcanonical caloric curve in the phase transition region may either converge towards the Maxwell construction or keep a backbending behavior, since a negative heat capacity system can be thermodynamically stable even in the thermodynamical limit if it is isolated (see Chap. 5). In particular we have shown in Chap. 5 that if the interaction is long-ranged the topological anomaly leading to the convex intruder in the entropy is not cured by increasing the number of particles [20], [30], [31].

Within our approach based on the topology of the probability distribution of observables [25] we have just shown that ensemble inequivalence arises from fluctuations of the order parameter. Ensembles putting different constraints on the fluctuations of the order parameter leads to different thermodynamics. In the case of phase transitions with non-zero latent heat , the total energy usually plays the role of an order parameter except in the microcanonical ensemble. Therefore, the microcanonical ensemble which forbids energy fluctuation is expected to presents a different thermodynamics than the other (canonical) ensembles. This different behavior may remain at the thermodynamical limit depending upon the specific properties of the considered system. In such a case, it may happen that the energy of a subsystem may become an order parameter when the total energy is constrained by a conservation law or a microcanonical sorting. This frequently occurs for Hamiltonians containing a kinetic energy contribution: if the kinetic heat capacity is large enough we will now show that the kinetic energy becomes an order parameter in the microcanonical ensemble. This is almost a paradox since in any other ensembles in which no total energy conservation is imposed the kinetic energy has a trivial perfect gas behavior while in the microcanonical ensemble it becomes an order parameter with the specific bimodal structure at the phase transition. Then, the microcanonical caloric curve presents at the thermodynamical limit a temperature jump in complete disagreement with the canonical ensemble.

Let us consider a finite system for which the Hamiltonian can be separated into two components $E = E_1 + E_2$, that are statistically independent $(W(E_1, E_2) = W_1(E_1)W_2(E_2))$ and such that the associated degrees of freedom scale in the same way with the number of particles; we will also consider the case where $S_1 = \log W_1$ has no anomaly while $S_2 = \log W_2$ presents a convex intruder which is preserved at the thermodynamical limit ($V \to \infty$, $N \to \infty$ with $N/V = cst$). Typical examples of E_1 are given by the kinetic energy for a classical system with velocity independent interactions. For other similar one-body operators see [30].

The probability to get a partial energy $E_1$when the total energy is E is given by

$$P_E\left(E_1\right) = \exp\left(S_1\left(E_1\right) + S_2\left(E - E_1\right) - S\left(E\right)\right) \tag{47}$$

The extremum of $P_E\left(E_1\right)$ is obtained for the partitioning of the total energy E between the kinetic and potential components that equalizes the two partial temperatures $\overline{T}_1{}^{-1} = \partial_{E_1} S_1(\overline{E}_1) = \partial_{E_2} S_2(E - \overline{E}_1) = \overline{T}_2{}^{-1}$. If $\overline{E}_1$is unique, $P_E\left(E_1\right)$ is mono-modal and we can use a saddle point approximation around this solution to compute the entropy $S\left(E\right) = \log \int_{-\infty}^{E} dE_1 \exp\left(S_1\left(E_1\right) + S_2\left(E - E_1\right)\right)$. At the lowest order, the entropy is simply additive so that the microcanonical temperature of the global system $\partial_E S(E) = \overline{T}^{-1}$ is the one of the most probable energy partition. Therefore, the most probable partial energy $\overline{E}_1$ acts as a microcanonical thermometer. If $\overline{E}_1$is always unique, the kinetic thermometer in the backbending region will follow the whole decrease of temperature as the total energy increases. Therefore, the total caloric curve will present the same anomaly as the potential one.

If conversely the partial energy distribution is double humped , then the equality of the partial temperatures admits three solutions one of them $\overline{E}_1^{(0)}$ being a minimum. At this point the partial heat capacities $C_1^{-1} = -\overline{T}^2\, \partial_{E_1}^2 S_1(\overline{E}_1^{(0)})$ and $C_2^{-1} = -\overline{T}^2\, \partial_{E_2}^2 S_2(E - \overline{E}_1^{(0)})$ fulfill the relation

$$C_1^{-1} + C_2^{-1} < 0 \tag{48}$$

This happens when the potential heat capacity is negative and the kinetic energy is large enough $(C_1 > -C_2)$ to act as an approximate heat bath: the partial energy distribution $P_E\left(E_1\right)$ in the microcanonical ensemble is then bimodal as the total energy distribution $P_\beta\left(E\right)$ in the canonical ensemble. The bimodality of $P_E\left(E_1\right)$ implies that the kinetic energy is an order parameter of the transition in the microcanonical ensemble. In this case, performing a double saddle point approximation around the two maxima leads to a microcanonical temperature given by a weighted average of the two estimations from the two maxima of the kinetic energy distribution

$$T = \partial_E S(E) = \frac{\overline{P}^{(1)}\sigma^{(1)}/\overline{T}^{(1)} + \overline{P}^{(2)}\sigma^{(2)}/\overline{T}^{(2)}}{\overline{P}^{(1)}\sigma^{(1)} + \overline{P}^{(2)}\sigma^{(2)}} \tag{49}$$

where $\overline{T}^{(i)} = T_1(\overline{E}_1^{(i)})$ are the kinetic temperatures calculated at the two maxima, $\overline{P}^{(i)} = P_E(\overline{E}_1^{(i)})$ are the probabilities of the two peaks and $\sigma^{(i)}$ their widths. At the thermodynamical limit (48) reads $c_1^{-1} + c_2^{-1} < 0$, with $c = \lim_{N\to\infty} C/N$. If this condition is fulfilled the probability distribution $P_\beta(E)$ presents two maxima for all finite sizes and only the highest peak survives at $N = \infty$. Let E_t be the energy at which $P_{E_t}(\overline{E}^{(1)}) = P_{E_t}(\overline{E}^{(2)})$. Because of (49) at the thermodynamical limit the caloric curve will follow the high (low) energy maximum of $P_E(E_1)$ for all energies below (above) E_t; there will be a temperature jump at the transition energy E_t.

This patent violation of ensemble equivalence means that, contrary to the physical intuition based on macro-systems, the equations of state are expected to explicitly depend on the characteristics of the considered ensemble of events i.e. the state variables: the fluctuating observables and the conserved quantities imposed by the dynamics or by the sorting variables used in the data analysis. This implies the impossibility to define a unique thermodynamics, i.e. a unique EoS, for systems undergoing a first order phase transition.

Let us illustrate the above results with two examples for a classical gas of interacting particles. For the kinetic energy contribution we have $S_1(E) = c_1 \ln(E/N)^N$ with a constant kinetic heat capacity per particle $c_1 = 3/2$. For the potential part we will take two polynomial parametrizations of the interaction caloric curve presenting a back bending which are displayed in the left part of Fig. 25. If the decrease of the partial temperature $T_2(E_2)$ is steeper than $-2/3$ (Fig. 25a) [20] (48) is verified and the kinetic caloric curve $T_1(E - E_1)$ (dot-dashed line) crosses the potential one $T_2(E_2)$ (full line) in three different points for all values of the total energy lying inside the coexistence region. The resulting caloric curve for the whole system is shown in Fig. 25b (symbols) together with the thermodynamical limit (lines) evaluated from the double saddle point approximation (49). In this case one observes a temperature jump at the transition energy. If the temperature decrease is smoother (Fig. 25c) the shape of the interaction caloric curve is preserved at the thermodynamical limit (Fig. 25d).

The occurrence of a temperature jump in the thermodynamical limit is easily spotted by looking at the bidimensional canonical event distribution $P_\beta(E_1, E_2)$ in the partial energies plane. This density of states is just the product of the independent kinetic and potential canonical probabilities as shown in the left part of Fig. 26 for the two model equation of states of Fig. 25 at the transition temperature $\beta = \beta_t$. In order to discuss the microcanonical ensemble one has to introduce the total energy $E = E_1 + E_2$. Keeping E and E_1 as variables instead of (E_1, E_2) is nothing but a simple coordinate change with unit Jacobian. Thus we can look at the canonical distribution as a function of E and E_1, $P_\beta(E, E_1) \propto \exp S_1(E_1) \exp S_2(E - E_1) \exp(-\beta E)$ which is shown in the right

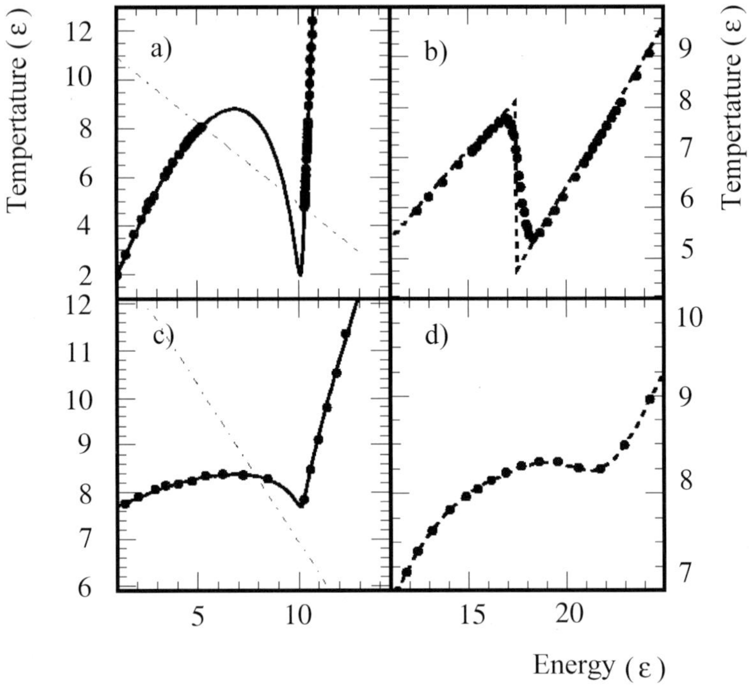

Fig. 25. Left panels: temperature as a function of the potential energy E_2(full lines) and of the kinetic energy $E - E_2$ (dot-dashed lines) for two model equation of states of classical systems showing a first order phase transition. Symbols: temperatures extracted from the most probable kinetic energy thermometer from (5). Right panels: total caloric curves (symbols) corresponding to the left panels and thermodynamical limit of (7) (dashed lines).

part of Fig. 26. The deformation of the event distribution induced by the microcanonical constraint does not cause a topological difference between our two model cases; this explains why both converge to the Maxwell construction for $N \to \infty$ in the canonical ensemble. If we now study the microcanonical ensemble we have to look at constant energy cuts of $P_\beta(E, E_1)$ leading to the microcanonical distribution $P_E(E_1)$ within a normalization constant. If the anomaly in the potential equation of state is sufficiently important, the distortion of events is such that one can still see the two phases coexist even after a sorting in energy as shown in the same Fig. 26 for two cuts of $P_\beta(E, E_1)$ at an energy close to the transition energy.

7 Observables: Melting of Metallic Clusters [32]

In the year 2000 the first experimental signature of a back bending caloric curve has been reported in the melting of metallic clusters. The experiment is rather simple. The clusters are first produced and selected. Then the clusters get ther-

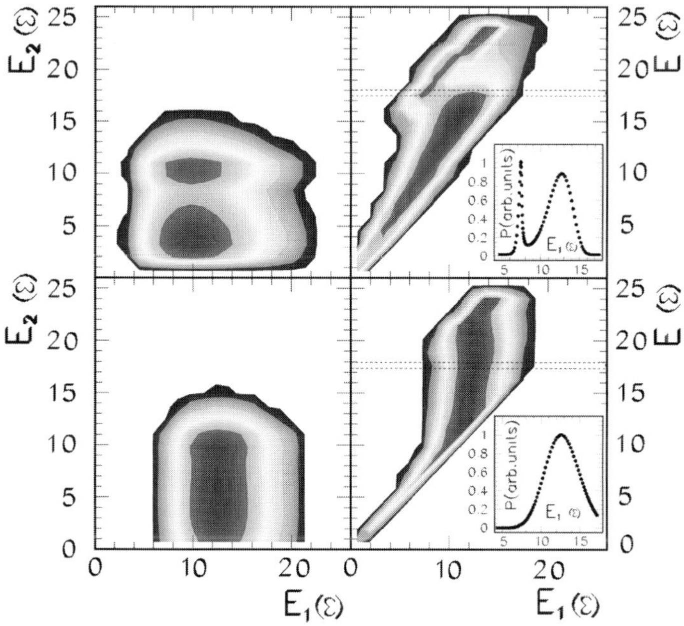

Fig. 26. Canonical event distributions in the potential versus kinetic energy plane (left panels) and total versus kinetic energy plane (right panels) at the transition temperature for the two model equations of state of Fig. 1. The inserts show two constant total energy cuts of the distributions.

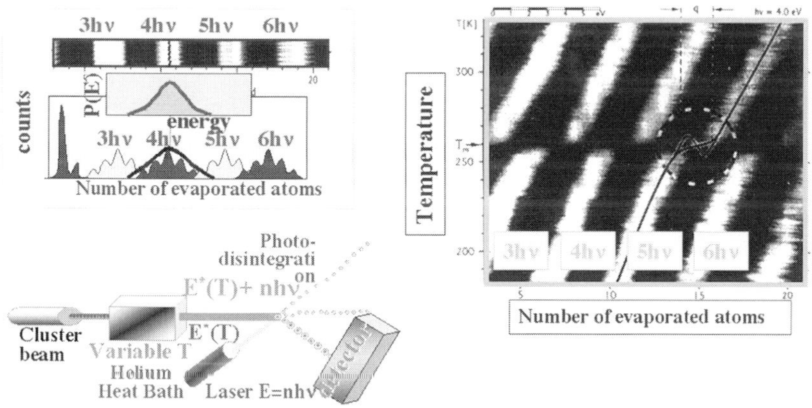

Fig. 27. Left bottom: a schematic picture of the experimental set-up. Left top: illustration of the shift of the cluster internal energy distribution by the photons' energy leading to a measurable distribution of evaporated fragments. Right: the observed correlation between the temperature of the Helium bath and the fragmentation pattern induced by the absorption of several photons. The "accident" at a given temperature indicates the melting point [32].

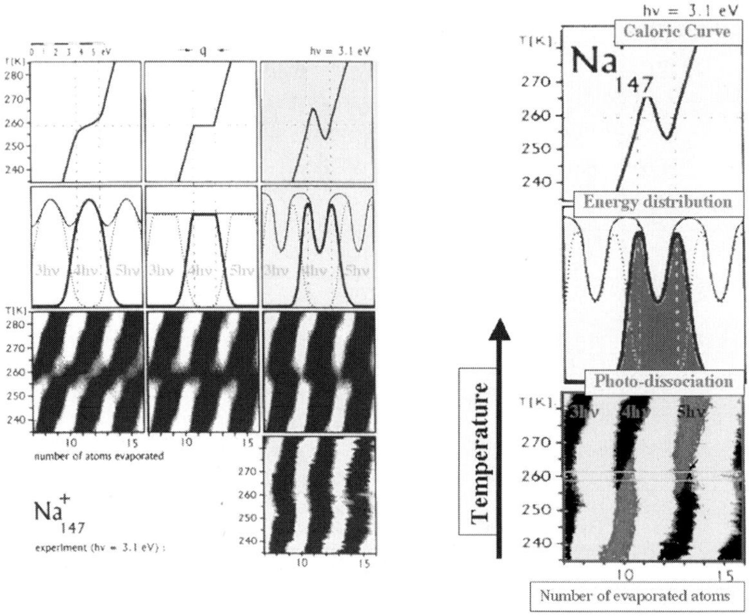

Fig. 28. Left side: several scenarii for the transition between two phases, a broad monomodal energy distribution, a flat one and a bimodal one, are compared with the experimental results. The observed pattern (last row) is only compatible with the existence of a bimodal energy distribution. Right side: enlarged diagram indicating the presence of a bimodality which looks like a sudent jump in the energy distribution. [32].

malized in the melting temperature region in an helium heat bath. After thermalization they are further excited by a laser beam absorbing several photons, thanks to the plasmon vibration. The average energy is then such that clusters have time to evaporate atoms within the experiment time scale. The number of evaporated atoms provides a measure of the cluster excitation energy. Since the photon energy is fixed, this is a measure of the excitation energy at the exit of the heat bath. Changing the temperature the thermal excitation changes and the distribution of evaporated atoms is shifted. The obtained bidimensional pictures of the number of evaporated atoms as a function of the oven temperature clearly show an anomaly corresponding to the melting point. (see Fig. 27).

Instead of looking for a back bending of the microcanonical caloric curve the authors of reference [32] have tried to observe the associated bimodal canonical energy distribution. Since the number of evaporated atoms is a measure of the cluster excitation energy before the photons' absorption, this distribution is a measure of the canonical distribution of excitation energy. The difference between a back bending and a monotonous caloric curve is that the energy distribution has a bi-modal or a mono-modal shape (see Chap. 6). This induces a modification of the fragmentation pattern. The only difficulty is that the system may absorb different numbers of photons. Therefore one observe replicas of the same distribution every $h\nu$. The trick is thus to correctly tune the laser energy so that it will not blur the jump. On can see in figure28 that the observed pattern is only compatible with a negative heat capacity system. The results can be interpreted if one assumes a 10 degree decrease of the temperature at the phase transition energy.

8 Observables: The Nuclear Physics Case

8.1 Liquid Gas Phase Transition in Nuclear Systems

Since nuclear forces resemble to Van der Waals interactions the nuclear phase diagram (see Fig. 29) is expected to present a liquid gas phase transition. Our present knowledge of the nuclear equation of state is limited. The main reason is the difficulty to treat the nuclear many-body problem and to define a reliable in medium interaction. The saturation energy and density, i.e. the ground state of nuclear matter, are well established but the compressibility, i.e. the variation of the energy as a function of the density around the saturation point, is still under discussion. As far as the temperature dependence of nuclear properties is concerned very little is also know in an absolute way. Only the entropy variation, i.e. the level density parameter $a = S/T$, of a finite nucleus as an open system has been clearly established through evaporation studies.

An important research activity is now devoted to the extraction of reliable information of the nuclear equation of states and the associated phase diagram. Heavy ion reactions are routinely used to test mechanical and thermodynamical properties of nuclei. In particular in the recent years the multifragmentation regime has been tentatively associated with the occurrence of a liquid-gas phase transition.

8.2 Negative Heat Capacities and Abnormal Fluctuations

At the same time of the observation of negative heat capacity in the melting transition (see Chap. 7), a $C < 0$ signal has also been reported for the nuclear multifragmentation transition using the fluctuations of the energy partition [33].

The investigation method [34] can be easily explained for a classical fluid and tested in the framework of the lattice-gas model. The total energy E of the considered system can be decomposed into two independent components, its kinetic and potential energy: $E = E_k + E_p$. In a microcanonical ensemble with a total energy E the total degeneracy factor $W(E) = \exp(S(E))$ is thus simply given by the folding product of the individual degeneracy factors $W_i(E_i) = \exp(S_i(E_i))$ of the two subsystems $i = k, p$. One can then define for the total system as well as for the two subsystems the microcanonical temperatures T_i and the associated heat capacities C_i. If we now look at the kinetic energy distribution when the total energy is E we get

$$P_k^E(E_k) = \exp(S_k(E_k) + S_p(E - E_k) - S(E)) \tag{50}$$

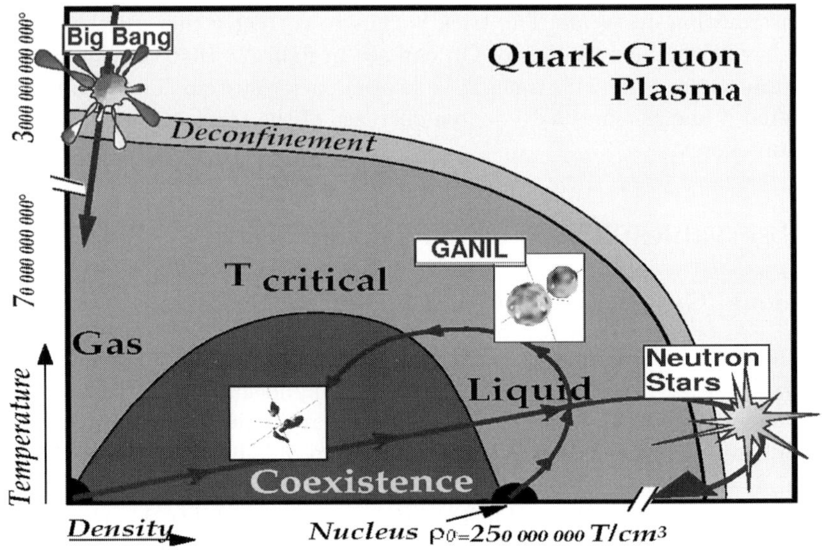

Fig. 29. Nuclear Phase diagram in the temperature density plane. At low temperature and below the saturation density a liquid gas phase transition is expected.

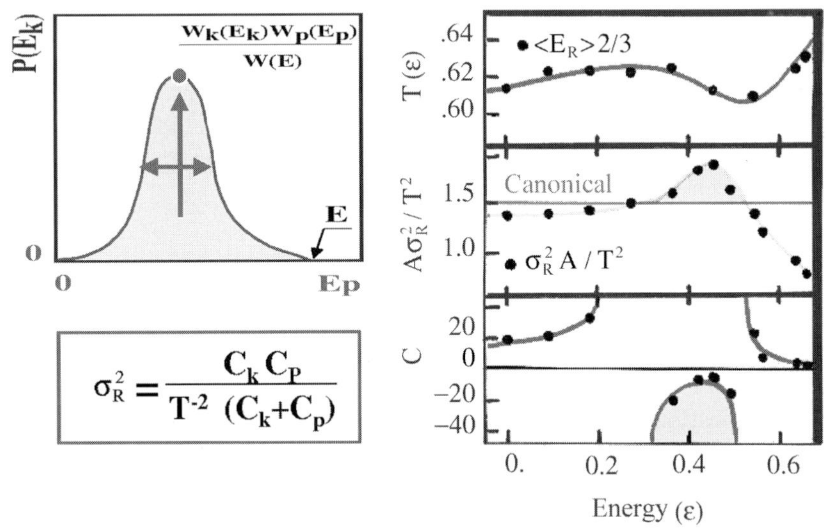

Fig. 30. Left part: the schematic distribution of partial energy for a fixed total energy. Right: the comparison of the various measurements (dots) with the exact results of the lattice-gas model (lines).

Using (50) we directly get that the most probable partitioning of the total energy E between the potential and kinetic components is characterized by a unique microcanonical temperature $\bar{T} \equiv T_k\left(\bar{E}_k^E\right) = T_p\left(E - \bar{E}_k^E\right)$. Therefore the most probable kinetic energy $\bar{E}_k^E$ can be used as a microcanonical thermometer

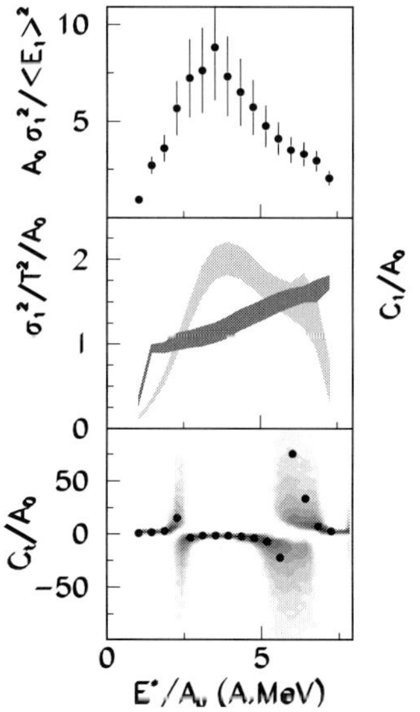

Fig. 31. Configurational energy fluctuations (top), comparison with the canonical expectation (middle) and deduced heat capacity as a function of the excitation energy (bottom) for the Au quasi projectile formed in Au+Au collisions at 35 MeV/nucleon. Multics-Miniball data from [33]

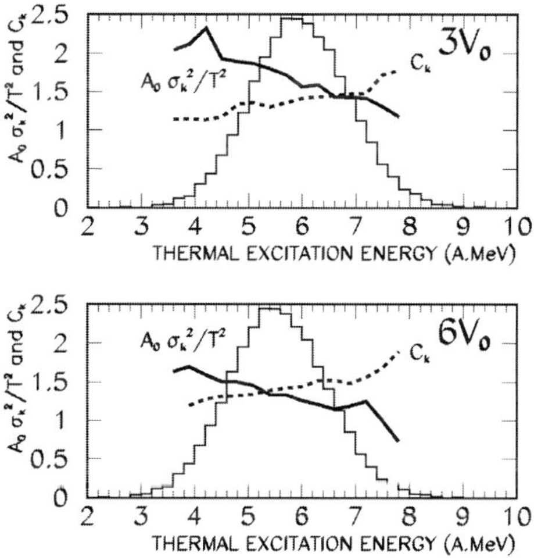

Fig. 32. Configurational energy fluctuations (full line) compared to the canonical expectation (dashed line) and excitation energy distribution (histogram) for central Xe+Sn collisions at 32 MeV/nucleon. The two pictures correspond to two different hypotheses in the data reconstruction. Indra data from [33]

as shown in Fig. 30. Using a Gaussian approximation for $P_k^E(E_k)$ the kinetic energy variance can be calculated as [34]

$$\sigma_k^2 = \bar{T}^2 \frac{C_k C_p}{C_k + C_p} \tag{51}$$

where C_k and C_p are the microcanonical heat capacities calculated for the most probable energy partition.

As shown in Fig. 30 when C_p diverges and then becomes negative, σ_k^2 remains positive but overcomes the canonical expectation $\sigma_k^2 = \bar{T}^2 C_k$. This anomalously large kinetic energy fluctuation is a signature of the first order phase transition. Equation (51) can be inverted to extract from the observed fluctuations the heat capacity

$$C \simeq C_k + C_p = \frac{\bar{T}^2 C_k^2}{\bar{T}^2 C_k - \sigma_k^2} \tag{52}$$

Figure 30 shows that the heat capacity extracted from the kinetic energy fluctuations is in very good agreement with the exact one. This means that kinetic energy fluctuations are an experimentally accessible measure of the heat capacity which allows to sign divergences and negative branches characteristic of the phase transition.

8.3 Experimental Results

The negative heat capacity signal of a phase transition has been looked for in experiments. In such a case an easy splitting of the energy is between the thermal excitation and agitation on one side and the partition Q-value plus the Coulomb interaction on the other side. The expected canonical prediction can be inferred from the relation between the average kinetic energy and the temperature since this provides C_k. Figures 31 and 32 show the first experimental results of a fluctuation overcoming the canonical expectation with the corresponding deduced heat capacity for excited nuclei in the gold mass range [33].

It is important to know that these measurements need a very sophisticated and accurate technique to correctly reconstruct the configurational energy fluctuations at the time of fragment formation. These reconstructions often need hypotheses such as the volume of the freeze-out and the origin of emitted particles. Additional measurements to control these hypotheses have to be performed. However, kinetic energy fluctuations are a very promising way to infer thermodynamical properties and to signal phase transitions.

References

1. E.T.Jaynes, Phys.Rev.106(1957)620 and Phys.Rev.108(1957)171
2. A.Katz, 'Principles of statistical mechanics' Freeman (1967)
3. S.Abe, Y.Okamoto, 'Nonextensive statistical mechanics and its applications', Lecture Notes in Physics vol.560 (2001)

4. R. Balian, 'From microphysics to macrophysics', Springer Verlag (1982)
5. Ph. Chomaz, Ann. Phys. Fr. 21 (1996) 669
6. L. Onsager, Phys.Rev. 65 (1944) 117
7. C.N.Yang, Phys.Rev. 85 (1952)809
8. E.A.Guggenheim, J.Chem.phys.13 (1945) 253.
9. L.D.Landau, E.M.Lifshitz, 'Statistical Physics', Pergamon Press (1980)
10. M.E.Fisher,'Critical phenomena', Academic (1971)
11. Ph.Chomaz, F.Gulminelli, Phys.Rev.Lett.82 (1999)1402
12. Ph.Chomaz, F.Gulminelli, Int.Journ.Mod.Phys.E8 (1999) 527
13. M.S.Challa, J.H.Hetherington, Phys.Rev.Lett.60 (1988)77 and Phys.Rev.A38 (1988) 6324
14. D.H.E.Gross, Phys.Rep.279 (1997)119 and "Microcanonical thermodynamics: phase transitions in finite systems", Lecture notes in Physics vol.66, World Scientific (2001)
15. F.Gulminelli, Ph.Chomaz, Europhys.Lett.50 (2000) 434
16. F.Gulminelli, Ph.Chomaz, Phys.Rev.Lett.85 (2000) 3587
17. T.L.Beck, R.S.Berry, Journ.Chem.Phys. 88 (1988) 3910
18. L.Moretto et al., Phys.Rep.287 (1997) 249
19. V.Antonov, Vest.Leningrad Univ.7(1962)135
20. D.Lynden-Bell, Monthly Not.Astr.Soc. 138 (1968) 495
21. W.Thirring, Z.Phys.235 (1970) 339
22. T.Padmanabhan, Phys.Rep.188 (1990) 285
23. K. Binder, D. P. Landau, Phys. Rev. B 30 (1984) 1477.
24. P. Labastie and R. L. Whetten, Phys. Rev. Lett. 65 (1990) 1567.
25. Ph. Chomaz, F. Gulminelli and V. Duflot, Phys. Rev. E 64 (2001) 046114 and arXiv:cond-mat/0112222.
26. T.Poston, I.Stewart, "Catastrophe Theory and its applications", Pitman (1978)
27. C.N.Yang and T.D.Lee, Phys.Rev.87 (1952) 404.
28. K.C. Lee, Phys. Rev. 53 E (1996) 6558.
29. M.S.S.Challa et al. Phys. Rev. 34B (1986) 1841.
30. J. Barré, D. Mukamel and S. Ruffo, Phys. Rev. Lett.87 (2001)030601-1.
31. L. Casetti, M. Pettini and E. G. D. Cohen, Phys. Rep. 337 (2000) 237.
32. M. Schmidt et al, Nature 393 (1998) 238 and Phys.Rev.Lett.86 (2001) 1191
33. M. D'Agostino et al., Nucl.Phys. A 650 (1999) 329, Phys.Lett.B 473 (2000) 219, Nucl.Phys.A 699 (2002)795.
34. P.Chomaz and F.Gulminelli, Nucl. Phys. A647 (1999) 153.

Phase Transitions in Systems with $1/r^\alpha$ Attractive Interactions

Eddie G.D. Cohen[1] and Iaroslav Ispolatov[2]

[1] Center for Studies in Physics and Biology, The Rockefeller University, 1230 York Avenue, New York, NY 10021 USA
[2] Departamento de Física, Universidad de Santiago de Chile, Casilla 302, Correo 2, Santiago, Chile

Abstract. We study the gravitational collapse – like phase transitions in the microcanonical ensemble for a class of systems of N particles with nonintegrable attractive pair potentials $1/r^\alpha$, with $0 < \alpha < 3$ in three dimensions. Introducing a continuum description and a mean field approximation in the limit N to infinity, an integral equation for the density profile of the particles is obtained, which is solved numerically. For all $0 < \alpha < 3$ and sufficiently small short range cutoffs r_0, there appears to be cutoff-independent critical energy, below which the system jumps discontinuously from a normal uniform to a collapsed core – halo state. This gravitational-like transition will become a normal first order transition when the cutoff range r_0 becomes larger than some critical value r_{0c} which is an increasing function of α.

1 Introduction

We consider systems with particles interacting with an attractive potential $\sim 1/r^\alpha$ with $0 < \alpha < 3$ in $d = 3$ dimensions. The Newtonian gravitation potential is then a special case with $\alpha = 1$. The main question we want to discuss is whether this class of systems behaves like the gravitational ($\alpha = 1$) system, especially as far as their collapse and phase transitions are concerned.

In order to investigate this question we studied systems of N particles in a microcanonical ensemble (ME) with an interparticle pair potential of the form $V(r_{ij}) \equiv V_{ij} = -G/|\boldsymbol{r}_i - \boldsymbol{r}_j|^\alpha$ in a spherical container of radius R and total energy E and α in the above mentioned range. Here $\boldsymbol{r}_i$ is the position of particle i and G a coupling constant which determines the strength of the interaction between two particles.

Using for the ME entropy the expression:

$$S(E) = \log \frac{1}{N!} \int \dots \int \prod_{k=1}^{N} \frac{d\boldsymbol{r}_k d\boldsymbol{p}_k}{h^3} \quad \delta \left[E - \sum_{i=1}^{N} \frac{p_i^2}{2M} - \sum_{i<j=1}^{N} V_{ij} \right] \tag{1}$$

where we set Boltzmann's constant $k_B = 1$ and $\boldsymbol{p}_k$ is the momentum of particle k. Here the factor h^3, with h Planck's constant, has been inserted to make the argument of the logarithm on the right hand side (r.h.s.) of (1) dimensionless and assure the conventional connection with quantum mechanics.

2 Mathematics

The evaluation of (1) proceeds in a number of steps, which we will summarize in the following, referring for all details to the literature [1,7]. The first steps appear to be for all α essentially the same as for the usual $\alpha = 1$ case.

1. a) Integrate over the $\boldsymbol{p}_k$, with $k = 1, ..., N$;
   b) Introduce a dimensionless energy $\epsilon = ER^\alpha/GN^2$;
   c) Introduce dimensionless coordinates by the scaling: $\boldsymbol{x}_i = \boldsymbol{r}_i/R \ \ (i = 1, ..., N)$;
   d) For $N \gg 1$, use for the evaluation of (1) a continuum approximation, replacing particles by (particle) density profiles:

$$\int ... \int \prod_{k=1}^{N} d\boldsymbol{x}_k ... \to \int D\rho(\boldsymbol{x}) .... \tag{2}$$

   where the functional integral on the r.h.s. of (2) is over density profiles. This introduces an additional factor constraining the number of particles, $\delta[\int \rho(\boldsymbol{x}) d\boldsymbol{x} - 1]$.
   e) Use Fourier representations of the delta function above and the theta function which appears after integration of the delta function in (1) over the momenta and ensures the positiveness of the kinetic energy.

   These steps lead then to the following asymptotically $N \to \infty$ exact expression for $S(\epsilon)$:
2.

$$S(\epsilon) \approx S_0 + \log \int D\rho \cdot \int_{-i\infty}^{i\infty} \frac{d\gamma}{2\pi i} \int_{-i\infty}^{i\infty} \frac{d\beta}{2\pi i} \exp\{N\tilde{s}[\rho(\cdot), \epsilon, \gamma, \beta]\} \tag{3}$$

   Here S_0 is an energy-independent constant, the integrals over γ and β originate from step e) and $\tilde{s}$ is an "effective action" functional, given by:

$$\tilde{s}[\rho(\cdot), \epsilon, \gamma, \beta] = \beta[\epsilon + \Phi(\alpha)] + \gamma\left[\int \rho(\boldsymbol{x}) d\boldsymbol{x} - 1\right] - \frac{3}{2}\log\beta - \int \rho(\boldsymbol{x}) \log[\rho(\boldsymbol{x})/e] d\boldsymbol{x} \tag{4}$$

   Also, β can be interpreted as the inverse temperature of the system and the magnitude of the potential energy $\Phi(\alpha)$ is given by:

$$\Phi(\alpha) = \frac{1}{2} \int \int d\boldsymbol{x}_1 d\boldsymbol{x}_2 \frac{\rho(\boldsymbol{x}_1)\rho(\boldsymbol{x}_2)}{|\boldsymbol{x}_1 - \boldsymbol{x}_2|^\alpha} \tag{5}$$

   implying that $\epsilon + \Phi(\alpha)$ is directly related to the kinetic energy of the system.
3. $S(\epsilon)$ is evaluated approximately by its saddle point, using that $N \to \infty$.
4. The equations for the extrema of $\tilde{s}[\rho(\cdot), \epsilon, \gamma, \beta]$ needed for the saddle point approximation, give:

   a) unique values γ_s and β_s and
   b) a non-linear integral equation for $\rho_s(\boldsymbol{x})$, where the subscript s refers to the saddle point approximation.
5. The four equations, which determine the extrema of $\tilde{s}$ are:

$$\rho_s(\boldsymbol{x}) = \exp[\gamma_s + \beta_s \varphi_s(\boldsymbol{x})] \tag{6}$$

where

$$\gamma_s = -\log \int \exp[\beta_s \varphi_s(\boldsymbol{x})] d\boldsymbol{x} \tag{7}$$

$$\begin{aligned} \beta_s &= \frac{3}{2}[\epsilon + \frac{1}{2}\int \rho_s(\boldsymbol{x}_2)\varphi_s(\boldsymbol{x}_2)d\boldsymbol{x}_2]^{-1} \\ &= \frac{3}{2}\ [\text{kinetic energy}]^{-1} = \frac{1}{T} \end{aligned} \tag{8}$$

$$\varphi_s(\boldsymbol{x}) = \int \frac{\rho_s(\boldsymbol{x}_1)d\boldsymbol{x}_1}{|\boldsymbol{x}_1 - \boldsymbol{x}|^\alpha} \tag{9}$$

which is the negative of the (negative) potential energy at the position $\boldsymbol{x}$ due to $\rho_s(\boldsymbol{x}_1)$. For $\alpha = 1$, the Poisson-Boltzmann-Emden (PBE) equation for the potential $\varphi_s(\boldsymbol{x})$ is obtained. This follows immediately from (6) by taking the Laplacian and using Poisson's equation for the $1/r$-potential. The evaluation of $S(\epsilon)$ is then reduced to a non-linear second order differential equation, which is the starting point of most further developments in the theory of the gravitational potential [2,7]. Due to the nonlinearity of the PBE equation, numerical methods have to be used for its solution, but some analytical results can be obtained as well. Clearly this is not the case for $\alpha \neq 1$, so that we had to take recourse exclusively to numerical results.
6. From the (3–9), an expression for the entropy per particle $s(\epsilon) = S(\epsilon)/N$, at the saddle point can be obtained as:

$$s(\epsilon) = \frac{3}{2}\log[\epsilon + \Phi(\alpha)] - \int \rho_s(\boldsymbol{x}) \log \rho_s(\boldsymbol{x}) d\boldsymbol{x} + O\left(\frac{1}{N}\right), \tag{10}$$

if terms independent of ϵ are ignored.
7. a) To proceed further we look for spherically symmetric solutions, assuming that $\rho(\boldsymbol{x}) = \rho(x)$, as is also done for $\alpha = 1$.
   b) The angular integration in the integral equation (6) for $\rho_s(\boldsymbol{x})$ can then be performed.
   c) We will consider here potentials $\propto 1/r^\alpha$ with $0 < \alpha < 3$, with or without short range regularizations which we will specify more precisely below.
8. The numerical solution of the integral equation for the density profile $\rho(x)$ from (6), together with (7–9), can then be obtained for $\alpha \neq 2$ from:

$$\begin{aligned} \rho(x) &= \rho_0 \beta_s \exp\left[\frac{2\pi}{(2-\alpha)x}\int_0^1 \rho(x_1)x_1 \Delta(x; x_1) dx_1\right] \\ &= \rho_0 \exp B[\rho(.), x] \end{aligned} \tag{11}$$

where

$$\rho_0 = \left[\int_0^1 (\exp B[\rho(.), x]) dx \right]^{-1} \tag{12}$$

is the normalization of $\rho(x)$;

$$\beta_s = \frac{3}{2} \left[\epsilon + \frac{4\pi^2}{2-\alpha} \int_0^1 \rho(x_1)\rho(x_2) x_1 x_2 \Delta(x_2; x_1) dx_1 dx_2 \right]^{-1} \tag{13}$$

with

$$\Delta(x; x_1) = (x + x_1)^{2-\alpha} - |x - x_1|^{2-\alpha} \tag{14}$$

while for $d = 2$ one has:

$$\frac{1}{2-\alpha} \Delta(x; x_1) \to L(x; x_1) \tag{15}$$

where

$$L(x; x_1) = \ln(x + x_1) - \ln|x - x_1| \tag{16}$$

9. The actual solution for $\rho_s(x)$ as a function of ϵ is then obtained by an iterative procedure

$$\rho^{\{i+1\}}(x) = \alpha \rho_0^{\{i\}} \exp B[\rho^{\{i\}}(.), x] + (1-\alpha)\rho^{\{i\}}(x), \tag{17}$$

where i is the number of the iteration. For details we refer to [1].

3 Physics

The physics of the solutions obtained as described above for $s(\epsilon)$ appears to behave, for all $0 < \alpha < 3$, like those for $\alpha = 1$. When the range of regularization is sufficiently short, we observe the following features which are usually associated with a gravitational phase transition:

1. There exists an $\epsilon^*(\alpha)$ such that:

$$s_{n-st}(\epsilon^*) = s_{c-st}(\epsilon^*) \tag{18}$$

   where the entropy derivative $ds(\epsilon)/d\epsilon$, i.e. the inverse temperature $\beta(\epsilon)$, undergoes a discontinuous jump (see Fig. 1). In (18) n, c refer to normal (uniform) and collapsed (core-halo) states, respectively and st to a stable state.
2. There exists an $\epsilon_c(\alpha)$ such that:
   a) the uniform state collapses for energies $\epsilon < \epsilon_c(\alpha)$, when there is no convergent iterative solution of (6–9) anymore and the system is unstable.
   b) We find numerically $\epsilon_c(\alpha = 1) = -0.3346 \approx \epsilon_c^{PBE}(\alpha = 1) = -0.335$, the value obtained from the PBE theory for $\alpha = 1$. A plot of $\epsilon = \epsilon_c(\alpha)$ is given in Fig. 2. When decreasing the energy ϵ starting from large ϵ-values to lower ϵ-values, the impossibility of a phase transition with

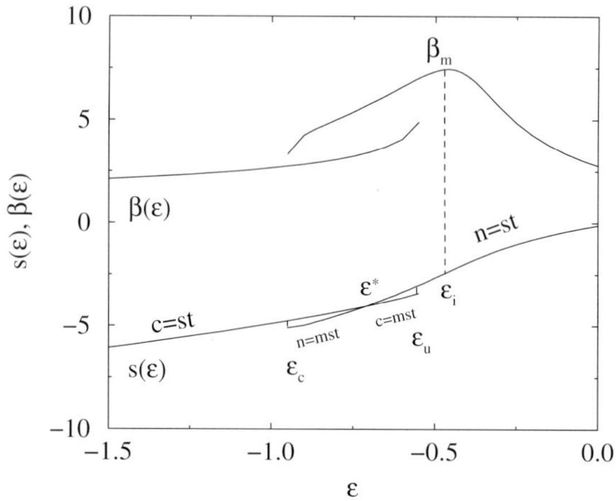

Fig. 1. Plots of entropy $s(\epsilon)$ and inverse temperature $\beta(\epsilon) = ds/d\epsilon$ for $\alpha = 1/2$. The radius of excluded central volume $r_0 = 5 \times 10^{-4}$

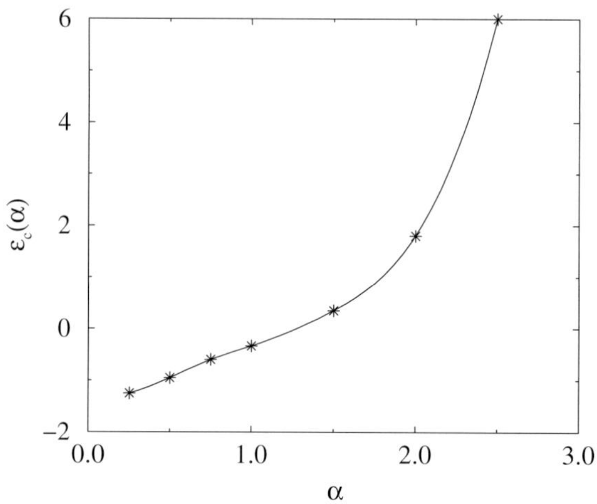

Fig. 2. Plot of collapse energy $\epsilon_c(\alpha)$ vs. potential exponent α

two coexisting phases at ϵ^*, because of their different temperatures $(\beta(\epsilon) = ds(\epsilon)/d\epsilon)$ and macroscopic structures, forces the stable system to bypass ϵ^* ("undercool") and follow a metastable path, till it reaches $\epsilon = \epsilon_c < \epsilon^*$, when the system finally jumps from a normal metastable phase to a collapsed stable phase, a transition which "should" have occurred at ϵ^*.

3. As in the $\alpha = 1$ case, the collapse at $\epsilon_c(\alpha)$ is fundamentally different, when there is or is not a short-range regularization. For:
   a) If there is no regularization, there is a phase transition with an infinitely large entropy jump from the metastable to a stable state

with infinite entropy. Then all normal (i.e. non-collapsed) states are metastable with respect to this collapsed state at the origin $r = 0$.

b) If there is a short-range regularization , then there is a phase transition with a finite jump from a metastable to a collapsed state (cf. Fig. 1). For increasing energies ϵ, starting from small ϵ-values and increasing to larger ϵ-values, a similar discontinuous jump occurs at $\epsilon = \epsilon_u$, bypassing again ϵ^* ("overheat") and follow a metastable path till it reaches $\epsilon = \epsilon_u > \epsilon^*$, when the system jumps from a collapsed metastable phase to a normal stable phase.

4. The above is illustrated in Fig. 1, which we summarize in the following.
   a) Starting at large ϵ, $s(\epsilon)$ goes first through an inflection point ϵ_i, where $d^2s/d\epsilon^2 = 0$;
   b) for $\epsilon < \epsilon_i$, the specific heat, $c_v(\epsilon) = -[(ds(\epsilon)/d\epsilon)^2/ds^2(\epsilon)/d\epsilon^2] < 0$ is negative, which, in the ME, is indicative for the occurrence of a phase transition [2,4].
   c) There is no iterative solution for a metastable collapsed phase for $\epsilon > \epsilon_u$ since the system is then unstable.
   d) the appearance of ϵ^*, where the entropies of the normal and collapsed are equal, but no phase transition occurs, except when the cut-off r_0 is sufficiently large (see below).
   e) Similarly as at $\epsilon = \epsilon_u$, there is no iterative solution for a metastable normal phase for $\epsilon < \epsilon_c$, since the system is then unstable. Instead, it jumps to a stable collapsed solution for $r_0 \geq 0$, where the jump in $s(\epsilon) = \infty$ if $r_0 = 0$ and finite if $r_0 > 0$. The observed "undercooled – overheated" phase transition behavior is usually called a gravitational or sometimes a "zeroth order" phase transition. We note that these phase transitions at ϵ_c and ϵ_u are unusual, in that they take place between two different kinds of phases: a metastable and a stable phase. As mentioned before, a normal first order phase transition, which should take place at $\epsilon = \epsilon^*$ where the entropy of the two phases is identical, does not occur, because of a discontinuity in its ϵ-derivative at $\epsilon = \epsilon^*$, i.e. of $\beta(\epsilon^*)$. See, however, below, sub points 3 and 4 in the next section.

5. a) The short range regularizations (which prevent entropy and density singularities in the collapsed state) taken here can be of two types: that of an excluded shell impenetrable to the particles in the center of the system so that the interpareticle potential is "bare", $V_{ij} = -G/|\boldsymbol{r}_i - \boldsymbol{r}_j|^\alpha$ but particle coordinates are restricted by $r_0 < r < R$ [8], or of a "soft" interparticle potential like $-G/[(\boldsymbol{r}_i - \boldsymbol{r}_j)^2 + r_0^2]^{\alpha/2}$. Despite their obvious physical difference they give rise to very similar thermodynamic behavior; both regularizations are characterized by a lengthscale r_0 sometimes called a "cutoff radius" below.
   b) The difference between the normal and the collapsed phase can be observed in their density profiles $\rho(x)$. In Fig. 3, $\rho(x)$ for $\alpha = \frac{1}{2}$ has been plotted for the normal as well as for the collapsed state. The

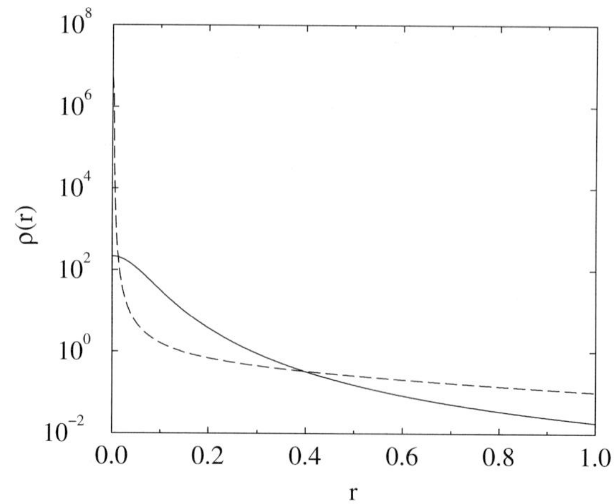

Fig. 3. Density profiles $\rho(x)$ for $\alpha = 1/2$ for normal (solid line) and collapsed (dashed line) phases for the energy $\epsilon^* = -0.708$, when entropies of both phases are the same. The radius of the excluded central volume is $r_0 = 5 \times 10^{-4}$

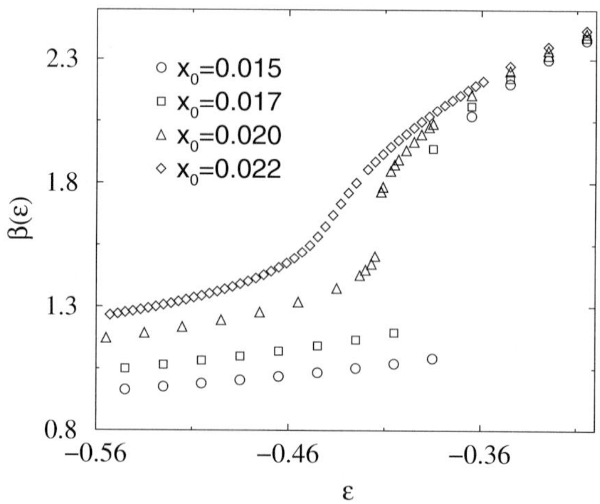

Fig. 4. Plots of inverse temperature $\beta(\epsilon) = ds/d\epsilon$ vs. energy ϵ for the potential with exponent $\alpha = 1$ and for different rescaled soft potential radii $x_0 = r_0/R$. A crossover from the gravitational to the first order phase transition occurs when x_0 is increased from 0.020 to 0.022, where the jump in β disappears

sharp maximum of the density near $x = 0$ clearly indicates a collapsed phase.

c) When the cutoff radius exceeds some critical value $r_{0c}(\alpha)$ the gravitational phase transition for $r_0 < r_{0c}$ goes over into a normal first order transition for $r_0 > r_{0c}$ (see Fig. 4 for the softcore cutoff) as has been discussed in more detail in [5]. This critical cutoff radius $r_{0c}(\alpha)$

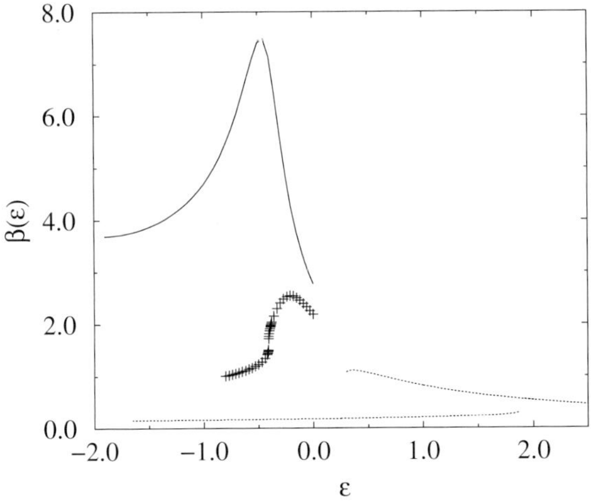

Fig. 5. Plots of inverse temperature $\beta(\epsilon) = ds/d\epsilon$ vs. energy ϵ for the rescaled soft potential radius $x_0 = r_0/R = 0.02$ and different potential exponents α. For $\alpha = 1/2$ (solid line), a typical maximum in $\beta(\epsilon)$ indicates the presence of a normal first-order phase transition meaning that $x_0 > x_{0c}(\alpha = 1/2)$; for $\alpha = 1$ (+), $\beta(\epsilon)$ experiences a vanishing jump which indicates that $x_0 \approx x_{0c}(\alpha = 1)$ (see also Fig. 4); finally for $\alpha = 3/2$ (dotted line), $\beta(\epsilon)$ experiences a noticeable jump revealing that $x_0 < x_{0c}(\alpha = 3/2)$. All together this suggests that r_{0c} increases with α

depends not only on the potential exponent α, but also on the nature of the cutoff itself and is different for the central excluded volume and the softcore cutoffs considered here. Therefore the dependence of $r_{0c}(\alpha)$ on α does not have such universal character as that of $\epsilon_c(\alpha)$ (Fig. 2) which appears to be virtually insensitive to the nature and range of a cutoff, provided that the cutoff is sufficiently small. We observe that the critical cutoff radius $r_{0c}(\alpha)$ is an increasing function of α (Fig. 5).

d) We remark that the normal first order transition that replaces the gravitational phase transition for $r_0 > r_{0c}$ is similar to first order phase transitions in other long-range interacting systems, e.g. in the mean field Potts model [3], where $\beta(\epsilon)$ is continuous with a $c_v = -[\beta)(\epsilon)]^2/[d\beta/d\epsilon] < 0$, since $d\beta/d\epsilon > 0$ and therefore $c_v < 0$ in an ϵ-interval. As said before, the relation between a negative specific heat and a first order phase transition in the ME has been emphasized in the gravitational literature for $\alpha = 1$ [2] and in the statistical mechanical literature only recently, especially through the work D. Gross [4].

4 Conclusions and Discussion

1. For $0 < \alpha < 3$, systems with attractive interparticle pair potentials $1/r^{\alpha}$, show collapse and a discontinuous behavior as a function of their energy in the microcanonical ensemble;
2. For $\alpha > 3$, the continuum approximation for the nonintegrable potentials $1/r^{\alpha}$ cannot be used, because of the importance then of short-range density fluctuations, and formally due to the divergence of $\int d\boldsymbol{r} r^{-\alpha}$ near $r = 0$;
3. The phase transition behavior depends crucially on the magnitude of the short range cut-off r_0. In fact, for all $0 < \alpha < 3$, there is a critical cut-off r_{0c}, so that for $r_0 > r_{0c}(\alpha)$ the $\beta(\epsilon)$ is continuous at $\epsilon = \epsilon^*$ and a (normal) first order transition can occur. The critical cutoff r_{0c} is an increasing function of α (see Fig. 5).
4. A qualitative discussion of the phase space behavior as a function of a cut-off can be found in a recent paper by Chavanis and Ispolatov [5]. In addition, we remark here that it seems physically reasonable that $r_{0c}(\alpha)$ increases with α, since with increasing α, i.e. with a deeper short range attractive well, a larger cut-off $r_{0c}(\alpha)$ is needed to diminish the effect of the collapse caused by the short range attraction. The $\epsilon_c(r_0)$ and $\epsilon_u(r_0)$, where the gravitational phase transition occurs, will approach each other with increasing r_0 and merge with ϵ^* at r_{0c}.
5. In fact, recent extensive Monte Carlo calculations for a number of systems with finite N [6] appear to confirm the validity of the use of the continuum and the saddle point approximations made in the basic approximate theory used for this class of systems. They also extend the results obtained in the $N \to \infty$ case to finite N systems, showing numerically that no essential changes in the phase behavior then occurs. These Monte Carlo results are also consistent with some earlier calculations by de Vega and Sanchéz [7].

Acknowledgment

This work was presented at the Winter School "Dynamics and Thermodynamics of Systems With Long-Range Interactions" in Les Houches, France. The authors gratefully acknowledge financial support from the School as well as from the Office of Basic Engineering Science of the US Department of Energy under grant number DE-FG02-88-ER13847, while I.I. is also thankful to FONDECYT, grant #1020052, for additional support.

References

1. I. Ispolatov, E. G. D. Cohen: Phys. Rev. Lett. **87**, 210601 (2000); Phys. Rev. E **64**, 056103 (2000)
2. See, e.g., T. Padmanabhan, Phys. Rep. **188**, 285 (1990)
3. J. Barré, D. Mukamel, S. Ruffo, Phys. Rev. Lett. **87**, 030601 (2001), I. Ispolatov, E. G. D. Cohen, Physica A **295**, 475 (2001)

4. D. Gross, *Microcanonical Thermodynamics*, World Scientific Lecture Notes in Physics, Vol.66, World Scientific Singapore (2001)
5. P.H. Chavanis, I. Ispolatov: Phys. Rev. E **66**, 036109 (2002)
6. L. Casetti, E. G. D. Cohen, in preparation
7. H. J. de Vega, N. Sanchéz, Nucl. Phys. B **625**, 409 (2002)
8. V. P. Youngkins, B. N. Miller, Phys. Rev. E. **6**2, 4583 (2000)

Nonextensivity: From Low-Dimensional Maps to Hamiltonian Systems

Constantino Tsallis[1], Andrea Rapisarda[2], Vito Latora[2], and Fulvio Baldovin[1]

[1] Centro Brasileiro de Pesquisas Físicas,
Rua Xavier Sigaud 150, 22290-180 Rio de Janeiro, RJ, Brazil
[2] Dipartimento di Fisica e Astronomia and Infn Universitá di Catania,
Corso Italia 57, 95129 Catania, Italy

Abstract. We present a brief pedagogical guided tour of the most recent applications of non-extensive statistical mechanics to well defined nonlinear dynamical systems, ranging from one-dimensional dissipative maps to many-body Hamiltonian systems.

1 Introduction

Thermodynamics and Boltzmann-Gibbs (BG) statistical mechanics constitute central formalisms of contemporary Physics. It is therefore of extreme importance to clearly establish under what exact conditions they are expected to apply, especially in what concerns the important and ubiquitous stationary state referred to as *thermal equilibrium*. This wisdom is more subtle than it looks at first sight. For instance, surprisingly enough, it is not yet exactly known the necessary and sufficient conditions for the celebrated BG distribution to correctly describe the thermostatistical state of a system (see, for instance, [1]). In order to settle such conditions, it is necessary to turn onto the microscopic dynamics of the system [2], would that be classical, quantum, relativistic, or any other one that appropriately describes the system under study.

Some of the central ingredients of BG statistical mechanics are the well known logarithmic entropy ($S_{BG} = -k\sum_{i=1}^{W} p_i \ln p_i$ in general, and $S_{BG}(\{p_i = 1/W\}) = k \ln W$ for equiprobability, where for simplicity we are referring here to the case of discrete distributions of probabilities), and its exponential weight $p_i \propto e^{-\beta E_i}$ ($\beta \equiv 1/kT$, $\{E_i\}$ being the energy spectrum) at equilibrium. This entropy has the remarkable property of extensivity for systems which are independent. More precisely, if we have a system composed by subsystems A and B such that $p_{ij}^{A+B} = p_i^A p_j^B$ $(\forall(i,j))$, then $S_{BG}(A+B) = S_{BG}(A) + S_{BG}(B)$. This property is at the basis of the standard understanding of thermodynamics. Such a property is not necessarily universal because it cannot be justified on general arguments (see, for instance, [3]). Strong evidence exists nowadays that it should not be generically assumed, contrarily to a widely spread, textbook belief among not few physicists. Very specifically, the *Boltzmann principle* $S_{BG}(\{p_i = 1/W\}) = k \ln W$ should not be used for any given specific system unless properly justified. It was Einstein [2] the first who warned about this. However, many physicists (see, for instance, [4]) have not yet done their way through this important concept. To be more precise, let us focus on N-body classical Hamiltonian d-dimensional systems including two-body interactions which do not present any mathematical complexity (such as an infinitely attractive potential) at short distances. It is acquired by now that the usual BG thermal equilibrium correctly describes the $t \to \infty$ stationary state for all systems whose interactions are *short-ranged* (e.g., if a two-body attractive potential decreases like $1/r^\alpha$ with $\alpha/d > 1$). In particular, in this case, it is irrelevant to consider first the $t \to \infty$ limit and then the $N \to \infty$ limit, or the other way around. The situation is much more subtle if we have *long-ranged* interactions (e.g., $0 \leq \alpha/d \leq 1$). In this case, the $\lim_{N\to\infty} \lim_{t\to\infty}$ still leads to the BG equilibrium, whereas the $\lim_{t\to\infty} \lim_{N\to\infty}$ generically does *not*. It happens, typically, that more than one basin of attraction exists in the space of the initial conditions. For some initial conditions the system directly goes to the BG statioanary state; for others it first goes to a non-BG state, remains there for a time which diverges with N, and only eventually goes to the BG state. In other words, depending on the size of the system and on its specific physical time scale, it might happen that the system remains in a quasi-stationary state different from the BG one for times longer than the age of the universe. In such cases, the only physically interesting situation is the non-BG one, which, perhaps for a vast class of systems, might be the one emerging within the nonextensive statistical mechanics we shall describe in the next Section. The present paper is a tutorial review of these concepts as they naturally emerge for three simple but important classes of systems, namely, (i) one-dimensional dissipative maps (e.g., the logistic map [5,6]), (ii) two-dimensional conservative maps (e.g., the standard map [7]), and (iii) classical many-body long-range-interacting Hamiltonians (e.g., the HMF model [8–12] and one of its generalizations, the so called $\alpha - XY$ model [13,14]). We shall not address here the many applications of nonextensive statistical mechanics available in the literature. The interested reader may refer to [15] for various reviews. Just as an illustration, let us mention some of those applications: turbulence [16,17], electron-positron annihilation [18], diffusion of *Hydra viridissima* [19], diffusion of quarks in gluon plasma [20], Levy and correlated anomalous diffusions [21], linguistics [22], economics [23,24], fluxes of cosmic rays [25], solar neutrinos [26], high energy particle collisions [27], self-organized criticality [28], among others. In one way or another, these phenomena seem to share long-range correlations in space/time, either long-range microscopic interactions, or long-range microscopic memory (nonmarkovian processes), or (multi)fractal boundary conditions, or, generically speaking, some mechanism which creates a scale-invariant hierarchical structure of some sort. Such possibility appears to emerge in classical many-body Hamiltonians and other nonlinear dynamical systems, everytime the Lyapunov spectrum approaches zero, hence chaos becomes impossible in the sense that the sensitivity to the initial conditions diverges less than exponentially in time, though a weak mixing can remain.

2 Mathematical Formalism

The first and main step in order to go from BG to nonextensive thermostatistics is to propose the use of an entropic form which generalizes that of BG, as follows [29]:

$$S_q \equiv k\frac{1-\sum_{i=1}^{W} p_i^q}{q-1} \quad (\sum_{i=1}^{W} p_i = 1;\ q \in \mathbb{R})\ . \tag{1}$$

(for simplicity, and without loss of generality, we shall adopt $k = 1$ from now on). This nonnegative form can be conveniently rewritten as

$$S_q = \langle \ln_q \frac{1}{p_i} \rangle\ , \tag{2}$$

where $\langle ... \rangle \equiv \sum_{i=1}^{W}(...)p_i$, and the *q-logarithm* function is defined as

$$\ln_q \equiv \frac{x^{1-q}-1}{1-q} \quad (\ln_1 x = \ln x;\ x \geq 0)\ . \tag{3}$$

The inverse function is the *q-exponential* one, given by

$$e_q^x \equiv [1+(1-q)\,x]^{\frac{1}{1-q}} \;\; (e_1^x = e^x)\ , \tag{4}$$

solution of $dy/dx = y^q$ with $y(0) = 1$. S_q satisfies several remarkable properties, such as *concavity* for all $\{p_i\}$ ($\forall q > 0$), *stability* with regard to $\{p_i\}$ ($\forall q > 0$) [30], and, for A and B independent subsystems, the following *pseudo-extensivity*:

$$S_q(A+B) = S_q(A) + S_q(B) + (1-q)S_q(A)S_q(B)\ , \tag{5}$$

from which the denomination *nonextensive* comes. For completeness, it is convenient to mention at this point that similar entropies are the Renyi one, defined as $S_q^R = [\ln \sum_{i=1}^{W} p_i^q]/[1-q]$, and the normalized one $S_q^N = S_q/[\sum_{i=1}^{W} p_i^q]$ (introduced independently by Abe and Rajagopal [31] and by Landsberg and Vedral [32]. Renyi entropy is very useful in the context of multifractal analysis (of nonlinear chaotic systems, for example). However, S_q^R and S_q^N are *not* concave for all positive values of q, *nor* are they stable [30]. These facts constitute severe drawbacks for basing a thermodynamical formalism on them. It is however convenient to have in mind that, since they are related to each other through monotonic functions $S_q^R = \ln[1+(1-q)S_q]/[1-q]$ and $S_q^N = S_q/[1+(1-q)S_q]$, they become extremized by the *same* probability distributions (assuming the supplementary constraints for optimization to be the same).

3 Applications to Low-Dimensional Maps

Low-dimensional systems are a field in which the nonextensive statistical mechanics has found one of its first and most proficuous applications. What makes low-dimensional systems extremely attractive is the simplicity of their dynamics and the possibility to have direct access to the phase space, sometimes even to plot a visual image of the entire phase space. In fact, the iterative rules of low-dimensional maps can be easily implemented on a computer and have provided useful methods to perform "numerical experiments" to check for the validity of the nonextensive formalism. As mentioned in the previous section, the BG statistics is expected to fail when the sensitivity to initial conditions is not exponential. Therefore, different authors have looked and found applications of the nonextensive formalism to cases in which the maps have a nonstandard sensitivity to initial conditions, as indicated by the vanishing of the Lyapunov exponents. Different techniques have been produced in order to calculate the 'entropic parameter' q, and it turns out that this parameter characterizes 'classes of universality' of systems, at least for some of their statistical behaviors. For pedagogical reasons here we will discuss only some of the results obtained for the logistic map, as representative of a dissipative system, and for the standard map, as an example of a conservative one.

3.1 Dissipative Maps: The Logistic Map

The logistic map is one of the simplest one-dimensional dissipative system one can imagine. It has been intensively studied and it has led to important developments in chaos theory [5]. Since it has all fundamental characteristics of non-conservative systems, the logistic map is often taken as a paradigmatic example; in other words it can be considered as the Ising model of non-linear dynamical systems. The logistic map can be described by the following iterative rule[1]

$$x_{t+1} = f_a(x_t) = 1 - ax_t^2 \quad -1 \leq x_t \leq 1;\ 0 \leq a \leq 2; \tag{6}$$

and it shows different regimes according to the value of the control parameter a [5,33]. For small values of a the map is regular since it has non-positive Lyapunov exponents (see Fig. 1). In particular, for $a < \frac{3}{4}$ the iterates of the logistic map converge to a single fixed point $x^* = f_a(x^*)$, whose value depends on a. At $a = \frac{3}{4}$ we have $\lambda = 0$ and we observe the first pitchfork bifurcation: for a value slightly larger than $\frac{3}{4}$ the iterates converge to a cycle-2. The number of points of the attractor $f_a(x)$ keeps doubling at distinct, increasing values of the parameter a, until a critical value $a_c = 1.401155198...$ is reached, where the attractor becomes infinite. This is known in the literature as the period-doubling route to chaos. In fact, beyond a_c the behavior of the logistic map is chaotic (positive Lyapunov exponent λ) for most of the values of a. At the chaos threshold $a = a_c$, λ is once again zero, and this point is the well known *edge of chaos* which has been the subject of intensive debates about the definition of what a complex systems is

[1] An equivalent, perhaps more popular version of the logistic map is $x_{t+1} = rx_t(1-x_t)$, with $0 \leq x_t \leq 1$ and $1 \leq r \leq 4$. However, we notice that form (6) is easily generalized to other universality classes of order $z \geq 1$: $x_{t+1} = 1 - a|x_t|^z$. In this way, a z-logistic map is in fact representative of the universality class of unimodal maps of order z; i.e., of one-dimensional maps with a single maximum of order z.

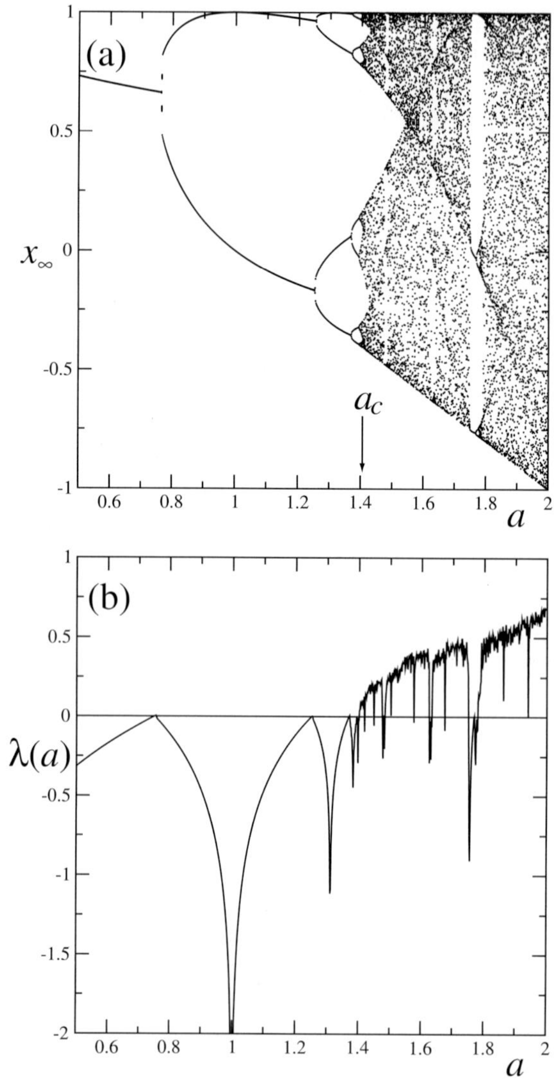

Fig. 1. (a) Attractor of the logistic map as a function of a. The edge of chaos is at the critical value $a_c = 1.401155198...$ (b) Lyapunov exponent λ as a function of a.

[34,35]. It is exactly at the edge of chaos and at all other points where λ is zero (as at period doublings, but also for instance at inverse tangent bifurcations [5]), that a generalization of the standard BG entropy will be fruitful, as we show below.

Sensitivity to initial conditions. Let us start by explaining better the meaning of the crucial concept of sensitivity to initial conditions. For all the cases in which the Lyapunov exponent λ is positive (negative) we expect on the average

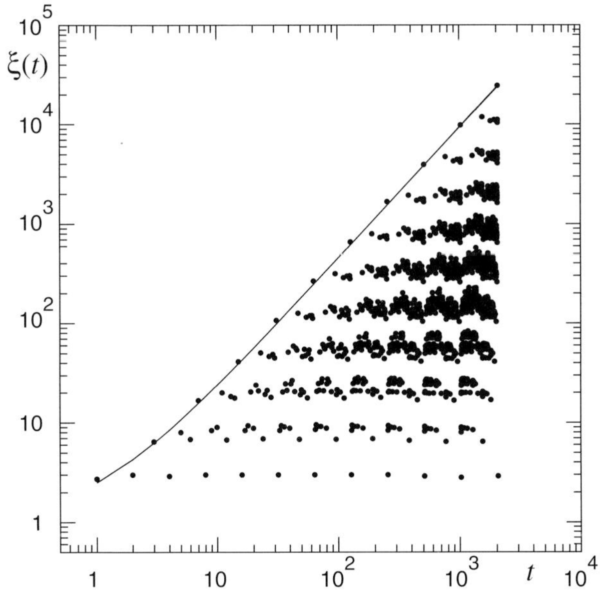

Fig. 2. Sensitivity to initial conditions for the logistic map at $a = a_c$. The dots represent $\xi(t)$ for two initial data started at $x_0 = 0$ and $x'_0 \sim 10^{-8}$. The solid line is the function in (8), with $q = 0.2445...$ and $\lambda_q = \ln \alpha / \ln 2 = 1.3236....$

an exponential increase (decrease) of any small initial distance $\xi(t) \equiv \frac{|\mathbf{x}_t - \mathbf{x}'_t|}{|\mathbf{x}_0 - \mathbf{x}'_0|}$ to hold. We typically have

$$\xi(t) = \exp(\lambda t), \tag{7}$$

where $\mathbf{x}_t$ and $\mathbf{x}'_t$ are the positions at time t of two initially close trajectories that may be considered, for the sake of generality, in a d-dimensional phase space. In the case of the logistic map $d = 1$. This case is referred in the literature as *strongly sensitive* (*insensitive*) to initial conditions. What is not usually reported in textbooks is what happens when λ, as defined by (7), is equal to zero. In [36] was proposed that in such situations the system still exhibits sensitivity to initial conditions, though in a form that is the q-generalization of (7):

$$\xi(t) = \exp_q(\lambda_q t) \equiv [1 + (1-q)\lambda_q t]^{\frac{1}{1-q}} \quad (q \in \mathbb{R}). \tag{8}$$

This equation recovers (7) for $q = 1$, and displays a quite rich spectrum of possiblities according to the sign of λ_q and q. Let us concentrate on the case $\lambda_q > 0$, $q < 1$. In this case the system presents a *weak sensitivity* to initial conditions, in the sense that two initially close trajectories diverge in time as a power-law instead than exponentially. A weakly sensitive system is well described once q and λ_q are given. In [36] q was calculated numerically for the logistic map at the edge of chaos, observing that the upper bound of $\xi(t)$ lies on a line that is precisely a power-law (see Fig. 2); furthermore, it was conjectured that a value $q = 0.2445....$ could be deduced from the Feigenbaum's constant α. These results were recently confirmed in [37], where it was proved that, at the edge of chaos the upper bound of $\xi(t)$ has exactly the form of (8), with $q = 0.2445...$ and $\lambda_q = \ln \alpha / \ln 2 = 1.3236....$ In fact, as a consequence of the Feigenbaum's renormalization group recursion relation, two close initial conditions starting in the neighborhood of the origin $x = 0$, produce, each $t = 2^n - 1$ $(n = 0, 1, 2, ...)$, a dominant power-law separation of their iterates that has the form (8). After these particular steps, the iterates reproduce self-similar sequences, that are also power-laws. The effect of starting the initial points in other regions of the phase space instead that in the vicinity of $x = 0$ is just a shifting in time of Fig. 2.

The renormalization group approach has proven powerful also at other points where the Lyapunov exponent is zero. In [38] the same q-exponential form (8) has been exactly obtained for the sensitivity to initial conditions at pitchfork and inverse tangent bifurcations, now with other values for q and λ_q. It is important to notice that the previous scheme of application of (8) generalizes to the other universality classes of unimodal maps of order[2] $z > 1$.

Entropy production. The analysis of the entropy production provides an alternative approach to the same problem and a second way to characterize the q-index at the edge of chaos. Instead of considering two trajectories initially close, we take a distribution of initial conditions localized in a tiny region of the phase space, we let this distribution evolve in time according to (6), and we study the increase of the entropy [39,6]. In the case of the logistic map, we can partition the phase space interval $-1 \leq x \leq 1$ into W equal cells, and consider an initial distribution of N points placed at random inside one of the cells; the initial cell is also chosen randomly in the partition. The normalized number of points that occupy a cell defines a probability distribution: $p_i \equiv n_i/N$ $(\sum_{i=1}^{W} p_i = 1)$, and as the system evolves, from the probabilities $p_i(t)$ we can calculate the entropy production $S_q(t)$, via (1). In the chaotic regime the BG entropy ($q = 1$) exhibits, before a saturation due to the finiteness of W, a linear increase in time, and in this stage the entropy production rate (that is expected to be equal to the Kolmogorov-Sinai entropy [40]) coincides with the positive Lyapunov exponent [41]. This is shown in Fig. 3 for the case $a = 2$, where $S_q(t)$ is reported for three different values of q. As t evolves, $S_q(t)$ increases (in all cases bounded by $\frac{W^{1-q}-1}{1-q}$, or $\ln W$ when $q = 1$), but only the curve for $q = 1$ shows a clear linear behavior with a slope equal to the Lyapunov exponent $\lambda = \ln 2$ [39]. For $q < 1$ the curve is convex, while for $q > 1$ the curve is concave. The slope in the linear stage does not depend on the dimension of the cells of the partition and on the size of the initial distribution. Therefore, in the chaotic regime the standard B-G entropy is the only q-entropy that displays a finite entropy production rate $\lim_{t\to\infty\, W\to\infty} S_q(t)/t$. We wish now to repeat the same analysis at the edge of chaos, where λ is zero. Due to the power-law sensitivity to initial conditions and to the fractality of phase space (the attractor now is a multifractal) we expect that anomalous behavior may be observed. In Fig. 4 we consider $a = a_c$ and we plot $S_q(t)$ for four different values of q; the curves are obtained with a large W

[2] See previous footnote.

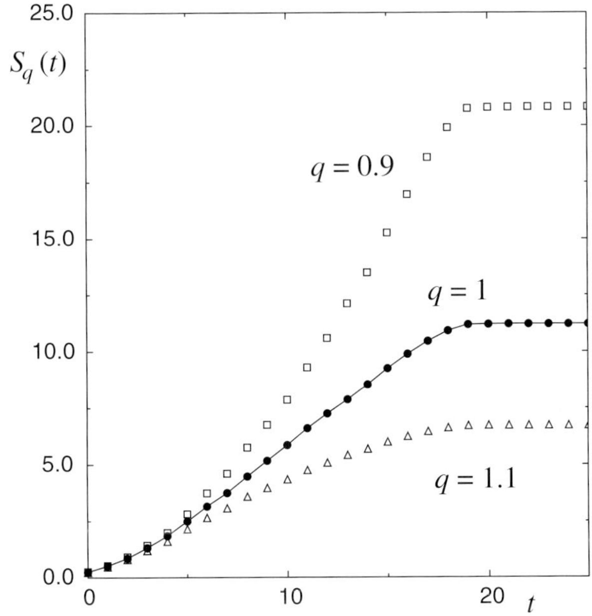

Fig. 3. Time evolution of S_q for the logistic map with $a = 2$. We consider three different values of q. The results are averages over 500 runs with $N = 10^6$ and $W = 10^5$.

and with an average over many different runs since the complexity of the phase space originates many fluctuations. Consistently with the value of q extracted for the sensibility to initial conditions, the result is that the growth of $S_q(t)$ is found to be linear when $q = 0.2445...$, while for $q < q_c$ ($q > q_c$) the curve is convex (concave). This behavior is similar to the one of Fig. 3, with a major difference: the linear growth is not at $q = 1$, but at $q = 0.2445...$ [6].

Multifractal analysis. There is also a third, geometrical method that allows us to calculate the q of the logistic map. This method is based only on the description of the multifractal attractor existing at $a = a_c$, and gives exactly the same value we have already obtained. The multifractal attractor can be characterized by using the multifractal function $f(\alpha)$ [33]. This function is defined in the interval $[\alpha_{min}, \alpha_{max}]$, and its maximum equals the fractal or Hausdorff dimension d_f. The value of q can be calculated from [36]:

$$\frac{1}{1-q} = \frac{1}{\alpha_{min}} - \frac{1}{\alpha_{max}} , \tag{9}$$

where α_{min} (α_{max}) are the two values for which $f(\alpha)$ is defined. In particular for the logistic map at the edge of chaos $\alpha_{min} = 0.380...$, $\alpha_{max} = 0.755...$ and we have again $q = 0.2445...$.

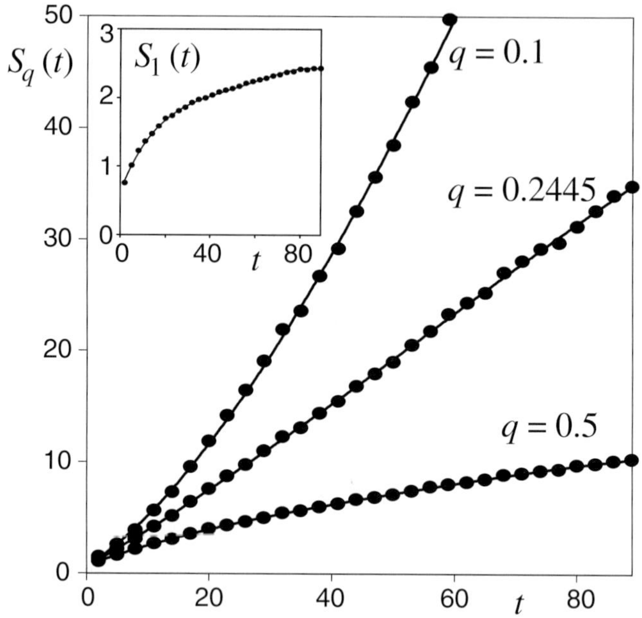

Fig. 4. Time evolution of S_q for the logistic map with $a = a_c$. We consider four different values of q. The case $q = 1$ is in the inset. $W = N = 2.5\ 10^6$ and results are averages over 15115 runs.

3.2 Conservative Maps: The Standard Map

Conservative systems present quite different statistical properties than dissipative ones. One of these differences is that the transition from chaoticity to regularity happens in the phase space without an 'edge of chaos', but in a rather complicated way as described by the Kolmogorov-Arnold-Moser (KAM) theorem [42], see Fig. 5e-g). Conservative maps may be derived, for example, from Hamiltonian systems. In this case the original system is characterized by an even number of dimensions $d = 2n$, where n is the number of degrees of freedom. If the Hamiltonian H is time-independent the dimension can be reduced by one. Moreover, as in statistical mechanics we are interested mostly in *recurrent trajectories* (i.e., in those trajectories that come back again and again, indefinitely, to any part of the phase space they have once visited), we can take a *Poincaré section* of the phase space cutting transversally the constant-energy hypersurface and considering the successive intersections of each orbit with this transversal surface. In this way we reduce the dimension of the phase space to $d_M = 2n - 2$ and we obtain a great numerical simplification, as we can now analyze a discrete-time iteration map instead of a continuous-time system of differential equations. The map thus obtained may be shown to be *simplectic* (see, e.g. [42]); this implies that the d_M Lyapunov exponents are coupled in pairs, where each member of the pair has the opposite of the other.

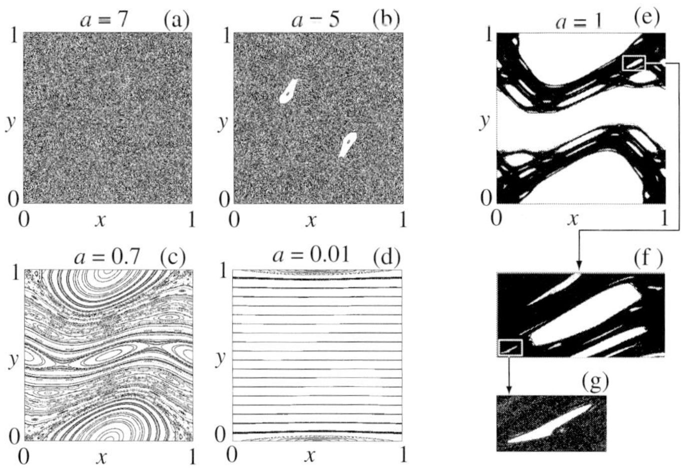

Fig. 5. (a)–(d) Phase portrait of the standard map for typical values of a. $N = 20 \times 20$ orbits (black dots) were started with a uniform distribution in the unit square and traced for $0 \leq t \leq 200$. (e)–(g) Islands-around-islands. (e): $N = 100 \times 100$ initial data were started inside a small square of side $\sim 10^{-2}$ and traced for $0 \leq t \leq 5000$. (f) is a magnification of the island inside the rectangle in (e). (g) is a magnification of the island inside the rectangle in (f).

The minimum dimension that allows for the application of this scheme is $d_M = 2$ ($n = 4$), and a paradigmatic example of this kind, that plays for conservative systems a similar role the logistic map plays for dissipative ones, is the standard map, that may be characterized by the equations[3]:

$$
\begin{aligned}
x_{t+1} &= y_t + \frac{a}{2\pi} \sin(2\pi x_t) + x_t \pmod{1}, \\
y_{t+1} &= y_t + \frac{a}{2\pi} \sin(2\pi x_t) \pmod{1},
\end{aligned}
\tag{10}
$$

where $a \in \mathbb{R}$ ($t = 0, 1, \ldots$). The system is integrable when $a = 0$, while, roughly speaking, chaoticity increases when $|a|$ increases (see Fig. 5a-d).

As for the logistic map, we expect that anomalous effects may be produced as the complexity of the phase space increases (i.e., when we pass from the chaotic to the regular regime reducing $|a|$), and as this happens, we would like to see how the nonextensive formalism generalizes the BG one, at least for some of the statistical variables we have previously described.

Sensitivity to initial conditions. In the case of the standard map, the sensitivity to initial condition depends dramatically on the position in the phase

[3] The map is also known as the kicked rotator map.

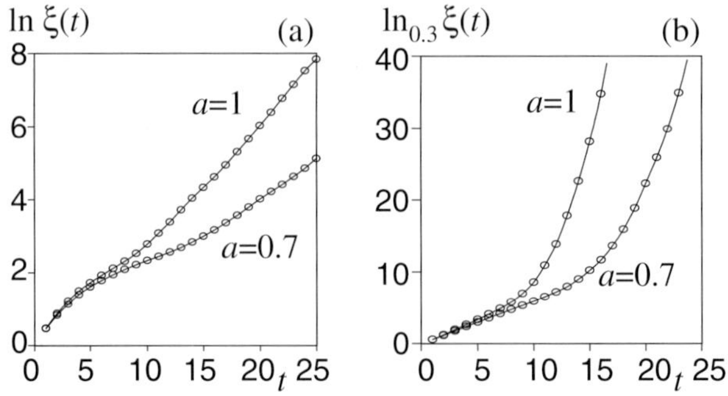

Fig. 6. Average sensitivity to initial conditions of the standard map for small values of a. $|\mathbf{x}_0 - \mathbf{x}'_0| \sim 10^{-9}$ and results are averages over 5000 runs. An initial separation of order In (a) $\ln(\xi(t))$ displays a linear increase only after a certain time. Correspondingly, in (b) $\ln_{0.3}(\xi(t))$ exhibits a linear increase for the initial steps.

space. Chaotic regions produces exponential separation of initially close trajectories, while regular regions exhibit linear separation of initially close orbits. At the border between these regions, appear fractal-like structures of islands-around-islands (see, e.g., [43], see Fig. 5e-g). We notice however that in statistical mechanics we are mainly interested in extracting global averages behaviors, so that we can try to calculate an average $\xi(t)$ by sampling uniformily the whole phase space. When this experiment is performed for rather small values of $|a|$, as we show in Fig. 6, a *crossover* in time between two different regimes occur [44]. Fig. 6a displays in fact a linear increase of $\ln(\xi(t))$ only after a certain iteration step. Correspondingly, Fig. 6b exhibits a linear increase of $\ln_q(\xi(t))$ precisely for those steps where $\ln(\xi(t))$ is concave (numerically it was found $q \simeq 0.3$). Moreover, the crossover time increases when $|a|$ decreases. This means that two different behaviors dominates the average sensitivity to initial conditions for different time. Initially, because of the predominance of fractal-like structures in the phase space, the sensitivity to initial conditions is power-law; then, after a certain time, the sensitivity to initial conditions becomes exponential due to the rapidity of the exponential growth if compared with the power-law. If we wish to model such a behavior with a single function, we can first observe that exponential and power-law sensitivity to initial conditions may be viewed as solution of the differential equations $\dot{\xi} = \lambda\xi$ and $\dot{\xi} = \lambda_q \xi^q$ respectively. A differential equation describing a crossover in time is then: $\dot{\xi} = \lambda\xi + (\lambda_q - \lambda)\xi^q$ $(q < 1,\ \ \lambda, \lambda_q > 0)$. The solution [45]

$$\xi(t) = \left[1 - \frac{\lambda_q}{\lambda} + \frac{\lambda_q}{\lambda} e^{(1-q)\lambda t}\right]^{\frac{1}{1-q}}, \tag{11}$$

presents, in the case $0 < \lambda \ll \lambda_q$, three asymptotic behaviors, namely (i) linear: $\xi \sim 1 + \lambda_q t$, for $0 \leq t \ll t_{cross1} \equiv \frac{1}{(1-q)\lambda_q}$; (ii) power-law: $\xi \sim [(1-q)\lambda_q]^{\frac{1}{1-q}}\, t^{\frac{1}{1-q}}$, for $t_{cross1} \ll t \ll t_{cross2} \equiv \frac{1}{(1-q)\lambda}$; (iii) exponential: $\xi \sim (\frac{\lambda_q}{\lambda})^{\frac{1}{1-q}} e^{\lambda t}$, for $t \gg t_{cross2}$.

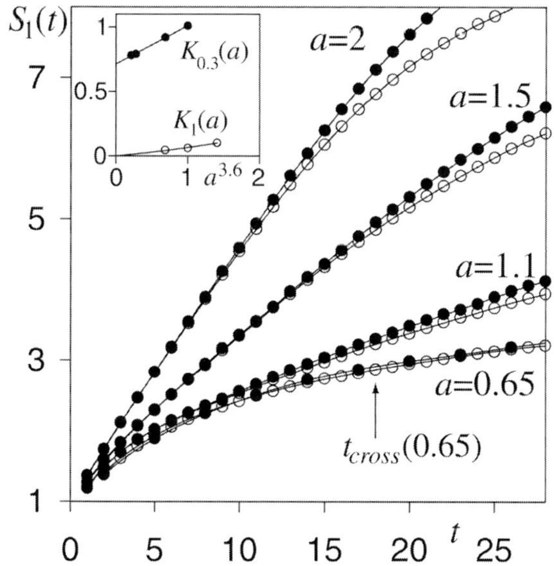

Fig. 7. For the standard map we show $S_1(t)$ for different values of $a \leq 2$ (5000 to 7000 runs were averaged). Full circles correspond to $N = W = 1000 \times 1000$ for $a = 2, 1.5, 1.1$ ($N = W = 5000 \times 5000$ for $a = 0.65$); empty circles correspond to $N = W = 448 \times 448$ for $a = 2, 1.5, 1.1$ ($N = W = 2236 \times 2236$ for $a = 0.65$). Inset: Slopes of $S_1(t)$ ($K_1(a)$) and of $S_{0.3}(t)$ =20 ($K_{0.3} \simeq 0.71 + 0.30\ a^{3.6}$) in their linear regimes (see also Fig. 8). The lines are guides to the eye.

Entropy production. If the connection between sensitivity to initial condition and entropy production that we have observed for the logistic map is correct, we should observe a crossover between different statistical regimes even when we analyze the entropy production of the standard map. This has been done in [46], with a numerical experiment analogous to the one previously described for the logistic map and that is difficult to perform because of the complexity of the phase space and the related effects of rapid saturation of the entropy for W finite. In Figs. 7 and 8 we display how, for small values of $|a|$ and before saturation begins, the entropy production of $S_1(t)$ becomes linear only after a stage where $S_q(t)$ is linear, with, once again, a value $q \simeq 0.3$. Notice that the crossover time tends to infinity when $|a|$ goes to zero, so that in this limit the stage characterized by the power-law sensitivity to initial conditions and by this nonstandard entropy production lasts for infinite time.

4 Applications to Many-Particle Systems

4.1 Hamiltonian Nonextensive Systems: The HMF Model

In the last years there has been an intense research activity in investigating microscopic dynamical features in connection with macroscopic features. In par-

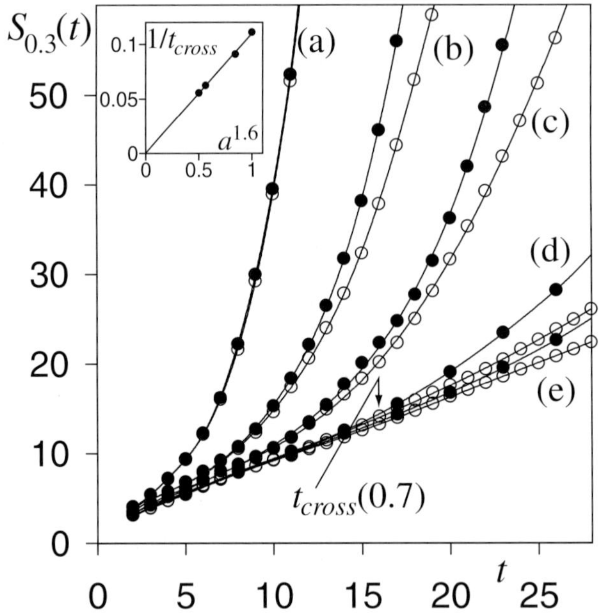

Fig. 8. For the standard map we show $S_{0.3}(t)$ for: $a = 1.5$ (a); $a = 1.1$ (b); $a = 0.9$ (c); $a = 0.7$ (d); $a = 0.65$ (e) (5000 to 7000 runs were averaged). Full circles correspond to $N = W = 1000 \times 1000$ for (a), (b), (c), $N = W = 5000 \times 5000$ for (d), (e); empty circles correspond to $N = W = 448 \times 448$ for (a), (b), (c), $N = W = 2236 \times 2236$ for (d), (e). Inset: $t_{cross}(a)$ defined as the intersection of the linear part (before it starts bending) and a standard extrapolation of the bended part of the curves $S_{0.3}(t)$. Notice that $t_{cross}(a)$ diverges for $a \to 0$. The lines are guides to the eye. Our results suggest $\lim_{t\to\infty} \lim_{a\to 0} \lim_{W\to\infty} \lim_{N\to\infty} \frac{S_{0.3}(t)}{t} \simeq 0.71$ for $q = q^* \simeq 0.3$, whereas this limit vanishes (diverges) for $q > q^*$ $(q < q^*)$.

ticular several authors have focused their attention in characterizing from a dynamical point of view macroscopic phase transitions [47–51]. It has been found, for example, that in correspondence of a phase transition there is a peak in the chaoticity of the microscopic dynamics. In general chaoticity is of fundamental importance for any system to reach the standard equilibration. Anomalies should be expected when this does not occur, since ergodicity breaking might happen and the system can remain trapped in some region of phase space. In this respect a model whose features have been particularly interesting is the Hamiltonian Mean Field model, from here on referred as HMF model, and its generalizations where the range of the interaction can be changed. The model was introduced in ref [8] and extensively studied in refs [8–11,13,14]. In this volume one chapter is dedicated to its equilibrium properties both from a dynamical and from a thermodynamical point of view [12]. In the present section we consider only the anomalies found in the out-of-equilibrium regime before equilibration to the Boltzmann-Gibbs regime and how these behavior can be related to nonextensive statistics [29].

The HMF model describes a system of N planar classical spins interacting through an infinite-range potential [8]. The Hamiltonian can be written as:

$$H = K + V = \sum_{i=1}^{N} \frac{{p_i}^2}{2} + \frac{1}{2N} \sum_{i,j=1}^{N} [1 - cos(\theta_i - \theta_j)] \ , \tag{12}$$

where θ_i is the ith angle and p_i the conjugate variable representing the angular momentum or the rotational velocity since unit mass is assumed. The interaction is the same as in the ferromagnetic XY model [3], though the summation is extended to all couples of spins and not restricted to first neighbors. Following the Kac procedure [12], the coupling constant in the potential is divided by N. This makes H only formally extensive ($V \sim N$ when $N \to \infty$) [29,15,13,14], since the energy remains non-additive, i.e. the system cannot be trivially divided in two independent sub-systems. The canonical analytical solution of the model predicts a second-order phase transition. At low energy density the system is in a ferromagnetic phase characterized by a magnetization $M \sim 1$, where M is the modulus of

$$\mathbf{M} = \frac{1}{N} \sum_{i=1}^{N} \mathbf{m}_i \ , \tag{13}$$

with $\mathbf{m}_i = [cos(\theta_i), sin(\theta_i)]$. Increasing the energy beyond a critical value, the spins become homogeneously oriented on the unit circle and $M \sim 0$. The *caloric curve*, i.e. the dependence of the energy density $U = E/N$ on the temperature T, is given by $U = \frac{T}{2} + \frac{1}{2}(1 - M^2)$ [12] and shown in Fig. 9 as a full curve. The critical point is at energy density $U_c = 0.75$ corresponding to a critical temperature $T_c = 0.5$ [8]. The dynamical behavior of HMF can be investigated in the microcanonical ensemble by starting the system with water bag initial conditions, i.e. $\theta_i = 0$ for all i ($M = 1$) and velocities uniformly distributed, and integrating numerically the equations of motion [9]. As shown in Fig. 9, microcanonical simulations are in general in good agreement with the canonical ensemble, except for a region below U_c, where it has also been found a dynamics characterized by Lévy walks, anomalous diffusion [10] and a negative specific heat [11]. Ensemble inequivalence and negative specific heat have also been found in self-gravitating systems [52], nuclei and atomic clusters [53–56]. In our case such anomalies emerge as dynamical transient features, although in a generalized version of the HMF model [51,12] it seems that they remain also in the BG equilibrium regime. In order to understand better this disagreement we focus on a particular energy value, namely $U = 0.69$ below the critical energy, and we study the time evolution of temperature, magnetization, and velocity distributions.

In Fig. 10 we report the temporal evolution of $2\langle K\rangle/N$ for $U = 0.69$ and $N = 500$. This quantity, at equilibrium, is expected to coincide with the temperature ($\langle\cdot\rangle$ denotes time averages). The figure shows that when the system is started with out of equilibrium initial conditions (water bag in our case) rapidly reaches a metastable quasi-stationary state (QSS) which does not coincide with the canonical prediction. After a short transient time,not reported here, $2\langle K\rangle/N$ shows a plateau corresponding to a N-dependent temperature $T_{QSS}(N)$ (and

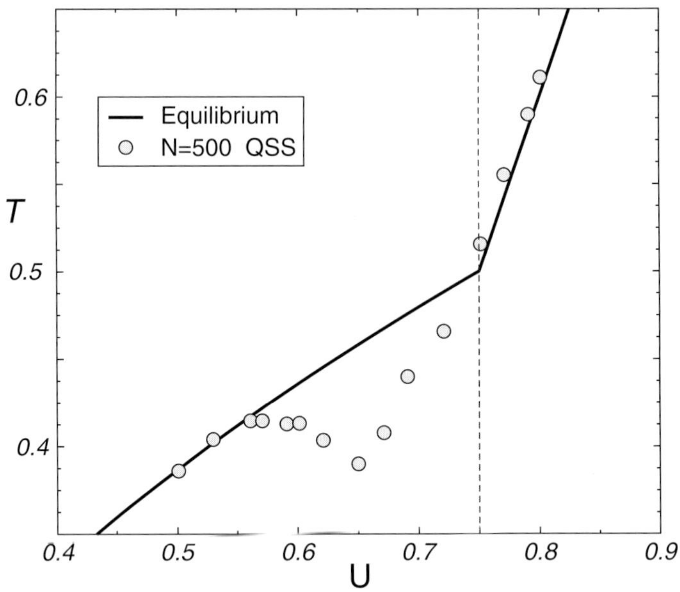

Fig. 9. We compare the equilibrium caloric curve of the HMF model (full curve) with the numerical simulation for a system of size N=500. The numerical integration is followed up to a short time, when the system is in the metastable quasi-stationary state (QSS).

$M_{QSS} \sim 0$, $M \to 0$ if $N \to \infty$) lower than the canonical temperature. This metastable state needs a long time to relax to the canonical equilibrium state with temperature $T_{can} = 0.476$ and magnetization $M_{can} = 0.307$. In the same figure we plot a typical single-particle motion in the two insets. In panel (b) the motion is characterized by anomalous diffusion and Lévy walks with constant velocity [10]. This motion is well described by a model introduced in [57]. These features change at equilibration, see panel (c) where the motion is a standard random walk and diffusion becomes normal. In [10] it has been found that the lifetime of the metastable QSS and the crossover time from anomalous to normal diffusion do coincide within the numerical accuracy adopted. On the other hand, the duration of the plateau increases with the size of the system. In fact the lifetime of QSS has a linear dependence on N as shown in Fig. 11 where simulations for different out-of-equilibrium initial conditions are plotted. These numerical results indicate that the two limits $t \to \infty$ and $N \to \infty$ do not commute: if the thermodynamic limit is performed before the infinite time limit, the system does not relax to the BG equilibrium and the anomalies discussed above remain forever . Indeed similar features have also been found for spin glasses [58,59] and granular matter [64,60] and could probably be interpreted within a general theoretical scenario. When N increases $T_{QSS}(N)$ tends to a definite value which is $T_\infty = 0.38$. The latter is obtained analytically as the metastable prolongation, at energies just below $U_c = 0.75$, of the high-energy branch charaterized

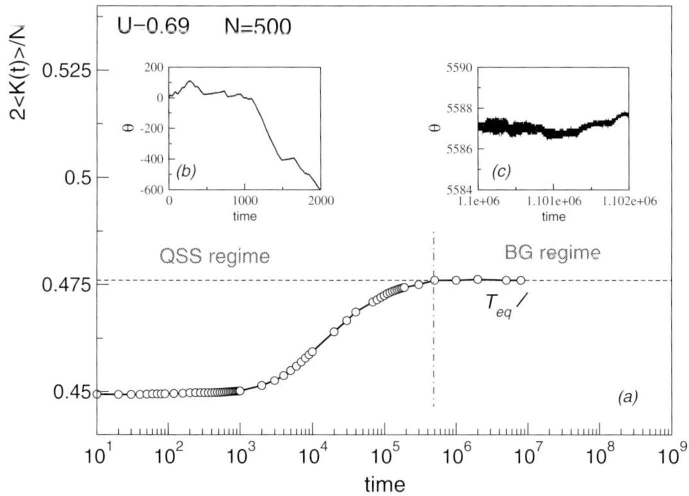

Fig. 10. For the HMF model at U=0.69 and N=500, we plot the temperature, calculated by means of the average kinetic energy, as a function of time. Two plateaux emerge from the calculation. The first one is the metastable quasi-stationary state (QSS), while the second one corresponds to the Boltzmann-Gibbs (BG) equilibrium regime. In the inset (b) we show the angle vs time of a typical rotor in the metastable regime: the motion shows Lévy walks and anomalous diffusion [10]. In the inset (c) we show the single-particle motion in the BG equilibrium regime: the motion is a random walk.

by $M = 0$. In [61] we have also found that $[T_{QSS}(N) - T_\infty] \propto N^{-1/3}$. This scaling, exploiting the expression for the caloric curves, implies also a scaling of the magnetization, which goes to zero like $M_{QSS} \propto N^{-1/6}$. A further numerical check of this law was obtained in [62].

It is important at this point to discuss the behavior of the Lyapunov exponents as a function of N and of time in order to investigate if also this measure shows some anomaly. In fact this is just the case: also the largest Lyapunov exponent λ has two temporal regimes and in particular two plateaux. We plot in Fig. 12 the time evolution of λ for the case $U = 0.69$ and $N = 500$. The numerical simulation averaged over 350 events shows that in the metastable QSS regime the largest Lyapunov exponent is smaller than at the BG equilibrium. In the inset we report the scaling of λ in the QSS regime as a function of the size: increasing N, λ_{QSS} tends to zero as $N^{-1/9}$ [62]. The scaling law for the Lyapunov exponent can be related to the scaling of the magnetization as discussed in [62]. The fact that $\lambda \to 0$ for $N \to \infty$ implies that mixing is negligible as N increases and one expects anomalies in the relaxation process in the sense advanced by Krylov [63] and in a very similar way to the case of the maps discussed in the previous section.

In order to check relaxation in HMF not only the temperature, the second moment, but also the higher moments of the velocity distribution should coincide with those of the BG equilibrium one. In Fig. 13 we focus our attention on the ve-

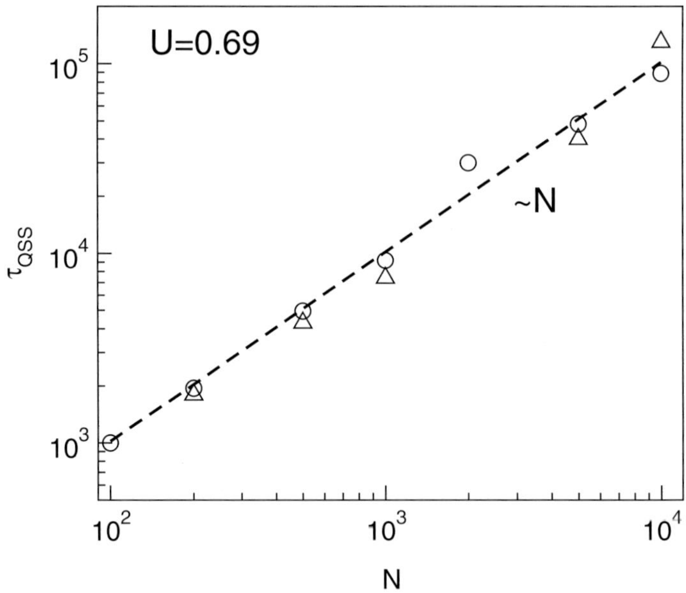

Fig. 11. For the HMF model we plot the lifetime of the metastable quasi-stationary state, τ_{QSS}, for U=0.69, as a function of N. Different out-of-equilibrium initial conditions are indicated by different symbols. The figure shows that the duration of these states diverges with N.

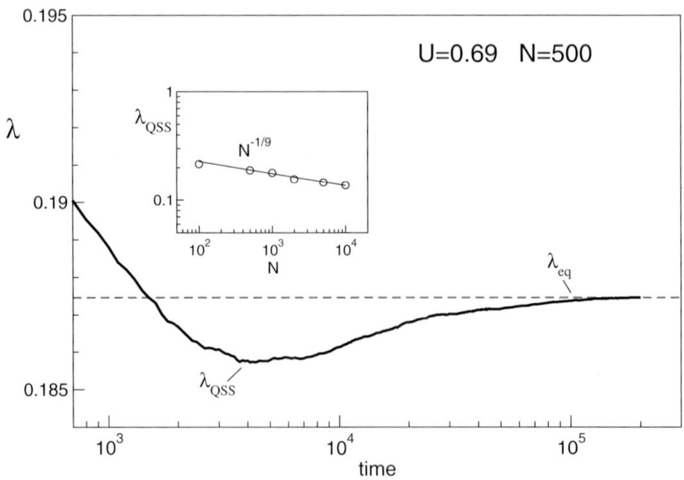

Fig. 12. We plot the largest Lyapunov exponent, λ, vs time for the HMF model at $U = 0.69$ and $N = 500$. Two plateaux are found also in this case. In the inset we show that λ_{QSS} goes to zero as $N^{-1/9}$. This behavior can be explained with the anomalous fluctuations of the magnetization. See text for further details.

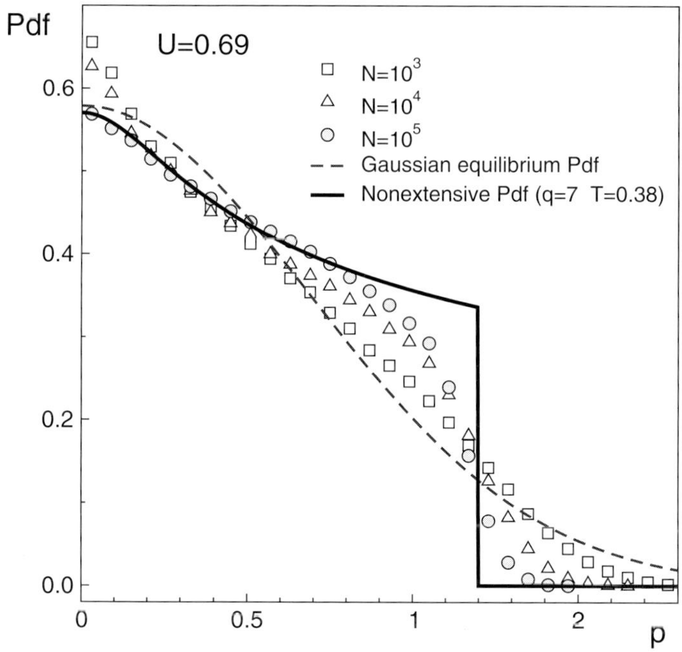

Fig. 13. We plot the velocity probability distribution functions of the HMF model for U=0.69 and different N values (open symbols). The Gaussian equilibrium shape is also plotted (dashed curve) together with the fit with the nonextensive pdf formula obtained for $q = 7$ and $T = 0.38$ (full curve).

locity probability distribution functions (pdfs). The figure shows that the initial uniform velocity pdfs, quickly acquire and mantain for the complete duration of the metastable state a *non-Gaussian shape* . In particular the velocity pdf of the QSS is wider than a Gaussian (drawn as dashed curve) for small velocities, but shows a faster decrease for $p > 1.2$. The enhancement for velocities around $p \sim 1$ is consistent with the anomalous diffusion and the Lévy walks (with average velocity $p \sim 1$) observed in the QSS regime [10]. The following rapid decrease for $p > 1.2$ is due to conservation of total energy. The stability of the QSS velocity pdf can be explained by the fact that, since for $N \to \infty$, $M_{QSS} \to 0$ also the force on the spins tends to zero with N, being $F_i = -M_x sin\theta_i + M_y cos\theta_i$. If N is finite, one has only a small random force, which makes the system eventually evolve into the usual Maxwell-Boltzmann distribution after some time. When this happens, Lévy walks disappear and anomalous diffusion leaves place to Brownian diffusion [10], see insets in Fig. 10. Nonextensive thermodynamics is able to reproduce the non-Gaussian pdf reported in Fig. 13. In fact the nonextensive formalism provides, for the canonical ensemble, a q-dependent power law distribution in the variables p_i , θ_i. This distribution has to be integrated over all θ_i and all but one p_i in order to obtain the one-momentum pdf, $P_q(p)$, to be compared with the numerical one, $P_{num}(p)$, obtained by considering, within the present molecular dynamical frame, increasingly large N-sized subsystems of an increasingly large M-system. Within the $M \gg N \gg 1$ numerical limit, we expect to go from the microcanonical ensemble to the canonical one (the cut-off is then expected to gradually disappear as indeed occurs in the usual short-range Hamiltonians), thus justifying the comparison between $P_q(p)$ and $P_{num}(p)$. The enormous complexity of this procedure made us to turn instead onto a naive, but tractable, comparison, namely that of our present numerical results with the following one-free-particle pdf [29,15] $P(p) \propto [1-(\frac{1}{2T})(1-q)p^2]^{1/(1-q)}$, which recovers the Maxwell-Boltzmann distribution for $q = 1$. The same type of formula has been recently used to describe successfully turbulent Couette-Taylor flow [16] and non-Gaussian pdfs related to anomalous diffusion of *Hydra* cells in cellular aggregates. In the case of HMF the best fit is obtained by a curve with entropic index $q = 7$, and temperature $T = 0.38$. The agreement between numerical results and theoretical curve improves with the size of the system. A finite-size scaling confirming the validity of the fit was reported in [61]. Since in our case we get $q > 3$, the theoretical curve does not have a finite integral and thus it needs to be truncated with a sharp cut-off in order to make the total probability equal to one. It is however clear that, the fitting value $q = 7$ is only an effective nonextensive entropic index. In [61] we verified, through the calculation of the fractal dimension D_2, that a dynamical correlation emerges in the μ-space before the final arrival to a quasi-uniform distribution. During intermediate times some filamentary structures appear, a similar feature has recently been found also in self-gravitating systems [50], which might be closely related to the plateaux observed in Fig. 10. We have found that these correlations do not sensibly depend on N, thus likely the possible connection does not concern the entire μ-space, but perhaps only the small sticky regions between the "chaotic sea" and the quasi-orbits [65]. Recently another application of the generalized nonextensive formalism has been observed for this model in the relaxation regime. It has been found that the time relaxation of macroscopic quantities, like the temperature for example, obey power-laws and more specifically q-exponential curves. To illustrate this, in Fig. 14, we plot the time relaxation of the temperature to its final value for $U = 0.69$ and for two sizes N=500,1000. In particular we show $y(t) = |T - T_{eq}|$, with $T_{eq} = 0.476$ as a function of the time t. The system relaxes to equilibrium according to the q-exponential formula[4]

$$y(t) = y_0 \left[1 - (1-q)\,\frac{t}{\tau}\right]^{1/(1-q)} , \tag{14}$$

where q is the entropic index, τ a characteristic time and y_0 a saturation value [29]. For $N = 500$ the same data shown in Fig. 10 are used. For this particular size, one finds a value of the entropic index equal to $q = 1.71$ and a characteristic relaxation time $\tau = 18000$. A slightly greater value of q equal to 1.81 is found for N=1000. In the inset we report the q-logarithm of same data [66]. This entropic index q is related to the relaxation and there is no reason to be exactly the

[4] The minus in front of the term $(1-q)$ at variance with formula (4) is due to the fact we are considering a decreasing q-exponential and $q > 1$.

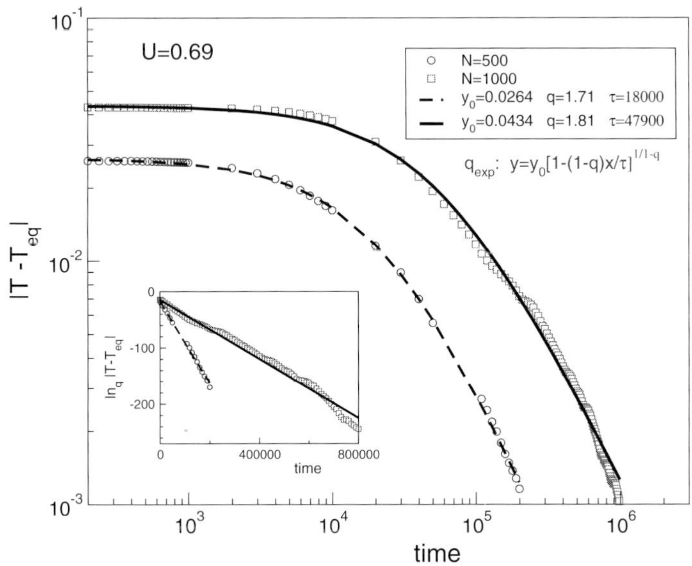

Fig. 14. For th HMF model, we plot the difference between the temperature T and that one at equilibrium T_{eq}, for U=0.69 and N=500,1000, vs time (open symbols). The points are the result of an ensemble average over 1000 runs. The figure shows that the relaxation process is a power law in time and can be very well reproduced with a decreasing q-exponential. We report the q-exponential fits for comparison (full and dashed curves). In the inset we plot the q-logarithm of the same data. The entropic index q is related to relaxation and should not necessarily be equal to that one of velocity pdfs.

same as the one obtained for the velocity pdfs. It is interesting to notice that similar relaxation process characterized by a q-exponential has been recently discovered also in echo experiment on the Internet [67]. Two final considerations before concluding this section. The first one is that in [68] it has been found that the HMF model shows aging in correspondence of the metastable QSS. Thus it is a very intriguing challenge left for future investigation, the study of the possible connections between spin glasses, hamiltonian mean field models and nonextensive statistics. Second, the anomalies we have found are not a peculiar characteristic of the HMF model. In fact the same behavior has been found also in a generalization of the HMF model, the $\alpha - XY$ model [13,14], in which the spins are placed on a d-dimensional lattice, and the interaction energy between two generic spins i and j (of formula (12)) is modulated by an extra factor $1/r_{ij}^{\alpha}$. Here r_{ij} is the distance between spin i and spin j [14]. If $\alpha/d < 1$ nonextensivity induces anomalies similar to those described for the HMF model. Conversely, if $\alpha/d > 1$, i.e. when the potential decreases rapidly enough with r, the $\alpha - XY$ model behaves in the standard way and does not exibit any of the anomalies discussed in this section. Similar features induced by nonextensivity have been found also for long-range Lennard-Jones-like potentials [69].

5 Conclusions

We have finally ended our guided tour in the realm of nonextensive systems. Our choice was to focus only on a few simple examples, chosen in order to represent the most instructive classes of systems. On one hand we have considered *low-dimensional systems*, in particular low-dimensional maps as the logistic map (representative of a dissipative system), and the standard map (representative of conservative systems). On the other hand we have considered *high-dimensional systems* discussing the anomalies observed in the HMF model, a classical many-body long-range-interacting Hamiltonian. Though representing very different kind of systems, all the examples reported share some common properties: long-range correlations in space/time, either long-range microscopic interactions or long-range microscopic memory (nonmarkovian processes), or (multi)fractal boundary conditions, or, generically speaking, some mechanism which creates a scale-invariant hierarchical structure of some sort. Such properties appear in classical many-body Hamiltonians and low-dimensional nonlinear dynamical systems everytime that the Lyapunov exponents approach zero, i.e. when chaos (strong mixing) is impossible and the sensitivity to the initial conditions is not exponential in time. In most cases of weak sensitivity to initial conditions, a generalized nonextensive formalism may replace standard thermostatistics.

References

1. F. Takens, in *Structures and Dynamics - Finite Dimensional Deterministic Studies*, eds. H.W. Broer, F. Dumortier, S.J. van Strien and F. Takens (North Holland, Amsterdam, 1991), p. 253; A.C.D. van Enter, R. Fernandez and A.D. Sokal, J. Stat. Phys. **72**, 879 (1993).
2. A. Einstein, Annalen der Physik **33** (1910) 1275. E.G.D. Cohen, Physica A **305**, 19 (2002); M. Baranger, Physica A **305**, 27 (2002); C. Beck and E.G.D. Cohen, [cond-mat/0205097].
3. L. Tisza, *Generalized Thermodynamics*, MIT Press (1966); P.T. Landsberg, *Thermodynamics and Statistical Mechanics*, Dover (1991).
4. D.H.E. Gross, *Thermo-Statistics or Topology of the Microcanonical Entropy Surface*, in "Dynamics and Thermodynamics of Systems with Long-Range Interactions", T. Dauxois, S. Ruffo, E. Arimondo, M. Wilkens Eds., Lecture Notes in Physics Vol. 602, Springer (2002), (in this volume)
5. H.G. Schuster, *Deterministic Chaos. An Introduction*, 3rd Revised Edition (VCH Publishers, Weinheim, 1995).
6. V. Latora, M. Baranger, A. Rapisarda, C. Tsallis, Phys. Lett. A **273**, 97 (2000); M. Baranger, V. Latora, and A. Rapisarda, Chaos Solitons and Fractals **13**, 471 (2001); U. Tirnakli, Physica A **305**, 119 (2002).
7. M. Tabor, *Chaos and Integrability in Nonlinear Dynamics*, Wiley, New York (1989).
8. M. Antoni and S.Ruffo, Phys. Rev. E **52**, 2361 (1995).
9. V. Latora , A. Rapisarda and S. Ruffo, Phys. Rev. Lett. **80**, 692 (1998); Physica D **131**, 38 (1999) and Progr. Theor. Phys. Suppl. **139**, 204 (2000).
10. V. Latora, A. Rapisarda and S. Ruffo, Phys. Rev. Lett. **83**, 2104 (1999) and Physica A **280**, 81 (2000).

11. V. Latora and A. Rapisarda, Nucl. Phys. A **681**, 331c (2001).
12. T. Dauxois, V. Latora, A. Rapisarda, S. Ruffo, A. Torcini, *The Hamiltonian Mean Field Model: from Dynamics to Statistical Mechanics and Back*, in "Dynamics and Thermodynamics of Systems with Long-Range Interactions", T. Dauxois, S. Ruffo, E. Arimondo, M. Wilkens Eds., Lecture Notes in Physics Vol. 602, Springer (2002), (in this volume)
13. C. Anteneodo and C. Tsallis , Phys. Rev. Lett. **80**, 5313 (1998); F. Tamarit and C. Anteneodo, Phys. Rev. Lett. **84**, 208 (2000).
14. A. Campa, A. Giansanti and D. Moroni, Phys. Rev. E **62**,303 (2000) and Physica A **305**, 137 (2002).
15. C. Tsallis, in *Nonextensive Statistical Mechanics and Thermodynamics*, eds. S.R.A. Salinas and C. Tsallis, Braz. J. Phys. **29**, 1 (1999); in *Nonextensive Statistical Mechanics and Its Applications*, eds. S. Abe and Y. Okamoto, Series *Lecture Notes in Physics* (Springer-Verlag, Heidelberg, 2001); , in *Classical and Quantum Complexity and Nonextensive Thermodynamics*, eds. P. Grigolini, C. Tsallis and B.J. West, Chaos , Solitons and Fractals **13**, Number 3, 371 (Pergamon-Elsevier, Amsterdam, 2002); in *Non Extensive Statistical Mechanics and Physical Applications*, eds. G. Kaniadakis, M. Lissia and A. Rapisarda, Physica A **305**, 1 (2002); in *Nonextensive Entropy - Interdisciplinary Applications*, eds. M. Gell-Mann and C. Tsallis (Oxford University Press, Oxford, 2002), to appear; in Annals of the Brazilian Academy of Sciences (2002), to appear [http://www.scielo.br/scielo.php] [cond-mat/0205571].
16. C. Beck, G.S. Lewis and H.L. Swinney , Phys. Rev. E **63**, 035303(R) (2001). See also C. Beck, Physica A **295**, 195 (2001) and *Phys. Rev. Lett.* **87**, 180601 (2001).
17. T. Arimitsu and N. Arimitsu, Physica A **305**, 218 (2002).
18. I. Bediaga, E.M.F. Curado and J. Miranda, Physica A **286**, 156 (2000).
19. A. Upadhyaya, J.-P. Rieu, J.A. Glazier and Y. Sawada, Physica A **293**, 549 (2001).
20. D.B. Walton and J. Rafelski, Phys. Rev. Lett. **84**, 31 (2000).
21. D.H. Zanette and P.A. Alemany, Phys. Rev. Lett. **75**, 366 (1995); C. Tsallis, S.V.F Levy, A.M.C. de Souza and R. Maynard, Phys. Rev. Lett. **75**, 3589 (1995) [Erratum: **77**, 5442 (1996)]; D. Prato and C. Tsallis, Phys. Rev. E **60**, 2398 (1999); C. Tsallis and D.J. Bukman, Phys. Rev. E **54**, R2197 (1996); E.K. Lenzi, C. Anteneodo and L. Borland, Phys. Rev. E **63**, 051109 (2001).
22. M.A. Montemurro, Physica A **300**, 567 (2001).
23. L. Borland, Phys. Rev. Lett. (2002), in press.
24. C. Anteneodo, C. Tsallis and A.S. Martinez, Europhys. Lett. (2002), in press [cond-mat/0109203].
25. C. Tsallis, J.C. Anjos and E.P. Borges, [astro-ph/0203258].
26. P. Quarati, A. Carbone, G. Gervino, G. Kaniadakis, A. Lavagno and E. Miraldi, Nucl. Phys. A **621**, 345c (1997).
27. D.B. Ion and M.L.D. Ion, Phys. Rev. Lett. **81**, 5714 (1998); Phys. Rev. Lett. **83**, 463 (1999); Phys. Rev. E **60**, 5261 (1999); Phys. Lett. B **474**, 395 (2000); Phys. Lett. B **482**, 57 (2000); Phys. Lett. B **503**, 263 (2001).
28. F.A. Tamarit, S.A. Cannas and C. Tsallis, Eur. Phys. J. B **1**, 545 (1998).
29. C. Tsallis, J. Stat. Phys. **52**, 479 (1988); E.M.F. Curado and C. Tsallis, J. Phys. A **24**, L69 (1991) [Corrigenda: **24**, 3187 (1991) and **25**, 1019 (1992)]; C. Tsallis, R.S. Mendes and A.R. Plastino, Physica A **261**, 534 (1998).
30. S. Abe, [cond-mat/0206078].
31. A.K. Rajagopal and S. Abe, Phys. Rev. Lett. **83**, 1711 (1989).
32. P.T. Landsberg and V. Vedral, Phys. Lett. A **247**, 211 (1998); P.T. Landsberg, in *Nonextensive Statistical Mechanics and Thermodynamics*, eds. S.R.A. Salinas and C. Tsallis, Braz. J. Phys. **29**, 46 (1999).

33. C. Beck and F. Schlogl, *Thermodynamics of chaotic systems* (Cambridge University Press, Cambridge, 1993).
34. P. Bak, *How Nature Works : The Science of Self-Organized Criticality*, (Springer–Verlag, New York 1996); A. Bhowal, Physica A **247**, 327 (1997).
35. M. Baranger, *Chaos, Complexity, and Entropy: a Physics Talk for Non-Physicists.* http://www.necsi.org/projects/baranger/cce.html
36. C. Tsallis, A.R. Plastino and W.-M Zheng, Chaos, Solitons and Fractals **8**, 885 (1997). U.M.S. Costa, M.L. Lyra, A.R. Plastino and C. Tsallis, Phys. Rev. E **56**, 245 (1997), and M.L. Lyra and C. Tsallis, Phys. Rev. Lett. **80**, 53 (1998).
37. F. Baldovin and A. Robledo, [cond-mat/0205371].
38. F. Baldovin and A. Robledo, [cond-mat/0205356].
39. V. Latora and M. Baranger, Phys. Rev. Lett. **82**, 520 (1999).
40. A. N. Kolmogorov, Dokl. Akad. Nauk SSSR **119**, 861 (1958); **124**, 754 (1959).
41. Ya. B. Pesin, Russian Math. Surveys **32:4**, 55 (1977).
42. E. Ott, *Chaos in dynamical systems* (Cambridge University Press, 1993).
43. G.M. Zaslavsky and B.A. Niyazov, Phys. Rep. **283**, 73 (1997).
44. F. Baldovin and C. Tsallis, in progress.
45. C. Tsallis, G. Bemski, and R.S. Mendes, Phys. Lett. A **257**, 93 (1999).
46. F. Baldovin, C. Tsallis and B. Schulze, [cond-mat/0203595].
47. A. Bonasera, V. Latora and A. Rapisarda, Phys. Rev. Lett. **75**, 3434 (1995).
48. L. Casetti, M. Pettini and E.G.D. Cohen, Phys. Rep. **337**, 237 (2000).
49. L. Milanovic, H.A.Posch and W. Thirring, Phys. Rev. E **57**, 2763 (1998).
50. H. Koyama and T. Konishi, Phys. Lett. A **279**, 226 (2001); Y. Sota, O. Iguchi, M. Morikawa, T. Tatekawa and K. Maeda, Phys. Rev. E **64**, 056133 (2001).
51. M. Antoni and A. Torcini, Phys. Rev. E **57**, R6233 (1998); A. Torcini and M. Antoni, Phys. Rev. E **59**, 2746 (1999).
52. D. Lynden-Bell , Physica A **263**, 293 (1999).
53. D.H.E. Gross, *Microcanonical thermodynamics: Phase transitions in Small systems*, 66 Lectures Notes in Physics, World scientific, Singapore (2001).
54. M. D'Agostino et al., Phys. Lett. B **473**, 219 (2000).
55. P. Chomaz, F. Gulminelli, *Phase Transitions in finite systems*, in "Dynamics and Thermodynamics of Systems with Long-Range Interactions", T. Dauxois, S. Ruffo, E. Arimondo, M. Wilkens Eds., Lecture Notes in Physics Vol. 602, Springer (2002), (in this volume)
56. M. Schmidt, R. Kusche, T. Hippler, J. Donges and W. Kronmueller, Phys. Rev. Lett. **86**, 1191 (2001).
57. J. Klafter and G. Zumofen, Phys. Rev. E **49**, 4873 (1994).
58. G. Parisi, Physica A **280**, 115 (2000).
59. P.G. Benedetti and F.H. Stillinger, Nature **410**, 259 (2001).
60. A. Coniglio, Physica A **306**, 76 (2002).
61. V. Latora, A. Rapisarda, C. Tsallis, Phys. Rev. E **64**, 056134 (2001)
62. V. Latora, A. Rapisarda, C. Tsallis, Physica A **305** , 129 (2002)
63. N.S.Krylov, *Works on the Foundations of Statistical Physics*, translated by A.B. Migdal, Ya. G. Sinai and Yu. L. Zeeman, Princeton University Press (1979) and Nature **153**, 709 (1944)
64. A. Kudrolli, J. Henry, Phys. Rev. E **62**, R1489 (2000); Y-H. Taguchi, H. Takayasu, Europhys. Lett. **30**, 499 (1995)
65. M.F. Shlesinger, G.M. Zaslavsky, J. Klafter, Nature **363**, 31 (1993)
66. A. Rapisarda, C. Tsallis, A. Giansanti in preparation
67. S. Abe, N. Suzuki, [cond-mat/0204336]
68. M.A. Montemurro, F. Tamarit C. Anteneodo [cond-mat/0205355]
69. E.P. Borges, C. Tsallis, Physica A **305**, 148 (2002)

Part II

Astrophysics

Statistical Mechanics of Gravitating Systems in Static and Cosmological Backgrounds

Thanu Padmanabhan

IUCAA, Pune University Campus, Ganeshkhind, Pune – 7, India,
nabhan@iucaa.ernet.in

Abstract. This pedagogical review addresses several issues related to statistical description of gravitating systems in both static and expanding backgrounds, focusing on the latter. After briefly reviewing the results for the static background, I describe the key issues and recent progress in the context of gravitational clustering of collision-less particles in an expanding universe. The questions addressed include: (a) How does the power injected into the system at a given wave number spread to smaller and larger scales? (b) How does the power spectrum of density fluctuations behave asymptotically at late times? (c) What are the universal characteristics of gravitational clustering that are independent of the initial conditions and manifest at the late time evolution of the system? The review is intended for non cosmologists and will be of interest to people working in fluid mechanics, non linear dynamics and condensed matter physics.

1 Introduction

The statistical mechanics of systems dominated by gravity is of interest both from the theoretical and "practical" perspectives. Theoretically this field has close connections with areas of condensed matter physics, fluid mechanics, renormalization group etc. and poses an incredible challenge as regards the basic formulation. From the practical point of view, the ideas find application in many different areas of astrophysics and cosmology, especially in the study of globular clusters, galaxies and gravitational clustering in the expanding universe. (For a general review of statistical mechanics of gravitating systems, see [1]; textbook description of the subject is available in [2], [3]. Review of gravitational clustering in expanding background is available in [4] and in several textbooks in cosmology [5]. There have been many attempts to understand these phenomena by different groups; see [6], [7], [8], [9], [10], [11] and the references cited therein.) Given the diversity of the subject, it will be useful to begin with a broad overview and a description of the issues which will be addressed in this review.

I will concentrate mostly on the problem of gravitational clustering in the expanding universe which is one of the most active research areas in cosmology. However, to place this problem in context, it is necessary to discuss statistical mechanics of isolated gravitating systems (without any cosmological expansion) in some detail. In Part I of this review, I cover this aspect highlighting the important features but referring the reader to existing previous literature for details. Part II presents a more detailed description of gravitational clustering in the context of cosmology. The rest of the introduction will be devoted to a summary of different issues which will be expanded upon in the later sections.

Let me begin with the issues which arise in the study of isolated gravitating systems (like, say, a cluster of stars) treated as a collection of structure-less point particles. In Newtonian theory, the gravitational force can be described as a gradient of a scalar potential and the evolution of a set of particles under the action of gravitational forces can be described the equations

$$\ddot{\mathbf{x}}_i = -\nabla\phi(\mathbf{x}_i, t); \quad \nabla^2\phi = 4\pi G \sum_i m_i \delta_D(\mathbf{x} - \mathbf{x}_i) \tag{1}$$

where $\mathbf{x}_i$ is the position of the i–th particle, m_i is its mass. For sufficiently large number of particles, it is useful to investigate whether some kind of statistical description of such a system is possible. Such a description, however, is complicated by the long-range, unscreened, nature of gravitational force. The force acting on any given particle receives contribution even from particles which are far away. If a self gravitating system is divided into two parts, the total energy of the system cannot be expressed as the sum of the gravitational energy of the components. Many of the conventional results in statistical physics are based on the extensivity of the energy which is clearly not valid for gravitating systems. To make progress, we have to use different techniques which are appropriate for each situation.

To construct the statistical description of such a system, one should begin with the construction of the micro-canonical ensemble describing such a system. If the Hamiltonian of the system is $H(p_i, q_i)$ then the volume $g(E)$ of the constant energy surface $H(p_i, q_i) = E$ will be of primary importance in the micro-canonical ensemble. The logarithm of this function will give the entropy $S(E) = \ln g(E)$ and the temperature of the system will be $T(E) \equiv \beta(E)^{-1} = (\partial S/\partial E)^{-1}$.

Systems for which a description based on canonical ensemble is possible, the Laplace transform of $g(E)$ with respect to a variable β will give the partition function $Z(\beta)$. It is, however, trivial to show that gravitating systems of interest in astrophysics cannot be described by a canonical ensemble [1], [12], [13]. Virial theorem holds for such systems and we have $(2K+U) = 0$ where K and U are the total kinetic and potential energies of the system. This leads to $E = K + U = -K$; since the temperature of the system is proportional to the total kinetic energy, the specific heat will be negative: $C_V \equiv (\partial E/\partial T)_V \propto (\partial E/\partial K) < 0$. On the other hand, the specific heat of any system described by a canonical ensemble $C_V = \beta^2 \left\langle (\Delta E)^2 \right\rangle$ will be positive definite. Thus one cannot describe self gravitating systems of the kind we are interested in by canonical ensemble.

One may attempt to find the equilibrium configuration for self gravitating systems by maximizing the entropy $S(E)$ or the phase volume $g(E)$. It is again easy to show that no global maximum for the entropy exists for classical point particles interacting via Newtonian gravity. To prove this, we only need to construct a configuration with arbitrarily high entropy which can be achieved as follows: Consider a system of N particles initially occupying a region of finite volume in phase space and total energy E. We now move a small number of these particles (in fact, a pair of them, say, particles 1 and 2 will do) arbitrarily close to each other. The potential energy of interaction of these two particles,

Gm_1m_2/r_{12}, will become arbitrarily high as $r_{12} \to 0$. Transferring some of this energy to the rest of the particles, we can increase their kinetic energy without limit. This will clearly increase the phase volume occupied by the system without bound. This argument can be made more formal by dividing the original system into a small, compact core and a large diffuse halo and allowing the core to collapse and transfer the energy to the halo.

The absence of the global maximum for entropy – as argued above – depends on the idealization that there is no short distance cut-off in the interaction of the particles, so that we could take the limit $r_{12} \to 0$. If we assume, instead, that each particle has a minimum radius a, then the typical lower bound to the gravitational potential energy contributed by a pair of particles will be $-Gm_1m_2/2a$. This will put an upper bound on the amount of energy that can be made available to the rest of the system.

We have also assumed that part of the system can expand without limit – in the sense that any particle with sufficiently large energy can move to arbitrarily large distances. In real life, no system is completely isolated and eventually one has to assume that the meandering particle is better treated as a member of another system. One way of obtaining a truly isolated system is to confine the system inside a spherical region of radius R with, say, reflecting wall.

The two cut-offs a and R will make the upper bound on the entropy finite, but even with the two cut-offs, the primary nature of gravitational instability cannot be avoided. The basic phenomenon described above (namely, the formation of a compact core and a diffuse halo) will still occur since this is the direction of increasing entropy. Particles in the hot diffuse component will permeate the entire spherical cavity, bouncing off the walls and having a kinetic energy which is significantly larger than the potential energy. The compact core will exist as a gravitationally bound system with very little kinetic energy. A more formal way of understanding this phenomena is based on the virial theorem for a system with a short distance cut-off confined to a sphere of volume V. In this case, the virial theorem will read as [2]

$$2T + U = 3PV + \Phi \tag{2}$$

where P is the pressure on the walls and Φ is the correction to the potential energy arising from the short distance cut-off. This equation can be satisfied in essentially three different ways. If T and U are significantly higher than $3PV$, then we have $2T + U \approx 0$ which describes a self gravitating systems in standard virial equilibrium but not in the state of maximum entropy. If $T \gg U$ and $3PV \gg \Phi$, one can have $2T \approx 3PV$ which describes an ideal gas with no potential energy confined to a container of volume V; this will describe the hot diffuse component at late times. If $T \ll U$ and $3PV \ll \Phi$, then one can have $U \approx \Phi$ describing the compact potential energy dominated core at late times. In general, the evolution of the system will lead to the production of the core and the halo and each component will satisfy the virial theorem in the form (2). Such an asymptotic state with two distinct phases [14] is quite different from what would have been expected for systems with only short range interaction.

Considering its importance, I shall briefly describe in Sect. 2 a toy model which captures the essential physics of the above system in an exactly solvable context.

The above discussion focussed on the existence of global maximum to the entropy and we proved that it does not exist in the absence of two cut-offs. It is, however, possible to have *local* extrema of entropy which are not global maxima. Intuitively, one would have expected the distribution of matter in the configuration which is a local extrema of entropy to be described by a Boltzmann distribution, with the density given by $\rho(\mathbf{x}) \propto \exp[-\beta\phi(\mathbf{x})]$ where ϕ is the gravitational potential related to ρ by Poisson equation. This is indeed true and a formal proof will be given in Sect. 3. This configuration is usually called the isothermal sphere (because it can be shown that, among all solutions to this equation, the one with spherical symmetry maximizes the entropy) and since it is a local maximum of entropy, it deserves careful study. I will describe briefly some of the interesting features of the isothermal spheres in Sect. 4 and this configuration will play a dominant role throughout our review. The second (functional) derivative of the entropy with respect to the configuration variables will determine whether the local extremum of entropy is a local maximum or a saddle point [15], [16] and some of these results are described at the end of Sect. 4. The relevance of the long-range of gravity in all the above phenomena can be understood by studying model systems with an attractive potential varying as $r^{-\alpha}$ with different values for α. Such studies confirm the results and interpretation given above; (see [17] and references cited therein).

Let us now consider the situation in the context of an expanding background described in Part II which is the main theme of the review. There is considerable amount of observational evidence to suggest that one of the dominant energy densities in the universe is contributed by self gravitating point particles. The smooth average energy density of these particles drive the expansion of the universe while any small deviation from the homogeneous energy density will cluster gravitationally. One of the central problems in cosmology is to describe the non linear phases of this gravitational clustering starting from a initial spectrum of density fluctuations. It is often enough (and necessary) to use a statistical description and relate different statistical indicators (like the power spectra, nth order correlation functions ...) of the resulting density distribution to the statistical parameters (usually the power spectrum) of the initial distribution. The relevant scales at which gravitational clustering is non linear are less than about 10 Mpc (where 1 Mpc $= 3 \times 10^{24}$ cm is the typical separation between galaxies in the universe) while the expansion of the universe has a characteristic scale of about few thousand Mpc. Hence, non linear gravitational clustering in an expanding universe can be adequately described by Newtonian gravity provided the rescaling of lengths due to the background expansion is taken into account. This is easily done by introducing a *proper* coordinate for the $i-$th particle $\mathbf{r}_i$, related to the *comoving* coordinate $\mathbf{x}_i$, by $\mathbf{r}_i = a(t)\mathbf{x}_i$ with $a(t)$ describing the stretching of length scales due to cosmic expansion. The Newtonian dynamics works with the proper coordinates $\mathbf{r}_i$ which can be translated to the behaviour of the comoving coordinate $\mathbf{x}_i$ by this rescaling. (Some basic results in cosmology are summarized in Appendix A.)

As to be expected, cosmological expansion completely changes the nature of the problem because of several new factors which come in: (a) The problem has now become time dependent and it will be pointless to look for equilibrium solutions in the conventional sense of the word. (b) On the other hand, the expansion of the universe has a civilizing influence on the particles and acts counter to the tendency of gravity to make systems unstable. (c) In any small local region of the universe, one would assume that the conclusions describing a finite gravitating system will still hold true approximately. In that case, particles in any small sub region will be driven towards configurations of local extrema of entropy (say, isothermal spheres) and towards global maxima of entropy (say, core-halo configurations).

An extra feature comes into play as regards the expanding halo from any sub region. The expansion of the universe acts as a damping term in the equations of motion and drains the particles of their kinetic energy – which is essentially the lowering of temperature of any system participating in cosmic expansion. This, in turn, helps gravitational clustering since the potential wells of nearby sub regions can capture particles in the expanding halo of one region when the kinetic energy of the expanding halo has been sufficiently reduced.

The actual behaviour of the system will, of course, depend on the form of $a(t)$. However, for understanding the nature of clustering, one can take $a(t) \propto t^{2/3}$ which describes a matter dominated universe with critical density (see Appendix A). Such a power law has the advantage that there is no intrinsic scale in the problem. Since Newtonian gravitational force is also scale free, one would expect some scaling relations to exist in the pattern of gravitational clustering. Incredibly enough, converting this intuitive idea into a concrete mathematical statement turns out to be non trivial and difficult. I shall discuss several attempts to give concrete shape to this idea in Sects. 5.1, 6 and 8 but there is definite scope for further work in this direction.

To make any progress we need a theoretical formulation which will relate statistical indicators in the non linear regime of clustering to the initial conditions. In particular, we need a robust prescription which will allow us to obtain the two-point correlation function and the nonlinear power spectrum from the initial power spectrum. Fortunately, this problem has been solved to a large extent and hence one can use this as a basis for attacking several other key questions. There are four key theoretical questions which are of considerable interest in this area that I will focus on:

- If the initial power spectrum is sharply peaked in a narrow band of wavelengths, how does the evolution transfer the power to other scales? This is, in some sense, analogous to determining the Green function for the gravitational clustering except that superposition will not work in the non linear context.
- What is the asymptotic nature of evolution for the self gravitating system in an expanding background? In particular, how can one connect up the local

behaviour of gravitating systems to the overall evolution of clustering in the universe? (If we assume that the isothermal spheres play an important role in the local description of gravitating system, we would expect a strong trace of it to survive even in the context of cosmological clustering. This is indeed true as I shall show in Sects. 6, 7 and 8 but only in the asymptotic limit, under certain assumptions.)

- Does the gravitational clustering at late stages wipe out the memory of initial conditions or does the late stage evolution depend on the initial power spectrum of fluctuations?
- Do the virialized structures formed in an expanding universe due to gravitational clustering have any invariant properties? Can their structure be understood from first principles?

All the above questions are, in some sense, open and thus are good research problems. I will highlight the progress which has been made and give references to original literature for more detailed discussion.

Part I: Gravitational Clustering in Static Backgrounds

2 Phases of the Self Gravitating System

As described in section 1 the statistical mechanics of finite, self gravitating, systems have the following characteristic features: (a) They exhibit negative specific heat while in virial equilibrium. (b) They are inherently unstable to the formation of a core-halo structure and global maximum for entropy does not exist without cut-offs at short and large distances. (c) They can be broadly characterized by two phases – one of which is compact and dominated by potential energy while the other is diffuse and behaves more or less like an ideal gas. The purpose of this section is to describe a simple toy model which exhibits all these features and mimics a self gravitating system [1].

Consider a system with two particles described by a Hamiltonian of the form

$$H\left(\mathbf{P},\mathbf{Q};\mathbf{p},\mathbf{r}\right)=\frac{\mathbf{P}^2}{2M}+\frac{\mathbf{p}^2}{2\mu}-\frac{Gm^2}{r} \tag{3}$$

where $(\mathbf{Q},\mathbf{P})$ are coordinates and momenta of the center of mass, $(\mathbf{r},\mathbf{p})$ are the relative coordinates and momenta, $M=2m$ is the total mass, $\mu=m/2$ is the reduced mass and m is the mass of the individual particles. This system may be thought of as consisting of two particles (each of mass m) interacting via gravity. We shall assume that the quantity r varies in the interval (a,R). This is equivalent to assuming that the particles are hard spheres of radius $a/2$ and that the system is confined to a spherical box of radius R. We will study the "statistical mechanics" of this simple toy model.

To do this, we shall start with the volume $g(E)$ of the constant energy surface $H = E$. Straightforward calculation gives

$$g(E) = AR^3 \int_a^{r_{\max}} r^2 dr \left[E + \frac{Gm^2}{r}\right]^2 . \tag{4}$$

where $A = (64\pi^5 m^3/3)$. The range of integration in (4) should be limited to the region in which the expression in the square brackets is positive. So we should use $r_{\max} = (Gm^2/|E|)$ if $(-Gm^2/a) < E < (-Gm^2/R)$, and use $r_{\max} = R$ if $(-Gm^2/R) < E < +\infty$. Since $H \geq (-Gm^2/a)$, we trivially have $g(E) = 0$ for $E < (-Gm^2/a)$. The constant A is unimportant for our discussions and hence will be omitted from the formulas hereafter. The integration in (4) gives the following result:

$$\frac{g(E)}{(Gm^2)^3} = \begin{cases} \frac{R^3}{3}(-E)^{-1}\left(1+\frac{aE}{Gm^2}\right)^3, & (-Gm^2/a) < E < (-Gm^2/R) \\ \frac{R^3}{3}(-E)^{-1}\left[\left(1+\frac{RE}{Gm^2}\right)^3 - \left(1+\frac{aE}{Gm^2}\right)^3\right], & (-Gm^2/R) < E < \infty. \end{cases} \tag{5}$$

This function $g(E)$ is continuous and smooth at $E = (-Gm^2/R)$. We define the entropy $S(E)$ and the temperature $T(E)$ of the system by the relations

$$S(E) = \ln g(E); \quad T^{-1}(E) = \beta(E) = \frac{\partial S(E)}{\partial E}. \tag{6}$$

All the interesting thermodynamic properties of the system can be understood from the $T(E)$ curve.

Consider first the case of low energies with $(-Gm^2/a) < E < (-Gm^2/R)$. Using (5) and (6) one can easily obtain $T(E)$ and write it in the dimensionless form as

$$t(\epsilon) = \left[\frac{3}{1+\epsilon} - \frac{1}{\epsilon}\right]^{-1} \tag{7}$$

where we have defined $t = (aT/Gm^2)$ and $\epsilon = (aE/Gm^2)$.

This function exhibits the peculiarities characteristic of gravitating systems. At the lowest energy admissible for our system, which corresponds to $\epsilon = -1$, the temperature t vanishes. This describes a tightly bound low temperature phase of the system with negligible random motion. The $t(\epsilon)$ is clearly dominated by the first term of (7) for $\epsilon \sim -1$. As we increase the energy of the system, the temperature *increases*, which is the normal behaviour for a system. This trend continues up to

$$\epsilon = \epsilon_1 = -\frac{1}{2}(\sqrt{3}-1) \simeq -0.36 \tag{8}$$

at which point the $t(\epsilon)$ curve reaches a maximum and turns around. As we increase the energy further the temperature *decreases*. The system *exhibits negative specific heat in this range.*

Equation (7) is valid from the minimum energy $(-Gm^2/a)$ all the way up to the energy $(-Gm^2/R)$. For realistic systems, $R \gg a$ and hence this range is quite wide. For a small region in this range, [from $(-Gm^2/a)$ to $(-0.36Gm^2/a)$] we have positive specific heat; for the rest of the region the specific heat is negative. *The positive specific heat region owes its existence to the nonzero short distance cutoff.* If we set $a = 0$, the first term in (7) will vanish; we will have $t \propto (-\epsilon^{-1})$ and negative specific heat in this entire domain.

For $E \geq (-Gm^2/R)$, we have to use the second expression in (5) for $g(E)$. In this case, we get:

$$t(\epsilon) = \left[\frac{3\left[(1+\epsilon)^2 - \frac{R}{a}(1+\frac{R}{a}\epsilon)^2\right]}{(1+\epsilon)^3 - (1+\frac{R}{a}\epsilon)^3} - \frac{1}{\epsilon} \right]^{-1} . \tag{9}$$

This function, of course, matches smoothly with (7) at $\epsilon = -(a/R)$. As we increase the energy, the temperature continues to decrease for a little while, exhibiting negative specific heat. However, this behaviour is soon halted at some $\epsilon = \epsilon_2$, say. The $t(\epsilon)$ curve reaches a minimum at this point, turns around, and starts increasing with increasing ϵ. We thus enter another (high-temperature) phase with positive specific heat. From (9) it is clear that $t \simeq (1/2)\epsilon$ for large ϵ. (Since $E = (3/2)NkT$ for an ideal gas, we might have expected to find $t \simeq (1/3)\epsilon$ for our system with $N = 2$ at high temperatures. This is indeed what we would have found if we had defined our entropy as the logarithm of the volume of the phase space with $H \leq E$. With our definition, the energy of the ideal gas is actually $E = [(3/2)N - 1]kT$; hence we get $t = (1/2)\epsilon$ when $N = 2$). The form of the $t(\epsilon)$ for $(a/R) = 10^{-4}$ is shown in Fig. 1 by the dashed curve. The specific heat is positive along the portions AB and CD and is negative along BC.

The overall picture is now clear. Our system has two natural energy scales: $E_1 = (-Gm^2/a)$ and $E_2 = (-Gm^2/R)$. For $E \gg E_2$, gravity is not strong enough to keep $r < R$ and the system behaves like a gas confined by the container; we have a high temperature phase with positive specific heat. As we lower the energy to $E \simeq E_2$, the effects of gravity begin to be felt. For $E_1 < E < E_2$, the system is unaffected by either the box or the short distance cutoff; this is the domain dominated entirely by gravity and we have negative specific heat. As we go to $E \simeq E_1$, the hard core nature of the particles begins to be felt and the gravity is again resisted. This gives rise to a low temperature phase with positive specific heat.

We can also consider the effect of increasing R, keeping a and E fixed. Since we imagine the particles to be hard spheres of radius $(a/2)$, we should only consider $R > 2a$. It is amusing to note that, if $2 < (R/a) < (\sqrt{3} + 1)$, there is no region of negative specific heat. As we increase R, this negative specific heat region appears and increasing R increases the range over which the specific heat is negative. Suppose a system is originally prepared with some E and R values such that the specific heat is positive. If we now increase R, the system may find itself in a region of negative specific heat. *This suggests the possibility that an instability may be triggered in a constant energy system if its radius increases beyond a critical value.* We will see later that this is indeed true.

Since systems described by canonical distribution cannot exhibit negative specific heat, it follows that canonical distribution will lead to a very different

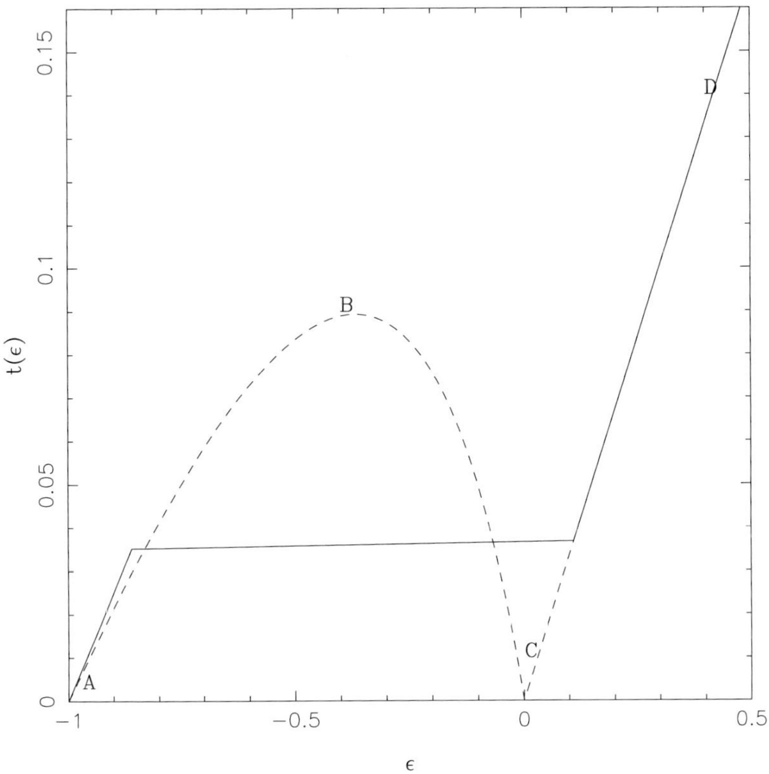

Fig. 1. The relation between temperature and energy for a model mimicking self gravitating systems. The dashed line is the result for micro-canonical ensemble and the solid line is for canonical ensemble. The negative specific heat region, BC, in the micro-canonical description is replaced by a phase transition in the canonical description. See text for more details

physical picture for this range of (mean) energies $E_1 < E < E_2$. It is, therefore, of interest to look at our system from the point of view of canonical distribution by computing the partition function. In the partition function

$$Z(\beta) = \int d^3P d^3p d^3Q d^3r \exp(\ \beta H) \tag{10}$$

the integrations over P, p and Q can be performed trivially. Omitting an overall constant which is unimportant, we can write the answer in the dimensionless form as

$$Z(t) = t^3 \left(\frac{R}{a}\right)^3 \int_1^{R/a} dx x^2 \exp\left(\frac{1}{xt}\right) \tag{11}$$

where t is the dimensionless temperature defined in (7). Though this integral cannot be evaluated in closed form, all the limiting properties of $Z(\beta)$ can be easily obtained from (11).

The integrand in (11) is large for both large and small x and reaches a minimum for $x = x_m = (1/2t)$. At high temperatures, $x_m < 1$ and hence the minimum falls outside the domain of integration. The exponential contributes very little to the integral and we can approximate Z adequately by

$$Z \approx t^3 \left(\frac{R}{a}\right)^3 \int_1^{R/a} dx x^2 \left[1 + \frac{2x_m}{x}\right] = \frac{t^3}{3} \left(\frac{R}{a}\right)^6 \left(1 + \frac{3a}{2Rt}\right). \tag{12}$$

On the other hand, if $x_m > 1$ the minimum lies between the limits of the integration and the exponential part of the curve dominates the integral. We can easily evaluate this contribution by a saddle point approach, and obtain

$$Z \approx \left(\frac{R}{a}\right)^3 t^4 (1 - 2t)^{-1} \exp\left(\frac{1}{t}\right). \tag{13}$$

As we lower the temperature, making x_m cross 1 from below, the contribution switches over from (12) to (13). The transition is exponentially sharp. The critical temperature at which the transition occurs can be estimated by finding the temperature at which the two contributions are equal. This occurs at

$$t_c = \frac{1}{3} \frac{1}{\ln(R/a)}. \tag{14}$$

For $t < t_c$, we should use (13) and for $t > t_c$ we should use (12).

Given $Z(\beta)$ all thermodynamic functions can be computed. In particular, the mean energy of the system is given by $E(\beta) = -(\partial \ln Z/\partial\beta)$. This relation can be inverted to give the $T(E)$ which can be compared with the $T(E)$ obtained earlier using the micro-canonical distribution. From (12) and (13) we get,

$$\epsilon(t) = \frac{aE}{Gm^2} = 4t - 1 \tag{15}$$

for $t < t_c$ and

$$\epsilon(t) = 3t - \frac{3a}{2R} \tag{16}$$

for $t > t_c$. Near $t \approx t_c$, there is a rapid variation of the energy and we cannot use either asymptotic form. The system undergoes a phase transition at $t = t_c$ absorbing a large amount of energy

$$\Delta\epsilon \approx \left(1 - \frac{1}{3\ln(R/a)}\right). \tag{17}$$

The specific heat is, of course, positive throughout the range. This is to be expected because canonical ensemble cannot lead to negative specific heats.

The $T - E$ curves obtained from the canonical (unbroken line) and micro-canonical (dashed line) distributions are shown in Fig. 1. (For convenience, we have rescaled the $T - E$ curve of the micro-canonical distribution so that $\epsilon \simeq 3t$ asymptotically.) At both very low and very high temperatures, the canonical and micro-canonical descriptions match. The crucial difference occurs at the intermediate energies and temperatures. Micro-canonical description predicts negative specific heat and a reasonably slow variation of energy with temperature. Canonical description, on the other hand, predicts a phase transition with rapid variation of energy with temperature. Such phase transitions are accompanied by large fluctuations in the energy, which is the main reason for the disagreement between the two descriptions [1], [12], [13].

Numerical analysis of more realistic systems confirm all these features. Such systems exhibit a phase transition from the diffuse virialized phase to a core dominated phase when the temperature is lowered below a critical value [14]. The transition is very sharp and occurs at nearly constant temperature. The energy released by the formation of the compact core heats up the diffuse halo component.

3 Mean Field Description of Gravitating Systems

The analysis in the previous two sections shows that there is no *global* maximum for the entropy for a self gravitating system of point particles and the evolution will proceed towards the formation of a core halo configuration and will continue inexorably in the absence of cut-offs. It is, however, possible to find configurations for these systems which are *local* maxima of the entropy. This configuration, called the isothermal sphere, will be of considerable significance in our discussions.

Consider a system of N particles interacting with each other through the two-body potential $U(\mathbf{x},\mathbf{y})$. The entropy S of this system, in the micro-canonical description, is defined through the relation

$$\mathrm{e}^S = g(E) = \frac{1}{N!}\int d^{3N}x d^{3N}p\delta(E-H) = \frac{A}{N!}\int d^{3N}x\left[E-\frac{1}{2}\sum_{i\neq j}U(\mathbf{x}_i,\mathbf{x}_j)\right]^{\frac{3N}{2}} \tag{18}$$

wherein we have performed the momentum integrations and replaced $(3N/2-1)$ by $(3N/2)$. We shall approximate the expression in (18) in the following manner.

Let the spatial volume V be divided into M (with $M \ll N$) cells of equal size, large enough to contain many particles but small enough for the potential to be treated as a constant inside each cell. Instead of integrating over the particle coordinates $(\mathbf{x}_1, \mathbf{x}_2, ..., \mathbf{x}_N)$ we shall sum over the number of particles n_a in the cell centered at $\mathbf{x}_a$ (where $a = 1, 2, ..., M$). Using the standard result that the integration over $(N!)^{-1}d^{3N}x$ can be replaced by

$$\sum_{n_1=1}^{\infty}\left(\frac{1}{n_1!}\right)\sum_{n_2=1}^{\infty}\left(\frac{1}{n_2!}\right)\cdots\sum_{n_M=1}^{\infty}\left(\frac{1}{n_M!}\right)\delta\left(N-\sum_a n_a\right)\left(\frac{V}{M}\right)^N \tag{19}$$

and ignoring the unimportant constant A, we can rewrite (18) as

$$
\begin{aligned}
\mathrm{e}^S &= \sum_{n_1=1}^{\infty}\left(\frac{1}{n_1!}\right)\sum_{n_2=1}^{\infty}\left(\frac{1}{n_2!}\right)\cdots\sum_{n_M=1}^{\infty}\left(\frac{1}{n_M!}\right)\delta\left(N-\sum_a n_a\right)\left(\frac{V}{M}\right)^N \\
&\qquad\times\left[E-\frac{1}{2}\sum_{a\neq b}^{M} n_a U_{ab} n_b\right]^{\frac{3N}{2}} \\
&\approx \sum_{n_1=1}^{\infty}\sum_{n_2=1}^{\infty}\cdots\sum_{n_M=1}^{\infty}\delta\left(N-\sum_a n_a\right)\exp S[\{n_a\}]
\end{aligned}
\tag{20}
$$

where

$$
S[\{n_a\}] = \frac{3N}{2}\ln\left[E-\frac{1}{2}\sum_{a\neq b}^{M} n_a U(\mathbf{x}_a,\mathbf{x}_b) n_b\right] - \sum_{a=1}^{M} n_a \ln\left(\frac{n_a M}{\mathrm{e}V}\right). \tag{21}
$$

In arriving at the last expression we have used the Sterling's approximation for the factorials. The mean field limit is now obtained by retaining in the sum in (20) only the term for which the summand reaches the maximum value, subject to the constraint on the total number. That is, we use the approximation

$$
\sum_{\{n_a\}} \mathrm{e}^{S[n_a]} \approx \mathrm{e}^{S[n_{a,\max}]} \tag{22}
$$

where $n_{a,\max}$ is the solution to the variational problem

$$
\left(\frac{\delta S}{\delta n_a}\right)_{n_a=n_{a,\max}} = 0 \qquad \text{with} \quad \sum_{a=1}^{M} n_a = N. \tag{23}
$$

Imposing this constraint by a Lagrange multiplier and using the expression (21) for S, we obtain the equation satisfied by $n_{a,\max}$:

$$
\frac{1}{T}\sum_{b=1}^{M} U(\mathbf{x}_a,\mathbf{x}_b) n_{b,\max} + \ln\left(\frac{n_{a,\max} M}{V}\right) = \text{constant} \tag{24}
$$

where we have defined the temperature T as

$$
\frac{1}{T} = \frac{3N}{2}\left(E-\frac{1}{2}\sum_{a\neq b}^{M} n_a U(\mathbf{x}_a,\mathbf{x}_b) n_b\right)^{-1} = \beta. \tag{25}
$$

We see from (21) that this expression is also equal to $(\partial S/\partial E)$; therefore T is indeed the correct thermodynamic temperature. We can now return back to the continuum limit by the replacements

$$
n_{a,\max}\frac{M}{V} = \rho(\mathbf{x}_a); \qquad \sum_{a=1}^{M} \to \frac{M}{V}\int . \tag{26}
$$

In this limit, the extremum solution (24) is given by

$$\rho(\mathbf{x}) = A\exp(-\beta\phi(\mathbf{x})); \quad \text{where} \quad \phi(\mathbf{x}) = \int d^3\mathbf{y}U(\mathbf{x},\mathbf{y})\rho(\mathbf{y}) \tag{27}$$

which, in the case of gravitational interactions, becomes

$$\rho(\mathbf{x}) = A\exp(-\beta\phi(\mathbf{x})); \quad \phi(\mathbf{x}) = -G\int \frac{\rho(\mathbf{y})d^3\mathbf{y}}{|\mathbf{x}-\mathbf{y}|}. \tag{28}$$

Equation (28) represents the configuration of extremal entropy for a gravitating system in the mean field limit. The constant β is already determined through (25) in terms of the total energy of the system. The constant A has to be fixed in terms of the total number (or mass) of the particles in the system.

The various manipulations in (18) to (24) tacitly assume that the expressions we are dealing with are finite. But for gravitational interactions *without* a short distance cut-off, the quantity e^S – and hence all the terms we have been handling – are divergent. We should, therefore, remember that a short distance cut-off is needed to justify the entire procedure [9], [17], [18] and that (28) – which is based on a strict r^{-1} potential and does not incorporate any such cutoff – can only be approximately correct. We will now study some of the properties of the solution of (28) which is a local extremum of the entropy.

4 Isothermal Sphere

The extremum condition for the entropy, (28), is equivalent to the differential equation for the gravitational potential:

$$\nabla^2\phi = 4\pi G\rho_c e^{-\beta[\phi(\mathbf{x})-\phi(0)]} \tag{29}$$

Given the solution to this equation, all other quantities can be determined. As we shall see, this system shows several peculiarities.

It is convenient to introduce the length, mass and energy scale by the definitions

$$L_0 \equiv (4\pi G\rho_c\beta)^{1/2}, \quad M_0 = 4\pi\rho_c L_0^3, \quad \phi_0 \equiv \beta^{-1} = \frac{GM_0}{L_0} \tag{30}$$

where $\rho_c = \rho(0)$. All other physical variables can be expressed in terms of the dimensionless quantities

$$x \equiv \frac{r}{L_0}, \quad n \equiv \frac{\rho}{\rho_c}, \quad m = \frac{M(r)}{M_0}, \quad y \equiv \beta\left[\phi - \phi(0)\right]. \tag{31}$$

In terms of $y(x)$ the isothermal equation (29) becomes

$$\frac{1}{x^2}\frac{d}{dx}(x^2\frac{dy}{dx}) = e^{-y} \tag{32}$$

with the boundary condition $y(0) = y'(0) = 0$. Let us consider the nature of solutions to this equation.

By direct substitution, we see that $n = (2/x^2)\,, m = 2x, y = 2\ln x$ satisfies these equations. This solution, however, is singular at the origin and hence is not physically admissible. The importance of this solution lies in the fact that other (physically admissible) solutions tend to this solution [1], [19] for large values of x. This asymptotic behavior of all solutions shows that the density decreases as $(1/r^2)$ for large r implying that the mass contained inside a sphere of radius r increases as $M(r) \propto r$ at large r. To find physically useful solutions, it is necessary to assume that the solution is cutoff at some radius R. For example, one may assume that the system is enclosed in a spherical box of radius R. In what follows, it will be assumed that the system has some cutoff radius R.

Equation (32) is invariant under the transformation $y \to y + a\ ;\ x \to kx$ with $k^2 = \mathrm{e}^a$. This invariance implies that, given a solution with some value of $y(0)$, we can obtain the solution with any other value of $y(0)$ by simple rescaling. Therefore, only one of the two integration constants in (32) is really non-trivial. Hence it must be possible to reduce the degree of the equation from two to one by a judicious choice of variables [19]. One such set of variables are:

$$v \equiv \frac{m}{x}; \quad u \equiv \frac{nx^3}{m} = \frac{nx^2}{v}. \tag{33}$$

In terms of v and u, (29) becomes

$$\frac{u}{v}\frac{dv}{du} = -\frac{(u-1)}{(u+v-3)}. \tag{34}$$

The boundary conditions $y(0) = y'(0) = 0$ translate into the following: v is zero at $u = 3$, and $(dv/du) = -5/3$ at (3,0). The solution $v\,(u)$ has to be obtained numerically: it is plotted in Fig. 2 as the spiraling curve. The singular points of this differential equation are given by the intersection of the straight lines $u = 1$ and $u + v = 3$ on which, the numerator and denominator of the right hand side of (34) vanishes; that is, the singular point is at $u_s = 1$, $v_s = 2$ corresponding to the solution $n = (2/x^2), m = 2x$. It is obvious from the nature of the equations that the solutions will spiral around the singular point.

The nature of the solution shown in Fig. 2 allows us to put an interesting bounds on physical quantities including energy. To see this, we shall compute the total energy E of the isothermal sphere. The potential and kinetic energies are

$$\begin{aligned} U &= -\int_0^R \frac{GM(r)}{r}\frac{dM}{dr}dr = -\frac{GM_0^2}{L_0}\int_0^{x_0} mnx dx \\ K &= \frac{3}{2}\frac{M}{\beta} = \frac{3}{2}\frac{GM_0^2}{L_0}m(x_0) = \frac{GM_0^2}{L_0}\frac{3}{2}\int_0^{x_0} nx^2 dx \end{aligned} \tag{35}$$

where $x_0 = R/L_0$. The total energy is, therefore,

$$\begin{aligned} E = K + U &= \frac{GM_0^2}{2L_0}\int_0^{x_0} dx(3nx^2 - 2mnx) \\ &= \frac{GM_0^2}{2L_0}\int_0^{x_0} dx \frac{d}{dx}\{2nx^3 - 3m\} = \frac{GM_0^2}{L_0}\{n_0x_0^3 - \frac{3}{2}m_0\} \end{aligned} \tag{36}$$

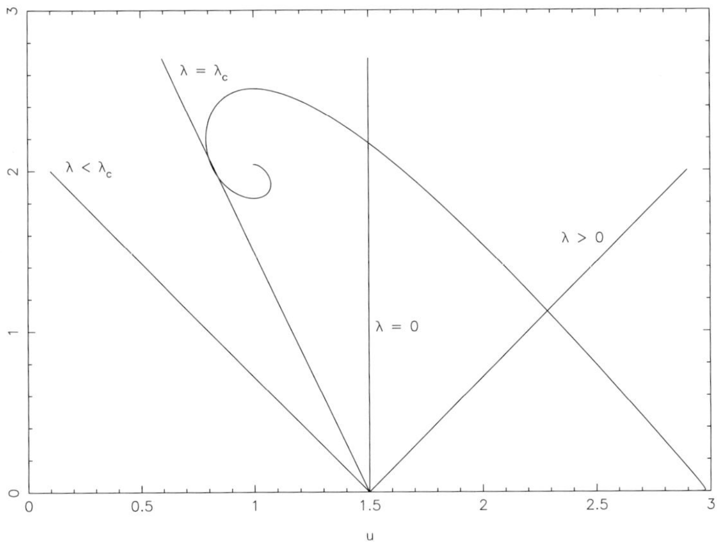

Fig. 2. Bound on RE/GM^2 for the isothermal sphere

where $n_0 = n(x_0)$ and $m_0 = m(x_0)$. The dimensionless quantity (RE/GM^2) is given by

$$\lambda = \frac{RE}{GM^2} = \frac{1}{v_0}\{u_0 - \frac{3}{2}\}. \tag{37}$$

Note that the combination (RE/GM^2) *is a function of* (u, v) *alone.* Let us now consider the constraints on λ. Suppose we specify some value for λ by specifying R, E and M. Then such an isothermal sphere *must* lie on the curve

$$v = \frac{1}{\lambda}\left(u - \frac{3}{2}\right); \qquad \lambda \equiv \frac{RE}{GM^2} \tag{38}$$

which is a straight line through the point $(1.5, 0)$ with the slope λ^{-1}. On the other hand, since *all* isothermal spheres must lie on the $u-v$ curve, *an isothermal sphere can exist only if the line in (38) intersects the* $u - v$ *curve.*

For large positive λ (positive E) there is just one intersection. When $\lambda = 0$, (zero energy) we still have a unique isothermal sphere. (For $\lambda = 0$, (38) is a vertical line through $u = 3/2$.). When λ is negative (negative E), the line can cut the $u-v$ curve at more than one point; thus more than one isothermal sphere can exist with a given value of λ. [Of course, specifying M, R, E individually will remove this non-uniqueness]. But as we decrease λ (more and more negative E) the line in (38) will slope more and more to the left; and when λ is smaller than a critical value λ_c, the intersection will cease to exist. *Thus no isothermal sphere can exist if* (RE/GM^2) *is below a critical value* λ_c.[1] This fact follows

[1] This derivation is due to the author [16]. It is surprising that Chandrasekhar, who has worked out the isothermal sphere in uv coordinates as early as 1939, missed

immediately from the nature of $u - v$ curve and (38). The value of λ_c can be found from the numerical solution in figure. It turns out to be about (-0.335).

The isothermal sphere has a special status as a solution to the mean field equations. Isothermal spheres, however, cannot exist if $(RE/GM^2) < -0.335$. Even when $(RE/GM^2) > -0.335$, the isothermal solution need not be stable. The stability of this solution can be investigated by studying the second variation of the entropy. Such a detailed analysis shows that the following results are true [15], [20], [16]. (i) Systems with $(RE/GM^2) < -0.335$ cannot evolve into isothermal spheres. Entropy has no extremum for such systems. (ii) Systems with $((RE/GM^2) > -0.335)$ and $(\rho(0) > 709\,\rho(R))$ can exist in a meta-stable (saddle point state) isothermal sphere configuration. Here $\rho(0)$ and $\rho(R)$ denote the densities at the center and edge respectively. The entropy extrema exist but they are not local maxima. (iii) Systems with $((RE/GM^2) > -0.335)$ and $(\rho(0) < 709\,\rho(R))$ can form isothermal spheres which are local maximum of entropy.

Part II. Gravitational Clustering in Expanding Universe

Let us next consider the gravitational clustering of a system of collision-less point particles *in an expanding universe* which poses several challenging theoretical questions. Though the problem can be tackled in a 'practical' manner using high resolution numerical simulations, such an approach hides the physical principles which govern the behaviour of the system. To understand the physics, it is necessary that we attack the problem from several directions using analytic and semi analytic methods. These sections will describe such attempts and will emphasize the semi analytic approach and outstanding issues, rather than more well established results. Some basic results in cosmology are summarized in Appendix A.

5 Gravitational Clustering at Nonlinear Scales

The expansion of the universe sets a natural length scale (called the Hubble radius) $d_H = c(\dot{a}/a)^{-1}$ which is about 4000 Mpc in the current universe. Since the non linear effects due to gravitational clustering occur at significantly smaller length scales, it is possible to use Newtonian gravity to describe these phenomena. In any region small compared to $d_{\rm H}$ one can set up an unambiguous coordinate system in which the *proper* coordinate of a particle $\mathbf{r}(t) = a(t)\mathbf{x}(t)$ satisfies the Newtonian equation $\ddot{\mathbf{r}} = -\nabla_{\mathbf{r}}\Phi$ where Φ is the gravitational potential. Expanding $\ddot{\mathbf{r}}$ and writing $\Phi = \Phi_{\rm FRW} + \phi$ where $\Phi_{\rm FRW}$ is due to the smooth

---

discovering the energy bound shown in Fig. 2. Chandrasekhar [19] has the uv curve but does not over-plot lines of constant λ. If he had done that, he would have discovered Antonov instability decades before Antonov did [15].

(mean) density of matter and ϕ is due to the perturbation in the density, we get

$$\ddot{a}\mathbf{x} + 2\dot{a}\dot{\mathbf{x}} + a\ddot{\mathbf{x}} = -\nabla_{\mathbf{r}}\Phi_{\text{FRW}} - \nabla_{\mathbf{r}}\phi = -\nabla_{\mathbf{r}}\Phi_{\text{FRW}} - a^{-1}\nabla_{\mathbf{x}}\phi \tag{39}$$

The first terms on both sides of the equation ($\ddot{a}\mathbf{x}$ and $-\nabla_{\mathbf{r}}\Phi_{\text{FRW}}$) should match since they refer to the global expansion of the background FRW universe (see (115) of Appendix A). Equating them individually gives the results

$$\ddot{\mathbf{x}} + 2\frac{\dot{a}}{a}\dot{\mathbf{x}} = -\frac{1}{a^2}\nabla_x\phi\,; \qquad \Phi_{\text{FRW}} = -\frac{1}{2}\frac{\ddot{a}}{a}r^2 = -\frac{2\pi G}{3}\rho_b r^2 \tag{40}$$

where ϕ is the gravitational potential generated by the perturbed, Newtonian, mass density through

$$\nabla_x^2\phi = 4\pi G a^2(\delta\rho) = 4\pi G\rho_b a^2\delta. \tag{41}$$

If $\mathbf{x}_i(t)$ is the trajectory of the i–th particle, then equations for gravitational clustering in an expanding universe, in the Newtonian limit, can be summarized as

$$\ddot{\mathbf{x}}_i + \frac{2\dot{a}}{a}\dot{\mathbf{x}}_i = -\frac{1}{a^2}\nabla_{\mathbf{x}}\phi; \quad \nabla_x^2\phi = 4\pi G a^2\rho_b\delta \tag{42}$$

where $\rho_b(t)$ is the smooth background density of matter. We stress that, in the non-relativistic limit, the perturbed potential ϕ satisfies the usual Poisson equation.

Usually one is interested in the evolution of the density contrast $\delta(t,\mathbf{x}) \equiv [\rho(t,\mathbf{x}) - \rho_b(t)]/\rho_b(t)$ rather than in the trajectories. Since the density contrast can be expressed in terms of the trajectories of the particles, it should be possible to write down a differential equation for $\delta(t,\mathbf{x})$ based on the equations for the trajectories $\mathbf{x}_i(t)$ derived above. It is, however, somewhat easier to write down an equation for $\delta_{\mathbf{k}}(t)$ which is the spatial Fourier transform of $\delta(t,\mathbf{x})$. To do this, we begin with the fact that the density $\rho(\mathbf{x},t)$ due to a set of point particles, each of mass m, is given by

$$\rho(\mathbf{x},t) = \frac{m}{a^3(t)}\sum_i \delta_D[\mathbf{x} - \mathbf{x}_i(t)] \tag{43}$$

where $\mathbf{x}_i(t)$ is the trajectory of the ith particle. To verify the a^{-3} normalization, we can calculate the average of $\rho(\mathbf{x},t)$ over a large volume V. We get

$$\rho_b(t) \equiv \int\frac{d^3\mathbf{x}}{V}\rho(\mathbf{x},t) = \frac{m}{a^3(t)}\left(\frac{N}{V}\right) = \frac{M}{a^3V} = \frac{\rho_0}{a^3} \tag{44}$$

where N is the total number of particles inside the volume V and $M = Nm$ is the mass contributed by them. Clearly $\rho_b \propto a^{-3}$, as it should. The density contrast $\delta(\mathbf{x},t)$ is related to $\rho(\mathbf{x},t)$ by

$$1 + \delta(\mathbf{x},t) \equiv \frac{\rho(\mathbf{x},t)}{\rho_b} = \frac{V}{N}\sum_i \delta_D[\mathbf{x} - \mathbf{x}_i(t)] = \int d\mathbf{q}\,\delta_D[\mathbf{x} - \mathbf{x}_T(t,\mathbf{q})]. \tag{45}$$

In arriving at the last equality we have taken the continuum limit by replacing: (i) $\mathbf{x}_i(t)$ by $\mathbf{x}_T(t,\mathbf{q})$ where $\mathbf{q}$ stands for a set of parameters (like the initial position, velocity etc.) of a particle; for simplicity, we shall take this to be initial position. (ii) (V/N) by $d^3\mathbf{q}$ since both represent volume per particle. Fourier transforming both sides we get

$$\delta_{\mathbf{k}}(t) \equiv \int d^3\mathbf{x} e^{-i\mathbf{k}\cdot\mathbf{x}} \delta(\mathbf{x},t) = \int d^3\mathbf{q} \exp[-i\mathbf{k}.\mathbf{x}_T(t,\mathbf{q})] - (2\pi)^3 \delta_D(\mathbf{k}) \tag{46}$$

Differentiating this expression, and using the equation of motion (42) for the trajectories give, after straightforward algebra, the equation (see [21], [22], [24]):

$$\ddot{\delta}_{\mathbf{k}} + 2\frac{\dot{a}}{a}\dot{\delta}_{\mathbf{k}} = \frac{1}{a^2}\int d^3\mathbf{q} e^{-i\mathbf{k}.\mathbf{x}_T(t,\mathbf{q})} \left\{ i\mathbf{k}\cdot\nabla\phi - a^2(\mathbf{k}\cdot\dot{\mathbf{x}}_T)^2 \right\} \tag{47}$$

which can be further manipulated to give

$$\ddot{\delta}_{\mathbf{k}} + 2\frac{\dot{a}}{a}\dot{\delta}_{\mathbf{k}} = 4\pi G\rho_b \delta_{\mathbf{k}} + A_{\mathbf{k}} - B_{\mathbf{k}} \tag{48}$$

with

$$A_{\mathbf{k}} = 4\pi G\rho_b \int \frac{d^3\mathbf{k}'}{(2\pi)^3} \delta_{\mathbf{k}'}\delta_{\mathbf{k}-\mathbf{k}'} \left[\frac{\mathbf{k}.\mathbf{k}'}{k'^2}\right] \tag{49}$$

$$B_{\mathbf{k}} = \int d^3\mathbf{q} \left(\mathbf{k}.\dot{\mathbf{x}}_T\right)^2 \exp\left[-i\mathbf{k}.\mathbf{x}_T(t,\mathbf{q})\right]. \tag{50}$$

This equation is exact but involves $\dot{\mathbf{x}}_T(t,\mathbf{q})$ on the right hand side and hence cannot be considered as closed.

The structure of (48) and (50) can be simplified if we use the perturbed gravitational potential (in Fourier space) $\phi_{\mathbf{k}}$ related to $\delta_{\mathbf{k}}$ by

$$\delta_{\mathbf{k}} = -\frac{k^2\phi_{\mathbf{k}}}{4\pi G\rho_b a^2} = -\left(\frac{k^2 a}{4\pi G\rho_0}\right)\phi_{\mathbf{k}} = -\left(\frac{2}{3H_0^2}\right) k^2 a \phi_{\mathbf{k}} \tag{51}$$

and write the integrand for $A_{\mathbf{k}}$ in the symmetrised form as

$$\begin{aligned} \delta_{\mathbf{k}'}\delta_{\mathbf{k}-\mathbf{k}'}\left[\frac{\mathbf{k}.\mathbf{k}'}{k'^2}\right] &= \frac{1}{2}\delta_{\mathbf{k}'}\delta_{\mathbf{k}-\mathbf{k}'}\left[\frac{\mathbf{k}.\mathbf{k}'}{k'^2} + \frac{\mathbf{k}.(\mathbf{k}-\mathbf{k}')}{|\mathbf{k}-\mathbf{k}'|^2}\right] \\ &= \frac{1}{2}\left(\frac{\delta'_{\mathbf{k}}}{k'^2}\right)\left(\frac{\delta_{\mathbf{k}-\mathbf{k}'}}{|\mathbf{k}-\mathbf{k}'|^2}\right)\left[(\mathbf{k}-\mathbf{k}')^2\mathbf{k}.\mathbf{k}' + k'^2\left(k^2 - \mathbf{k}.\mathbf{k}'\right)\right] \\ &= \frac{1}{2}\left(\frac{2a}{3H_0^2}\right)^2 \phi_{\mathbf{k}'}\phi_{\mathbf{k}-\mathbf{k}'}\left[k^2(\mathbf{k}.\mathbf{k}' + k'^2) - 2(\mathbf{k}.\mathbf{k}')^2\right] \end{aligned} \tag{52}$$

In terms of $\phi_{\mathbf{k}}$ (with $\mathbf{k}' = (\mathbf{k}/2) + \mathbf{p}$), (48) becomes,

$$\begin{aligned} \ddot{\phi}_{\mathbf{k}} + 4\frac{\dot{a}}{a}\dot{\phi}_{\mathbf{k}} &= -\frac{1}{2a^2}\int \frac{d^3\mathbf{p}}{(2\pi)^3}\phi_{\frac{1}{2}\mathbf{k}+\mathbf{p}}\phi_{\frac{1}{2}\mathbf{k}-\mathbf{p}}\left[\left(\frac{k}{2}\right)^2 + p^2 - 2\left(\frac{\mathbf{k}.\mathbf{p}}{k}\right)^2\right] \\ &+ \left(\frac{3H_0^2}{2}\right)\int\frac{d^3\mathbf{q}}{a}\left(\frac{\mathbf{k}.\dot{\mathbf{x}}}{k}\right)^2 e^{i\mathbf{k}.\mathbf{x}} \end{aligned} \tag{53}$$

where $\mathbf{x} = \mathbf{x}_T(t,\mathbf{q})$. We shall now consider several applications of this equation.

5.1 Application 1: 'Renormalizability' of Gravity

Gravitational clustering in an expanding universe brings out an interesting feature about gravity which can be described along the following lines. Let us consider a large number of particles which are interacting via gravity in an expanding background and form bound gravitating systems. At some time t, let us assume that a fraction f of the particles are in virialized, self-gravitating clusters which are reasonably immune to the effect of expansion. Imagine that we replace each of these clusters by single particles at their centers of mass with masses equal to the total mass of the corresponding clusters. (The total number of particles have now been reduced but, if the original number was sufficiently large, we may assume that the resulting number of particles is again large enough to carry on further evolution with a valid statistical description.) We now evolve the resulting system to a time t' and compare the result with what would have been obtained if we had evolved the original system directly to t'. Obviously, the characteristics of the system at small scales (corresponding to the typical size R of the clusters at time t) will be quite different. However, at large scales ($kR \ll 1$), the characteristics will be the same both the systems. In other words, the effect of a bunch of particles, in a virialized cluster, on the rest of the system is described, to the lowest order, by just the monopole moment of the cluster – which is taken into account by replacing the cluster by a single particle at the center of mass having appropriate mass. In this sense, gravitational interactions are "renormalizable" – where the term is used in the specific sense defined above.

The result has been explicitly verified in simulations [23] but one must emphasize that the whole idea of numerical simulations of such systems tacitly assumes the validity of this result. If the detailed non linear behaviour at small scales, say within galaxies, influences very large scale behaviour of the universe (say, at super cluster scales), then it will be impossible to simulate large scale structure in the universe with finite resolution.

One may wonder how this feature (renormalizability of gravity) is taken care of in (48). Inside a galaxy cluster, for example, the velocities $\dot{\mathbf{x}}_T$ can be quite high and one might think that this could influence the evolution of $\delta_{\mathbf{k}}$ at all scales. This does not happen and, to the lowest order, the contribution from virialized bound clusters cancel in $A_{\mathbf{k}} - B_{\mathbf{k}}$. We shall now provide a proof of this result [24].

We begin by writing the right hand side $\mathcal{R}$ of (47) concentrating on the particles in a given cluster.

$$\mathcal{R} = \int d^3\mathbf{q}\, e^{-i\mathbf{k}\cdot\mathbf{Q}} ik^a \left\{ \frac{\partial_a \phi}{a^2} + i\dot{Q}^a(\mathbf{k}\cdot\dot{\mathbf{Q}}) \right\} \tag{54}$$

where we have used the notation $\mathbf{Q} = \mathbf{x}_T$ for the trajectories of the particles and the subscripts $a, b, ... = 1, 2, 3$ denote the components of the vector. For a set of particles which form a bound virialized cluster, we have from (42) the equation of motion

$$\ddot{Q}^i + 2\frac{\dot{a}}{a}\dot{Q}^i = -\frac{1}{a^2}\frac{\partial\phi}{\partial Q^i} \tag{55}$$

We multiply this equation by Q^j, sum over the particles in the particular cluster and symmetrize on i and j, to obtain the equation

$$\frac{d^2}{dt^2}\sum Q^iQ^j - 2\sum \dot{Q}^i\dot{Q}^j + 2\frac{\dot{a}}{a}\frac{d}{dt}\sum Q^iQ^j = -\frac{1}{a^2}\sum\left(Q^j\frac{\partial\phi}{\partial Q^i} + Q^i\frac{\partial\phi}{\partial Q^j}\right) \tag{56}$$

We use the summation symbol, rather than integration over $\mathbf{q}$ merely to emphasize the fact that the sum is over particles of a given cluster. Let us now consider the first term in the right hand side of (54) with the origin of the coordinate system shifted to the center of mass of the cluster. Expanding the exponential as $e^{-i\mathbf{k}\cdot\mathbf{Q}} \approx (1 - i\mathbf{k}\cdot\mathbf{Q}) + \mathcal{O}(k^2R^2)$ where R is the size of the cluster, we find that in the first term, proportional to $\nabla\phi$, the sum of the forces acting on all the particles in the cluster (due to self gravity) vanishes. The second term gives, on symmetrization,

$$\sum ia^{-2}(\mathbf{k}\cdot\nabla\phi)e^{-i\mathbf{k}\cdot\mathbf{Q}} \approx \frac{k^ak^b}{2a^2}\sum\left(Q^b\frac{\partial\phi}{\partial Q^a} + Q^a\frac{\partial\phi}{\partial Q^b}\right) \tag{57}$$

Using (56) we find that

$$\sum ia^{-2}(\mathbf{k}\cdot\nabla\phi)e^{-i\mathbf{k}\cdot\mathbf{Q}} = +\sum(\mathbf{k}\cdot\dot{\mathbf{Q}})^2 + \frac{1}{2}\left(\frac{d^2}{dt^2} + 2\frac{\dot{a}}{a}\frac{d}{dt}\right)\sum(\mathbf{k}\cdot\mathbf{Q})^2 \tag{58}$$

The second term is of order $\mathcal{O}(k^2R^2)$ and can be ignored, giving

$$\sum ia^{-2}(\mathbf{k}\cdot\nabla\phi)e^{-i\mathbf{k}\cdot\mathbf{Q}} \approx +\sum(\mathbf{k}\cdot\dot{\mathbf{Q}})^2 + \mathcal{O}(k^2R^2) \tag{59}$$

Consider next the second term in the right hand side of (54) with the same expansion for the exponential. We get

$$\begin{aligned}\sum(ik^a)e^{-i\mathbf{k}\cdot\mathbf{Q}}\left[i\dot{Q}^ak^b\dot{Q}^b\right] &\cong -\sum k^ak^b\dot{Q}^a\dot{Q}^b(1 - i\mathbf{k}\cdot\mathbf{Q}) + \mathcal{O}(k^2R^2)\\ &= -\sum(\mathbf{k}\cdot\dot{\mathbf{Q}})^2 + \sum(\mathbf{k}\cdot\dot{\mathbf{Q}})^2k^aQ^a + \mathcal{O}(k^2R^2)\end{aligned} \tag{60}$$

The second term is effectively zero for any cluster of particles for which $\mathbf{Q} \to -\mathbf{Q}$ is a symmetry. Hence the two terms on the right hand side of (54) cancel each other for all particles in the same virialized cluster; that is, the term $(A_\mathbf{k} - B_\mathbf{k})$ receives contribution only from particles which are not bound to any of the clusters to the order $\mathcal{O}(k^2R^2)$. If the typical size of the clusters formed at time t is R, then for wave-numbers with $k^2R^2 \ll 1$, we can ignore the contribution from the clusters. Hence, in the limit of $k \to 0$ we can ignore $(A_\mathbf{k} - B_\mathbf{k})$ term and treat (48) as linear in $\delta_\mathbf{k}$; large spatial scales in the universe can be described by linear perturbation theory even when small spatial scales are highly non linear.

There is, however, an important caveat to this claim. In the right hand side of (48) one is comparing the first term (which is linear in $\delta_\mathbf{k}$) with the contribution $(A_\mathbf{k} - B_\mathbf{k})$. If, at the relevant wavenumber, the first term $4\pi G\rho_b\delta_\mathbf{k}$ is negligibly small, then the *only* contribution will come from $(A_\mathbf{k} - B_\mathbf{k})$ and, of course, we cannot ignore it in this case. The above discussion shows that this contribution will scale as k^2R^2 and will lead to a development of $\delta_{\mathbf{k}} \propto k^2$ if originally (in linear theory) $\delta_k \propto k^n$ with $n > 2$ as $k \to 0$.

We shall next describe the linear evolution; the development of $\delta_{\mathbf{k}} \propto k^2$ tail at large spatial scales will be taken up in section 5.4.

5.2 Application 2: Evolution at Large Scales

If the density contrasts are small and linear perturbation theory is to be valid, we should be able to ignore the terms $A_{\mathbf{k}}$ and $B_{\mathbf{k}}$ in (48). Thus the linear perturbation theory in Newtonian limit is governed by the equation

$$\ddot{\delta}_{\mathbf{k}} + 2\frac{\dot{a}}{a}\dot{\delta}_{\mathbf{k}} = 4\pi G\rho_b\delta_{\mathbf{k}} \tag{61}$$

From the structure of (48) it is clear that we will obtain the linear equation if $A_{\mathbf{k}} \ll 4\pi G\rho_b\delta_{\mathbf{k}}$ and $\mathbf{B}_{\mathbf{k}} \ll 4\pi G\rho_b\delta_{\mathbf{k}}$. A *necessary* condition for this $\delta_{\mathbf{k}} \ll 1$ but this is *not* a sufficient condition – a fact often ignored or incorrectly treated in literature. As we saw in the last section, if $\delta_{\mathbf{k}} \to 0$ for certain range of $\mathbf{k}$ at $t = t_0$ (but is nonzero elsewhere) then $(A_{\mathbf{k}} - B_{\mathbf{k}}) \gg 4\pi G\rho_b\delta_{\mathbf{k}}$ and the growth of perturbations around $\mathbf{k}$ will be entirely determined by nonlinear effects.

For the present, we shall assume that $A_{\mathbf{k}} - B_{\mathbf{k}}$ is ignorable and study the resulting system. In that case, (61) has the growing solution $\delta_{\mathbf{k}}(t) = [a(t)/a(t_i)]\delta_{\mathbf{k}}(t_i)$ in the matter dominated universe with $a(t) \propto t^{2/3}, \rho_b \propto a^{-3}$. This shows that when linear perturbation theory is applicable the density perturbations grow as $a(t)$. The power spectrum $P(\mathbf{k}, t) = |\delta_{\mathbf{k}}(t)|^2$ and the correlation function $\xi(\mathbf{x}, t)$ [which is the Fourier transform of the power spectrum] both grow as $a^2(t)$ while $\phi_{\mathbf{k}} \propto k^{-2}(\delta_{\mathbf{k}}/a)$ remains constant in time. This analysis allows us to fix the evolution of clustering at sufficiently large scales (that is, for sufficiently small k) uniquely. The clustering at these scales which is well described by linear theory, and the power spectrum grows as a^2.

5.3 Application 3: Formation of First Non Linear Structures

A useful insight into the nature of linear perturbation theory (as well as nonlinear clustering) can be obtained by examining the nature of particle trajectories which lead to the growth of the density contrast $\delta_L(a) \propto a$ obtained above. To determine the particle trajectories corresponding to the linear limit, let us start by writing the trajectories in the form

$$\mathbf{x}_T(a, \mathbf{q}) = \mathbf{q} + \mathbf{L}(a, \mathbf{q}) \tag{62}$$

where $\mathbf{q}$ is the Lagrangian coordinate (indicating the original position of the particle) and $\mathbf{L}(a, \mathbf{q})$ is the displacement. The corresponding Fourier transform of the density contrast is given by the general expression

$$\delta_{\mathbf{k}}(a) = \int d^3\mathbf{q}\, e^{-i\mathbf{k}\cdot\mathbf{q} - i\mathbf{k}\cdot\mathbf{L}(a,\mathbf{q})} - (2\pi)^3\delta_{\text{Dirac}}[\mathbf{k}] \tag{63}$$

In the linear regime, we expect the particles to have moved very little and hence we can expand the integrand in the above equation in a Taylor series in $(\mathbf{k}\cdot\mathbf{L})$. This gives, to the lowest order,

$$\delta_{\mathbf{k}}(a) \cong -\int d^3\mathbf{q}\, e^{-i\mathbf{k}\cdot\mathbf{q}}(i\mathbf{k}\cdot\mathbf{L}(a,\mathbf{q})) = -\int d^3\mathbf{q}\, e^{-i\mathbf{k}\cdot\mathbf{q}}\,(\nabla_{\mathbf{q}}\cdot\mathbf{L}) \tag{64}$$

showing that $\delta_{\mathbf{k}}(a)$ is Fourier transform of $-\nabla_{\mathbf{q}}.\mathbf{L}(a,\mathbf{q})$. This allows us to identify $\nabla\cdot\mathbf{L}(a,\mathbf{q})$ with the original density contrast in real space $-\delta_{\mathbf{q}}(a)$. Using the Poisson equation we can write $\delta_{\mathbf{q}}(a)$ as a divergence; that is

$$\nabla\cdot\mathbf{L}(a,\mathbf{q}) = -\delta_{\mathbf{q}}(a) = -\frac{2}{3}H_0^{-2}a\nabla\cdot(\nabla\phi) \tag{65}$$

which, in turn, shows that a consistent set of displacements that will lead to $\delta(a)\propto a$ is given by

$$\mathbf{L}(a,\mathbf{q}) = -(\nabla\psi)a \equiv a\mathbf{u}(\mathbf{q}); \qquad \psi = (2/3)H_0^{-2}\phi \tag{66}$$

The trajectories in this limit are, therefore, linear in a:

$$\mathbf{x}_T(a,\mathbf{q}) = \mathbf{q} + a\mathbf{u}(\mathbf{q}) \tag{67}$$

A useful approximation to describe the quasi linear stages of clustering is obtained by using the trajectory in (67) as an ansatz valid *even at quasi linear epochs.* In this approximation, called Zeldovich approximation, the proper Eulerian position $\mathbf{r}$ of a particle is related to its Lagrangian position $\mathbf{q}$ by

$$\mathbf{r}(t) \equiv a(t)\mathbf{x}(t) = a(t)[\mathbf{q} + a(t)\mathbf{u}(\mathbf{q})] \tag{68}$$

where $\mathbf{x}(t)$ is the comoving Eulerian coordinate. If the initial, unperturbed, density is $\overline{\rho}$ (which is independent of $\mathbf{q}$), then the conservation of mass implies that the perturbed density will be

$$\rho(\mathbf{r},t)d^3\mathbf{r} = \bar{\rho}d^3\mathbf{q}. \tag{69}$$

Therefore

$$\rho(\mathbf{r},t) = \bar{\rho}\left[\det\left(\frac{\partial q_i}{\partial r_j}\right)\right]^{-1} = \frac{\bar{\rho}/a^3}{\det(\partial x_j/\partial q_i)} = \frac{\rho_b(t)}{\det(\delta_{ij} + a(t)(\partial u_j/\partial q_i))} \tag{70}$$

where we have set $\rho_b(t) = [\bar{\rho}/a^3(t)]$. Since $\mathbf{u}(\mathbf{q})$ is a gradient of a scalar function, the Jacobian in the denominator of (70) is the determinant of a real symmetric matrix. This matrix can be diagonalised at every point $\mathbf{q}$, to yield a set of eigenvalues and principal axes as a function of $\mathbf{q}$. If the eigenvalues of $(\partial u_j/\partial q_i)$ are $[-\lambda_1(\mathbf{q}),\, -\lambda_2(\mathbf{q}),\, -\lambda_3(\mathbf{q})]$ then the perturbed density is given by

$$\rho(\mathbf{r},t) = \frac{\rho_b(t)}{(1-a(t)\lambda_1(\mathbf{q}))(1-a(t)\lambda_2(\mathbf{q}))(1-a(t)\lambda_3(\mathbf{q}))} \tag{71}$$

where $\mathbf{q}$ can be expressed as a function of $\mathbf{r}$ by solving (68). This expression describes the effect of deformation of an infinitesimal, cubical, volume (with the faces of the cube determined by the eigenvectors corresponding to λ_n) and the consequent change in the density. A positive λ denotes collapse and negative λ signals expansion.

In a over dense region, the density will become infinite if one of the terms in brackets in the denominator of (71) becomes zero. In the generic case, these eigenvalues will be different from each other, so that we can take, say, $\lambda_1 \geq \lambda_2 \geq \lambda_3$. At any particular value of $\mathbf{q}$ the density will diverge for the first time when $(1 - a(t)\lambda_1) = 0$; at this instant the material contained in a cube in the $\mathbf{q}$ space gets compressed to a sheet in the $\mathbf{r}$ space, along the principal axis corresponding to λ_1. Thus sheet like structures, or 'pancakes', will be the first nonlinear structures to form when gravitational instability amplifies density perturbations.

5.4 Application 4: A Non Linear Tail at Small Wavenumber

There is an interesting and curious result which is characteristic of gravitational clustering that can be obtained directly from (53). Consider an initial power spectrum which has very little power at large scales; more precisely, we shall assume that $P(k)$ dies faster than k^4 for small k. If these large scales are described by linear theory – as one would have normally expected, then the power at these scales can only grow as a^2 and it will always be sub dominant to k^4. It turns out that this conclusion is incorrect. As the system evolves, small scale nonlinearities will develop in the system and – if the large scales have too little power intrinsically (i.e. if n is large) – then the long wavelength power will soon be dominated by the "tail" of the short wavelength power arising from the nonlinear clustering. This occurs because, in (48), the nonlinear term ($A_{\mathbf{k}}$ - $B_{\mathbf{k}}$) $= \mathcal{O}(k^2R^2)$ can dominate over $4\pi G\rho_b\delta_{\mathbf{k}}$ at long wavelengths (as $\mathbf{k} \to 0$) and lead to the development of a k^4 power spectrum at large scales. This is a purely non linear effect which we shall now describe.

To do this, we shall use the Zeldovich approximation to obtain [4] a closed equation for $\phi_{\mathbf{k}}$. The trajectories in Zeldovich approximation, given by (67) can be used in (53) to provide a *closed* integral equation for $\phi_{\mathbf{k}}$. In this case,

$$\mathbf{x}_T(\mathbf{q}, a) = \mathbf{q} + a\nabla\psi; \quad \dot{\mathbf{x}_{\mathrm{T}}} = \left(\frac{2a}{3t}\right)\nabla\psi; \quad \psi = \frac{2}{3H_0^2}\varphi \tag{72}$$

and, to the same order of accuracy, $B_{\mathbf{k}}$ in (50) becomes:

$$\int d^3\mathbf{q}\,(\mathbf{k}\cdot\dot{\mathbf{x}_{\mathrm{T}}})^2\,e^{-i\mathbf{k}\cdot(\mathbf{q}+\mathbf{L})} \simeq \int d^3\mathbf{q}(\mathbf{k}\cdot\dot{\mathbf{x}_{\mathrm{T}}})^2 e^{-i\mathbf{k}\cdot\mathbf{q}} \tag{73}$$

Substituting these expressions in (53) we find that the gravitational potential is described by the closed integral equation:

$$
\begin{aligned}
\ddot{\phi}_{\mathbf{k}} + 4\frac{\dot{a}}{a}\dot{\phi}_{\mathbf{k}} &= -\frac{1}{3a^2}\int \frac{d^3\mathbf{p}}{(2\pi)^3}\phi_{\frac{1}{2}\mathbf{k}+\mathbf{p}}\phi_{\frac{1}{2}\mathbf{k}-\mathbf{p}}\mathcal{G}(\mathbf{k},\mathbf{p}) \\
\mathcal{G}(\mathbf{k},\mathbf{p}) &= \frac{7}{8}k^2 + \frac{3}{2}p^2 - 5\left(\frac{\mathbf{k}\cdot\mathbf{p}}{k}\right)^2
\end{aligned} \tag{74}
$$

This equation provides a powerful method for analyzing non linear clustering since estimating $(A_{\mathbf{k}} - B_{\mathbf{k}})$ by Zeldovich approximation has a very large domain of applicability.

A formal way of obtaining the k^4 tail is to solve (74) for long wavelengths [4]; i.e. near $\mathbf{k} = 0$. Writing $\phi_{\mathbf{k}} = \phi_{\mathbf{k}}^{(1)} + \phi_{\mathbf{k}}^{(2)} + ...$ where $\phi_{\mathbf{k}}^{(1)} = \phi_{\mathbf{k}}^{(L)}$ is the time *independent* gravitational potential in the linear theory and $\phi_{\mathbf{k}}^{(2)}$ is the next order correction, we get from (74), the equation

$$
\ddot{\phi}_{\mathbf{k}}^{(2)} + 4\frac{\dot{a}}{a}\dot{\phi}_{\mathbf{k}}^{(2)} \cong -\frac{1}{3a^2}\int \frac{d^3\mathbf{p}}{(2\pi)^3}\phi^L_{\frac{1}{2}\mathbf{k}+\mathbf{p}}\phi^L_{\frac{1}{2}\mathbf{k}-\mathbf{p}}\mathcal{G}(\mathbf{k},\mathbf{p}) \tag{75}
$$

The solution to this equation is the sum of a solution to the homogeneous part [which decays as $\dot{\phi} \propto a^{-4} \propto t^{-8/3}$ giving $\phi \propto t^{-5/3}$] and a particular solution which grows as a. Ignoring the decaying mode at late times and taking $\phi_{\mathbf{k}}^{(2)} = aC_{\mathbf{k}}$ one can determine $C_{\mathbf{k}}$ from the above equation. Plugging it back, we find the lowest order correction to be,

$$
\phi_{\mathbf{k}}^{(2)} \cong -\left(\frac{2a}{21H_0^2}\right)\int \frac{d^3\mathbf{p}}{(2\pi)^3}\phi^L_{\frac{1}{2}\mathbf{k}+\mathbf{p}}\phi^L_{\frac{1}{2}\mathbf{k}-\mathbf{p}}\mathcal{G}(\mathbf{k},\mathbf{p}) \tag{76}
$$

Near $\mathbf{k} \simeq 0$, we have

$$
\begin{aligned}
\phi_{\mathbf{k}\simeq 0}^{(2)} &\cong -\frac{2a}{21H_0^2}\int \frac{d^3\mathbf{p}}{(2\pi)^3}|\phi^L_{\mathbf{p}}|^2\left[\frac{7}{8}k^2 + \frac{3}{2}p^2 - \frac{5(\mathbf{k}\cdot\mathbf{p})^2}{k^2}\right] \\
&= \frac{a}{126\pi^2 H_0^2}\int_0^\infty dp p^4 |\phi_{\mathbf{p}}^{(L)}|^2
\end{aligned} \tag{77}
$$

which is independent of $\mathbf{k}$ to the lowest order. Correspondingly the power spectrum for density $P_\delta(k) \propto a^2k^4P_\varphi(k) \propto a^4k^4$ in this order.

The generation of long wavelength k^4 tail is easily seen in simulations if one starts with a power spectrum that is sharply peaked in $|\mathbf{k}|$. Figure 3 shows the results of such a simulation [25] in which the y-axis is $[\Delta(k)/a(t)]$ where $\Delta^2(k) \equiv k^3P/2\pi^2$ is the power per logarithmic band in k. In linear theory $\Delta \propto a$ and this quantity should not change. The curves labelled by $a = 0.12$ to $a = 20.0$ show the effects of nonlinear evolution, especially the development of k^4 tail.

5.5 Application 5: Generation of Small Scale Power

Figure 3 also shows that, as the clustering proceeds, power is generated at spatial scales smaller than the scale k_0^{-1} at which the power is injected. This feature can

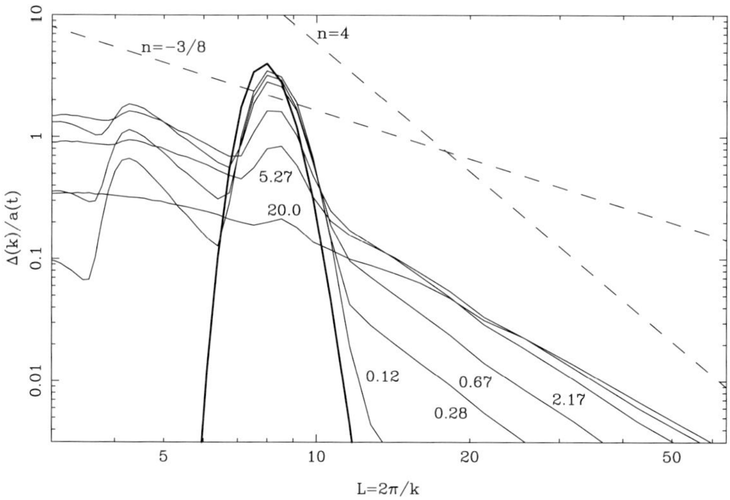

Fig. 3. The transfer of power to long wavelengths forming a k^4 tail is illustrated using simulation results. Power is injected in the form of a narrow peak at $L = 8$. Note that the y–axis is (Δ/a) so that there will be no change of shape under linear evolution with $\Delta \propto a$. As time goes on a k^4 tail is generated which itself evolves according to the nonlinear scaling relation discussed later on

also be easily understood [23] from (76). Let the initial gravitational potential and the density contrast (in the linear theory) be sharply peaked at the wave number k_0, say, with:

$$\phi_{\mathbf{k}}^L = \mu \frac{H_0^2}{k_0^4} \delta_D[|\mathbf{k}| - k_0]; \quad k_0^3 \delta_{\mathbf{k}}^L = -\frac{2}{3}(\mu a) k_0 \delta_D[|\mathbf{k}| - k_0] \tag{78}$$

where μ is dimensionless constant indicating the strength of the potential and the other factors ensure the correct dimensions. Equation (76) shows that, the right hand side is nonzero only when the *magnitudes* of both the vectors $[(1/2)\mathbf{k} + \mathbf{p}]$ and $[(1/2)\mathbf{k} - \mathbf{p}]$ are k_0. This requires $\mathbf{k} \cdot \mathbf{p} = 0, (k/2)^2 + p^2 = k_0^2$. (This constraint has a simple geometric interpretation: Given any $\mathbf{k}$, with $k \leq 2k_0$ one constructs a vector $\mathbf{k}/2$ inside a sphere of radius k_0 and a vector $\mathbf{p}$ perpendicular to $\mathbf{k}/2$ reaching up to the shell at radius k_0 where the initial power resides. Obviously, this construction is possible only for $k < 2k_0$.) Performing the integration in (76) we find that

$$\phi_{\mathbf{k}}^{(2)} = \frac{\mu^2}{56\pi^2} \frac{H_0^2}{k_0^5} a \left(1 - \frac{k^2}{4k_0^2}\right) \left(1 + \frac{k^2}{3k_0^2}\right) \tag{79}$$

[We have again ignored the decaying mode which arises as a solution to the homogeneous part.] The corresponding power spectrum for the density field $P(k) = |\delta_k|^2 \propto a^2 k^4 |\phi_k|^2$ will evolve as

$$P^{(2)}(k) \propto (\mu a)^4 q^4 \left(1 - \frac{1}{4} q^2\right)^2 \left(1 + \frac{1}{3} q^2\right)^2; \quad q = \frac{k}{k_0} \tag{80}$$

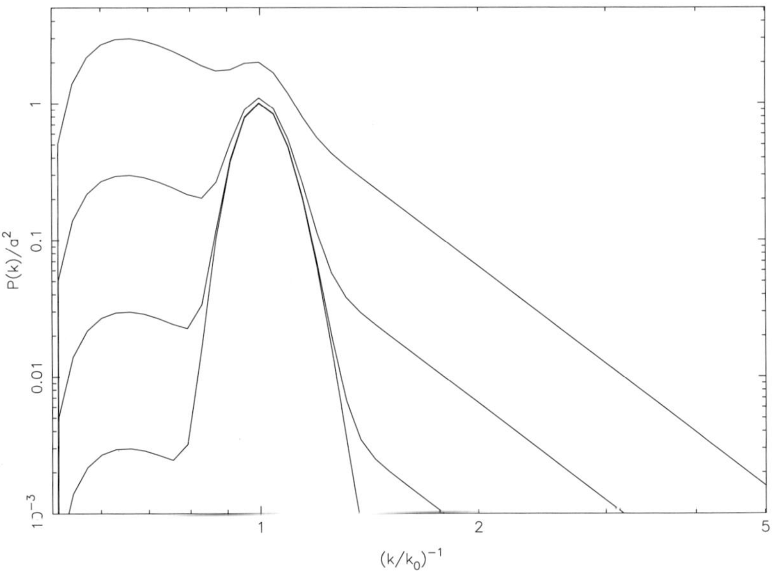

Fig. 4. Analytic model for transfer of power in gravitational clustering. The initial power was injected at the wave number k_0 with a Gaussian window of width $\Delta k/k_0 = 0.1$. First order calculation shows that the power is transfered to larger spatial scales with a k^4 tail and to the shorter spatial scales, all the way down to $(1/2)k_0^{-1}$. The plot gives the total power spectrum divided by a^2 (with y−axis normalized arbitrarily) at different times with a^2 changing by factor 10 between any two curves

The power at large spatial scales ($k \to 0$) varies as k^4 as discussed before. The power has also been generated at smaller scales in the range $k_0 < k < 2k_0$ with $P^{(2)}(k)$ being a maximum at $k_m \approx 1.54k_0$ corresponding to the length scale $k_m^{-1} \approx 0.65k_0^{-1}$. Figure 4 shows the power spectrum for density field (divided by a^2 to eliminate linear growth) computed analytically for a narrow Gaussian initial power spectrum centered at $k_0 = 1$. The curves are for $(\mu a/56\pi^2)^2 = 10^{-3}, 10^{-2}, 10^{-1}$ and 1. The similarity between Figs. 3 and 4 is striking and allows us to understand the simulation results. The key difference is that, in the simulations, newly generated power will further produce power at $4k_0, 8k_0, ...$ and each of these will give rise to a k^4 tail to the right. The resultant power will, of course, be more complicated than predicted by our analytic model. The generation of power near this maximum at $k_m^{-1} = 0.65k_0^{-1}$ is clearly visible as a second peak in Fig. 4 and around $2\pi/k_0 \approx 4$ in Fig. 3.

If we had taken the initial power spectrum to be Dirac delta function in the wave vector $\mathbf{k}$ (rather than on the *magnitude* of the wave vector, as we have done) the right hand side of (76) will contribute only when $(\frac{1}{2}\mathbf{k} \pm \mathbf{p}) = \mathbf{k}_0$. This requires $\mathbf{p} = 0$ and $\mathbf{k} = 2\mathbf{k}_0$ showing that the power is generated exactly at the second harmonic of the wave number. Spreading the initial power on a shell of radius k_0, spreads the power over different vectors leading to the result obtained above.

Equation (78) shows that $k_0^3\delta_{\mathbf{k}}$ will reach non linearity for $\mu a \approx (3/2)$. The situation is different as regards the gravitational potential due to the large numerical factor $56\pi^2$; the gravitational potential fluctuations are comparable to the original fluctuations only when $\mu a \approx 56\pi^2$. We shall say more about power transfer in gravitational clustering in section 7.

5.6 Application 6: Spherical Approximation

In the nonlinear regime – when $\delta \gtrsim 1$ – it is not possible to solve (48) exactly. Some progress, however, can be made if we assume that the trajectories are homogeneous; i.e. $\mathbf{x}(t,\mathbf{q}) = f(t)\mathbf{q}$ where $f(t)$ is to be determined. In this case, the density contrast is

$$\delta_{\mathbf{k}}(t) = \int d^3\mathbf{q} e^{-if(t)\mathbf{k}.\mathbf{q}} - (2\pi)^3\delta_D(\mathbf{k}) = (2\pi)^3\delta_D(\mathbf{k})[f^{-3} - 1] \equiv (2\pi)^3\delta_D(\mathbf{k})\delta(t) \tag{81}$$

where we have defined $\delta(t) \equiv \left[f^{-3}(t) - 1\right]$ as the amplitude of the density contrast for the $\mathbf{k} = 0$ mode. It is now straightforward to compute A and B in (48). We have

$$A = 4\pi G\rho_b\delta^2(t)[(2\pi)^3\delta_D(\mathbf{k})] \tag{82}$$

and

$$\begin{aligned} B &= \int d^3\mathbf{q}(k^a q_a)^2 \dot{f}^2 e^{-if(k_a q^a)} = -\dot{f}^2\frac{\partial^2}{\partial f^2}[(2\pi)^3\delta_D(f\mathbf{k})] \\ &= -\frac{4}{3}\frac{\dot{\delta}^2}{(1+\delta)}[(2\pi)^3\delta_D(\mathbf{k})] \end{aligned} \tag{83}$$

so that (48) becomes

$$\ddot{\delta} + 2\frac{\dot{a}}{a}\dot{\delta} = 4\pi G\rho_b(1+\delta)\delta + \frac{4}{3}\frac{\dot{\delta}^2}{(1+\delta)} \tag{84}$$

To understand what this equation means, let us consider, at some initial epoch t_i, a spherical region of the universe which has a slight constant over-density compared to the background. As the universe expands, the over-dense region will expand more slowly compared to the background, will reach a maximum radius, contract and virialize to form a bound nonlinear system. Such a model is called "spherical top-hat". It is possible to show that (84) is the same as the equation governing density evolution in a spherical model [4]. The detailed analysis of the spherical model [26] shows that the virialized systems formed at any given time have a mean density which is typically 200 times the background density of the universe at that time. Hence, it is often convenient to divide the growth of structures into three regimes: linear regime with $\delta \ll 1$; quasi-linear regime with $1 \lesssim \delta \lesssim 200$; and nonlinear regime with $\delta \gtrsim 200$.

6 Nonlinear Scaling Relations

Given an initial density contrast, one can trivially obtain the two point correlation function at any later epoch in the *linear* theory. If there is a procedure for relating the nonlinear correlation function and linear correlation function (even approximately) then one can make considerable progress in understanding nonlinear clustering. It is actually possible to do this [27], [28], [29], [30] and relate the exact mean correlation function

$$\bar{\xi}(t,x) = \frac{3}{x^3}\int_0^x \xi(t,y)y^2 dy \tag{85}$$

to the one computed in the linear theory, along the following lines:

The mean number of neighbours within a distance x of any given particle is

$$N(x,t) = (na^3)\int_o^x 4\pi y^2 dy[1+\xi(y,t)] \tag{86}$$

when n is the comoving number density. Hence the conservation law for pairs implies

$$\frac{\partial\xi}{\partial t} + \frac{1}{ax^2}\frac{\partial}{\partial x}[x^2(1+\xi)v] = 0 \tag{87}$$

where $v(t,x)$ denotes the mean relative velocity of pairs at separation x and epoch t. Using

$$(1+\xi) = \frac{1}{3x^2}\frac{\partial}{\partial x}[x^3(1+\bar{\xi})] \tag{88}$$

in (87), we get

$$\frac{1}{3x^2}\frac{\partial}{\partial x}[x^3\frac{\partial}{\partial t}(1+\bar{\xi})] = -\frac{1}{ax^2}\frac{\partial}{\partial x}\left[\frac{v}{3}\frac{\partial}{\partial x}[x^2(1+\bar{\xi})]\right]. \tag{89}$$

Integrating, we find:

$$x^3\frac{\partial}{\partial t}(1+\bar{\xi}) = -\frac{v}{a}\frac{\partial}{\partial x}[x^3(1+\bar{\xi})]. \tag{90}$$

[The integration would allow the addition of an arbitrary function of t on the right hand side. We have set this function to zero so as to reproduce the correct limiting behaviour]. It is now convenient to change the variables from t to a, thereby getting an equation for $\bar{\xi}$:

$$a\frac{\partial}{\partial a}[1+\bar{\xi}(a,x)] = \left(\frac{v}{-\dot{a}x}\right)\frac{1}{x^2}\frac{\partial}{\partial x}[x^3(1+\bar{\xi}(a,x))] \tag{91}$$

or, defining $h(a,x) = -(v/\dot{a}x)$

$$\left(\frac{\partial}{\partial\ln a} - h\frac{\partial}{\partial\ln x}\right)(1+\bar{\xi}) = 3h\left(1+\bar{\xi}\right). \tag{92}$$

This equation shows that the behaviour of $\bar{\xi}(a,x)$ is essentially decided by h, the dimensionless ratio between the mean relative velocity v and the Hubble velocity $\dot{a}x = (\dot{a}/a)r$, both evaluated at scale x.

In the extreme nonlinear limit $\left(\bar{\xi} \gg 1\right)$, we may expect bound structures not to expand with Hubble flow. To maintain a stable structure, the relative pair velocity $v\,(a,x)$ of particles separated by x should balance the Hubble velocity $Hr = \dot{a}x$; hence, $v = -\dot{a}x$ or $h\,(a,x) \cong 1$. The behaviour of $h\,(a,x)$ for $\bar{\xi} \ll 1$ is more complicated and can be shown [4],[22] that $h = (2/3)\bar{\xi}$ in the limit of $\bar{\xi} \ll 1$. Thus $h(a,x)$ depends on (a,x) only through $\bar{\xi}(a,x)$ in the linear limit, while $h \cong -1$ is the nonlinear limit. This suggests the ansatz that h depends on a and x only through $\bar{\xi}(a,x)$; that is, we assume that $h(a,x) = h[\bar{\xi}(a,x)]$. It is then possible to find a solution to (92) which reduces to the form $\bar{\xi} \propto a^2$ for $\bar{\xi} \ll 1$ using the method of characteristics [28]. The final result can be given as:

$$\bar{\xi}_L(a,l) = \exp\left(\frac{2}{3}\int^{\bar{\xi}(a,x)} \frac{d\mu}{h(\mu)(1+\mu)}\right); \quad l = x[1+\bar{\xi}(a,x)]^{1/3}. \qquad (93)$$

Given the function $h(\bar{\xi})$, this relates $\bar{\xi}_L$ and $\bar{\xi}$ or – equivalently – gives the mapping $\bar{\xi}(a,x) = U[\bar{\xi}_L(a,l)]$ between the nonlinear and linear correlation functions evaluated at different scales x and l. The lower limit of the integral is chosen to give $\ln\bar{\xi}$ for small values of $\bar{\xi}$ on the linear regime. [The (2/3) factor in the exponent becomes $(2/D)$ in D-dimensions and $l = x[1+\bar{\xi}(a,x)]^{1/D}$.]

The following points need to be stressed regarding this result: (i) Among all statistical indicators, it is *only* $\bar{\xi}$ which obeys a nonlinear scaling relation (NSR) of the form $\bar{\xi}_{\rm NL}(a,x) = U\left[\bar{\xi}_L(a,l)\right]$. Attempts to write similar relations for ξ or $P(k)$ have no fundamental justification. (ii) The non locality of the relation represents the transfer of power in gravitational clustering and cannot be ignored – or approximated by a local relation between $\bar{\xi}_{NL}(a,x)$ and $\bar{\xi}_L(a,x)$.

Given the form of $h(\bar{\xi})$, (93) determines the relation $\bar{\xi} = U[\bar{\xi}_L]$. It is, however, easier to determine the form of U, directly from theory and this was done in [29]. Here, I shall provide a more intuitive but physically motivated derivation along the following lines:

In the linear regime $\left(\bar{\xi} \ll 1, \bar{\xi}_L \ll 1)\right)$ we clearly have $U(\bar{\xi}_L) \simeq \bar{\xi}_L$. We can divide the non linear phases of evolution conveniently into two parts, which I will call quasi-linear $(1 \lesssim \bar{\xi} \lesssim 200)$ and non linear $(200 \lesssim \bar{\xi})$. In the quasi-linear phase, regions of high density contrast will undergo collapse and in the non linear phase more and more virialized systems will get formed. We recall that, in the study of finite gravitating systems made of point particles and interacting via Newtonian gravity, isothermal spheres play an important role and are the local maxima of entropy. Hence dynamical evolution drives the system towards an $(1/x^2)$ density profile. Since one expects similar considerations to hold at small scales, during the late stages of evolution of the universe, we may hope that isothermal spheres with $(1/x^2)$ profile may still play a role in the late stages of evolution of clustering in an expanding background. However, while converting the density profile to correlation function, we need to distinguish between two cases. In the quasi-linear regime, dominated by the collapse of high density peaks, the density profile around any peak will scale as the correlation function and we will have $\bar{\xi} \propto (1/x^2)$. On the other hand, in the nonlinear end, we will be probing the structure inside a single halo and $\xi(\mathbf{x})$ will vary as $\langle \rho(\mathbf{x}+\mathbf{y})\rho(\mathbf{y})\rangle$. If $\rho \propto |x|^{-\epsilon}$, then $\xi \propto |x|^{-\gamma}$ with $\gamma = 2\epsilon - 3$. This gives $\bar{\xi} \propto (1/x)$ for $\epsilon = 2$. Thus, if isothermal spheres are the generic contributors, then we expect the correlation function to vary as $(1/x)$ and nonlinear scales, steepening to $(1/x^2)$ at intermediate scales. Further, since isothermal spheres are local maxima of entropy, a configuration like this should remain undistorted for a long duration. This argument suggests that a $\bar{\xi}$ which goes as $(1/x)$ at small scales and $(1/x^2)$ at intermediate scales is likely to grow approximately as a^2 at all scales. The form of $U[\xi_L]$ must be consistent with this feature.

This criterion allows us to determine the form of U in the quasi-linear and non linear phases. In the quasi-linear regime, we want a linear correlation function of the form $\bar{\xi}_L(a,l) = a^2/l^2$ to be mapped to $\bar{\xi}_{\rm NL}(a,x) \propto a^2/x^2$. We thus demand $U[a^2l^{-2}] \propto a^2x^{-2}$ with $l \approx x\bar{\xi}_{\rm NL}^{1/3}$. It is trivial to see that this requires $U(z) \propto z^3$. Similarly, in the non linear end, we expect a linear correlation function of the form $\bar{\xi}_L(a,l) = a^2/l$ to be mapped to $\bar{\xi}_{\rm NL}(a,x) \propto a^2/x$ requiring $U[a^2l^{-1}] \propto a^2x^{-1}$ with $l \approx x\bar{\xi}_{\rm NL}^{1/3}$. This gives $U(z) \propto z^{3/2}$.

Combining all the results we find that the nonlinear mean correlation function can be expressed in terms of the linear mean correlation function by the relation:

$$\bar{\xi}(a,x) = \begin{cases} \bar{\xi}_L(a,l) & \text{for } \bar{\xi}_L < 1,\ \bar{\xi} < 1 \\ \bar{\xi}_L(a,l)^3 & \text{for } 1 < \bar{\xi}_L < 5.85,\ 1 < \bar{\xi} < 200 \\ 14.14\bar{\xi}_L(a,l)^{3/2} & \text{for } 5.85 < \bar{\xi}_L,\ 200 < \bar{\xi} \end{cases} \tag{94}$$

The numerical coefficients have been determined by continuity arguments. We have assumed the linear result to be valid up to $\bar{\xi} = 1$ and the virialisation to occur at $\bar{\xi} \approx 200$ which is result arising from the spherical model. The true test of such a model, of course, is N-body simulations and remarkably enough, simulations are very well represented by relations of the above form. [The fact that numerical simulations show a correlation between $\bar{\xi}(a,x)$ and $\bar{\xi}_L(a,l)$ was originally pointed out in [27]. These authors, however, gave a multi parameter fit to the data which has the virtue of representing the numerical work accurately. Equation (94), on the other hand, portrays the clear physical interpretation behind the result.] The *exact* values of the numerical coefficients can be obtained from simulations and it changes 14.14 to 11.7 and 200 to 125 in the above relations.

The result we have obtained for the non linear end corresponds to an assumption called *stable clustering* . If we ignore the effect of mergers, then it seems reasonable that virialized systems should maintain their densities and sizes in proper coordinates, i.e. the clustering should be "stable". This would require the correlation function to have the form $\bar{\xi}_{NL}(a,x) = a^3F(ax)$. [The factor a^3 arising from the decrease in background density]. One can easily show that this will lead to $U \propto \xi_L^{3/2}$ in the non linear regime. Another way deriving this result is to note that if the *proper* size of the objects do not change with time, then

$\dot{r} = 0$ which implies, statistically, $v = -\dot{a}x$ requiring $h = 1$. Integrating (93) with appropriate boundary condition leads to $U \propto \bar{\xi}_L^{3/2}$.

In case mergers of structures are important, one would consider this assumption to be suspect [31]. We can, however, generalize the above argument in the following manner: If the virialized systems have reached stationarity in the statistical sense, then it seems reasonable to assume that the function h – which is the ratio between two velocities – should reach some constant value. In that case, one can integrate (93) and obtain the result $\bar{\xi}_{NL} = a^{3h}F(a^h x)$ where h now denotes the asymptotic value. A similar argument will now show that

$$\bar{\xi}_{NL}(a,x) \propto [\bar{\xi}_L(a,l)]^{3h/2} \tag{95}$$

in the general case. If the linear power spectrum has an index n (with $P_L(k) \propto k^n, \bar{\xi}_L \propto x^{-(n+3)}$) then one would get

$$\bar{\xi}(a,x) \propto a^{(3-\gamma)h}x^{-\gamma}; \qquad \gamma = \frac{3h(n+3)}{2+h(n+3)} \tag{96}$$

Simulations are not accurate enough to fix the value of h; in particular, the asymptotic value of h could depend on n within the accuracy of the simulations.

It is possible to obtain similar relations between $\bar{\xi}(a,x)$ and $\bar{\xi}_L(a,l)$ in two dimensions as well by repeating the above analysis [32]. In 2-D the scaling relations turn out to be

$$\bar{\xi}(a,x) \propto \begin{cases} \bar{\xi}_L(a,l) & (\mathrm{Linear}) \\ \bar{\xi}_L(a,l)^2 & (\mathrm{Quasi-linear}) \\ \bar{\xi}_L(a,l)^h & (Nonlinear) \end{cases} \tag{97}$$

where h again denotes the asymptotic value. For power law spectrum the nonlinear correction function will $\bar{\xi}_{NL}(a,x) = a^{2-\gamma}x^{-\gamma}$ with $\gamma = 2(n+2)/(n+4)$. If we generalize the concept of stable clustering to mean constancy of h in the nonlinear epoch, then the correlation function will behave as $\bar{\xi}_{NL}(a,x) = a^{2h}F(a^h x)$. In this case, if the spectrum is a power law then the nonlinear and linear indices are related to

$$\gamma = \frac{2h(n+2)}{2+h(n+2)} \tag{98}$$

[In general, $\bar{\xi} \propto \bar{\xi}_L^D$ in the quasi-linear regime and $\bar{\xi} \propto \bar{\xi}_L^{Dh/2}$ in the nonlinear regime [33] where D is the dimension of space]. All the features discussed in the case of 3 dimensions are present here as well. For example, if the asymptotic value of h scales with n such that $h(n+2) =$ constant then the nonlinear index will be independent of the linear index. Fig. 5 shows the results of numerical simulation in 2D, which suggests that $h = 3/4$ asymptotically [34]

The ideas presented here can be generalized in two obvious directions [35]: (i) By considering peaks of different heights, drawn from an initial Gaussian random field, and averaging over the probability distribution one can obtain a more precise NSR. (ii) By using a generalized ansatz for higher order correlation functions, one can attempt to compute different statistical parameters in the quasi linear and nonlinear regimes.

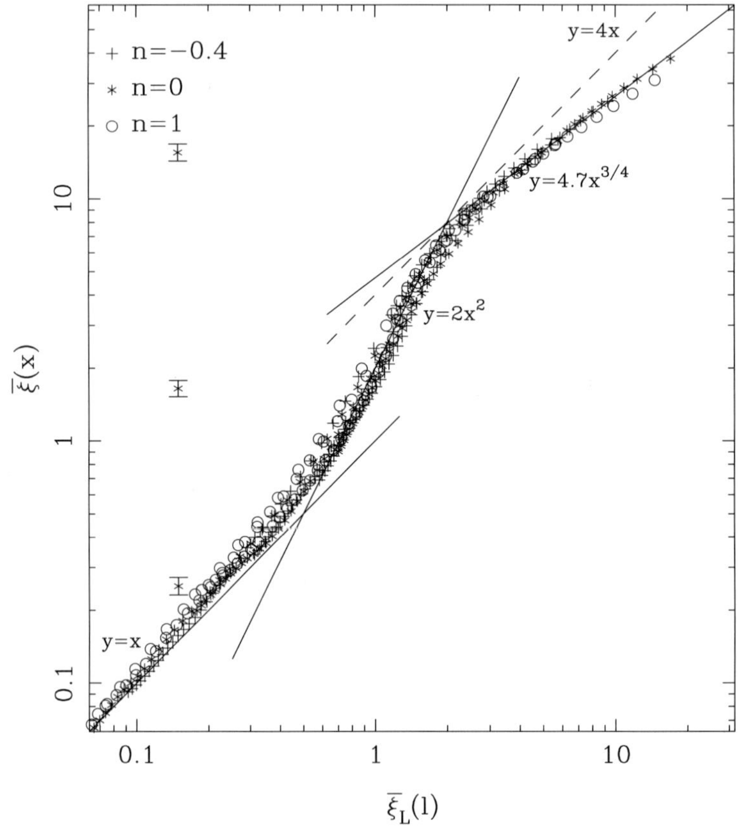

Fig. 5. The NSR in 2D giving $\bar{\xi}(x)$ as a function of the linear mean correlation function $\bar{\xi}_L(l)$. The theoretical predictions at the three regimes are shown by solid lines of slopes $1, 2, 3/4$. The data points are from simulations for different power spectra and different epochs. The overlapping of data points shows the existence of an NSR and its agreement with the theoretical prediction lends support to the model discussed in the text. The error bars indicate the typical accuracy of the result

The entire approach side steps three important issues which needs to be investigated more thoroughly than has been done in the literature.

- There has been no "first-principle" derivation of the non linear scaling relations in spite of several attempts by different groups. (The NSR does not depend on G, for example!) For example, I have not obtained the NSR from the basic equation (53) derived earlier in Sect. 5 because I (or anyone else) do not know how to go about doing this.
- A strong constraint any correlation function must satisfy is that its Fourier transform must be positive definite. The necessary and sufficient condition for a function $f(x)$ to have a positive definite Fourier transform is that it must be a convolution, viz., it must be expressible as an integral over y of a product $g(x+y)g(y)$ where g is some other function. In the NSR, one starts

with a linear correlation function and maps it by a functional transform to obtain a non linear correlation function. There is no guarantee that this functional transform will actually lead to a valid correlation function in the sense that it will have a positive definite Fourier transform. In fact, it will not except approximately; it is not clear what are the implications of this result.

- Both ξ and $\bar{\xi}$ can be negative for a range of scale. The NSR described above is not applicable to power spectra with negative correlation function. It is possible to generalize the formalism to cover this situation but such an extension leads to results which are fairly counter intuitive [30].

7 Critical Indices and Power Transfer in Gravitational Clustering

Given a model for the evolution of the power spectra in the quasi linear and nonlinear regimes, one could explore whether evolution of gravitational clustering possesses any universal characteristics. The derivation of NSR in the previous section has encoded in it the feature that $n = -1$ and $n = -2$ (in the power spectrum with n defined through the relation $P \propto k^n$) will appear as "critical indices" in the quasi-linear and non linear regimes respectively, in the sense that $\bar{\xi} \propto a^2$ for these indices (which matches with the linear evolution, $\xi_L \propto a^2$).

This suggests even a stronger result: any generic initial power spectrum will be driven to a particular form of power spectrum in the late stages of the evolution. Such a possibility arises because of the following reason: We see from (94) that [in the quasi linear regime] spectra with $n < -1$ grow faster than a^2 while spectra with $n > -1$ grow slower than a^2. This feature could drive the spectral index to $n = n_c \approx -1$ in the quasi linear regime irrespective of the initial index. Similarly, the index in the nonlinear regime could be driven to $n \approx -2$ during the late time evolution. So the spectral indices -1 and -2 are some kind of "fixed points" in the quasi linear and nonlinear regimes.

This effect can be understood better by studying the "effective" index for the power spectra at different stages of the evolution [25]. Let us define a *local* index for rate of clustering by

$$n_a(a, x) \equiv \frac{\partial \ln \bar{\xi}(a, x)}{\partial \ln a} \tag{99}$$

which measures how fast $\bar{\xi}(a, x)$ is growing. When $\bar{\xi}(a, x) \ll 1$, then $n_a = 2$ irrespective of the spatial variation of $\bar{\xi}(a, x)$ and the evolution preserves the shape of $\bar{\xi}(a, x)$. However, as clustering develops, the growth rate will depend on the spatial variation of $\bar{\xi}(a, x)$. Defining the effective spatial slope by

$$-[n_x(a, x) + 3] \equiv \frac{\partial \ln \bar{\xi}(a, x)}{\partial \ln x} \tag{100}$$

one can rewrite (92) as

$$n_a = h(\frac{3}{\bar{\xi}(a, x)} - n_x) \tag{101}$$

At any given scale of nonlinearity, decided by $\bar{\xi}(a,x)$, there exists a critical spatial slope n_c such that $n_a > 2$ for $n_x < n_c$ [implying rate of growth is faster than predicted by linear theory] and $n_a < 2$ for $n_x > n_c$ [with the rate of growth being slower than predicted by linear theory]. The critical index n_c is fixed by setting $n_a = 2$ in (101) at any instant. This requirement is established from the physically motivated desire to have a form of the two point correlation function that remains invariant under time evolution. Since the linear end of the two point correlation function scales as a^2, the required invariance of form constrains the index n_a to be 2 at *all* scales. The fact that $n_a > 2$ for $n_x < n_c$ and $n_a < 2$ for $n_x > n_c$ will tend to "straighten out" correlation functions towards the critical slope. [We are assuming that $\bar{\xi}(a,x)$ has a slope that is decreasing with scale, which is true for any physically interesting case]. From the NSR it is easy to see that in the range $1 \lesssim \bar{\xi} \lesssim 200$, the critical index is $n_c \approx -1$ and for $200 \lesssim \bar{\xi}$, the critical index is $n_c \approx -2$. This clearly suggests that the local effect of evolution is to drive the correlation function to have a shape with $(1/x)$ behaviour at nonlinear regime and $(1/x^2)$ in the intermediate regime. Such a correlation function will have $n_a \approx 2$ and hence will grow at a rate close to a^2.

To go from the scalings in two limits to an actual profile, we can use some fitting function. One possible interpolation [36] between the two limits is given by:

$$\bar{\xi}(a,x) = \left(\frac{Ba}{2}\left(\sqrt{1+\frac{L}{x}} - 1\right)\right)^2 \tag{102}$$

with L, B being constants. If we evolve this profile (with the optimum value of $B = 38.6$) from $a^2 = 1$ to $a^2 \approx 1000$ using the NSR, and plot $[\bar{\xi}(a,x)/a^2]$ against x, then the curves virtually fall on top of each other within about 10 per cent [36]. This shows that the profile does grow approximately as a^2.

These considerations also allow us to predict the nature of power transfer in gravitational clustering. Suppose that, initially, the power spectrum was sharply peaked at some scale $k_0 = 2\pi/L_0$ and has a small width Δk. When the peak amplitude of the spectrum is far less than unity, the evolution will be described by linear theory and there will be no flow of power to other scales. But once the peak approaches a value close to unity, power will be generated at other scales due to nonlinear couplings even though the amplitude of perturbations in these scales are less than unity. For modes with no initial power [or exponentially small power], nonlinear coupling provides the only driving term in (48). These generate power at the scale $\mathbf{k}$ through mode-coupling, provided power exists at some other scale. [We saw this effect in section 5.5 for a simple model.]

In $\mathbf{x}$–space, this arises along the following lines: If the initial spectrum is sharply peaked at some scale L_0, first structures to form are voids with a typical diameter L_0. Formation and fragmentation of sheets bounding the voids lead to generation of power at scales $L < L_0$. First bound structures will then form at the mass scale corresponding to L_0. In such a model, $\bar{\xi}_{\rm lin}$ at $L < L_0$ is nearly constant with an effective index of $n \approx -3$. Assuming we can use (94) with the local index in this case, we expect the power to grow very rapidly as compared to the linear rate of a^2. [The rate of growth is a^6 for $n = -3$ and a^4

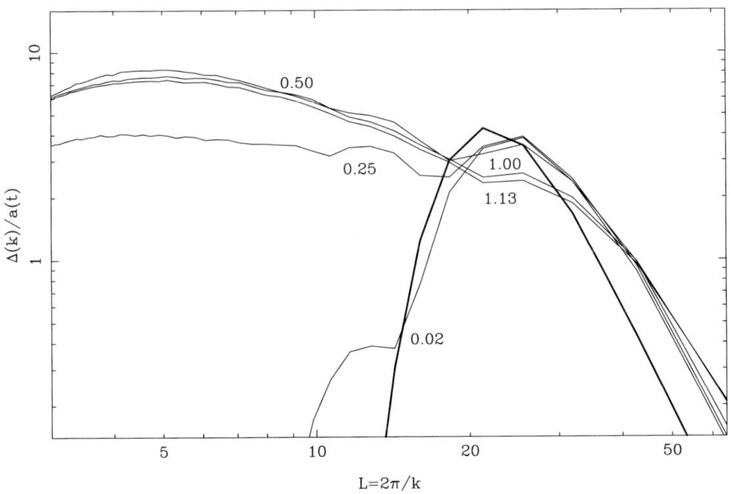

Fig. 6. The transfer of power in gravitational clustering

for $n = -2.5$.] Different rate of growth for regions with different local index will lead to steepening of the power spectrum and an eventual slowing down of the rate of growth. In this process, which is the dominant one, the power transfer is mostly from large scales to small scales. [Except for the generation of the k^4 tail at large scales which we have discussed earlier in subsection 5.4.]

From our previous discussion, we would have expected such an evolution to lead to a "universal" power spectrum with some critical index $n_c \approx -1$ for which the rate of growth is that of linear theory – viz., a^2. In fact, the same results should hold even when there exists small scale power; numerical simulations confirm this prediction and show that – in the quasi linear regime, with $1 < \delta < 100$ – power spectrum indeed has a universal slope (see Figs. 6, 7; for more details, see [25]).

The initial power spectrum for Fig. 6 was a Gaussian peaked at the scale $k_0 = 2\pi/L_0; L_0 = 24$ and having a spread $\Delta k = 2\pi/128$. The amplitude of the peak was chosen so that $\Delta_{lin}(k_0 = 2\pi/L_0, a = 0.25) = 1$, where $\Delta^2(k) = k^3 P(k)/(2\pi^2)$ and $P(k)$ is the power spectrum. The y-axis is $\Delta(k)/a$, the power per logarithmic scale divided by the linear growth factor. This is plotted as a function of scale $L = 2\pi/k$ for different values of scale factor $a(t)$ and the curves are labeled by the value of a. As we have divided the power spectrum by its linear rate of growth, the change of shape of the spectrum occurs strictly because of non-linear mode coupling. It is clear from this figure that power at small scales grows rapidly and saturates at the growth rate close to the linear rate [shown by crowding of curves] at later epochs. The effective index for the power spectrum approaches $n = -1$ within the accuracy of the simulations. This figure clearly demonstrates the general features we expected from our understanding of scaling relations.

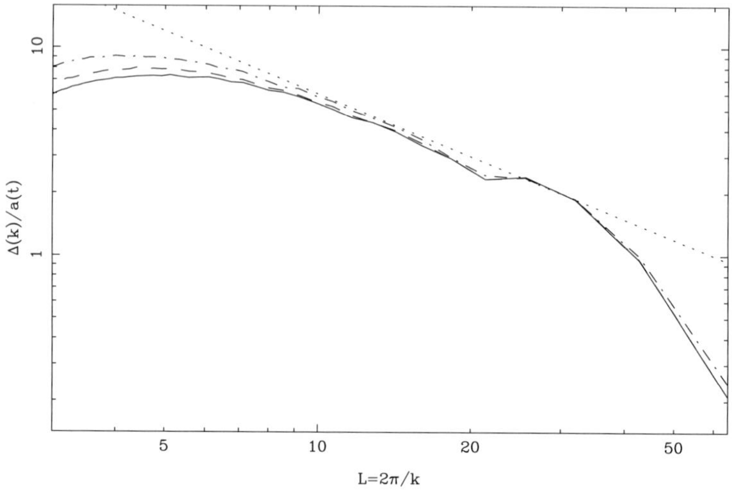

Fig. 7. The growth of gravitational clustering towards a universal power spectrum $P(k) \propto k^{-1}$

Figure 7 compares power spectra of three different models at a late epoch. Model I was described in the last para; Model II had initial power concentrated in *two* narrow windows in k-space. In addition to power around $L_0 = 24$ as in model I, we added power at $k_1 = 2\pi/L_1; L_1 = 8$ using a Gaussian with same width as that used in model I. Amplitude at L_1 was chosen five times higher than that at $L_0 = 24$, thus $\Delta_{lin}(k_1, a = 0.05) = 1$. Model III was similar to model II, with the small scale peak shifted to $k_1 = 2\pi/L_1; L_1 = 12$. The amplitude of the small scale peak was the same as in Model II. At this epoch $\Delta_{lin}(k_0) = 4.5$ and it is clear from this figure that the power spectra of these models are very similar to one another.

8 Universal Behaviour of Gravitational Clustering in the Asymptotic Limit

In the study of isolated gravitating systems, it is easy to arrive at a broad picture of the late time structure of the system. We have seen that it will be made of a very compact core and a diffuse halo and – if the system is confined by a reflecting wall – then this state could be thought of as being made of two distinct phases as described in section 2. By and large, all memory of initial stage could have been wiped out in the asymptotic steady state.

The situation is much more complicated in the case of gravitational clustering in an expanding background. To begin with, defining an asymptotic state in a universe which is expanding according to $a(t) \propto t^{2/3}$ itself is problematic. Taking a strict $t \to \infty$ limit is meaningless and hence one is interested in time-scales which are finite but much longer than other time scales in the problem. We need to make this notion more precise. Secondly, we saw in section 5.1 that the virialized cores and the diffuse halo are only weakly coupled in the cosmological context. Hence the memory of initial conditions cannot be wiped out as easily in this case as, for example, in the context of systems with short range interaction or even in the context of gravitating systems in static backgrounds. We will now discuss some of these issues.

Let us start with a time $t = t_i$ at which the density fluctuations δ_k is much smaller than unity at all scales. (If necessary, we shall assume that there are cut-offs at small and large scales.) The evolution in the initial stages can then be understood by linear theory developed in section 5.2 ; the power at all scales grow as a^2 and the power spectrum maintains its shape. As time goes on, the density contrast at some scale will hit unity and structures corresponding to that characteristic scale will begin to get formed. If the power spectrum decreases with k^{-1} (so that there is more power at small spatial scales), then small scales will go non linear first.

Once the non linear structure has formed, we need to study the evolution at different wave numbers differently. If $k_{\rm nl}$ is the scale at which $\delta \approx 1$, then our analysis in section 5.1 shows that the evolution is still linear at $k \ll k_{\rm nl}$ provided the initial power was more than k^4; if not, a k^4 tail develops rapidly at these large spatial scales. At $k \gg k_{\rm nl}$, virialized structures would have formed, which are fairly immune to overall expansion of the universe. The discussion in section 5.1 shows that these scales also do not affect the evolution at smaller k. In this sense, the universe decouples nicely into a clustered component and an unclustered one which do not affect each other to the lowest order. The situation is most complex for $k \approx k_{\rm nl}$ since neither of the approximations are possible.

The critical scale $k_{\rm nl}$ itself is a function of time and, in the popular cosmological models, keeps decreasing with time; that is, larger and larger spatial scales go non linear as time goes on. In such an intrinsically time dependent situation, one could ask for different kinds of universal behaviour and it is important to distinguish between them.

To begin with, one could examine whether the power spectrum (or the correlation function) has a universal shape and evolution at late times, independent of initial power spectrum. We have seen in section 7 that the late time power spectrum *does* have a universal behaviour if the initial spectrum was sharply peaked. In this case, at length scales smaller than the initial scale at which the power is injected, one obtains two critical indices $n = -1$ and -2 and the two point correlation function has an approximate shape of (102) or even simpler, $\bar{\xi}(a, x) \propto a^2 x^{-1}(l + x)^{-1}$ where l is the length scale at which $\bar{\xi} \approx 200$. At scales bigger than the scale at which power was originally injected, the spectrum develops a k^4 tail.

But if the initial spectrum is *not* sharply peaked, each band of power evolves by this rule and the final result is a lot messier. The NSR developed in section 6 allows one to tackle this situation and ask how the non linear scales will behave. Here one of the most popular assumptions used in the literature is that of stable clustering which requires $v = -\dot{a}x$ (or $h = 1$) for sufficiently large $\bar{\xi}(a, x)$. Integrating (93) with $h = 1$, we get $\bar{\xi}(a, x) = \bar{\xi}_L^{3/2}(a, x)$. If $P_{in}(k) \propto k^n$

so that $\bar{\xi}_L(a,x) \propto a^2 x^{-(n+3)}$, then $\bar{\xi}(a,x)$ at nonlinear scales will vary as

$$\bar{\xi}(a,x) \propto a^{\frac{6}{n+5}} x^{-\frac{3(n+3)}{n+5}}; \qquad (\bar{\xi} \gg 200) \tag{103}$$

if stable clustering is true. Clearly, the power law index in the nonlinear regime "remembers" the initial index. The same result holds for more general initial conditions. *If* stable clustering *is* valid, then the late time behaviour of $\bar{\xi}(a,x)$ is strongly dependent on the initial conditions. The two (apparently reasonable) requirements: (i) validity of stable clustering at highly nonlinear scales and (ii) the independence of late time behaviour from initial conditions, *are mutually exclusive.*

This is yet another peculiarity of gravity in the context of expanding background. The initial conditions are *not* forgotten – unlike in systems with short range interactions or even in the context of gravitating systems in static background. The physical reason for this feature is the weak coupling between the virialized "cores" and diffuse "halos", described in section 5.1. This weakness of the coupling keeps the memory of the initial conditions alive as the system evolves.

While this conclusion is most probably correct, the assumption of stable clustering may not be true. In fact, there is some evidence from simulations that this assumption is not true [31]. In that context, another natural assumption which could replace stable clustering, will be the following: *We assume that h reaches a constant value asymptotically which is not necessarily unity.* Then we get $\bar{\xi}(a,x) = a^{3h} F[a^h x]$ where h now denotes the constant asymptotic value of of the function. For an initial spectrum which is scale-free power law with index n, this result translates to

$$\bar{\xi}(a,x) \propto a^{\frac{2\gamma}{n+3}} x^{-\gamma} \tag{104}$$

where γ is given by

$$\gamma = \frac{3h(n+3)}{2+h(n+3)} \tag{105}$$

One can obtain a γ which is independent of initial power law index provided h satisfies the condition $h(n+3) = c$, a constant. Unfortunately, simulations are not good enough yet to test this conclusion in the really asymptotic domain.

The second question one could ask, concerns the density profiles of individual virialized halos which, of course, is related to the behaviour of the correlation function [36], [37], [38]. To focus on this relation, let us start with the simple assumption that the density field $\rho(a,\mathbf{x})$ at late stages can be expressed as a superposition of several halos, each with some density profile; that is, we take

$$\rho(a,\mathbf{x}) = \sum_i f(\mathbf{x}-\mathbf{x}_i, a) \tag{106}$$

where the i-th halo is centered at $\mathbf{x}_i$ and contributes an amount $f(\mathbf{x}-\mathbf{x}_i,a)$ at the location $\mathbf{x}_i$ [We can easily generalize this equation to the situation in which there are halos with different properties, like core radius, mass etc by summing over the number density of objects with particular properties; we shall not bother to do this. At the other extreme, the exact description merely corresponds to taking the f's to be Dirac delta functions. Hence there is no loss of generality in (106)]. The power spectrum for the density contrast, $\delta(a,\mathbf{x}) = (\rho/\rho_b - 1)$, corresponding to the $\rho(a,\mathbf{x})$ in (106) can be expressed as

$$P(\mathbf{k},a) = \left(a^3 \left|f(\mathbf{k},a)\right|\right)^2 \left|\sum_i \exp -i\mathbf{k}\cdot\mathbf{x}_i(a)\right|^2 = \left(a^3 \left|f(\mathbf{k},a)\right|\right)^2 P_{\rm cent}(\mathbf{k},a) \tag{107}$$

where $P_{\rm cent}(\mathbf{k},a)$ denotes the power spectrum of the distribution of centers of the halos. This equation (107) can be used directly to probe the nature of halo profiles along the following lines. If the correlation function varies as $\bar{\xi} \propto x^{-\gamma}$, the correlation function of the centers vary as $\bar{\xi} \propto x^{-\gamma_C}$ and the individual profiles are of the form $f(x) \propto x^{-\epsilon}$, then the relation $P(k) = |f(k)|^2 P_{\rm cent}(k)$ translates to

$$\epsilon = 3 + \frac{1}{2}(\gamma - \gamma_C) \tag{108}$$

At very non linear scales, the centers of the virialized clusters will coincide with the deep minima of the gravitational potential. Hence the power spectrum of the centers will be proportional to the power spectrum of the gravitational potential $P_\phi(k) \propto k^{n-4}$ if $P(k) \propto k^n$. Since the correlation functions vary as $x^{-(\alpha+3)}$ when the power spectra vary as k^α, it follows that $\gamma = \gamma_C - 4$. Substituting into (108) we find that $\epsilon = 1$ at the extreme non linear scales. On the other hand, in the quasi-linear regime, reasonably large density regions will act as cluster centers and hence one would expect $P_{\rm cent}(k)$ and $P(k)$ to scale in a similar fashion. This leads to $\gamma \approx \gamma_C$, giving $\epsilon \approx 3$. So we would expect the halo profile to vary as x^{-1} at small scales steepening to x^{-3} at large scales. A simple interpolation for such a density profile will be

$$f(x) \propto \frac{1}{x(x+l)^2} \tag{109}$$

Such a profile, usually called NFW profile [38], is often used in cosmology though I have not come across the simple theoretical argument given above in the literature. Unfortunately, it is not possible to get this result from more detailed and transparent arguments.

In general, at very nonlinear scales, the correlation function probes the profile of individual halos and $P_{cent} \approx$ constant implying $\gamma_C = 3$ in (108). At these scales, we get

$$\gamma = 2\epsilon - 3 \tag{110}$$

For the situation considered in (105), with $h(n+3) = c$, the halo profile will have the index

$$\epsilon = 3\left(\frac{c+1}{c+2}\right) \tag{111}$$

which is independent of initial conditions. Note that we are now demanding the asymptotic value of h to *explicitly depend* on the initial conditions though the *spatial* dependence of $\bar{\xi}(a, x)$ does not. In other words, the velocity distribution – which is related to h – still "remembers" the initial conditions. This is indirectly reflected in the fact that the growth of $\bar{\xi}(a, x)$ – represented by $a^{6c/((2+c)(n+3))}$ – does depend on the index n.

As an example of the power of such a – seemingly simple – analysis note the following: Since $c \geq 0$, it follows that $\epsilon > (3/2)$; invariant profiles with shallower indices (for e.g with $\epsilon = 1$) discussed above are not consistent with the evolution described above.

While the above arguments are suggestive, they are far from conclusive. It is, however, clear from the above analysis and it is not easy to provide *unique* theoretical reasoning regarding the shapes of the halos. The situation gets more complicated if we include the fact that all halos will not all have the same mass, core radius etc and we have to modify our equations by integrating over the abundance of halos with a given value of mass, core radius etc. This brings in more ambiguitics and depending on the assumptions we make for each of these components and the final results have no real significance. The issue is theoretically wide open.

Acknowledgement

I thank D. Lynden-Bell, R. Nityananda, J. P. Ostriker and K. Subramaniam for several discussions over the years. I thank organisers of the LesHouches School, especially Thierry Dauxois, for inviting me to the Les Houches school and for making this contribution.

Appendix A: Basic Cosmology

I summarize in this appendix some of the key equations in cosmology which are used in the review. More detailed discussions can be found in standard text books [5].

The cosmological model which we use in our discussion is described by a spacetime interval of the form

$$ds^2 = dt^2 - a^2(t)d\mathbf{x}^2 \tag{112}$$

where $a(t)$ is called the *expansion factor* which describes the rate of expansion of the universe and the units are chosen so that $c = 1$. The form of $a(t)$ is determined by the energy density present in the universe and is determined through the equations

$$\frac{1}{2}\dot{a}^2 - \frac{G}{a}\left(\frac{4\pi}{3}\rho_b a^3\right) = 0; \quad d(\rho_b a^3) = -pda^3 \tag{113}$$

where ρ_b is the energy density of matter and p is the pressure related to ρ_b by an equation of state of the form $p = p(\rho_b)$. This equation of state, along with the second equation in (113) determines ρ_b as a function of a. Substituting in the first equation in (113), one can determine $a(t)$. As a mnemonic (and only as a mnemonic), one can think of the first equation as giving the sum of kinetic energy and potential energy of the universe to be zero for the universe and the second equation as an equivalent of $dE = -pdV$. There are also equivalent to the relation $(\ddot{a}/a) = -(4\pi G/3)\rho_b(t)$

Non relativistic matter moving at speeds far less than speed of light ($v \ll c$) will have the pressure $p \approx \rho_b v^2$ negligibly small compared to the energy density $\rho_b c^2$. We can then set $p \approx 0$ obtaining $\rho_b \propto a^{-3}$ and $a \propto t^{2/3}$. Since an overall multiplication constant in a can be absorbed by rescaling the lengths, the normalization of $a(t)$ is arbitrary. It is conventional to take $a = (t/t_0)^{2/3}$ with t_0 denoting the current age of the universe. The rate of expansion at present is called the Hubble constant and is defined by

$$H_0 = \left(\frac{\dot{a}}{a}\right)_0 = \frac{2}{3t_0} \tag{114}$$

Observationally, $H_0 \approx 75$ km s^{-1} Mpc^{-1}. This defines a natural time scale $t_H = H_0^{-1} \approx 10^{10}$ years and a length scale $d_H = cH_0^{-1} \approx 4000$ Mpc. Note that H_0 and the energy density are related by $\rho_0 = (3H_0^2/8\pi G)$.

The general relativistic effects of gravity are felt over length and time scales comparable to t_H, d_H. Since the typical separation between galaxies is about 1 Mpc and the size of large superclusters in the universe is about 100 Mpc, the general relativistic effects are not important at the present epoch for most of the issues in structure formation. Equation (112) then suggests that if we use the coordinate $\mathbf{r} \equiv a(t)\mathbf{x}$, the physical laws in an expanding universe will take more familiar form at small scales. (The $\mathbf{r}$ is called the *proper coordinate*, while $\mathbf{x}$ is called the *comoving coordinate*.) For example, gravity can be described by Newtonian theory in proper coordinates with the cosmic background providing an extra gravitational potential $\Phi_{\mathrm{FRW}} = -(2\pi G/3)\rho_b(t)r^2$ corresponding to a uniform distribution of matter.

When matter expands with the universe homogeneously, the proper coordinate separation between any two points vary as $\dot{\mathbf{r}} = \dot{a}\mathbf{x} = (\dot{a}/a)\mathbf{r}$. (This is called Hubble expansion with $\mathbf{v} = H\mathbf{r}$.) In this case,

$$\ddot{\mathbf{r}} = -\frac{\ddot{a}}{a}\mathbf{r} = \frac{4\pi G}{3}\rho_b(t)\mathbf{r} = -\nabla_{\mathbf{r}}\Phi_{\mathrm{FRW}} \tag{115}$$

Gravitational clustering and growth of inhomogeneities require particles to move relative to the cosmic expansion with $\dot{\mathbf{x}} \neq 0$.

In the study of structure formation, the central quantity one uses is the density contrast defined as $\delta(t, \mathbf{x}) = [\rho(t, \mathbf{x}) - \rho_b(t)]/\rho_b(t)$ which characterizes the fractional change in the energy density compared to the background. (Here $\rho_b(t) = \langle\rho(t, \mathbf{x})\rangle$ is the mean background density.) Since one is often interested in the statistical behaviour of structures in the universe, it is conventional to assume that δ and other related quantities are elements of an ensemble. Many popular models of structure formation suggest that the initial density perturbations in the early universe can be represented as a Gaussian random variable with zero mean (that is, $\langle\delta\rangle = 0$) and a given initial power spectrum. The latter quantity is defined through the relation $P(t,k) = \langle|\delta_k(t)|^2\rangle$ where $\delta_{\mathbf{k}}$ is the Fourier transform of $\delta(t,\mathbf{x})$. It is also conventional to define the two-point correlation function $\xi(t,x)$ and the mean correlation function $\bar{\xi}(t,x)$ via the equations

$$\xi(t,x) = \int \frac{d^3k}{(2\pi)} P(t,k)\, e^{i\mathbf{k}\cdot\mathbf{x}}; \quad \bar{\xi}(t,x) = \frac{3}{x^2}\int_0^x \xi(t,y)\, y^2 dy \tag{116}$$

Though gravitational clustering will make the density contrast non Gaussian at late times, the power spectrum and the correlation functions continue to be of primary importance in the study of structure formation.

References

1. T. Padmanabhan: Physics Reports **188**, 285 (1990).
2. T. Padmanabhan: *Theoretical Astrophysics, Vol.I: Astrophysical Processes*, (Cambridge University Press, Cambridge, 2000), chapter 10.
3. J. Binney and S. Tremaine: *Galactic Dynamics*, (Princeton University Press, New Jersey, 1987).
4. T. Padmanabhan: 'Aspects of Gravitational Clustering', *Large Scale Structure Formation*, ed. by R. Mansouri and R. Brandenberger, (Astrophysics and Space Science Library, volume 247, Kluwer Academic, Dordrecht, 2000), astro-ph/9911374.
5. P.J.E. Peebles: *Principles of Physical Cosmology*, (Princeton University Press, New Jersey, 1993);
   J.A. Peacock: *Cosmological Physics*, (Cambridge University Press, Cambridge, 1999);
   T. Padmanabhan: *Structure Formation in the Universe*, (Cambridge University Press, Cambridge 1993);
   T. Padmanabhan: *Theoretical Astrophysics, Vol.III: Galaxies and Cosmology*, (Cambridge University Press, Cambridge, 2002, in press).
6. P.-H. Chavanis: *Statistical mechanics of two-dimensional vortices and stellar systems*, in "Dynamics and Thermodynamics of Systems with Long-Range Interactions", T. Dauxois, S. Ruffo, E. Arimondo, M. Wilkens Eds., Lecture Notes in Physics Vol. 602, Springer (2002), (in this volume)
7. H. J. de Vega, N. Sanchez, F. Combes: Chaos, Soliton and Fractals **10**, 329 (1999)
   H. J. de Vega, J. Siebert: Phys. Rev. E **66**, 016112 (2002)
8. P. Valageas: A &A, **382**, 477 (2001); A&A , **379**, 8 (2001).
9. E. Follana, V. Laliena: Phys. Rev. **E 61**, 6270 (2000).
10. M. Bottaccio, A. Amici, P. Miocchi, R. Capuzzo Dolcetta, M. Montuori, L. Pietronero: Europhys. Lett. **57**, 315 (2002)
11. Roman Scoccimarro: 'A New Angle on Gravitational Clustering'. To appear in the proceedings of the *15th Florida Workshop in Nonlinear Astronomy and Physics, "The Onset of Nonlinearity"*, astro-ph/0008277.
12. D. Lynden-Bell: Physica A **263**, 293 (1999)
13. D. Lynden-Bell and R. M. Lynden-Bell: Mon. Not. R. Astr. Soc. **181**, 405 (1977).
14. E.B. Aaronson and C.J. Hansen: ApJ **177**, 145 (1972);
    T. Padmanabhan and D. Narasimha, unpublished.

15. V.A. Antonov, V.A. : Vest. Leningrad Univ. **7**, 135 (1962); Translation: IAU Symposium **113**, 525 (1985).
16. T.Padmanabhan: Astrophys. Jour. Supp. , **71**, 651 (1989).
17. I. Ispolatov and E.G.D. Cohen: 'Collapse in $1/r^{\alpha}$ interacting systems', cond-mat/0106381.
18. G. Horowitz and J. Katz, Ap.J., **222**, 941 (1978).
19. S. Chandrasekhar: *An Introduction to the Study of Stellar Structure*, (Dover 1939).
20. D. Lynden-Bell and R. Wood: Mon. Not. R. Astr. Soc. **138**, 495 (1968).
21. D.Pines and D.Bohm, Phys.Rev. **85**, 338 (1952).
22. T. Padmanabhan: *Cosmology and Astrophysics through Problems*, (Cambridge University Press, Cambridge 1996).
23. T. Padmanabhan: unpublished.
24. P.J.E. Peebles: *Large Scale Structure of the Universe*, (Princeton University Press, New Jersey, 1980).
25. J.S. Bagla and T. Padmanabhan: Mon Not. R. Astr. Soc., **286**, 1023 (1997).
26. S. Engineer, N. Kanekar and T. Padmanabhan , Mon Not. R. Astr. Soc., **314** , 279 (2000).
27. A.J.S. Hamilton, P. Kumar, E. Lu and A. Mathews: ApJ, **374**, L1 (1991).
28. R. Nityananda, and T. Padmanabhan: Mon Not. R. Astr. Soc., **271**, 976 (1994).
29. T. Padmanabhan: Mon Not. R. Astr. Soc., **278**, L29 (1996).
30. N. Kanekar and T. Padmanabhan: Mon Not. R. Astr. Soc. , 324, 988 (2001).
31. T. Padmanabhan, R. Cen, J.P. Ostriker, F.J. Summers. Ap. J, **466**, 604 (1996).
32. T. Padmanabhan: in 'Gravitation and Cosmology', proceedings of the *ICGC- 95 conference*, eds. S. V. Dhurandhar and T. Padmanabhan, (Kluwer Academic Publishers, Dordrecht, 1997), p37.
33. T. Padmanabhan and N. Kanekar: Phys. Rev. D, **61**, 023515 (2000)
34. J.S. Bagla, S. Engineer and T. Padmanabhan: Ap.J., **495**, 25-28 (1998).
35. D. Munshi, T. Padmanabhan: Mon Not. R. Astr. Soc., **290**, 193 (1997).
36. T. Padmanabhan and S. Engineer: ApJ, *493*, 509 (1998).
37. K. Subramanian: 'Self-similar collapse and the structure of dark matter halos: A fluid approach', astro-ph/9909280.
    K. Subramanian, Renyue Cen, Jeremiah P. Ostriker: Ap.J, **538**, 528-542 (2000).
38. Julio F. Navarro, Carlos S. Frenk, Simon D.M. White: Ap.J. **462**, 563-575 (1996).

Statistical Mechanics of Two-Dimensional Vortices and Stellar Systems

Pierre-Henri Chavanis

Laboratoire de Physique Quantique, Université Paul Sabatier,
118, route de Narbonne, 31062 Toulouse, France

Abstract. The formation of large-scale vortices is an intriguing phenomenon in two-dimensional turbulence. Such organization is observed in large-scale oceanic or atmospheric flows, and can be reproduced in laboratory experiments and numerical simulations. A general explanation of this organization was first proposed by Onsager (1949) by considering the statistical mechanics for a set of point vortices in two-dimensional hydrodynamics. Similarly, the structure and the organization of stellar systems (globular clusters, elliptical galaxies,...) in astrophysics can be understood by developing a statistical mechanics for a system of particles in gravitational interaction as initiated by Chandrasekhar (1942). These statistical mechanics turn out to be relatively similar and present the same difficulties due to the unshielded long-range nature of the interaction. This analogy concerns not only the equilibrium states, i.e. the formation of large-scale structures, but also the relaxation towards equilibrium and the statistics of fluctuations. We will discuss these analogies in detail and also point out the specificities of each system.

1 Introduction

Two-dimensional flows with high Reynolds numbers have the striking property of organizing spontaneously into coherent structures (the vortices) which dominate the dynamics [93] (see Fig. 1). The robustness of Jupiter's Great Red Spot, a huge vortex persisting for more than three centuries in a turbulent shear between two zonal jets, is probably related to this general property. Some other coherent structures like dipoles (pairs of cyclone/anticyclone) and sometimes tripoles have been found in atmospheric or oceanic systems and can persist during several days or weeks responsible for atmospheric blocking. Some astrophysicists invoke the existence of organized vortices in the gaseous component of disk galaxies in relation with the emission of spiral density waves [99]. It has also been proposed that planetary formation might have begun inside persistent gaseous vortices born out of the protoplanetary nebula [5,121,15,60,33] (see Fig. 2). As a result, hydrodynamical vortices occur in a wide variety of geophysical or astrophysical situations and their robustness demands a general understanding.

Similarly, it is striking to observe that self-gravitating systems follow a kind of organization despite the diversity of their initial conditions and their environement [9] (see Fig. 3). This organization is illustrated by morphological classification schemes such as the Hubble sequence for galaxies and by simple rules which govern the structure of individual self-gravitating systems. For example, elliptical galaxies display a quasi-universal luminosity profile described

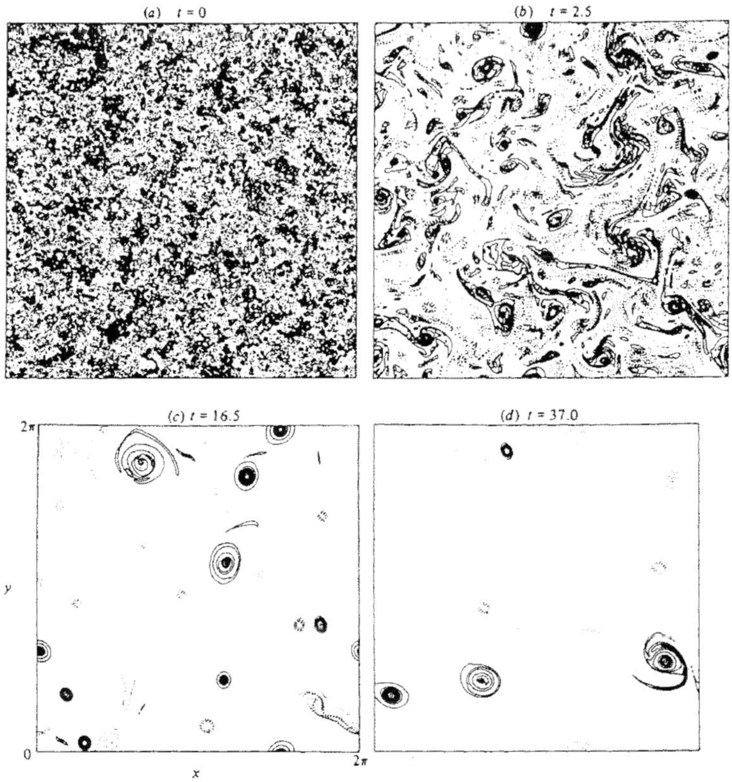

Fig. 1. Self-organization of two-dimensional turbulent flows into large-scale vortices [93]. These vortices are long-lived and dominate the dynamics.

by de Vaucouleur's $R^{1/4}$ law and most of globular clusters are well-fitted by the Michie-King model. On the other hand, the flat rotation curves of spiral galaxies can be explained by the presence of a dark matter halo with a density profile decreasing as r^{-2} at large distances. The fractal nature of the interstellar medium and the large scale structures of the universe also display some form of organization.

The question that naturally emerges is what determines the particular configuration to which a self-gravitating system or a large-scale vortex settles. It is possible that their actual configuration crucially depends on the conditions that prevail at their birth and on the details of their evolution. However, in view of their apparent regularity, it is tempting to investigate whether their organization can be favoured by some fundamental physical principles like those of thermodynamics and statistical physics. We ask therefore if the actual states of self-gravitating systems in the universe and coherent vortices in two-dimensional turbulent flows are not simply more probable than any other possible configuration, i.e. if they cannot be considered as *maximum entropy states.* This statistical

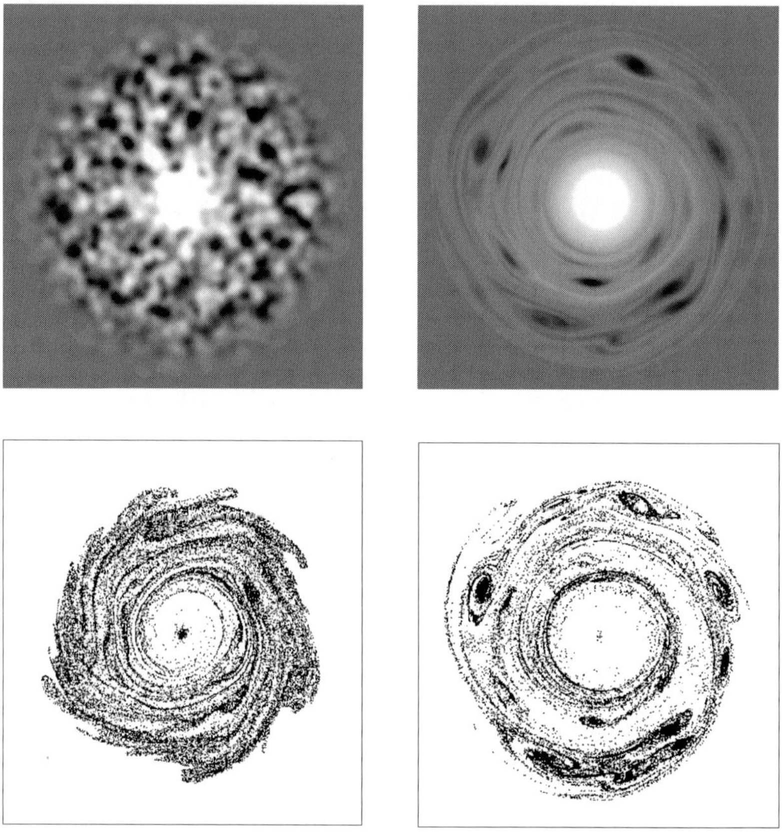

Fig. 2. A scenario of planet formation inside large-scale vortices presumably present in the Keplerian gaseous disk surrounding a star at its birth. Starting from a random vorticity field, a series of anticyclonic vortices appears spontaneously (upper panel). Due to the Coriolis force and to the friction with the gas, these vortices can efficiently trap dust particles passing nearby (lower pannel). The local increase of dust concentration inside the vortices can initiate the formation of planetesimals and planets by gravitational instability. This numerical simulation is taken from [15].

mechanics approach has been initiated by Onsager [101] for a system of point vortices and by Chandrasekhar [21] in the case of self-gravitating systems.

It turns out that the statistical mechanics of two-dimensional vortices and self-gravitating systems present a deep analogy despite the very different physical nature of these systems. This analogy was pointed out by Chavanis in [29,32,35] and further developed in [54,30,34,36,47,48]. In the following, we will essentially discuss the statistical mechanics of 2D vortices and refer to the review of Padmanabhan [103] (and his contribution in this book) for more details about the statistical mechanics of self-gravitating systems. We will see that the analogy between two-dimensional vortices and (three-dimensional) self-gravitating systems

Fig. 3. Large-scale structures in the universe as observed with the Hubble space telescope. The analogy with Fig. 1 is striking and will be discussed in detail in this paper.

concerns not only the prediction of the equilibrium state, i.e. the formation of large-scale structures, but also the statistics of fluctuations and the relaxation towards equilibrium.

This paper is organized as follows. In Sect. 2, we discuss the statistical mechanics of point vortices introduced by Onsager [101] and further developed by Joyce and Montgomery [70] and Pointin and Lundgren [107] among others (see a complete list of references in the book of Newton [98]). We discuss the existence of a thermodynamic limit in Sect. 2.7 and make the connexion with field theory. Statistical equilibrium states of axisymmetric flows are obtained analytically in Sect. 2.8-2.9. The relation with equilibrium states of self-gravitating systems is shown in Sect. 2.10. In Sect. 3, we discuss the statistics of velocity fluctuations produced by a random distribution of point vortices and use this stochastic approach to obtain an estimate of the diffusion coefficient of point vortices. Application to 2D decaying turbulence is considered in Sect. 3.4. In Sect. 4, we describe the relaxation of a point vortex in a thermal bath and analyze this relaxation in terms of a Fokker-Planck equation involving a diffusion and a drift. In Sect. 5, we develop a more general kinetic theory of point vortices. A new kinetic equation is obtained which satisfies all conservation laws of the point vortex system and increases the Boltzmann entropy (H-theorem). We mention the connexion with the kinetic theory of stars developed by Chandrasekhar [21]. In Sect. 6, we discuss the violent relaxation of 2D vortices and stellar systems. We mention the analogy between the Vlasov and the Euler equations and between the statistical approach developed by Lynden-Bell [90] for collisionless stellar systems and by Kuz'min [83], Miller [95] and Robert and Sommeria [111] for continuous vorticity fields. The concepts of "chaotic mixing" and "incomplete relaxation" are discussed in the light of a relaxation theory in Sect. 6.3. Application of statistical mechanics to geophysical flows and Jupiter's Great Red Spot are evocated in Sect. 6.4.

2 Statistical Mechanics of Point Vortices in Two-Dimensional Hydrodynamics

2.1 Two-Dimensional Perfect Flows

The equations governing the dynamics of an invisicid flow are the equation of continuity and the Euler equation:

$$\frac{\partial \rho}{\partial t} + \nabla(\rho \mathbf{u}) = 0, \tag{1}$$

$$\frac{\partial \mathbf{u}}{\partial t} + (\mathbf{u}\nabla)\mathbf{u} = -\frac{1}{\rho}\nabla p. \tag{2}$$

For an incompressible flow, the equation of continuity reduces to the condition

$$\nabla \mathbf{u} = 0. \tag{3}$$

If, in addition, the flow is two-dimensional, this last equation can be written $\partial_x u + \partial_y v = 0$, where (u, v) are the components of the velocity. According to the Schwarz theorem, there exists a streamfunction ψ such that $u = \partial_y \psi, v = -\partial_x \psi$, or, equivalently

$$\mathbf{u} = -\mathbf{z} \times \nabla\psi, \tag{4}$$

where $\mathbf{z}$ is a unit vector normal to the flow. The vorticity

$$\boldsymbol{\omega} = \nabla \times \mathbf{u} = \omega \mathbf{z}, \qquad \text{with} \qquad \omega = \partial_x v - \partial_y u, \tag{5}$$

is directed along the vertical axis. According to the Stokes formula, the circulation of the velocity along a closed curve (C) delimiting a domain area (S) is

$$\Gamma = \oint_{(C)} \mathbf{u}\, d\mathbf{l} = \int_{(S)} \omega d^2\mathbf{r}. \tag{6}$$

Taking the curl of (4), we find that the vorticity is related to the stream function by a Poisson equation

$$\Delta\psi = -\omega, \tag{7}$$

where $\Delta = \partial_{xx}^2 + \partial_{yy}^2$ is the Laplacian operator. In an unbounded domain, this equation can be written in integral form as

$$\psi(\mathbf{r}, t) = -\frac{1}{2\pi}\int \omega(\mathbf{r}', t) \ln|\mathbf{r} - \mathbf{r}'| d^2\mathbf{r}', \tag{8}$$

and the velocity field can be expressed in terms of the vorticity as

$$\mathbf{u}(\mathbf{r},t) = \frac{1}{2\pi}\mathbf{z} \times \int \omega(\mathbf{r}',t)\frac{\mathbf{r}-\mathbf{r}'}{|\mathbf{r}-\mathbf{r}'|^2}d^2\mathbf{r}'. \tag{9}$$

In a bounded domain, (8) must be modified so as to take into account vortex images. The impermeability condition implies that ψ is constant on the boundary and we shall take $\psi = 0$ by convention. Taking the curl of (2), the pressure term disapears and the Euler equation becomes

$$\frac{\partial \omega}{\partial t} + \mathbf{u}\nabla\omega = 0. \tag{10}$$

This corresponds to the transport of the vorticity ω by the velocity field $\mathbf{u}$. It is easy to show that the flow conserves the kinetic energy

$$E = \int \frac{\mathbf{u}^2}{2} d^2\mathbf{r}. \tag{11}$$

Using (4), (7), one has successively

$$E = \frac{1}{2}\int (\nabla\psi)^2 d^2\mathbf{r} = \frac{1}{2}\int \psi(-\Delta\psi)d^2\mathbf{r} = \frac{1}{2}\int \omega\psi d^2\mathbf{r}, \tag{12}$$

where the second equality is obtained by a part integration with the condition $\psi = 0$ on the boundary. Therefore, E can be interpreted either as the kinetic energy of the flow (see (11)) or as a potential energy of interaction between vortices (see (12)).

2.2 The Point Vortex Gas

We shall consider the situation in which the velocity is created by a collection of N point vortices. In that case, the vorticity field can be expressed as a sum of δ-functions in the form

$$\omega(\mathbf{r},t) = \sum_{i=1}^{N} \gamma_i \delta(\mathbf{r} - \mathbf{r}_i(t)), \tag{13}$$

where $\mathbf{r}_i(t)$ denotes the position of point vortex i at time t and γ_i is its circulation. According to (9)(13), the velocity of a point vortex is equal to the sum of the velocities $\mathbf{V}(j \to i)$ produced by the $N-1$ other vortices, i.e.

$$\mathbf{V}_i = \sum_{j \neq i} \mathbf{V}(j \to i) \qquad \text{with} \qquad \mathbf{V}(j \to i) = -\frac{\gamma_j}{2\pi}\mathbf{z} \times \frac{\mathbf{r}_j - \mathbf{r}_i}{|\mathbf{r}_j - \mathbf{r}_i|^2}. \tag{14}$$

As emphasized by Kirchhoff [79], the above dynamics can be cast in a Hamiltonian form

$$\gamma_i \frac{dx_i}{dt} = \frac{\partial H}{\partial y_i}, \qquad \gamma_i \frac{dy_i}{dt} = -\frac{\partial H}{\partial x_i}, \tag{15}$$

$$H = -\frac{1}{2\pi}\sum_{i<j}\gamma_i\gamma_j \ln|\mathbf{r}_i - \mathbf{r}_j|, \tag{16}$$

where the coordinates (x, y) of the point vortices are canonically conjugate. These equations of motion still apply when the fluid is restrained by boundaries, in which case the Hamiltonian (16) is modified so as to allow for vortex images, and may be constructed in terms of Green's functions depending on the geometry of the domain. Since H is not explicitly time dependant, it is a constant of the motion and it represents the "potential" energy of the point vortices. The other conserved quantities are the angular momentum $\sum_i \gamma_i r_i^2$ and the impulse $\sum_i \gamma_i \mathbf{r}_i$. Note that the Hamiltonian (16) does not involve a "kinetic" energy of the point vortices in the usual sense (i.e., a quadratic term $\sum_i \frac{p_i^2}{2m}$). This is related to the particular circumstance that a point vortex is not a material particle. Indeed, an isolated vortex remains at rest contrary to a material particle which has a rectilinear motion due to its inertia. Point vortices form therefore a very peculiar Hamiltonian system. Note also that the Hamiltonian of point vortices can be either positive or negative (in the case of vortices of different signs) whereas the kinetic energy of the flow is necessarily positive. This is clearly a drawback of the point vortex model.

2.3 The Microcanonical Approach of Onsager (1949)

The statistical mechanics of point vortices was first considered by Onsager [101] who showed the existence of negative temperature states at which point vortices of the same sign cluster into "supervortices". He could therefore explain the formation of large, isolated vortices in nature. This was a remarkable anticipation since observations were very scarce at that time.

Let us consider a liquid enclosed by a boundary, so that the vortices are confined to an area A. Since the coordinates (x, y) of the point vortices are canonically conjugate, the phase space coincides with the configuration space and is *finite*:

$$\int dx_1dy_1...dx_Ndy_N = \left(\int dxdy\right)^N = A^N. \tag{17}$$

This striking property contrasts with most classical Hamiltonian systems considered in statistical mechanics which have unbounded phase spaces due to the presence of a kinetic term in the Hamiltonian.

As is usual in the microcanonical description of a system of N particles, we introduce the density of states

$$g(E) = \int dx_1dy_1...dx_Ndy_N\delta\Big(E - H(x_1, y_1, ..., x_N, y_N)\Big), \tag{18}$$

which gives the phase space volume per unit interaction energy E. The equilibrium N-body distribution of the system, satisfying the normalization condition $\int \mu(\mathbf{r}_1, ..., \mathbf{r}_N)d^2\mathbf{r}_1...d^2\mathbf{r}_1 = 1$, is given by

$$\mu(\mathbf{r}_1, ..., \mathbf{r}_N) = \frac{1}{g(E)}\delta(E - H(\mathbf{r}_1, ..., \mathbf{r}_N)). \tag{19}$$

This formula simply means that, in the microcanonical ensemble, all accessible microstates (having the required energy E) are equiprobable at statistical equilibrium.

The phase space volume which corresponds to energies $H(\mathbf{r}_1, ..., \mathbf{r}_N)$ less than a given value E can be written

$$\Phi(E) = \int_{E_{min}}^{E} g(E)dE. \tag{20}$$

It increases monotonically from zero to A^N when E goes from E_{min} to $+\infty$. Therefore, $g(E) = d\Phi(E)/dE$ will have a maximum value at some $E = E_m$, say, before decreasing to zero when $E \to +\infty$.

In the microcanonical ensemble, the entropy and the temperature are defined by

$$S = \ln g(E), \qquad \beta = \frac{1}{T} = \frac{dS}{dE}. \tag{21}$$

For $E > E_m$, $S(E)$ is a decreasing function of energy and consequently the temperature is *negative*. Now, high energy states $E \gg E_m$ are clearly those in which the vortices of the same sign are crowded as close together as possible. For energies only slighlty greater than E_m, the concentration will not be so dramatic but there will be a tendency for the vortices to group themselves together on a macroscopic scale and form "clusters" or "supervortices". By contrast, for $E < E_m$, the temperature is positive and the vortices have the tendency to accumulate on the boundary of the domain in order to decrease their energy. For a system with positive and negative vortices, the negative temperature states, achieved for relatively high energies, consist of two large counter-rotating vortices physically well separated in the box. On the contrary when $E \to -\infty$, the temperature is positive and vortices of opposite circulation tend to pair off.

2.4 The Equation of State

For a two-dimensional gas of particles interacting via a Coulombian potential in $\ln r$, the equation of state can be derived exactly. Let us assume that the system is enclosed in a domain of surface $V = R^2$. The density of states can be written

$$g(E, V) = \int_0^R ... \int_0^R \prod_{i=1}^{N} d^2\mathbf{r}_i \delta\Big(E + \frac{1}{2\pi}\sum_{i<j}\gamma_i\gamma_j \ln|\mathbf{r}_i - \mathbf{r}_j|\Big). \tag{22}$$

Making the change of variable $\mathbf{x} = \mathbf{r}/R$, we find that $g(E, V) = V^N g(E', 1)$ with $E' = E + \frac{1}{4\pi}\ln V \sum_{i<j}\gamma_i\gamma_j$. Therefore, the entropy satisfies $S(E, V) = N \ln V + S(E', 1)$ and the pressure $P = T(\partial S/\partial V)_E$ is exactly given by

$$P = \frac{N}{\beta V}\Big(1 + \frac{\beta}{4\pi N}\sum_{i<j}\gamma_i\gamma_j\Big). \tag{23}$$

If the vortices have the same circulation γ, we obtain

$$P = \frac{N}{\beta V}\left(1 + \frac{\gamma^2}{8\pi}(N-1)\beta\right). \tag{24}$$

In particular the pressure vanishes for

$$\beta_c = -\frac{8\pi}{(N-1)\gamma^2}. \tag{25}$$

We shall see in Sect. 2.8 that this *negative* critical inverse temperature is the minimum inverse temperature that the system can achieve.

If, on the other hand, we consider a neutral system consisting of $N/2$ vortices of circulation γ and $N/2$ vortices of circulation $-\gamma$, we find

$$P = \frac{N}{\beta V}\left(1 - \frac{\beta\gamma^2}{8\pi}\right). \tag{26}$$

This result is well-known is plasma physics [113]. The critical temperature at which the pressure vanishes is now positive

$$\beta_c = \frac{8\pi}{\gamma^2}, \tag{27}$$

and independant on the number of point vortices in the system. For $\beta > \beta_c$ the pressure is negative so this range of temperatures is forbidden. For simplicity, we have not taken into account the contribution of images in the previous calculations; this can slightly change the results but this should not alter the existence of the critical inverse temperatures reported above.

2.5 The Mean-Field Approximation

It is easy to show that the exact distribution of point vortices (13) expressed in terms of δ-functions is solution of the Euler equation (10). This is proved as follows. Taking the derivative of (13) with respect to time, we obtain

$$\frac{\partial\omega}{\partial t} = -\sum_{i=1}^{N}\gamma\nabla\delta(\mathbf{r}-\mathbf{r}_i(t))\mathbf{V}_i. \tag{28}$$

Since $\mathbf{V}_i = \mathbf{u}(\mathbf{r}_i(t), t)$, we can rewrite the foregoing equation in the form

$$\frac{\partial\omega}{\partial t} = -\nabla\sum_{i=1}^{N}\gamma\delta(\mathbf{r}-\mathbf{r}_i(t))\mathbf{u}(\mathbf{r}, t). \tag{29}$$

Since the velocity is divergenceless, we obtain

$$\frac{\partial\omega}{\partial t} = -\mathbf{u}(\mathbf{r}, t)\nabla\sum_{i=1}^{N}\gamma\delta(\mathbf{r}-\mathbf{r}_i(t)) = -\mathbf{u}\nabla\omega. \tag{30}$$

Therefore, in the point vortex model, the Euler equation (10) contains exactly the same information as the Hamiltonian system (15).

This description in terms of δ-functions, while being technically correct, is useless for practical purposes because it requires the knowledge of the exact trajectories of the point vortices for an arbitrary initial condition or the solution of the Euler equation (10). When N is large, this task is impossibly difficult. Therefore, instead of the exact vorticity field expressed in terms of δ-functions, one is more interested by functions which are smooth. For that reason, we introduce a smooth vorticity field $\langle\omega\rangle(\mathbf{r},t)$ which is proportional to the average number of vortices contained in the cell $(\mathbf{r},\mathbf{r}+d\mathbf{r})$ at time t. This description requires that it is possible to divide the domain in a large number of cells in such a way that each cell is (a) large enough to contain a macroscopic number of point vortices but (b) small enough for all the particles in the cell can be assumed to possess the same average characteristics of the cell.

Formally, the average vorticity field is given by

$$\langle\omega\rangle(\mathbf{r},t)=\sum_{i=1}^{N}\gamma\langle\delta(\mathbf{r}-\mathbf{r}_i(t))\rangle, \tag{31}$$

where the statistical average of a function $X(\mathbf{r}_1,...,\mathbf{r}_N)$ is defined by

$$\langle X\rangle=\int\mu(\mathbf{r}_1,...,\mathbf{r}_N,t)X(\mathbf{r}_1,...,\mathbf{r}_N)d^2\mathbf{r}_1...d^2\mathbf{r}_N, \tag{32}$$

where $\mu(\mathbf{r}_1,...,\mathbf{r}_N,t)$ is the N-body distribution function of the system at time t. The average vorticity can be rewritten

$$\langle\omega\rangle(\mathbf{r},t)=N\gamma P(\mathbf{r},t)=\gamma\langle n\rangle(\mathbf{r},t), \tag{33}$$

where we have introduced the one-vortex distribution function

$$P(\mathbf{r}_1,t)=\int\mu(\mathbf{r}_1,...,\mathbf{r}_N,t)d^2\mathbf{r}_2...d^2\mathbf{r}_N, \tag{34}$$

and the local vortex density $\langle n\rangle=NP(\mathbf{r},t)$. In the foregoing relations, we have implicitly used the fact that the vortices are identical. The average energy of the system is given by

$$\begin{aligned}E=\langle H\rangle&=-\frac{1}{4\pi}\sum_{i\neq j}\gamma^2\langle\ln|\mathbf{r}_i-\mathbf{r}_j|\rangle\\&=-\frac{1}{4\pi}N(N-1)\gamma^2\int g(\mathbf{r},\mathbf{r}',t)\ln|\mathbf{r}-\mathbf{r}'|d^2\mathbf{r}d^2\mathbf{r}',\end{aligned} \tag{35}$$

where

$$g(\mathbf{r}_1,\mathbf{r}_2,t)=\int\mu(\mathbf{r}_1,...,\mathbf{r}_N,t)d^3\mathbf{r}_3...d^2\mathbf{r}_N, \tag{36}$$

is the two-body distribution function. In the mean-field approximation, which is exact in a properly defined thermodynamic limit with $N\to+\infty$ (see Sect. 2.7), we have

$$g(\mathbf{r}_1,\mathbf{r}_2,t)=P(\mathbf{r}_1,t)P(\mathbf{r}_2,t). \tag{37}$$

Accounting that $N(N-1) \simeq N^2$ for large N, the average energy takes the form

$$E = -\frac{1}{4\pi} \int \langle\omega\rangle(\mathbf{r},t)\langle\omega\rangle(\mathbf{r}',t) \ln|\mathbf{r}-\mathbf{r}'| d^2\mathbf{r} d^2\mathbf{r}'. \tag{38}$$

Using (8), the expression (38) for E can be rewritten

$$E = \frac{1}{2}\int \langle\omega\rangle\psi d^2\mathbf{r} = \int \frac{\langle\mathbf{u}\rangle^2}{2} d^2\mathbf{r}. \tag{39}$$

where ψ is the streamfunction created by the average vorticity $\langle\omega\rangle$ and where $\langle\mathbf{u}\rangle$ is the smooth velocity field.

2.6 The Maximum Entropy State

We now wish to determine the equilibrium distribution of vortices following a statistical mechanics approach. Using Boltzmann procedure, we divide the domain in macrocells with area Δ. Let n_i denote the number of point vortices in the cell Δ_i. We now decompose each macrocell into ν microcells with equal area. A *macrostate* is determined by the distribution $\{n_i\}$. Several configurations can lead to the same macrostate: each of them will be called a *microstate.* Using a combinatorial analysis, the number of microstates corresponding to the macrostate $\{n_i\}$ is

$$W(\{n_i\}) = N!\prod_i \frac{\nu^{n_i}}{n_i!}. \tag{40}$$

The logarithm of this number defines the Boltzmann entropy. Using Stirling formula and considering the continuum limit in which $\Delta, \nu \to 0$, we get the classical formula

$$S = -N \int P(\mathbf{r}) \ln P(\mathbf{r}) d^2\mathbf{r}, \tag{41}$$

where $P(\mathbf{r})$ is the density probability that a point vortex be found in the surface element centered on $\mathbf{r}$. At equilibrium, the system is in the most probable macroscopic state, i.e. the state that is the most represented at the microscopic level. This optimal state is obtained by maximizing the Boltzmann entropy (41) at fixed energy (39) and vortex number N, or total circulation

$$\Gamma = N\gamma = \int \langle\omega\rangle d^2\mathbf{r}. \tag{42}$$

Writing the variational principle in the form

$$\delta S - \beta\delta E - \alpha\delta\Gamma = 0, \tag{43}$$

where β and α are Lagrange multipliers, it is readily found that the maximum entropy state corresponds to the Boltzmann distribution

$$\langle\omega\rangle = A e^{-\beta\gamma\psi}, \tag{44}$$

with inverse temperature β. We can account for the conservation of angular momentum $L = \int \langle\omega\rangle r^2 d^2\mathbf{r}$ (in a circular domain) and impulse $P = \int \langle\omega\rangle y d^2\mathbf{r}$ (in a channel) by introducing appropriate Lagrange multipliers Ω and U for each of these constraints. In that case, (44) remains valid provided that we replace the streamfunction ψ by the relative streamfunction $\psi' = \psi + \frac{\Omega}{2}r^2 - Uy$. Substituting the Boltzmann relation between $\langle\omega\rangle$ and ψ in the Poisson equation (7), we obtain a differential equation for the streamfunction

$$-\Delta\psi = Ae^{-\beta\gamma\psi}, \tag{45}$$

which determines the statistical equilibrium distribution of vortices. This Boltzmann Poisson equation can be easily generalized to include a spectrum of circulations among the vortices.

The combinatorial analysis presented in this section was performed by Joyce and Montgomery [70]. The mean-field equation (45) was also obtained by Pointin and Lundgren [107] from the equilibrium BGK hierarchy. In fact, this mean-field approach was first developed by Onsager but his results were not published (U. Frisch, private communication).

2.7 Field Theory and Thermodynamic Limit

Let us consider a collection of N point vortices with equal circulation γ in a bounded domain of size R. We want to give a rational to the mean-field approximation considered previously by defining a proper thermodynamic limit for point vortices. Rigorous results have been established in [17,58]. In the following, we present a less rigorous, albeit equivalent, field theory approach inspired by the work of [65,103,57] for self-gravitating systems.

In the microcanonical ensemble, the quantity of fundamental interest is the density of states

$$g(E) = \int \delta\Big(E - H(\mathbf{r}_1, ..., \mathbf{r}_N)\Big) \prod_{i=1}^{N} d^2\mathbf{r}_i, \tag{46}$$

which is explicitly given by

$$g(E) = \int_0^R ... \int_0^R \prod_{i=1}^{N} d^2\mathbf{r}_i \delta\Big(E + \frac{\gamma^2}{2\pi}\sum_{i<j} \ln\frac{|\mathbf{r}_i - \mathbf{r}_j|}{R}\Big), \tag{47}$$

where the potential of interaction has been normalized by R. For simplicity, we have ignored the contribution of the images but this shall not affect the final results. We now introduce the change of variables $\mathbf{x} = \mathbf{r}/R$ and define the function

$$u(\mathbf{x}_1, ..., \mathbf{x}_N) = \frac{1}{N}\sum_{i<j} \ln|\mathbf{x}_i - \mathbf{x}_j|, \tag{48}$$

and the dimensionless energy

$$\Lambda = \frac{2\pi E}{N^2\gamma^2}. \tag{49}$$

In terms of these quantities, the density of states can be rewritten

$$g(E) = \frac{2\pi V^N}{N^2\gamma^2} \int_0^1 ... \int_0^1 \prod_{i=1}^{N} d^2\mathbf{x}_i \delta\Big(\Lambda + \frac{1}{N} u(\mathbf{x}_1, ..., \mathbf{x}_N)\Big). \tag{50}$$

The proper thermodynamic limit for a system of point vortices with equal circulation in the microcanonical ensemble is such that $N \to +\infty$ with fixed Λ. We see that the box size R does not enter in the normalized energy Λ. Therefore, the thermodynamic limit corresponds to $N \to +\infty$ with $\gamma \sim N^{-1} \to 0$ and $E \sim 1$. This is a very unusual thermodynamic limit due to the non-extensivity of the system. Note that the total circulation $\Gamma = N\gamma$ remains fixed in this process.

For sufficiently large N, the density of states can be written

$$g(E) \simeq \int \mathcal{D}\rho \; e^{NS[\rho]} \; \delta\Big(\Lambda - E[\rho]\Big)\delta\Big(1 - \int \rho(\mathbf{r}) d^2\mathbf{r}\Big), \tag{51}$$

with

$$S[\rho] = -\int \rho(\mathbf{r}) \ln \rho(\mathbf{r}) d^2\mathbf{r}, \tag{52}$$

$$E[\rho] = -\frac{1}{2} \int \rho(\mathbf{r})\rho(\mathbf{r}') \ln|\mathbf{r} - \mathbf{r}'| d^2\mathbf{r} d^2\mathbf{r}'. \tag{53}$$

In the above formula, $g(E)$ has been expressed as a functional integral over the macrostates recalling that $e^{NS[\rho]} = W(\{\rho\})$ gives the number of microstates corresponding to the macrostate $\rho(\mathbf{r})$ (see Sect. 2.6). The crucial point to realize is that the vortex number N appears explicitly in the exponential, all other terms being of order unity. Therefore, at the thermodynamic limit $N \to +\infty$, the functional integral is dominated by the distribution $\rho_*(\mathbf{r})$ which maximizes the Boltzmann entropy (52) under the constraints of fixed circulation and energy brought by the δ-functions. In this limit, the mean-field approximation is *exact* and we have

$$g(E) = e^{NS[\rho_*]}, \qquad S(E) = NS[\rho_*]. \tag{54}$$

In the canonical ensemble, the object of interest is the partition function

$$Z(\beta) = \int e^{-\beta H(\mathbf{r}_1, ..., \mathbf{r}_N)} \prod_{i=1}^{N} d^2\mathbf{r}_i, \tag{55}$$

which is the normalization factor in the Gibbs measure

$$\mu(\mathbf{r}_1, ..., \mathbf{r}_N) = \frac{1}{Z} e^{-\beta H(\mathbf{r}_1, ..., \mathbf{r}_N)}. \tag{56}$$

The free energy is defined by the relation

$$F(\beta) = -\frac{1}{\beta} \ln Z. \tag{57}$$

Using the notations introduced previously, we can rewrite the integral (55) in the form

$$Z(\beta) = V^N \int_0^N ... \int_0^N \prod_{i=1}^N d^2\mathbf{x}_i e^{\eta u(\mathbf{x}_1,...,\mathbf{x}_N)}, \tag{58}$$

where

$$\eta = \frac{\beta N \gamma^2}{2\pi}, \tag{59}$$

is the normalized inverse temperature. It plays the role of the "plasma parameter" in plasma physics. It can be shown that the partition function (58) is convergent only for $\eta > \eta_c = -4N/(N-1)$ which is consistent with the result (25). The proper thermodynamic limit for a canonical distribution of point vortices is such that $N \to +\infty$ with fixed η. In this limit, $\gamma \sim N^{-1}$ and $\beta \sim N$.

The partition function is related to the density of states by the Laplace transform

$$Z(\beta) = \int_{-\infty}^{+\infty} dE \; g(E) e^{-\beta E}. \tag{60}$$

Therefore, for large N one has, using (51),

$$Z(\beta) \simeq \int \mathcal{D}\rho \; e^{NJ[\rho]} \delta\left(1 - \int \rho(\mathbf{r}) d^2\mathbf{r}\right), \tag{61}$$

with

$$J[\rho] = S[\rho] - \eta E[\rho]. \tag{62}$$

At the thermodynamic limit, the functional integral is dominated by the distribution $\rho_*(\mathbf{r})$ which maximizes the Massieu function $J[\rho]$ under the constraint of a fixed circulation. In this limit, the mean-field approximation is exact and we have

$$Z(\beta) = e^{NJ[\rho_*]}, \qquad F = E - TS. \tag{63}$$

It can be noted that the critical points of entropy at fixed energy and circulation (microcanonical description) and the critical points of free energy at fixed temperature and circulation (canonical description) coincide. This is not necessarily the case for the second order variations of entropy and free energy. Therefore, a distribution $\rho(\mathbf{r})$ can be a maximum of $S[\rho]$ but a minimum (or a saddle point) of $J[\rho]$. In that case, the equilibrium is stable in the microcanonical ensemble but unstable in the canonical ensemble (and the procedure leading to (63) is clearly not valid). When this situation happens, the ensembles are *inequivalent* and phase transitions occur. This is the case, in particular, for the gravitational problem (see Sect. 2.10).

Finally, the grand canonical partition function is defined by

$$Z_{GC} = \sum_{N=0}^{+\infty} \frac{z^N}{N!} \int \prod_{i=1}^N d^2\mathbf{r}_i e^{-\beta H_N(\mathbf{r}_1,...,\mathbf{r}_N)}, \tag{64}$$

where z is the fugacity. Using the Hubbard-Stratanovich transformation

$$\int \mathcal{D}\xi \, e^{-\int \{\frac{1}{2}(\nabla\xi)^2 - \rho(\mathbf{r})\xi(\mathbf{r})\}d^2\mathbf{r}} = e^{-\frac{1}{2}\int \rho(\mathbf{r})G(\mathbf{r}-\mathbf{r}')\rho(\mathbf{r}')d^2\mathbf{r}d^2\mathbf{r}'}, \tag{65}$$

where G is the Green function of the Laplacian operator, we can rewrite the Boltzmann factor in the form

$$e^{-\beta H_N} = \int \mathcal{D}\xi \, e^{-\int \frac{1}{2}(\nabla\xi)^2 d^2\mathbf{r} + \sqrt{-\beta}\sum_{i=1}^{N}\gamma\xi(\mathbf{r}_i)}, \tag{66}$$

where we have assumed $\beta < 0$ which is the case of physical interest. Substituting this result in (64) and introducing the notation $\phi(\mathbf{r}) = \sqrt{-\beta}\gamma\xi(\mathbf{r})$, we can easily carry out the summation on N to obtain

$$Z_{GC} = \int \mathcal{D}\phi \, e^{-\frac{1}{T_{eff}}\int d^2\mathbf{r}\{\frac{1}{2}(\nabla\phi)^2 - \mu^2 e^{\phi}(\mathbf{r})\}}, \tag{67}$$

$$T_{eff} = -\beta\gamma^2, \qquad \mu^2 = -z\beta\gamma^2. \tag{68}$$

Therefore, the grand partition function of the point vortex gas corresponds to a Liouville field theory with an action

$$S[\phi] = \frac{1}{T_{eff}}\int d^2\mathbf{r}\{\frac{1}{2}(\nabla\phi)^2 - \mu^2 e^{\phi}(\mathbf{r})\}. \tag{69}$$

While the previous description is formally correct if we *define* Z and Z_{GC} by (55) and (64), it must be noted however that the canonical and grand canonical ensembles may not have a physical meaning for point vortices. In particular, it is not clear how one can impose a thermal bath at negative temperature. On the other hand, the usual procedure to derive the canonical ensemble from the microcanonical ensemble rests on a condition of additivity which is clearly lacking in the present case.

2.8 Axisymmetric Equilibrium States in a Disk

Let us consider a collection of N point vortices with circulation γ confined within a disk of radius R. At statistical equilibrium, the streamfunction ψ is solution of the Boltzmann-Poisson equation (45). If we work in a circular domain, we must in principle account for the conservation of angular momentum. This can lead to bifurcations between axisymmetric and off-axis solutions [117]. We shall, however, ignore this constraint for the moment in order to obtain analytical expressions for the thermodynamical parameters. This is a sufficient approximation to illustrate the structure of the problem, which is our main concern here.

If we confine our attention to axisymmetric solutions, the Boltzmann-Poisson equation (45) can be written

$$\frac{1}{r}\frac{d}{dr}\left(r\frac{d\psi}{dr}\right) = -\omega_0 e^{-\beta\gamma(\psi-\psi_0)}. \tag{70}$$

$$\psi'(0) = 0, \qquad \psi(R) = 0, \tag{71}$$

where ψ_0 and ω_0 are the values of streamfunction and vorticity at $r = 0$. Introducing the function $\phi = \beta\gamma(\psi - \psi_0)$ and the dimensionless radial distance $\xi = (|\beta|\gamma\omega_0)^{1/2}r$, (70) can be reduced to the form

$$\frac{1}{\xi}\frac{d}{d\xi}\left(\xi\frac{d\phi}{d\xi}\right) = \lambda e^{-\phi}, \tag{72}$$

$$\phi(0) = \phi'(0) = 0, \tag{73}$$

with $\lambda = 1$ if $\beta < 0$ and $\lambda = -1$ if $\beta > 0$. It turns out that this equation can be solved analytically as noticed by a number of authors. With the change of variables $t = \ln\xi$ and $\phi = 2\ln\xi - z$, (72) can be rewritten

$$\frac{d^2z}{dt^2} = -\lambda e^z = -\frac{d}{dz}(\lambda e^z). \tag{74}$$

This corresponds to the motion of a fictitious particle in a potential $V(z) = \lambda e^z$. This equation is readily integrated and, returning to original variables, we finally obtain

$$e^{-\phi} = \frac{1}{(1 + \frac{\lambda}{8}\xi^2)^2}. \tag{75}$$

From the circulation theorem (6) applied to an axisymmetric flow, we have

$$-\frac{d\psi}{dr}(r) = \frac{\Gamma(r)}{2\pi r}. \tag{76}$$

where $\Gamma(r) = \int_0^r \omega(r')2\pi r'dr'$ is the circulation within r. Taking $r = R$ and introducing the dimensionless variables previously defined, we obtain

$$\eta \equiv \frac{\beta\gamma\Gamma}{2\pi} = -\alpha\phi'(\alpha), \tag{77}$$

where $\alpha = (|\beta|\gamma\omega_0)^{1/2}R$ is the value of ξ at the box radius. Equation (77) relates α to the inverse temperature η. Using (75) and (77), the vorticity field can be written explicitly

$$\langle\omega\rangle(r) = \frac{4\Gamma}{\pi R^2(\eta+4)}\frac{1}{(1 - \frac{\eta}{\eta+4}\frac{r^2}{R^2})^2}. \tag{78}$$

At positive temperatures ($\eta > 0$), the vorticity is an increasing function of the distance and the vortices tend to accumulate on the boundary of the domain (Fig. 4). On the contrary at negative temperatures ($\eta < 0$), the vorticity is a decreasing function of the distance and the vortices tend to group themselves in the core of the domain to form a "supervortex" (Fig. 5). These results are consistent with Onsager's prediction [101]. We also confirm that statistical equilibrium states only exist for $\eta > \eta_c = -4$, as previously discussed. At this critical temperature, the central vorticity becomes infinite and the solution tends to a Dirac peak:

$$\langle\omega\rangle(\mathbf{r}) \to \Gamma\delta(\mathbf{r}), \qquad \text{for} \qquad \eta \to \eta_c = -4. \tag{79}$$

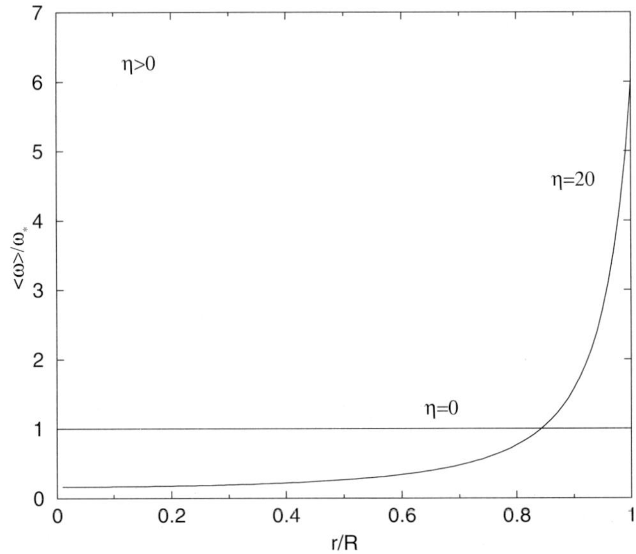

Fig. 4. Statistical equilibrium states of point vortices at positive temperatures ($\omega_* = \Gamma/\pi R^2$). The vortices are preferentially localized near the wall.

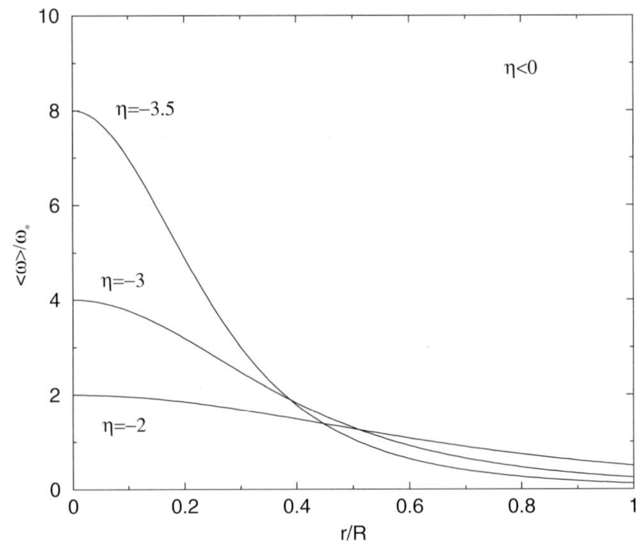

Fig. 5. Statistical equilibrium states of point vortices at negative temperatures showing a clustering. For $\eta = -4$, the vortices collapse at the center of the domain and the vorticity profile is a Dirac peak.

The energy defined by (39) can be written in the dimensionless form

$$\Lambda \equiv \frac{2\pi E}{\Gamma^2} = \frac{1}{2\eta^2}\int_0^{\alpha} \phi'(\xi)^2 \xi d\xi. \tag{80}$$

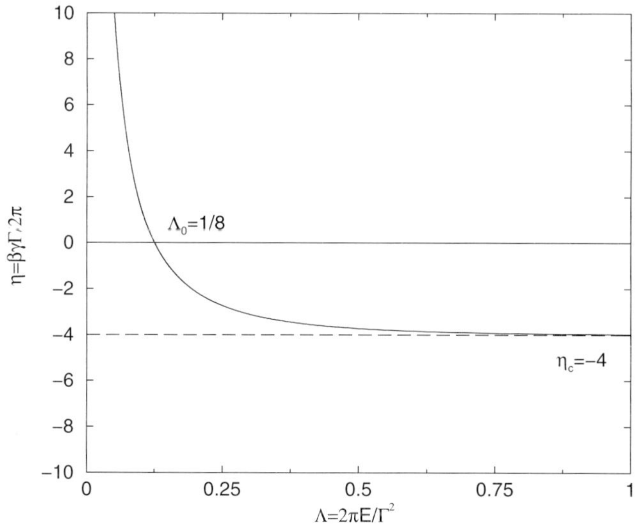

Fig. 6. Equilibrium phase diagram (caloric curve) for point vortices with equal circulation confined within a disk. For simplicity, the angular momentum has not been taken into account ($\Omega = 0$).

The integral can be carried out explicitly using (75). Eliminating α between (80) and (77), we find that the temperature is related to the energy by the equation of state

$$\Lambda = \frac{1}{\eta}\left[\frac{4}{\eta}\ln\left(\frac{4}{4+\eta}\right)+1\right]. \tag{81}$$

The corresponding equilibrium phase diagram is represented in Fig. 6. We check explicitly that the energy becomes infinite when $\eta \to \eta_c = -4$. The transition between positive and negative temperatures occurs for $\Lambda_0 = 1/8$.

The entropy (41) can also be calculated easily from the above results. Within an unimportant additive constant, it is given by

$$\frac{S}{N} = \frac{8}{\eta}\ln 4 - \left(1+\frac{8}{\eta}\right)\ln(4+\eta). \tag{82}$$

Using (81), we can easily check that $dS/dE = \beta$, as it should. For $\Lambda < \Lambda_0 = 1/8$, the entropy is an increasing function of energy (positive temperatures) and for $\Lambda > \Lambda_0$, it is a decreasing function of energy (negative temperatures). For $\Lambda \to 0$, $S/N \sim \ln\Lambda$ and for $\Lambda \to +\infty$, $S/N \sim -4\Lambda$ (Fig. 7).

It is amusing to note that if we define a local pressure by the relation $p(\mathbf{r}) = \langle\omega\rangle/\gamma\beta$ (which is similar to the local equation of state $p = \rho T$ for an ideal gas of material particles), we find that the pressure P at the boundary of the domain is exactly given by (24). Furthermore, using the Boltzmann distribution (44), we easily check that $\nabla p = -\langle\omega\rangle\nabla\psi$, which is similar to the equation of hydrostatic equilibrium for a fluid in a gravitational field. It is not clear whether these results bear more significance than is apparent at first sight.

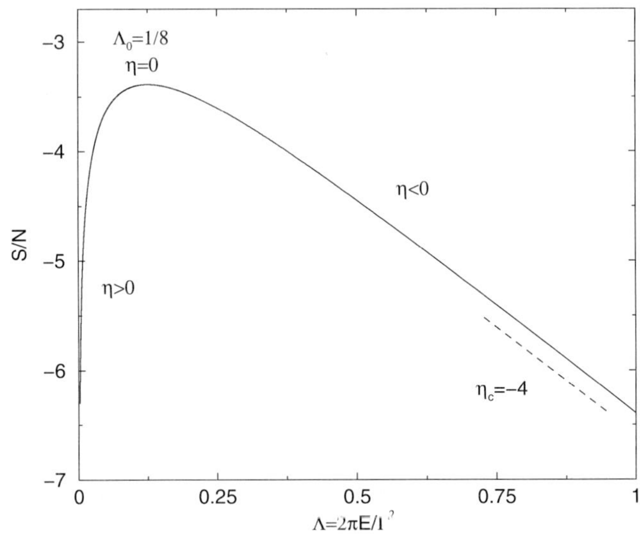

Fig. 7. Entropy vs energy plot for a system of point vortices with equal circulation confined within a disk.

2.9 Equilibrium States in an Unbounded Domain

In an infinite domain, it is necessary to take into account the conservation of angular momentum, because this constraint determines the typical size of the system. Furthermore, we shall assume that the vortices have the same sign otherwise positive and negative vortices will form dipoles and escape to infinity. Therefore, there is no equilibrium state in an infinite domain for a neutral system of point vortices. When the conservation of angular momentum is taken into account, the density of states is given by

$$g(E,L)=\int\delta\Big(E-H(\mathbf{r}_1,...,\mathbf{r}_N)\Big)\,\delta\Big(L-\sum_{i=1}^{N}\gamma r_i^2\Big)\prod_{i=1}^{N}d^2\mathbf{r}_i, \tag{83}$$

and the angular velocity of the flow by

$$\Omega=2T\left(\frac{\partial S}{\partial L}\right)_E. \tag{84}$$

Using the same trick as in Sect. 2.4 with the change of variables $\mathbf{x}=\mathbf{r}/\sqrt{L}$, one finds the exact result

$$\Omega=\frac{2N}{\beta L}\left(1+\frac{\beta N\gamma^2}{8\pi}\right). \tag{85}$$

Therefore, the vorticity field is determined by the Boltzmann distribution

$$\langle\omega\rangle=Ae^{-\beta\gamma\psi'}, \tag{86}$$

where ψ' is the relative streamfunction

$$\psi' \equiv \psi + \frac{\Omega}{2} r^2 = \psi + \frac{N}{4\beta L}(4+\eta) r^2. \tag{87}$$

For $\eta = 0$, one has

$$\langle \omega \rangle = \frac{\gamma N \Gamma}{\pi L} e^{-\frac{N\gamma}{L} r^2}. \tag{88}$$

For large r, the asymptotic behavior of (86) is

$$\langle \omega \rangle \sim \frac{1}{r^{-\eta}} e^{-\frac{N\gamma}{4L}(4+\eta) r^2} \qquad (r \to +\infty), \tag{89}$$

where we have used $\psi \sim -(\Gamma/2\pi)\ln r$ at large distances. From (89), one sees that $\eta \geq -4$ is required for the existence of an integrable solution.

Inserting the relation (86) in the Poisson equation (7), we get

$$-\Delta\psi' = A e^{-\frac{2\pi\eta}{N\gamma}\psi'} - \frac{N^2\gamma^2}{2\pi L \eta}(4+\eta). \tag{90}$$

With the change of variables

$$\boldsymbol{\xi} = \left(\frac{\gamma N}{L}\right)^{1/2} \mathbf{r}, \qquad \phi = \ln\left(\frac{LA}{N^2\gamma^2}\right) - \frac{2\pi\eta}{N\gamma}\psi', \tag{91}$$

the Boltzmann-Poisson equation (90) can be written

$$\frac{d^2\phi}{d\xi^2} + \frac{1}{\xi}\frac{d\phi}{d\xi} = 2\pi\eta e^{\phi} - (4+\eta), \tag{92}$$

and the vorticity (86) becomes

$$\langle \omega \rangle = \frac{N^2\gamma^2}{L} e^{\phi}. \tag{93}$$

Equation (92) has been obtained by Lundgren and Pointin [89] in a completely different manner. In the limit $\eta \to +\infty$, we have

$$\langle \omega \rangle = \frac{\Gamma^2}{2\pi L}, \qquad \text{if} \quad r \leq (2L/\Gamma)^{1/2}, \tag{94}$$

and $\langle \omega \rangle = 0$ otherwise. This vortex patch is the state of minimum energy at fixed circulation and angular momentum. For $\eta \to -4$, one has approximately

$$\langle \omega \rangle = \frac{N^2\gamma^2}{L} \frac{A}{(1 - A\pi\eta\frac{\gamma N}{4L} r^2)^2} e^{-\frac{\gamma N}{4L}(4+\eta) r^2}. \tag{95}$$

The first factor is an exact solution of (92) with the second term on the right hand side neglected (see Sect. 2.8). The second factor is a correction for large r, in agreement with the asymptotic result expressed by (89). The parameter A tends to infinity as $\eta \to -4$ and is determined from the condition $\int \langle \omega \rangle d^2\mathbf{r} = \Gamma$ by the formula

$$\pi A + \ln(\pi A) = -C - \ln\left(1 + \frac{\eta}{4}\right), \tag{96}$$

where $C = 0.577...$ is the Euler constant. For $\eta = -4$, the vorticity profile is a Dirac peak and the energy tends to $+\infty$.

2.10 The Gravitational N-Body Problem

It is interesting to compare the previous results with those obtained in the case of self-gravitating systems [103]. Formally, the structure of the N-vortex problem shares some analogies with the gravitational N-body problem. The force by unit of mass experienced by a star is given by

$$\mathbf{F}_i = \sum_{j\neq i} \mathbf{F}(j \to i), \qquad \mathbf{F}(j \to i) = Gm\frac{\mathbf{r}_j - \mathbf{r}_i}{|\mathbf{r}_j - \mathbf{r}_i|^3}, \tag{97}$$

where $\mathbf{F}(j \to i)$ is the force created by star j on star i. The force can be written as the gradient $\mathbf{F} = -\nabla\Phi$ of a gravitational potential Φ which is related to the stellar density

$$\rho(\mathbf{r}, t) = \sum_{i=1}^{N} m\delta(\mathbf{r} - \mathbf{r}_i), \tag{98}$$

by the Poisson equation

$$\Delta\Phi = 4\pi G\rho. \tag{99}$$

Furthermore, the equations of motion (Newton's equations) can be put in the Hamiltonian form

$$m\frac{d\mathbf{r}_i}{dt} = \frac{\partial H}{\partial \mathbf{v}_i}, \qquad m\frac{d\mathbf{v}_i}{dt} = -\frac{\partial H}{\partial \mathbf{r}_i},$$
$$H = \frac{1}{2}\sum_{i=1}^{N} mv_i^2 - \sum_{i<j}\frac{Gm^2}{|\mathbf{r}_i - \mathbf{r}_j|}. \tag{100}$$

In the analogy between stellar systems and two-dimensional vortices, the star density ρ plays the role of the vorticity ω, the force $\mathbf{F}$ the role of the velocity $\mathbf{V}$ and the gravitational potential Φ the role of the streamfunction ψ. The crucial point to realize is that, for the two systems, the interaction is a long-range unshielded Coulombian interaction (in $D = 3$ or $D = 2$ dimensions). This makes the connexion between point vortices and stellar systems deeper than between point vortices and electric charges for example. In particular, point vortices can organize into large scale clusters, like stars in galaxies, while the distribution of electric charges in a neutral plasma is uniform. There are, on the other hand, fundamental differences between stars and vortices. In particular, a star creates an acceleration while a vortex creates a velocity. On the other hand, the gravitational interaction is attractive and directed along the line joining the particles while the interaction between vortices is rotational and perpendicular to the line joining the vortices.

Despite these important differences, the statistical mechanics of 2D vortices and stellar systems are relatively similar. Like the point vortex gas, the self-gravitating gas is described at statistical equilibrium by the Boltzmann distribution

$$\langle\rho\rangle = Ae^{-\beta\Phi}, \tag{101}$$

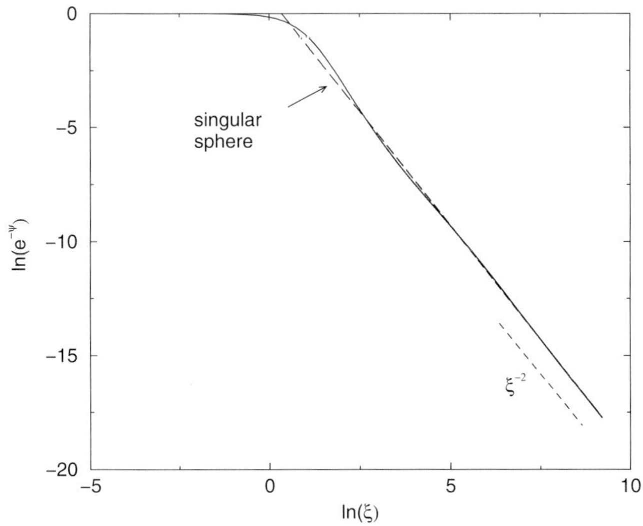

Fig. 8. Density profile of the self-gravitating gas at statistical equilibrium. The dashed line corresponds to the singular solution $\rho = 1/2\pi G\beta r^2$.

obtained by maximizing the Boltzmann entropy at fixed mass M and energy E . Its structure is therefore determined by solving the Boltzmann-Poisson equation

$$\Delta\Phi = 4\pi G A e^{-\beta\Phi}, \tag{102}$$

where A and $\beta > 0$ have to be related to M and E. This statistical mechanics approach has been developed principally for globular clusters relaxing towards equilibrium via two-body encounters [9]. It is clear that the Boltzmann-Poisson equation (102) is similar to the Boltzmann-Poisson equation (45) for point vortices at *negative* temperatures. The density profile determined by these equations is a decreasing function of the distance, which corresponds to a situation of *clustering* (see Figs. 5 and 8). The similarity of the maximum entropy problem for stars and vortices, and the Boltzmann-Poisson equations (102) (45), is a first manifestation of the formal analogy existing between these two systems.

However, due to the different dimension of space ($D = 3$ for stars instead of $D = 2$ for vortices), the mathematical problems differ in the details. First of all, the density profile determined by the Boltzmann-Poisson equation (102) in $D = 3$ decreases like r^{-2} at large distances leading to the so-called infinite mass problem since $M = \int_0^{+\infty} \rho \; 4\pi r^2 dr \to +\infty$ [19]. There is no such problem for point vortices in two dimensions: the vorticity decreases like r^{-4}, or even more rapidly if the conservation of angular momentum is accounted for, and the total circulation $\Gamma = \int_0^{+\infty} \omega \; 2\pi r dr$ is finite. The infinite mass problem implies that no statistical equilibrium state exists for open star clusters, even in theory. A system of particles in gravitational interaction tends to evaporate so that the final state is just two stars in Keplerian orbit. This evaporation process has been clearly identified in the case of globular clusters which gradually lose stars to

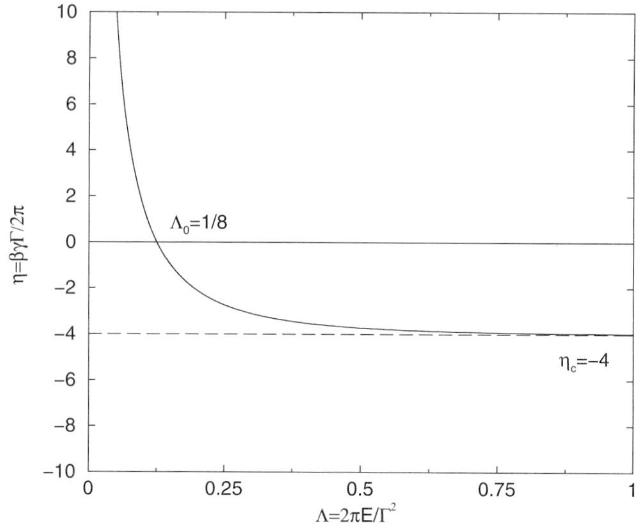

Fig. 9. Equilibrium phase diagram for self-gravitating systems confined within a box. For sufficiently low energy or temperature, there is no equilibrium state and the system undergoes gravitational collapse.

the benefit of a neighboring galaxy. In fact, the evaporation is so slow that we can consider in a first approximation that the system passes by a succession of quasiequilibrium states described by a truncated isothermal distribution function (Michie-King model) [9]. This justifies the statistical mechanics approach in that sense. Another way of solving the infinite mass problem is to confine the system within a box of radius R. However, even in that case, the notion of equilibrium poses problem regarding what now happens at the center of the configuration.

The equilibrium phase diagram (E, T) for bounded self-gravitating systems is represented in Fig. 9. The caloric curve has a striking spiral behavior parametrized by the density contrast $\mathcal{R} = \rho(0)/\rho(R)$ going from 1 (homogeneous system) to $+\infty$ (singular sphere) as we proceed along the spiral. There is no equilibrium state below $E_c = -0.335GM^2/R$ or $T_c = \frac{GMm}{2.52kR}$ [2,92]. In that case, the system is expected to collapse indefinitely. This is called *gravothermal catastrophe* in the microcanonical ensemble (fixed E) and *isothermal collapse* in the canonical ensemble (fixed T). Dynamical models show that the collapse is self-similar and develops a finite time singularity [105,85,56,91,84,46]. However, although the central density goes to $+\infty$, the shrinking of the core is so rapid that the core mass goes to zero. Therefore, the singularity contains no mass and this process alone cannot lead to a black hole.

Since the $T(E)$ curve has turning points, this implies that the microcanonical and canonical ensembles are not equivalent and that phase transitions will occur [103]. In the microcanonical ensemble, the series of equilibria becomes unstable after the first turning point of energy (MCE) corresponding to a density contrast of 709. At that point, the solutions pass from local entropy maxima to saddle points. In the canonical ensemble, the series of equilibria becomes unstable after the first turning point of temperature (CE) corresponding to a density contrast of 32.1. At that point, the solutions pass from minima of free energy $(F = E - TS)$ to saddle points. It can be noted that the region of negative specific heats between (CE) and (MCE) is stable in the microcanonical ensemble but unstable in the canonical ensemble, as expected on general physical grounds. The thermodynamical stability of isothermal spheres can be deduced from the topology of the $\beta - E$ curve by using the turning point criterion of Katz [75] who has extended Poincaré's theory of linear series of equilibria. The stability problem can also be reduced to the study of an eigenvalue equation associated with the second order variations of entropy or free energy as studied by Padmanabhan [102] in the microcanonical ensemble and by Chavanis [37] in the canonical ensemble. This study has been recently extended to other statistical ensembles [44]: grand canonical, grand microcanonical, isobaric.... The same stability limits as Katz are obtained but this method provides in addition the form of the density perturbation profiles that trigger the instability at the critical points. It also enables one to show a clear equivalence between thermodynamical stability in the canonical ensemble and dynamical stability with respect to the Navier-Stokes equations (Jeans problem) [37,44]. These analytical methods can be extended to general relativity [38]. It must be stressed, however, that the statistical equilibrium states of self-gravitating systems are at most metastable: there is no *global* maximum of entropy or free energy for a classical system of point masses in gravitational interaction [2].

Phase transitions in self-gravitating systems can be studied in detail by introducing a small-scale cut-off ϵ in order to regularize the potential. This can be achieved for example by considering a system of self-gravitating fermions (in which case an effective repulsion is played by the Pauli exclusion principle) [62,8,52,41,43] or a hard spheres gas [3,103,120]. Other forms of regularization are possible [59,128,45]. For these systems, there can still be gravitational collapse but the core will cease to shrink when it feels the influence of the cut-off. The result is the formation of a compact object with a large mass: a "fermion ball" or a hard spheres "condensate". The equilibrium phase diagram of self-gravitating fermions is represented in Fig. 10 and has been discussed at length by Chavanis [41] in the light of an analytical model. The introduction of a small-scale cut-off has the effect of unwinding the classical spiral of Fig. 9. For a small cut-off $\epsilon \ll 1$, the trace of the spiral is still visible and the $T(E)$ curve is multivalued (Fig. 11). This can lead to a gravitational first order phase transition between a gaseous phase with an almost homogeneous density profile (upper branch) and a condensed phase with a core-halo structure (lower branch). At a critical cut-off value $\epsilon = \epsilon_{MTP}$, the two phases merge and the gravitational phase transition disapears. This particular point is sometimes called a microcanonical tricritical point (MTP).

For $\epsilon > \epsilon_{MTP}$, the curve has the form of Fig. 12. The $T(E)$ curve is now univalued so that the equilibrium states are always stable in the microcanonical ensemble (they are global entropy maxima). In particular, the region of nega-

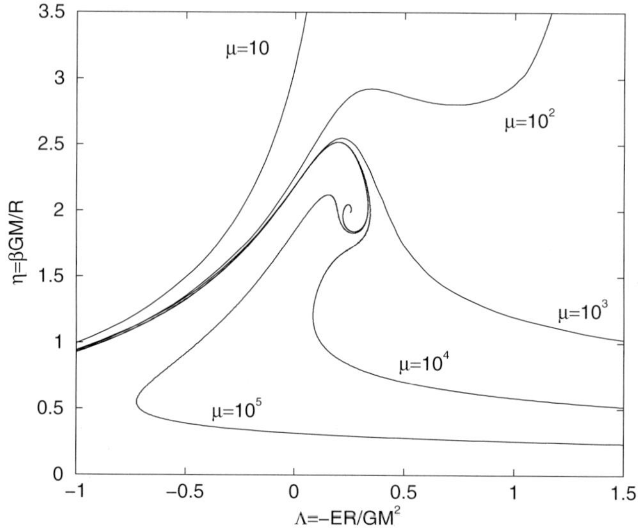

Fig. 10. Equilibrium phase diagram for self-gravitating fermions [41]. The degeneracy parameter μ plays the role of a small-scale cut-off $\epsilon \sim 1/\mu$. For $\epsilon \to 0$, the classical spiral of Fig. 9 is recovered.

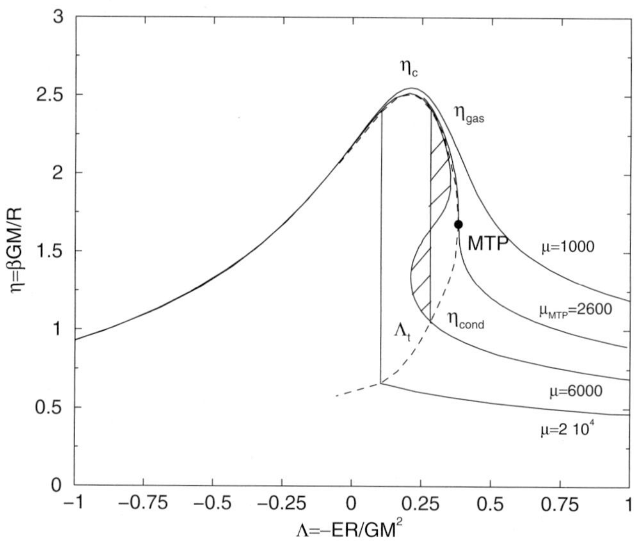

Fig. 11. Enlargement of the phase diagram near the tricritical point in the microcanonical ensemble. A priori, the phase transition should occur at the energy $E_t(\mu)$ at which the gaseous phase and the condensed phase have the same entropy. In fact, the entropic barrier played by the unstable solution on the wiggling branch is so hard to cross that the transition will not occur at E_t but rather at, or near, E_c the point of gravothermal catastrophe [77,45].

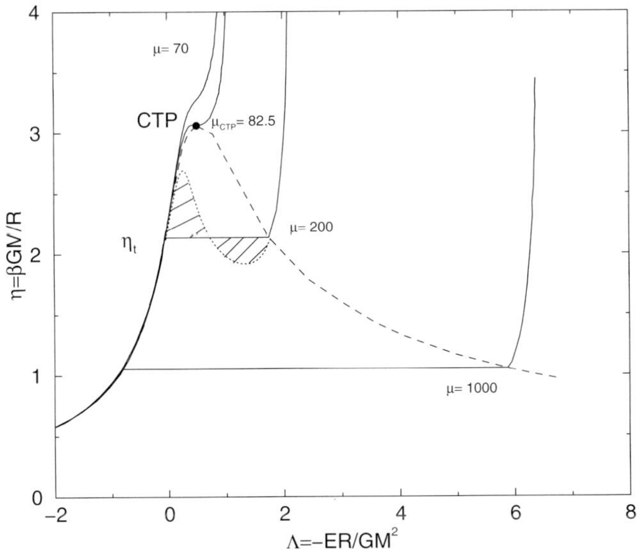

Fig. 12. Same as Fig. 11 near the tricritical point in the canonical ensemble. The region of negative specific heats, allowed in the microcanonical ensemble, is replaced by a phase transition in the canonical ensemble. The notion of metastability is discussed in [41,45].

tive specific heats (leading to a convex dip in the entropy vs energy plot [45]) is allowed in the microcanonical ensemble. By contrast, the $E(T)$ curve is multi-valued and this can lead to a normal first order phase transition in the canonical ensemble. The gaseous and condensed phases are thus connected by a Maxwell plateau which replaces the region of negative specific heats. At $\epsilon = \epsilon_{CTP}$, the two phases merge, the specific heat becomes infinite and the phase transition is second order. This particular point is sometimes called a canonical tricritical point (CTP). For $\epsilon > \epsilon_{CTP}$, the $T(E)$ curve is monotonous and the specific heat is positive. Therefore, as the cut-off parameter ϵ increases, the self-gravitating gas consecutively exhibits gravitational first order, normal first order, second-order and no phase transition at all [45]. A similar behavior exists for the Blume-Emery-Griffiths (BEG) model with infinite range interaction [6] and is probably representative of other systems presenting inequivalence of ensembles (small systems or systems with long-range interactions).

The equilibrium phase diagram of point vortices in two dimensions is completely different (Fig. 6). Since there is no turning point of energy or temperature, we can immediately infer that the equilibria are always stable and that the microcanonical and canonical ensembles are equivalent. In the microcanonical ensemble, an equilibrium state exists for all values of energy. Therefore, there is no "gravothermal catastrophe" in two dimensions [76]. The solutions of the Boltzmann-Poisson system for arbitrary dimension D and the disappearance of the spiral as we approach the critical dimension $D = 2$ have been studied in [116] (note that the spiral also disapears for $D \geq 10$). There exists, on the other hand, a critical inverse temperature $\eta_c = -4$ in two dimensions below which there is no equilibrium state in the canonical ensemble. This can lead to a situation of collapse [116] but, as indicated previously, it is not clear whether the canonical ensemble makes sense for a gas of point vortices.

3 Statistics of Velocity Fluctuations Arising from a Random Distribution of Point Vortices

3.1 The Marginal Gaussian Distribution

The aim of equilibrium statistical mechanics is to predict the final configuration of a system resulting from a complex evolution. We seek now to develop a kinetic theory of point vortices to determine how the system will reach this equilibrium configuration. The first problem to consider is the characterization of the velocity fluctuations experienced by a point vortex. These fluctuations are responsible for a diffusion process which is one of the driving source of the evolution. In this section, we describe a simple stochastic model [69,97,126,55,82,47,48,40] from which we can obtain an estimate of the diffusion coefficient of point vortices [47]. We follow the presentation of Chavanis and Sire [47,48].

Let us consider a collection of N point vortices randomly distributed in a disk of radius R. We assume that the vortices have a Poisson distribution, i.e. their positions are independent and uniformly distributed in average over the entire domain. In terms of statistical mechanics, this corresponds to an equilibrium state with $\beta = 0$ in a domain with no specific symmetry or $\beta \to +\infty$ (i.e. a state of minimum energy) if the angular momentum is conserved. We are particularly interested in the "thermodynamic limit" in which the number of vortices and the size of the domain go to infinity ($N \to \infty$, $R \to \infty$) in such a way that the vortex density $n = \frac{N}{\pi R^2}$ remains finite. In this limit, the Poisson distribution is stationary and is well-suited to the analysis of the fluctuations. For simplicity, we assume that the vortices have the same circulation γ. There is therefore a solid rotation of the system $\langle \mathbf{V} \rangle = \frac{1}{2} n \gamma \mathbf{r}_{\perp}$. We shall work in a rotating frame of reference so as to ignore this solid rotation. The generalization of our results to a spectrum of circulations among the vortices is relatively straightforward [47].

The velocity $\mathbf{V}$ occurring at the center O of the domain is the sum of the velocities $\mathbf{\Phi}_i$ ($i = 1, ..., N$) produced by the N vortices:

$$\mathbf{V} = \sum_{i=1}^{N} \mathbf{\Phi}_i, \qquad \mathbf{\Phi}_i = -\frac{\gamma}{2\pi} \frac{\mathbf{r}_{\perp i}}{r_i^2}, \tag{103}$$

where $\mathbf{r}_i$ denotes the position of the ith vortex relative to the point under consideration and, by definition, $\mathbf{r}_{\perp i}$ is the vector $\mathbf{r}_i$ rotated by $+\pi/2$. Since the vortices are randomly distributed, the velocity $\mathbf{V}$ fluctuates. It is therefore of interest to study the statistics of these fluctuations, i.e. the probability $W(\mathbf{V}) d^2\mathbf{V}$ that $\mathbf{V}$ occurs between $\mathbf{V}$ and $\mathbf{V} + d\mathbf{V}$. Basically, we have to determine the distribution of a sum of random variables. If the variance of the individual velocity $\mathbf{\Phi}$ were finite, we could immediately apply the central limit theorem and deduce that the distribution of $\mathbf{V}$ is Gaussian. Alternatively, if the variance of $\mathbf{\Phi}$ were rapidly diverging (i.e., algebraically), the distribution of $\mathbf{V}$ would be a Lévy law with an infinite variance and an algebraic tail. In the present case, the problem is intermediate between these two situations because the variance of the velocity created by a single vortex

$$\langle \Phi^2 \rangle = \int_{|\mathbf{r}|=0}^{R} \tau(\mathbf{r}) \Phi^2 d^2\mathbf{r} = \int_0^R \frac{1}{\pi R^2} \frac{\gamma^2}{4\pi^2 r^2} 2\pi r dr, \tag{104}$$

diverges *slowly*, i.e. logarithmically. For that reason, the distribution of $\mathbf{V}$ will be intermediate between Gaussian and Lévy laws.

The distribution $W_N(\mathbf{V})$ of the velocity can be expressed as

$$W_N(\mathbf{V}) = \int \prod_{i=1}^{N} \tau(\mathbf{r}_i) d^2\mathbf{r}_i \delta\left(\mathbf{V} - \sum_{i=1}^{N} \mathbf{\Phi}_i\right), \tag{105}$$

where $\tau(\mathbf{r}_i) d^2\mathbf{r}_i$ governs the probability of occurrence of the i-th point vortex at position $\mathbf{r}_i$. In writing this expression, we have assumed that the vortices are identical and uncorrelated. Now, using a method originally due to Markov, we express the δ-function appearing in (105) in terms of its Fourier transform

$$\delta(\mathbf{x}) = \frac{1}{(2\pi)^2} \int e^{-i\boldsymbol{\rho}\mathbf{x}} d^2\boldsymbol{\rho}. \tag{106}$$

With this transformation, $W_N(\mathbf{V})$ becomes

$$W_N(\mathbf{V}) = \frac{1}{4\pi^2} \int A_N(\boldsymbol{\rho}) e^{-i\boldsymbol{\rho}\mathbf{V}} d^2\boldsymbol{\rho}, \tag{107}$$

with

$$A_N(\boldsymbol{\rho}) = \left(\int_{|\mathbf{r}|=0}^{R} e^{i\boldsymbol{\rho}\mathbf{\Phi}} \tau(\mathbf{r}) d^2\mathbf{r} \right)^N, \tag{108}$$

where we have written

$$\mathbf{\Phi} = -\frac{\gamma}{2\pi} \frac{\mathbf{r}_\perp}{r^2}. \tag{109}$$

If we now suppose that the vortices are uniformly distributed on average, then

$$\tau(\mathbf{r}) = \frac{1}{\pi R^2}, \tag{110}$$

and (108) reduces to

$$A_N(\boldsymbol{\rho}) = \left(\frac{1}{\pi R^2} \int_{|\mathbf{r}|=0}^{R} e^{i\boldsymbol{\rho}\mathbf{\Phi}} d^2\mathbf{r} \right)^N. \tag{111}$$

Since

$$\frac{1}{\pi R^2}\int_{|\mathbf{r}|=0}^{R} d^2\mathbf{r} = 1, \tag{112}$$

we can rewrite our expression for $A_N(\boldsymbol{\rho})$ in the form

$$A_N(\boldsymbol{\rho}) = \left(1 - \frac{1}{\pi R^2}\int_{|\mathbf{r}|=0}^{R}(1-e^{i\boldsymbol{\rho}\boldsymbol{\Phi}})d^2\mathbf{r}\right)^N. \tag{113}$$

We now consider the limit in which the number of vortices and the size of the domain go to infinity in such a way that the density remains finite:

$$N \to \infty, \quad R \to \infty, \quad n = \frac{N}{\pi R^2} \quad \text{finite.}$$

If the integral occurring in (113) increases less rapidly than N, then

$$A(\boldsymbol{\rho}) = e^{-nC(\boldsymbol{\rho})}, \tag{114}$$

with

$$C(\boldsymbol{\rho}) = \int_{|\mathbf{r}|=0}^{R}(1-e^{i\boldsymbol{\rho}\boldsymbol{\Phi}})d^2\mathbf{r}. \tag{115}$$

We have dropped the subscript N to indicate that the limit $N \to \infty$, in the previous sense, has been taken. Note that $A(\boldsymbol{\rho})$ can still depend on N through logarithmic factors, so that (114) must be considered as an equivalent of (108) for large N, not a true limit. In fact, it is appropriate to consider that $N \to +\infty$ but not $\ln N$ since in physical situations the typical number of vortices does not exceed 10^4.

The characteristic function $C(\boldsymbol{\rho})$ can be calculated explicitly by taking $\boldsymbol{\Phi}$ as a variable of integration instead of $\mathbf{r}$ and transforming to polar coordinates [47]. The final result remains complicated but the following expression provides a sufficient approximation for our purposes:

$$C(\boldsymbol{\rho}) = \frac{\gamma^2\rho^2}{16\pi}\ln\left(\frac{4\pi N}{n\gamma^2\rho^2}\right). \tag{116}$$

Since $C(\boldsymbol{\rho})$ diverges weakly with N (logarithmically), the limiting process leading to formula (114) is permissible. The velocity distribution $W(\mathbf{V})$ is simply the Fourier transform of $A(\boldsymbol{\rho})$ with the expression (116) for $C(\boldsymbol{\rho})$. This leads to the following distribution for the velocity fluctuations:

$$W(\mathbf{V}) = \frac{4}{n\gamma^2\ln N}e^{-\frac{4\pi}{n\gamma^2\ln N}V^2} \quad (V \ll V_{crit}(N)), \tag{117}$$

$$W(\mathbf{V}) = \frac{n\gamma^2}{4\pi^2 V^4} \quad (V \gg V_{crit}(N)), \tag{118}$$

$$V_{crit}(N) \sim \left(\frac{n\gamma^2}{4\pi}\ln N\right)^{1/2} \ln^{1/2}(\ln N). \tag{119}$$

For small fluctuations, the core of the distribution is Gaussian as if the central limit theorem were applicable (this is due to the quadratic behavior of $C(\rho)$). For sufficiently large values of V, the velocity distribution $W(\mathbf{V})$ decays algebraically as for a Lévy law (this is due to the logarithmic term in $C(\rho)$). In the mathematical limit $\ln N \to +\infty$, the algebraic tail is rejected to infinity and the distribution is purely Gaussian. This is consitent with the generalized form of the central limit theorem described by Ibragimov and Linnik [67] and used by Min et al. [97] and Weiss et al. [126]. However, the convergence is so slow with N that the algebraic tail is always visible for point vortex systems [97,69,126,82]. This algebraic tail arises because we are on the frontier between Gaussian and Lévy laws (see Fig. 1.1 of Bouchaud and Georges [11]). Therefore, we proposed to call (117)-(119) the *Marginal Gaussian distribution* [48].

Since the distribution $W(\mathbf{V})$ decreases like V^{-4} for $V \to \infty$, the variance of the velocity diverges logarithmically. Since $V = \gamma/2\pi r$, this corresponds to a divergence at small scales ($r \to 0$). On the other hand, if we were to extend the Gaussian distribution (117) for all values of V, we would conclude that its variance

$$\langle V^2 \rangle - \frac{n\gamma^2}{4\pi} \ln N, \tag{120}$$

diverges logarithmically when $N \to \infty$. This corresponds to a divergence at large scales ($R \to +\infty$). We can recover these results more directly by calculating the variance of $\mathbf{V}$ from (103). Indeed,

$$\langle V^2 \rangle = N \int_{L_{min}}^{R} \frac{1}{\pi R^2} \frac{\gamma^2}{4\pi^2 r^2} 2\pi r dr = \frac{n\gamma^2}{2\pi} \ln\left(\frac{R}{L_{min}}\right), \tag{121}$$

where L_{min} is a lower cut-off and R an upper cut-off played here by the size of the domain.

We can give a physical interpretation of the algebraic tail of the velocity distribution in terms of the nearest neighbor approximation. Let us calculate the velocity distribution due to the nearest neighbor. For that purpose, we must first determine the probability $\tau_{n.n}(r)dr$ that the position of the nearest neighbor occurs between r and $r + dr$. Clearly, $\tau_{n.n}(r)dr$ is equal to the probability that no vortices exist interior to r times the probability that a vortex (any) exists in the annulus between r and $r + dr$. Therefore, it must satisfy an equation of the form

$$\tau_{n.n}(r)dr = \left(1 - \int_0^r \tau_{n.n}(r')dr'\right) n2\pi r dr, \tag{122}$$

where $n = \frac{N}{\pi R^2}$ denotes the mean density of vortices in the disk. Differentiating (122) with respect to r we obtain

$$\frac{d}{dr}\left[\frac{\tau_{n.n}(r)}{2\pi n r}\right] = -\tau_{n.n}(r). \tag{123}$$

This equation is readily integrated with the condition $\tau_{n.n}(r) \sim 2\pi n r$ as $r \to 0$, and we find

$$\tau_{n.n}(r) = 2\pi n r e^{-\pi n r^2}. \tag{124}$$

This is the distribution of the nearest neighbor in a random distribution of particles. From this formula, we can obtain the exact value for the "average distance" d between vortices. By definition,

$$d = \int_0^{+\infty} \tau_{n.n}(r) r dr. \tag{125}$$

Hence,

$$d = \frac{1}{2\sqrt{n}}. \tag{126}$$

If we assume that the velocity $\mathbf{V}$ is entirely due to the nearest neighbor, then

$$W_{n.n}(\mathbf{V}) d^2\mathbf{V} = \tau_{n.n}(\mathbf{r}) d^2\mathbf{r}, \tag{127}$$

with

$$\mathbf{V} = -\frac{\gamma}{2\pi} \frac{\mathbf{r}_\perp}{r^2}. \tag{128}$$

Using (124), we obtain

$$W_{n.n}(\mathbf{V}) = \frac{n\gamma^2}{4\pi^2 V^4} e^{-\frac{n\gamma^2}{4\pi V^2}}. \tag{129}$$

The nearest neighbor approximation is expected to give relevant results only for large values of the velocity. Thus, we can make the additional approximation

$$W_{n.n}(\mathbf{V}) = \frac{n\gamma^2}{4\pi^2 V^4}, \tag{130}$$

in perfect agreement with (118) valid for $V \gg V_{crit}$. This shows that the algebraic tail of the velocity distribution is essentially produced by the nearest neighbor as for a Lévy law.

Note also that the typical velocity due to the nearest neighbor is $V_{n.n} \sim \gamma/2\pi d$ where d is the inter-vortex distance (126). Hence $V_{n.n}^2 \sim n\gamma^2/\pi^2$. Comparing this result with (120), we see that the velocity due to all the vortices is, up to a logarithmic factor, of the same order as the velocity due to the nearest neighbor. We can say, in some sense, that the velocity fluctuation is marginally dominated by the nearest neighbor and that collective effects are responsible for logarithmic corrections. In fact, we can show [40] that the "effective velocity" created by a vortex at a distance r from the point under consideration is given by

$$V_{eff} = \frac{\gamma}{2\pi r} \frac{1}{1 + r/\Lambda}, \tag{131}$$

where $\Lambda \sim d$ is of the order of the inter-vortex distance. For $r \gg d$, the "effective" velocity decays as r^{-2} instead of the ordinary r^{-1} law recovered for $r \ll d$. This result indicates that, in a statistical sense, the velocity produced by a vortex is shielded by "cooperative" effects.

3.2 The Speed of Fluctuations

The function $W(\mathbf{V})$ does not provide us with all the necessary information concerning the fluctuations of $\mathbf{V}$. An important aspect of the problem concerns the *speed of fluctuations*, i.e. the typical duration $T(V)$ of the velocity fluctuation $\mathbf{V}$. This requires the knowledge of the bivariate probability $W(\mathbf{V},\mathbf{A})d^2\mathbf{V}d^2\mathbf{A}$ to measure simultaneously a velocity $\mathbf{V}$ with a rate of change

$$\mathbf{A}=\frac{d\mathbf{V}}{dt}=\sum_{i=1}^{N}\boldsymbol{\psi}_i, \tag{132}$$

$$\boldsymbol{\psi}_i=-\frac{\gamma}{2\pi}\left(\frac{\mathbf{v}_{\perp i}}{r_i^2}-\frac{2(\mathbf{r}_i\mathbf{v}_i)\mathbf{r}_{\perp i}}{r_i^4}\right), \tag{133}$$

where $\mathbf{v}_i=d\mathbf{r}_i/dt$ is the velocity of vortex i. Then, the duration $T(V)$ can be estimated by the formula

$$T(V)=\frac{|\mathbf{V}|}{\sqrt{\langle A^2\rangle_{\mathbf{V}}}}, \tag{134}$$

where

$$\langle A^2\rangle_{\mathbf{V}}=\frac{\int W(\mathbf{V},\mathbf{A})A^2d^2\mathbf{A}}{W(\mathbf{V})}, \tag{135}$$

is the mean square acceleration associated with a velocity fluctuation $\mathbf{V}$.

To determine the speed of fluctuations, we need therefore to calculate the bivariate probability $W_N(\mathbf{V},\mathbf{A})$ to measure simultaneously a velocity $\mathbf{V}$ with a rate of change $\mathbf{A}=d\mathbf{V}/dt$. According to (103), (132) and (133), $\mathbf{V}$ and $\mathbf{A}$ are the sum of N random variables $\mathbf{\Phi}_i$ and $\boldsymbol{\psi}_i$ depending on the positions $\mathbf{r}_i$ and on the velocities $\mathbf{v}_i$ of the point vortices. Unlike material particles, the variables $\{\mathbf{r}_i,\ \mathbf{v}_i\}$, for different i's, are not independent because the velocities of the vortices are determined by the configuration $\{\mathbf{r}_i\}$ of the system as a whole (see (103)). To be able to solve the problem analytically, we shall make a *decorrelation approximation* and treat $\{\mathbf{r}_i,\ \mathbf{v}_i\}$ $(i=1,...,N)$ as independent variables governed by the distribution

$$\tau(\mathbf{r},\mathbf{v})=\frac{1}{\pi R^2}\times\frac{4}{n\gamma^2\ln N}e^{-\frac{4\pi}{n\gamma^2\ln N}v^2}, \tag{136}$$

resulting from (110) and (117). It is remarkable that the distribution (136) is formally equivalent to the Maxwell-Boltzmann statistics of material particles at equilibrium. Owing to this analogy, we can interpret the variance

$$\overline{v^2}=\frac{n\gamma^2}{4\pi}\ln N, \tag{137}$$

as a kind of kinetic "temperature" of the point vortices. It should not be confused with the temperature β introduced in Sect. 2 and measuring the *clustering* of the vortices (in the present case, $\beta=0$ since the vortices are uniformly distributed in average).

When this decorrelation hypothesis is implemented, a straightforward generalization of the method used in Sect. 3.1 yields

$$W_N(\mathbf{V},\mathbf{A}) = \frac{1}{16\pi^4}\int A_N(\boldsymbol{\rho},\boldsymbol{\sigma})e^{-i(\boldsymbol{\rho}\mathbf{V}+\boldsymbol{\sigma}\mathbf{A})}d^2\boldsymbol{\rho}d^2\boldsymbol{\sigma}, \tag{138}$$

with

$$A_N(\boldsymbol{\rho},\boldsymbol{\sigma}) = \left(\int_{|\mathbf{r}|=0}^{R}\int_{|\mathbf{v}|=0}^{+\infty}\tau(\mathbf{r},\mathbf{v})e^{i(\boldsymbol{\rho}\boldsymbol{\Phi}+\boldsymbol{\sigma}\boldsymbol{\psi})}d^2\mathbf{r}d^2\mathbf{v}\right)^N, \tag{139}$$

where we have defined

$$\boldsymbol{\Phi} = -\frac{\gamma}{2\pi}\frac{\mathbf{r}_\perp}{r^2}, \tag{140}$$

$$\boldsymbol{\psi} = -\frac{\gamma}{2\pi}\left(\frac{\mathbf{v}_\perp}{r^2} - \frac{2(\mathbf{rv})\mathbf{r}_\perp}{r^4}\right). \tag{141}$$

Using an integration by parts, the conditional moment of the acceleration for a given velocity fluctuation $\mathbf{V}$ can be expressed as [47]:

$$W(\mathbf{V})\langle A^2\rangle_{\mathbf{V}} = -\frac{1}{\pi^2}\int\frac{\partial A}{\partial(\sigma^2)}(\boldsymbol{\rho},\mathbf{0})e^{-i\boldsymbol{\rho}\mathbf{V}}d^2\boldsymbol{\rho}. \tag{142}$$

We therefore need to Taylor expand the function $A(\boldsymbol{\rho},\boldsymbol{\sigma})$ for $|\boldsymbol{\sigma}| \to 0$ and carry out the integration in (142). The calculation is relatively tricky but can be done analytically [47]. Substituting the resulting expression for $\langle A^2\rangle_{\mathbf{V}}$ in (134), we obtain the following estimate for the speed of fluctuations:

$$T(V) = \frac{4\sqrt{\pi}V}{n^{3/2}\gamma^2\ln N}e^{-\frac{2\pi}{n\gamma^2\ln N}V^2} \quad (V \ll V_{crit}(N)), \tag{143}$$

$$T(V) = \frac{1}{\sqrt{\pi n}\ln N}\frac{1}{V} \quad (V \gg V_{crit}(N)). \tag{144}$$

For weak and large fluctuations, $T(V)$ decreases to zero like V and V^{-1} respectively. These asymptotic behaviors have a clear physical meaning in the nearest neighbor approximation. When $r = \gamma/2\pi V$ is small, corresponding to large velocities, it is highly improbable that another vortex will enter the disk of radius r before long. By contrast, on a short time scale $T \sim r/\overline{v} \sim (\gamma/\overline{v})V^{-1}$, the vortex will have left the disk. When $r = \gamma/2\pi V$ is large, corresponding to small velocities, the probability that the vortex will remain alone in the disk is low. The characteristic time before another vortex enters the disk varies like the inverse of the number of vortices expected to be present in the disk, i.e. $T \sim (r/\overline{v})1/n\pi r^2 \sim (1/n\gamma\overline{v})V$. The demarcation between weak and strong fluctuations corresponds to $V \sim \gamma n^{1/2}$, i.e. to the velocity produced by a vortex distant $n^{-1/2}$ from the point under consideration. These asymptotic behaviors are also consistent with the theory of Smoluchowski concerning the persistence of fluctuations (see discussion in [47]).

The average duration of a velocity fluctuation is defined by

$$\langle T\rangle = \int_0^{+\infty} T(V)W(V)2\pi V dV. \tag{145}$$

To leading order in $\ln N$, we obtain

$$\langle T \rangle = \frac{4}{3}\left(\frac{\pi}{6}\right)^{1/2} \frac{1}{n\gamma\sqrt{\ln N}}. \tag{146}$$

This formula shows that the typical duration of a velocity fluctuation scales like

$$T_{typ} \sim \frac{1}{n\gamma\sqrt{\ln N}}. \tag{147}$$

This corresponds to the typical time needed by a vortex moving with a typical velocity $\langle V^2\rangle^{1/2}$ (given by (120)) to cross the interparticle distance $d \sim n^{-1/2}$, as expected from general physical grounds [126].

3.3 The Diffusion Coefficient

According to the previous discussion, we can characterize the fluctuations of the velocity of a point vortex by two functions: a function $W(\mathbf{V})$ which governs the occurrence of the velocity $\mathbf{V}$ and a function $T(V)$ which determines the typical time during which the vortex moves with this velocity. Since the velocity fluctuates on a typical time $T_{typ} = d/\sqrt{\langle V^2\rangle}$ which is much smaller than the dynamical time $t_D = R/\sqrt{\langle V^2\rangle}$ needed by the vortex to cross the entire domain, the motion of the vortex will be essentially stochastic. If we denote by $P(\mathbf{r}, t)$ the probability density that the particle be found in $\mathbf{r}$ at time t, then $P(\mathbf{r}, t)$ will satisfy the diffusion equation

$$\frac{\partial P}{\partial t} = D\Delta P. \tag{148}$$

If the particle is at $\mathbf{r} = \mathbf{r}_0$ at time $t = 0$, the solution of (148) is clearly

$$P(\mathbf{r}, t|\mathbf{r}_0) = \frac{1}{4\pi D t} e^{-\frac{|\mathbf{r}-\mathbf{r}_0|^2}{4Dt}}, \tag{149}$$

where D is the diffusion coefficient. The mean square displacement that the particle is expected to suffer during an interval of time Δt large with respect to the fluctuation time T_{typ}, is

$$\langle (\Delta\mathbf{r})^2\rangle = 4D\Delta t. \tag{150}$$

We can obtain another expression for $\langle (\Delta\mathbf{r})^2\rangle$ in terms of the functions $W(\mathbf{V})$ and $T(V)$ defined in the previous sections. Indeed, dividing the interval

$$\Delta\mathbf{r} = \int_t^{t+\Delta t} \mathbf{V}(t')dt', \tag{151}$$

into a succession of discrete increments in position with amount $T(V_i)\mathbf{V}_i$, we readily establish that

$$\langle (\Delta\mathbf{r})^2\rangle = \langle T(V)V^2\rangle \Delta t. \tag{152}$$

Combining (150) and (152) we obtain an alternative expression for the diffusion coefficient in the form

$$D = \frac{1}{4}\int T(V)W(\mathbf{V})V^2 d^2\mathbf{V}. \tag{153}$$

Substituting for $T(V)$ and $W(\mathbf{V})$ in the foregoing expression, we obtain to leading order in $\ln N$ [47],

$$D = \frac{1}{72}\left(\frac{6}{\pi}\right)^{1/2}\gamma\sqrt{\ln N}. \tag{154}$$

We should not give too much credit to the numerical factor appearing in (154) since the definition (134) of $T(V)$ is just an order of magnitude. Note that the scaling form of D is consistent with the expression

$$D \sim T_{typ}\langle V^2\rangle \sim \gamma\sqrt{\ln N}, \tag{155}$$

that one would expect on general physical grounds.

3.4 Application to 2D Decaying Turbulence

The previous results have some direct applications to the context of 2D decaying turbulence. The relaxation of 2D decaying turbulence is a three-stage process: during an initial transient period, the fluid organizes itself from random fluctuations and a population of coherent vortices emerges. Then, when two like-sign vortices come into contact they merge and form a bigger structure. As time goes on, the vortex number decreases and their average size increases, in a process reminiscent of a coarsening stage. Finally, when only one dipole is left, it decays diffusively due to inherent viscosity. Direct numerical simulations [18] and experiments [61] show that the typical core vorticity ω remains constant during the course of the evolution and that the density decreases like $n \sim t^{-\xi}$ with $\xi \approx 0.7$. As the energy $E \sim \int \mathbf{u}^2 d^2\mathbf{r} \sim n\omega^2 a^4 \sim na^4$ is conserved throughout the evolution, the typical vortex radius increases like $a \sim t^{\xi/4}$.

In the *punctuated Hamiltonian model* [125], the flow is approximated by a collection of vortices with constant vorticity $\pm\omega$ (and circulation $\gamma \sim \pm\omega a_i^2$) whose centers follow the Hamiltonian dynamics (15)(16). When two like-sign vortices with radii a_1 and a_2 enter in collision, the resulting vortex keeps the same vorticity and takes a radius a such that $a^4 = a_1^4 + a_2^4$ which ensures the conservation of energy (as discussed above). This effective model reproduces the results of the numerical simulations and experiments, with again $\xi \simeq 0.7$. Since the average distance between vortices, of order $d \sim t^{\xi/2}$, increases more rapidly than their size, the point vortex model should provide increasing accuracy. We emphasize, however, that in the above mentioned studies the density only changes by a factor of order $4 \sim 5$ between the initial time and the final time so that the scaling exponent is measured on less than one decade.

The punctuated Hamiltonian model has been re-investigated recently by Sire and Chavanis [115] using a renormalization group procedure which allows for much longer time simulations that could otherwise be achieved. It is found that the scaling regime is achieved for very late times and is characterized by a decay exponent $\xi = 1$ (an effective exponent $\xi \simeq 0.7$ is recovered for shorter times). In addition, the decay of the total area occupied by the vortices results in a physical process by which merging occurs principally via 3-vortex collisions involving vortices of different sign. A simple kinetic theory based on an effective 3-vortex interaction returns this value $\xi = 1$. These theoretical results tend to be confirmed by recent direct numerical simulations [86].

During the decay, the vortices diffuse with a coefficient $D \sim \gamma$ given by (154), where $\gamma \sim \omega a^2$ is their circulation (we ignore here logarithmic corrections). If the diffusion coefficient were constant, then the dispersion of the vortices

$$\langle r^2 \rangle \sim Dt, \tag{156}$$

would increase linearly with time as in ordinary Brownian motion. However, since D varies with time according to

$$D \sim \omega a^2 \sim t^{\xi/2}, \tag{157}$$

we expect anomalous diffusion, i.e.

$$\langle r^2 \rangle \sim t^{\nu}, \tag{158}$$

with $\nu \neq 1$. Substituting (157) in (156), we obtain the following relation between ν and ξ:

$$\nu = 1 + \frac{\xi}{2}. \tag{159}$$

This formula is in perfect agreement with long time numerical simulations ($\nu = 3/2$ for $\xi = 1$) [115] and experiments ($\nu_{exp} \sim 1.3 - 1.4$ in the regime where $\xi = 0.7$) [61]. This hyperdiffusive behavior can be interpreted in terms of Lévy flights [115], with a large time flight distribution $P(\tau) \sim \tau^{-\mu}$ with $\mu = 3 - \xi/2$ in agreement with experiments ($\mu_{exp} \sim 2.6 \pm 0.2$ for $\xi = 0.7$)[61].

3.5 The Spatial Correlations in the Velocities Arising from a Random Distribution of Point Vortices

The stochastic approach described in Sect. 3.1 can be generalized to obtain exact results concerning the spatial correlations of the velocity fluctuations [48]. First of all, the distribution of the velocity increments between two neighboring points is given by the 2D Cauchy law [69,97,48]:

$$W(\delta \mathbf{V}) = \frac{2}{\pi n^2 \gamma^2 |\delta \mathbf{r}|^2} \left(1 + \frac{4|\delta \mathbf{V}|^2}{n^2 \gamma^2 |\delta \mathbf{r}|^2} \right)^{-3/2}, \tag{160}$$

which is a particular Lévy law. It is also possible to determine an analytical expression for the conditional moment $\langle \delta \mathbf{V} \rangle_{\mathbf{V}}$. We find:

$$\langle \delta \mathbf{V} \rangle_{\mathbf{V}} = -\frac{n\gamma}{2} B\left(\frac{4\pi V^2}{n\gamma^2 \ln N} \right) \left\{ \delta \mathbf{r}_{\perp} + 2\frac{(\mathbf{V}_{\perp} \delta \mathbf{r})}{V^2} \mathbf{V} \right\} \quad (V \ll V_{crit}(N)), \tag{161}$$

$$\langle \delta \mathbf{V} \rangle_{\mathbf{V}} = -\frac{2\pi}{\gamma} V^2 \left\{ \delta \mathbf{r}_\perp + 2 \frac{(\mathbf{V}_\perp \delta \mathbf{r})}{V^2} \mathbf{V} \right\} \quad (V \gg V_{crit}(N)), \tag{162}$$

where $B(x)$ denotes the function

$$B(x) = \frac{1}{x}(e^x - 1 - x). \tag{163}$$

On the other hand, the spatial auto-correlation function of the velocity between two points separated by an arbitrary distance is given by [48]:

$$\langle \mathbf{V}_0 \mathbf{V}_1 \rangle = \frac{n\gamma^2}{2\pi} \ln \left(\frac{R}{r_1} \right). \tag{164}$$

This leads to an energy spectrum

$$E(k) = \frac{n\gamma^2}{4\pi k}(1 - J_0(kR)), \tag{165}$$

which reduces to Novikov's result $E(k) = n\gamma^2/4\pi k$ for $k \to +\infty$ [100]. We observe that the velocity correlation function (164) diverges logarithmically as $r_1 \to 0$. This is consistent with the divergence of the variance of the velocity distribution (117)-(119). Therefore, it is more proper to characterize the spatial correlations of the velocity by the function $\langle V_{1\|} \rangle = \langle \mathbf{V}_0 \mathbf{V}_1 / |\mathbf{V}_0| \rangle$ which remains finite as $r_1 \to 0$. Its evaluation is more complex and leads to the result

$$\langle V_{1\|} \rangle = \langle |\mathbf{V}_0| \rangle - \left(\frac{n\gamma^2}{\pi \ln N} \right)^{1/2} \int_0^{+\infty} \frac{J_1(z)}{z^2} \Gamma_{s^2 z^2}\left(\frac{1}{2} \right) dz, \tag{166}$$

where

$$\Gamma_x(p+1) = \int_0^x e^{-t} t^p dt, \tag{167}$$

is the incomplete Γ-function and

$$\langle |\mathbf{V}_0| \rangle = \left(\frac{n\gamma^2}{16} \ln N \right)^{1/2}, \qquad s = \left(\frac{\pi n \ln N}{4} \right)^{1/2} r_1. \tag{168}$$

Similarly, the correlation function $K(r_1) = \langle \mathbf{V}_0 \mathbf{V}_1 / V_0^2 \rangle$ is given by

$$K(s) = 1 - \frac{4s^2}{\ln N} \left(1 - e^{-1/4s^2} + \frac{1}{4s^2} E_1 \left(\frac{1}{4s^2} \right) \right), \tag{169}$$

where $E_1(z)$ denotes the exponential integral

$$E_1(z) = \int_z^{+\infty} \frac{e^{-t}}{t} dt. \tag{170}$$

Other results concerning the characterization of the spatial velocity correlations can be found in [48].

3.6 Statistics of Fluctuations of the Gravitational Field

The stochastic approach developed in the preceding sections was inspired by the famous work of Chandrasekhar and von Neumann [20,27,28,24,25] concerning the fluctuations of the gravitational field created by a random distribution of stars. Let us consider a collection of N stars with mass m randomly distributed in a sphere of radius R with a uniform density n in average. The force by unit of mass created at the center O of the domain is

$$\mathbf{F} = \sum_{i=1}^{N} \boldsymbol{\Phi}_i, \qquad \boldsymbol{\Phi}_i = -\frac{Gm}{r_i^3}\mathbf{r}_i. \tag{171}$$

As before, the problem consists in determining the distribution of a sum of random variables. In the present case, the variance of the force created by one star diverges algebraically

$$\langle \Phi^2 \rangle \sim \frac{3}{4\pi R^3} \int \left(\frac{Gm}{r^2}\right)^2 4\pi r^2 dr \sim \frac{1}{r}, \tag{172}$$

so that the distribution of the total force is a Lévy law. This particular Lévy law is known as the *Holtsmark distribution* since it was first determined by Holtsmark in the context of the electric field created by a gas of simple ions [64]. In the thermodynamic limit $N, R \to +\infty$ with $n = \frac{3N}{4\pi R^3}$ finite, the distribution of $\mathbf{F}$ can be expressed by the Fourier transform

$$W(\mathbf{F}) = \frac{1}{2\pi^2 F} \int_0^{+\infty} k \sin(kF) e^{-ak^{3/2}} dk, \tag{173}$$

with

$$a = \frac{4}{15}(2\pi G)^{3/2} n m^{1/2}. \tag{174}$$

It has the asymptotic behavior

$$W(\mathbf{F}) \sim \frac{1}{2} G^{3/2} m^{1/2} n F^{-9/2} \qquad (F \to +\infty), \tag{175}$$

which can be shown to coincide with the distribution due to the nearest neighbor. The typical force due to the nearest neighbor $F_{n.n} \sim Gm/d^2 \sim Gmn^{2/3}$ is precisely of the same order as the average value of the force due to all the stars

$$\langle F \rangle = 4\Gamma\left(\frac{1}{3}\right)\left(\frac{8\sqrt{2}}{15}\right)^{2/3} Gmn^{2/3}, \tag{176}$$

determined from (173). This shows that only stars close to the star under consideration determine the fluctuations of the gravitational field. In fact, it is possible to show that the "effective" force created by a star at distance r from the star under consideration is [1]:

$$F_{eff} = \frac{Gm}{r^2} \frac{1}{1 + r^2/\Lambda^2}, \tag{177}$$

where $\Lambda \sim d$ is of the order of the interparticle distance. For weak separations, one has $F_{eff} \to Gm/r^2$ but for large separations $r \gg d$, the effect of individual stars compensate each other and the resulting force is reduced by a factor $(r/d)^2$. This corresponds to a shielding of the interaction, in a statistical sense, due to collective effects.

Chandrasekhar and von Neumann have used this stochastic model to determine the speed of fluctuations $T(F)$ [27], the diffusion coefficient of stars (a calculation completed by Kandrup [72]) and the spatial [28,24] and temporal [25] correlations of the gravitational field etc... There is a complete parallel with the results obtained by Chavanis and Sire [47,48,40] for point vortices and this is another manifestation of the deep formal analogy between the two systems.

4 Relaxation of a Point Vortex in a Thermal Bath

4.1 Analogy with Brownian Motion

We would like now to develop a kinetic theory of point vortices in order to describe their relaxation towards equilibrium. It has to be noted that an equilibrium state can be achieved in very different ways so that the kinetic theory of vortices is not unique and depends on the situation contemplated (see Sect. 6). In order to develop an intuition on the problem, we first propose a naive kinetic theory based on an analogy with Brownian theory [30,36]. The starting point of this analogy is to realize that the velocity of a point vortex can be decomposed in two terms: a smoothly varying function of position and time $\langle \mathbf{V}\rangle(\mathbf{r},t)$ and a function $\boldsymbol{\mathcal{V}}(t)$ taking into account the "granularity" of the system. The total velocity of a point vortex can therefore be written:

$$\mathbf{V} = \langle \mathbf{V}\rangle(\mathbf{r},t) + \boldsymbol{\mathcal{V}}(t). \tag{178}$$

The velocity $\langle \mathbf{V}\rangle(\mathbf{r},t)$ reflects the influence of the system as a *whole* and is generated by the mean vorticity $\langle\omega\rangle(\mathbf{r},t)$ according to the Biot and Savart formula

$$\langle \mathbf{V}\rangle(\mathbf{r},t) = -\frac{1}{2\pi}\mathbf{z}\times\int \frac{\mathbf{r}'-\mathbf{r}}{|\mathbf{r}'-\mathbf{r}|^2}\langle\omega\rangle(\mathbf{r}',t)d^2\mathbf{r}'. \tag{179}$$

The fluctuation $\boldsymbol{\mathcal{V}}(t)$ arises from the difference between the exact distribution $\omega_{exact}(\mathbf{r},t)$ of point vortices given by (13) and their "smoothed-out" distribution $\langle\omega\rangle(\mathbf{r},t) = n\gamma$. Indeed, if we consider a surface element σ, the number of point vortices actually present in this area will fluctuate around the mean number $n\sigma$. These fluctuations will be governed by a Poisson distribution with variance $n\sigma$. On account of these fluctuations, the velocity of a vortex will depart from its mean field value $\langle \mathbf{V}\rangle$. The velocity fluctuation $\mathcal{V}$, of order γ/d (where $d \sim n^{-1/2}$ is the inter-vortex distance), is much smaller than the average velocity $\langle V\rangle$, of order $n\gamma R$ (where R is the domain size), but this term has a cumulative effect which gives rise to a process of diffusion. It makes sense therefore to introduce a stochastic description of the vortex motion such as that for colloidal suspensions in a liquid [81,22] or stars in globular clusters [26]. However, contrary to the ideal Brownian motion, point vortex systems have relatively long correlation times so that $\mathcal{V}(t)$ is *not* a white noise. This makes the study much more complicated than usual and the technical developments of Sect. 5 are required. However, in order to gain some physical insights into the problem, we shall ignore this difficulty for the moment and describe the system by traditional stochastic processes.

According to (178), we would naively expect that the evolution of the density probability $P(\mathbf{r}, t)$ would be governed by a diffusion equation of the form

$$\frac{\partial P}{\partial t} + \langle \mathbf{V} \rangle \nabla P = D \Delta P, \tag{180}$$

where D is the diffusion coefficient. This would in fact be the case for a passive particle having no retroaction on the vortices or when the distribution of vortices is uniform in average like in Sect. 3. However, this diffusion equation cannot be valid when the system is inhomogeneous. Indeed, (180) does not converge towards the Boltzmann distribution (44) when $t \to +\infty$. It seems therefore that a term is missing to act against the diffusion.

This problem is similar to the one encountered by Chandrasekhar in his stochastic approach of stellar dynamics [23]. Chandrasekhar solved the problem by introducing a *dynamical friction* in order to compensate for the effect of diffusion. The occurence of this frictional force is a manifestation of the "fluctuation-dissipation" theorem in statistical mechanics. In the present context, the dynamical friction is replaced by a *systematic drift* of the vortices [30]. This drift appears when the distribution of vorticity is inhomogeneous. It can be understood in terms of a polarization process like in a phenomenon of induction. In Sect. 4.3, we shall derive the drift term directly from the Liouville equation by using a linear response theory. For the moment, we introduce this term *by hands* and rewrite the decomposition (178) in the form

$$\mathbf{V} = \langle \mathbf{V} \rangle - \xi \nabla \psi + \boldsymbol{\mathcal{V}}(t), \tag{181}$$

where ξ is the drift coefficient. Equation (181) must be viewed as a stochastic equation analogous to the Langevin equation in the ordinary Brownian theory. The corresponding Fokker-Planck equation can be written [30]:

$$\frac{\partial P}{\partial t} + \langle \mathbf{V} \rangle \nabla P = \nabla (D \nabla P + \xi P \nabla \psi). \tag{182}$$

The physical interpretation of each term is straightforward. The left hand side (which can be written dP/dt) is an advection term due to the smooth mean field velocity $\langle \mathbf{V} \rangle$. The right hand side can be written as the divergence of a current $-\nabla \mathbf{J}$ and is the sum of two terms: the first term is a diffusion due to the erratic motion of the vortices caused by the fluctuations of the velocity (see Sect. 3) and the second term accounts for the systematic drift of the vortices due to the inhomogeneity of the vortex cloud. At equilibrium, the drift precisely balances random scatterings and the Boltzmann distribution (44) is settled. More precisely, the condition that the Boltzmann distribution satisfies (182) identically requires that D and ξ be related according to the relation

$$\xi = D\beta\gamma, \tag{183}$$

which is a generalization of the Einstein formula to the case of point vortices. It is remarkable that we can obtain such a general relation without, at any point, being required to analyze the mechanism of "collisions". A more rigorous justification of this relation will be given in the next sections in which the diffusion coefficient and the drift term are calculated explicitly. Note that the diffusion current can be written $\mathbf{J} = \chi\nabla\alpha$ where $\alpha = \ln\langle\omega\rangle + \beta\gamma\psi$ is a "generalized potential" which is uniform at equilibrium, see (44). Therefore, the Fokker-Planck equation (182) is consistent with the linear thermodynamics of Onsager which relates the diffusion currents to the gradients of generalized potentials. The Fokker-Planck equation (182) can also be obtained from a variational formulation (the so-called Maximum Entropy Production Principle): it can be shown to maximize the rate of entropy production $\dot{S}$ under the constraints brought by the dynamics [36].

4.2 Diffusion Coefficient of Point Vortices in an Inhomogeneous Medium

In this section, we determine the value of the diffusion coefficient D which enters in the Fokker-Planck equation (182). The essential difference with the calculation of Sect. 3 is that the vorticity distribution is now *inhomogeneous* so that there exists a shear in the system. This shear considerably modifies the expression of the diffusion coefficient [30,36]. An explicit expression for D can be obtained in a "thermal bath approximation" which is valid if we are sufficiently close to equilibrium. To be specific, we consider a system of N point vortices at statistical equilibrium with an inverse temperature β_{eq}. These "field vortices" form the thermal bath. We now introduce a "test vortex" in the system an study its diffusion in the sea of "field vortices". For simplicity, we restrict ourselves to simple equilibrium flows which are either unidirectional or axisymmetric.

In the case of unidirectional flows in the x-direction, we define the diffusion coefficient in the y direction by

$$D = \lim_{t\to+\infty}\frac{\langle(\Delta y)^2\rangle}{2t} \quad \text{with} \quad \Delta y = \int_0^t V_y(t')dt', \tag{184}$$

where $\mathbf{V}(t)$ is given by (14). It is possible to put the diffusion coefficient in the form of a Kubo formula [36]:

$$D = \int_0^{+\infty} \langle V_y(t)V_y(t-\tau)\rangle d\tau, \tag{185}$$

where the quantity in bracket is the velocity auto-correlation function at different times. Using (14), we have explicitly

$$D = N\int_0^{+\infty} d\tau \int d^2\mathbf{r}_1 V_y(1\to 0,t)V_y(1\to 0,t-\tau)P_{eq}(y_1). \tag{186}$$

For axisymmetric equilibrium flows, one has similarly

$$D = \lim_{t\to+\infty} \frac{\langle(\Delta r)^2\rangle}{2t} \quad \text{with} \quad \Delta r = \int_0^t V_r(t')dt', \tag{187}$$

and

$$D = N\int_0^{+\infty} d\tau \int d^2\mathbf{r}_1 V_{r(t)}(1\to 0,t)V_{r(t-\tau)}(1\to 0,t-\tau)P_{eq}(r_1). \tag{188}$$

The velocity auto-correlation function is calculated in Appendix A under the assumption that between t and $t-\tau$ the point vortices follow the equilibrium streamlines, which is valid for sufficiently strong shears. It is found that the auto-correlation function decreases like t^{-2} at large times so that the diffusion coefficient is well-defined. It must be stressed, however, that the decorrelation is slow so that the fluctuations cannot be described by a white noise process, as indicated previously [1]. Using the results of Appendix A, the final expression for the diffusion coefficient in the presence of a shear can be put in the form [30,36]:

$$D = \frac{\gamma}{8}\frac{1}{|\Sigma(\mathbf{r})|}\ln N\langle\omega\rangle_{eq}, \tag{189}$$

where $|\Sigma(\mathbf{r})|$ is the local shear of the flow: $|\Sigma| = -d/dy\langle V\rangle_{eq}(y)$ for unidirectional flows and $|\Sigma| = rd/dr(\langle V\rangle_{eq}(r)/r)$ for axisymmetric flows. In Sect. 5.8, we shall calculate the diffusion coefficient in another manner and extend formula (189) to more general equilibrium flows with $|\Sigma(\mathbf{r})| = 2\sqrt{-\det(\Sigma)}$ where Σ is the stress tensor. The expression (189) for the diffusion coefficient can be written in the general form (see Sect. 3):

$$D = \frac{1}{4}\tau\langle V^2\rangle = \frac{\gamma\tau}{16\pi}\ln N\langle\omega\rangle_{eq}, \tag{190}$$

with a correlation time $\tau = 2\pi/|\Sigma(\mathbf{r})|$. Physically, it corresponds to the time needed by two vortices with relative velocity Σd (where d is the inter-vortex distance) to be stretched by the shear on a distance $\sim d$. Of course, when there is no shear, our approximations break down and the diffusion coefficient is given by (154) which involves a correlation time $\tau \sim 1/\langle\omega\rangle(\mathbf{r})\sqrt{\ln N}$ determined by the dispersion of the vortices [47]. These two results correspond to a limit of "strong shear" and "weak shear" respectively. Clearly, a general formula for D should take into account simultaneously the effect of the shear and the dispersion of the vortices.

4.3 Systematic Drift Experienced by a Point Vortex in an Inhomogeneous Medium: Linear Response Theory

In Sect. 4.1, we have argued on the basis of general considerations that when the vorticity distribution is inhomogeneous, a point vortex should experience a

[1] The problem is even more severe in stellar dynamics where the temporal correlation function of the gravitational force decreases like t^{-1}, responsible for logarithmic divergences in the diffusion coefficient [88].

systematic drift in addition to its diffusive motion. In this section, we justify this drift by a linear response theory, starting directly from the Liouville equation [30]. Consider a collection of N point vortices at statistical equilibrium. In the mean-field limit, the N-particle distribution function $\mu_{eq}(\{\mathbf{r}_k\})$ can be approximated by a product of N one-particle distribution functions P_{eq}, each of which at equilibrium with the same inverse temperature β_{eq}:

$$\mu_{eq}(\{\mathbf{r}_k\}) = \prod_{k=1}^{N} P_{eq}(\mathbf{r}_k) = \prod_{k=1}^{N} A e^{-\beta_{eq}\gamma\psi_{eq}(\mathbf{r}_k)}. \tag{191}$$

This distribution is stationary, in a statistical sense, since it corresponds to a maximum entropy state. The introduction of an additional point vortex, the "test vortex", will modify this equilibrium state. The distribution function of the field vortices becomes time dependant and can be written in the form

$$\mu(\{\mathbf{r}_k\}, t) = \mu_{eq}(\{\mathbf{r}_k\}) + \mu'(\{\mathbf{r}_k\}, t), \tag{192}$$

where the perturbation $\mu'(\{\mathbf{r}_k\}, t)$ reflects the influence of the test vortex on its neighbors, just like in a polarization process. The N-particle distribution function $\mu(\{\mathbf{r}_k\}, t)$ satisfies the Liouville equation

$$\frac{\partial \mu}{\partial t} + \sum_{i=1}^{N} \left[\sum_{j\neq i} \mathbf{V}(j \to i) + \mathbf{V}(0 \to i) \right] \frac{\partial \mu}{\partial \mathbf{r}_i} = 0. \tag{193}$$

Substituting (192) in (193), we obtain the evolution equation of the perturbation μ':

$$\frac{\partial \mu'}{\partial t} + \mathcal{L}\mu' = \beta_{eq}\gamma \sum_{i=1}^{N} \mathbf{V}_i \frac{\partial \psi_{eq}}{\partial \mathbf{r}_i}(\mathbf{r}_i)\mu_{eq}(\{\mathbf{r}_k\}), \tag{194}$$

where $\mathcal{L} \equiv \sum_{i=1}^{N} \mathbf{V}_i \frac{\partial}{\partial \mathbf{r}_i}$ is a Liouville operator and $\mathbf{V}_i = \sum_{j\neq i} \mathbf{V}(j \to i) + \mathbf{V}(0 \to i)$ denotes the total velocity of vortex i. This equation can be solved formally with the Greenian

$$G(t, t') \equiv \exp\left\{ - \int_{t'}^{t} \mathcal{L}(\tau) d\tau \right\}. \tag{195}$$

If $t = 0$ is the time at which the test vortex is introduced in the system, we have $\mu'(t = 0) = 0$. One then finds that

$$\mu'(t) = \beta_{eq}\gamma \int_0^t d\tau G(t, t - \tau) \sum_{i=1}^{N} \mathbf{V}_i \frac{\partial \psi_{eq}}{\partial \mathbf{r}_i}(\mathbf{r}_i)\mu_{eq}(\{\mathbf{r}_k\}). \tag{196}$$

The average velocity of the test vortex is expressed in term of the distribution function μ of the field vortices by

$$\langle \mathbf{V} \rangle = \int \prod_{k=1}^{N} d^2\mathbf{r}_k \mathbf{V} \mu(\{\mathbf{r}_k\}, t), \tag{197}$$

where $\mathbf{V} = \sum_{i=1}^{N} \mathbf{V}(i \to 0)$. Inserting (192) and (196) into (197), one obtains

$$\langle \mathbf{V} \rangle = \int \prod_{k=1}^{N} d^2\mathbf{r}_k \mathbf{V} \mu_{eq}(\{\mathbf{r}_k\}) + \beta_{eq}\gamma \int \prod_{k=1}^{N} d^2\mathbf{r}_k \mathbf{V}$$

$$\times \int_0^t d\tau G(t, t-\tau) \sum_{i=1}^{N} V_i^{\nu} \frac{\partial \psi_{eq}}{\partial r_i^{\nu}}(\mathbf{r}_i) \mu_{eq}(\{\mathbf{r}_k\}), \tag{198}$$

with summation over repeated greek indices. The two terms arising in this expression have a clear physical meaning. The first term is the mean field velocity $\langle \mathbf{V} \rangle_{eq} = -\mathbf{z} \times \nabla \psi_{eq}(\mathbf{r})$ created by the unperturbed distribution function $\mu_{eq}(\{\mathbf{r}_k\})$. The second term, arising from the perturbation μ', corresponds to the response of the system to the polarization induced by the test vortex. Because of this back reaction, the test vortex will experience a systematic drift $\langle \mathbf{V} \rangle_{drift} = \langle \mathbf{V} \rangle - \langle \mathbf{V} \rangle_{eq}$. Explicating the action of the Greenian (195), we obtain

$$\langle \mathbf{V} \rangle_{drift} = \beta_{eq}\gamma \int \prod_{k=1}^{N} d^2\mathbf{r}_k \sum_{i=1}^{N} \mathbf{V}(i \to 0, t) \int_0^t d\tau \sum_{i=1}^{N} \Bigg[\sum_{j \neq i} V^{\nu}(j \to i, t-\tau)$$

$$+ V^{\nu}(0 \to i, t-\tau) \Bigg] \frac{\partial \psi_{eq}}{\partial r_i^{\nu}}(\mathbf{r}_i(t-\tau)) \mu_{eq}(\{\mathbf{r}_k(t-\tau)\}), \tag{199}$$

where $\mathbf{r}_i(t-\tau)$ is the position at time $t-\tau$ of the point vortex i located at $\mathbf{r}_i(t) = \mathbf{r}_i$ at time t. This is obtained by solving the Kirchhoff-Hamilton equations of motion $d\mathbf{r}_i/dt = \mathbf{V}_i$ between t and $t-\tau$.

The exact expression of the drift (199) is completely inextricable in the general case. In order to enlighten its physical content, we have to make some approximations. We shall consider in the evaluation of the time integral that the point vortices are advected by the equilibrium mean-field velocity $\langle \mathbf{V} \rangle_{eq}$. This is reasonable in a first approximation because, when $N \to \infty$, the typical velocity fluctuations $\mathcal{V}$, of order $\gamma/d \sim \gamma N^{1/2}/R$ are much smaller than the mean field velocity $\langle \mathbf{V} \rangle_{eq}$, of order $N\gamma/R$. Of course, this approximation breaks up at scales smaller than $\delta \sim R/N$ when the velocity fluctuations become comparable to the average velocity. In that case, we cannot ignore the details of the discrete vortex interactions anymore and a specific treatment is necessary. For simplicity, we shall introduce a small-scale cut-off and replace the exact Greenian G by a smoother Greenian $\langle G \rangle_{eq}$ constructed with the averaged Liouville operator $\langle \mathcal{L} \rangle_{eq} \equiv \sum_{i=1}^{N} \langle \mathbf{V}^i \rangle_{eq} \frac{\partial}{\partial \mathbf{r}_i}$. In this approximation the correlations involving two different vortex pairs vanish and we obtain

$$\langle V^{\mu} \rangle_{drift} = \beta_{eq}\gamma \int \prod_{k=1}^{N} d^2\mathbf{r}_k \int_0^t d\tau \sum_{i=1}^{N} V^{\mu}(i \to 0, t)$$

$$\times V^{\nu}(0 \to i, t-\tau) \frac{\partial \psi_{eq}}{\partial r_i^{\nu}}(\mathbf{r}_i(t-\tau)) \prod_{i=k}^{N} P_{eq}(\mathbf{r}_k), \tag{200}$$

where we have used $P^{eq}(\mathbf{r}_k(t-\tau)) = P^{eq}(\mathbf{r}_k(t))$ since $P_{eq} = f(\psi_{eq})$ is constant along a streamline. Since the vortices are identical, we have equivalently

$$\langle V^\mu \rangle_{drift} = N\beta_{eq}\gamma \int d^2\mathbf{r}_1 \int_0^t d\tau V^\mu(1 \to 0, t) \times V^\nu(0 \to 1, t-\tau)\frac{\partial \psi_{eq}}{\partial r_1^\nu}(\mathbf{r}_1(t-\tau))P_{eq}(\mathbf{r}_1). \tag{201}$$

In the case of a unidirectional equilibrium flow in the x direction, the drift velocity (201) can be written

$$\langle V_y \rangle_{drift} = N\beta_{eq}\gamma \int d^2\mathbf{r}_1 \int_0^{+\infty} d\tau V_y(1 \to 0, t) V_y(0 \to 1, t-\tau)\frac{\partial \psi_{eq}}{\partial y_1}(y_1)P_{eq}(y_1), \tag{202}$$

where we have used $y_1(t-\tau) = y_1(t) = y_1$ and the time integration has been extended to $+\infty$. Since the space integration diverges as $\mathbf{r}_1 \to \mathbf{r}_0$, we can make a local approximation $\partial_y \psi(y_1) \simeq \partial_y \psi(y)$, neglect the contribution of images in the velocity Kernel and use $V_y(0 \to 1) = -V_y(1 \to 0)$. The local approximation reflects the strong influence of the nearest neighbor and is only marginally valid as discussed in Sect. 3. The expression of the drift can then be written

$$\langle V_y \rangle_{drift} = -\beta_{eq}\gamma D\frac{\partial \psi_{eq}}{\partial y}(y), \tag{203}$$

where D is the diffusion coefficient given by formula (186). The same type of relation is obtained for an axisymmetric equilibrium flow. More generaly, one can write the drift as [30,36]:

$$\langle \mathbf{V} \rangle_{drift} = -\beta_{eq}\gamma D \nabla \psi_{eq}, \tag{204}$$

where D is given by (189). We have thus derived, from a kinetic theory, an Einstein relation $\xi = D\gamma\beta_{eq}$ for the point vortex gas. This relation is of great conceptual importance and could be checked by direct numerical simulations of point vortex dynamics in the thermal bath approach.

The direction of the drift has also important physical implications. First, we note that the drift is always normal to the mean field velocity $\langle \mathbf{V} \rangle_{eq} = -\mathbf{z} \times \nabla \psi_{eq}$. Since $P_{eq} = f(\psi_{eq})$ for an equilibrium flow, this implies that the drift is directed along the vorticity gradient. This is clear at first sight if we write the expression of the drift in the form $\langle \mathbf{V} \rangle_{drift} = D\nabla \ln\langle \omega \rangle_{eq}$, using (204) and (44). Since $D \geq 0$, this formula indicates that the test vortex always *ascends* the vorticity gradient. In fact, we have assumed in the previous discussion that all the vortices have the same circulation. It is straightforward to generalize our calculations for a test vortex with negative circulation evolving in a bath of vortices with positive circulations. Similar results are obtained except that now the test vortex *descends* the vorticity gradient.

If we now take into account simultaneously the drift and the diffusion of the test vortex, we can argue that the evolution of the density probability $P(\mathbf{r}, t)$ is governed by the Fokker Planck equation

$$\frac{\partial P}{\partial t} + \langle \mathbf{V} \rangle_{eq} \nabla P = \nabla (D(\nabla P + \beta_{eq} \gamma P \nabla \psi_{eq})), \tag{205}$$

where we recall that ψ_{eq} is the *stationary* streamfunction generated by the equilibrium vorticity $\langle \omega \rangle_{eq}$ via the Poisson equation (7). In the case of a unidirectional flow, this Fokker-Planck equation can be transformed into a Schrödinger equation (with imaginary time) which can be solved analytically [116].

4.4 A Relaxation Equation for Point Vortices

If we are sufficiently close to equilibrium, we can try to apply our previous results to *all* vortices in the system, eliminating the somewhat arbitrary distinction between test and field vortices. We propose to describe the relaxation of a cloud of point vortices towards statistical equilibrium by the following set of equations

$$\frac{\partial P}{\partial t} + \langle \mathbf{V} \rangle \nabla P = \nabla (D(\nabla P + \beta \gamma P \nabla \psi)), \tag{206}$$

$$\Delta \psi = -N \gamma P, \tag{207}$$

consisting of the Fokker-Planck equation (205) *coupled* to the Poisson equation (7). This model is expected to be valid only close to equilibrium so that the inverse temperature β has a clear physical interpretation. In this model, a point vortex is assumed to undergo a diffusion process due to the fluctuations of the velocity and a *systematic drift* $\langle \mathbf{V} \rangle_{drift} = -\beta \gamma D \nabla \psi$ due to the inhomogeneity of the vortex cloud. At negative temperatures, the drift is directed inward and the vortices tend to *cluster* while at positive temperatures, the drift is directed outward and the vortices tend to *repell* each other and accumulate on the boundary. For $\beta = 0$, there is no drift and (206) reduces to the pure diffusion equation (180); in that case, the vorticity distribution is uniform in average. These results are consistent with the thermodynamical approach of Onsager [101] but the drift provides a physical mechanism to understand the clustering of point vortices at negative temperatures. It can also be noted that (206) is formally similar to the Smoluchowski equation describing the relaxation of colloidal suspensions in an external gravitational field [108]. In the present context, however, the field ψ is not fixed but is generated by the distribution of particles itself via the Poisson equation. The resulting Smoluchowski-Poisson system has been studied in [46,116] for various space dimensions.

The system of equations (206) and (207) is also similar to the model introduced by Chandrasekhar [23] in his stochastic description of stellar dynamics:

$$\frac{\partial f}{\partial t} + \mathbf{v} \frac{\partial f}{\partial \mathbf{r}} + \langle \mathbf{F} \rangle \frac{\partial f}{\partial \mathbf{v}} = \frac{\partial}{\partial \mathbf{v}} \left\{ D \left(\frac{\partial f}{\partial \mathbf{v}} + \beta m f \mathbf{v} \right) \right\}, \tag{208}$$

$$\Delta \Phi = 4\pi G \int f d^3 \mathbf{v}. \tag{209}$$

In that model, a star undergoes a diffusion process (in velocity space) due to the fluctuations of the gravitational force and a *dynamical friction* $\langle \mathbf{F} \rangle_{friction} = -D\beta m\mathbf{v}$ resulting from close encounters. Fundamentally, this friction is due to the inhomogeneity of the velocity distribution. The coefficient of dynamical friction is given by an Einstein formula $\xi = D\beta m$ in which the velocity dispersion $1/\beta$ of the stars enters explicitly.

The morphological similarity of the two models (206)-(207) and (208)-(209) is striking although the physical content of these equations is, of course, very different. In this analogy, we see that the systematic drift of the vortices is the counterpart of the dynamical friction of stars. In fact, Chandrasekhar's dynamical friction can be derived from a linear response theory [74] exactly like we have derived the systematic drift of a vortex. In addition, both terms can be understood physically as a result of a polarization process (see discussion in [36]). This is another mark of the formal analogy between point vortices and stars.

5 Kinetic Theory of Point Vortices

The previous relaxation equations are only valid for a test vortex evolving in a bath of field vortices, or for a cloud of point vortices close to statistical equilibrium. We would like now to relax this "thermal bath approximation" and describe more general situations which do not explicitly rely on the existence of a well-defined temperature or equilibrium state. Hence, we would like to develop a more complete kinetic theory of point vortices [36].

5.1 The Liouville Equation

Let us consider a collection of $N+1$ point vortices with identical circulation γ. Let $\mu(\mathbf{r}, \mathbf{r}_1, ..., \mathbf{r}_N, t)$ denote the $N+1$ particle distribution of the system, i.e. $\mu(\mathbf{r}, \mathbf{r}_1, ..., \mathbf{r}_N, t) d^2\mathbf{r} d^2\mathbf{r}_1 ... d^2\mathbf{r}_N$ represents the probability that point vortex 0 be in the cell $(\mathbf{r}, \mathbf{r}+d\mathbf{r})$, point vortex 1 in the cell $(\mathbf{r}_1, \mathbf{r}_1+d\mathbf{r}_1)$... and point vortex N in the cell $(\mathbf{r}_N, \mathbf{r}_N+d\mathbf{r}_N)$ at time t. The $(N+1)$-particle distribution function $\mu(t)$ satisfies the Liouville equation

$$\frac{\partial \mu}{\partial t} + \sum_{i=0}^{N} \mathbf{V}_i \frac{\partial \mu}{\partial \mathbf{r}_i} = 0, \tag{210}$$

where $\mathbf{V}_i$ is the velocity of vortex i produced by the other vortices according to (14). We also introduce the one- and N-particle distribution functions defined by

$$P(\mathbf{r}, t) = \int \mu(\{\mathbf{r}_k\}, t) \prod_{k=1}^{N} d^2\mathbf{r}_k, \tag{211}$$

$$\mu_{sys}(\mathbf{r}_1, ..., \mathbf{r}_N, t) = \int \mu(\{\mathbf{r}_k\}, t) d^2\mathbf{r}. \tag{212}$$

We write the distribution function μ in the suggestive form

$$\mu(\mathbf{r}, \mathbf{r}_1, ..., \mathbf{r}_N, t) = P(\mathbf{r}, t)\mu_{sys}(\mathbf{r}_1, ..., \mathbf{r}_N, t) + \mu_I(\mathbf{r}, \mathbf{r}_1, ..., \mathbf{r}_N, t), \tag{213}$$

where the quantity μ_I reflects the effect of correlations between vortices.

The Liouville equation (210) provides the correct starting point for the analysis of the dynamics of our vortex system. However, when N is large, this equation contains much more information than one can interpret. Consequently, what one would like to do is to describe the system in some average sense by a one-particle distribution function.

5.2 The Projection Operator Formalism

Our first objective is to derive some *exact* kinetic equations satisfied by $P(\mathbf{r}, t)$ and $\mu_{sys}(\mathbf{r}_1, ..., \mathbf{r}_N, t)$. This can be achieved by using the projection operator formalism developed by Willis and Picard [127]. This formalism was also used by Kandrup [73] in the context of stellar dynamics to derive a generalized Landau equation describing the time evolution of the distribution function of stars in an inhomogeneous medium. We shall just recall the main steps of the theory. More details can be found in the original paper of Willis and Picard [127] and in Kandrup [73]. To have similar notations, we set $x \equiv \{\mathbf{r}\}$ and $y \equiv \{\mathbf{r}_1, ..., \mathbf{r}_N\}$. Then, (213) can be put in the form

$$\mu(x, y, t) = \mu_R(x, y, t) + \mu_I(x, y, t), \tag{214}$$

with

$$\mu_R(x, y, t) = f(x, t)g(y, t), \tag{215}$$

where we have written $f(x, t) \equiv P(\mathbf{r}, t)$ and $g(y, t) \equiv \mu_{sys}(\mathbf{r}_1, ..., \mathbf{r}_N, t)$. The Liouville equation is also cast in the form

$$\frac{\partial \mu}{\partial t} = -iL\mu = -i(L_0 + L_{sys} + L')\mu, \tag{216}$$

where L_0 and L_{sys} act respectively only on the variables x and y, whereas the interaction Liouvillian L' acts upon both x and y (the complex number i is here purely formal and has been introduced only to have the same notations as in [127,73]).

Following Willis and Picard [127], we introduce the time-dependant projection operator

$$P(x, y, t) = g(y, t)\int dy + f(x, t)\int dx - f(x, t)g(y, t)\int dx \int dy. \tag{217}$$

We can easily check that

$$P(x, y, t)\mu(x, y, t) = \mu_R(x, y, t), \tag{218}$$

$$[1 - P(x, y, t)]\mu(x, y, t) = \mu_I(x, y, t). \tag{219}$$

We also verify that P is a projection in the sense that $P^2(t) = P(t)$. Applying P and $1-P$ on the Liouville equation (216), we obtain the coupled equations

$$\partial_t \mu_R(x,y,t) = -iPL\mu_R - iPL\mu_I, \tag{220}$$

and

$$\partial_t \mu_I(x,y,t) = -i(1-P)L\mu_R - i(1-P)L\mu_I. \tag{221}$$

These equations describe the separation between a "macrodynamics" and a "sub-dynamics".

Introducing the Greenian

$$\mathcal{G}(t,t') \equiv \exp\left\{-i\int_{t'}^{t} dt''[1-P(t'')]L\right\}, \tag{222}$$

we can immediately write down a formal solution of (221), namely

$$\mu_I(x,y,t) = -\int_0^t dt'\mathcal{G}(t,t')i[1-P(t')]L\mu_R(x,y,t'), \tag{223}$$

where we have assumed that, initially, the particles are uncorrelated so that $\mu_I(x,y,0) = 0$. Substituting for $\mu_I(x,y,t)$ from (223) in (220), we obtain

$$\partial_t \mu_R(x,y,t) = -iPL\mu_R - \int_0^t dt' P(t)L\mathcal{G}(t,t')[1-P(t')]L\mu_R(x,y,t'). \tag{224}$$

The integration over y will yield an equation describing the evolution of f. Using some mathematical properties of the projection operator (217), the final result can be put in the nice symmetrical form given by Willis and Picard [127],

$$\partial_t f(x,t) + iL_0 f + i\langle L'\rangle_{sys} f = -\int_0^t dt' \int dy \Delta_t L'\mathcal{G}(t,t')\Delta_{t'}L' g(y,t') f(x,t'), \tag{225}$$

where the notations stand for

$$\langle L'\rangle_{sys} = \int dy' L'(x,y')g(y',t), \tag{226}$$

$$\langle L'\rangle_0 = \int dx' L'(x',y)f(x',t), \tag{227}$$

$$\Delta_t L' = L' - \langle L'\rangle_{sys} - \langle L'\rangle_0. \tag{228}$$

Similarly, after integrating over x we find the equation satisfied by g,

$$\partial_t g(y,t) + iL_{sys} g - i\langle L'\rangle_1 g = -\int_0^t dt' \int dx \Delta_t L'\mathcal{G}(t,t')\Delta_{t'}L' g(y,t') f(x,t'). \tag{229}$$

5.3 Application to the Point Vortex System

The previous theory is completely general and we now consider its application to a system of point vortices [36]. Let us first rewrite the Liouville equation (210) in a form that separates the contribution of the test vortex from the contribution of the field vortices:

$$\frac{\partial\mu}{\partial t}+\sum_{i=1}^{N}\mathbf{V}(i\to 0)\frac{\partial\mu}{\partial\mathbf{r}}+\sum_{i=1}^{N}\mathbf{V}(0\to i)\frac{\partial\mu}{\partial\mathbf{r}_i}$$
$$+\sum_{i=1}^{N}\sum_{j=1,j\neq i}^{N}\mathbf{V}(j\to i)\frac{\partial\mu}{\partial\mathbf{r}_i}=0. \tag{230}$$

Applying the general theory of Willis and Picard [127], we obtain the following kinetic equation for the one-particle distribution function of a vortex system [36]:

$$\frac{\partial P}{\partial t}+\langle\mathbf{V}\rangle\frac{\partial P}{\partial\mathbf{r}}=\frac{\partial}{\partial r^{\mu}}\int_0^t d\tau\int\prod_{k=1}^{N}d^2\mathbf{r}_k\sum_{i=1}^{N}\sum_{j=1}^{N}\mathcal{V}^{\mu}(i\to 0)$$
$$\times\mathcal{G}(t,t-\tau)\left(\mathcal{V}^{\nu}(j\to 0)\frac{\partial}{\partial r^{\nu}}+\mathcal{V}^{\nu}(0\to j)\frac{\partial}{\partial r^{\nu}_j}\right)P(\mathbf{r},t-\tau)\mu_{sys}(\{\mathbf{r}_k\},t-\tau), \tag{231}$$

where the Greek indices refer to the components of $\boldsymbol{\mathcal{V}}$ in a fixed system of coordinates and $\boldsymbol{\mathcal{V}}(i\to 0)=\mathbf{V}(i\to 0)-\langle\mathbf{V}(i\to 0)\rangle$ denotes the velocity fluctuation. We can note that (231) already shares some analogies with the Fokker-Planck equation of Sect. 4. Indeed, the first term on the right hand side corresponds to a diffusion and the second term to a drift. For a passive particle $\mathcal{V}^{\nu}(0\to j)=0$ and the drift cancels out, as expected.

5.4 The Factorization Hypothesis

If the vortices are initially decorrelated then, for sufficiently short times, they will remain decorrelated. This means that the $(N+1)$-particle distribution function can be factorized in a product of $(N+1)$ one-particle distribution functions

$$\mu(\mathbf{r},\mathbf{r}_1,...,\mathbf{r}_N,t)=\prod_{k=0}^{N}P(\mathbf{r}_k,t). \tag{232}$$

If we integrate the Liouville equation (210) on the positions of the N vortices $1,...,N$ and use the factorization (232), we directly obtain [36]:

$$\frac{\partial P}{\partial t}+\langle\mathbf{V}\rangle\nabla P=0. \tag{233}$$

Therefore, for sufficiently short times, the average vorticity $\langle\omega\rangle$ satisfies the 2D Euler equation (called the Vlasov equation in other circumstances). However, at later times, the distribution function μ differs from the pure product (232) and the Euler equation does not provide a good approximation anymore. In Sect. 5.3 we have determined an exact equation (231) satisfied by the one-particle distribution function which is valid at any time. This equation is not closed, however, since it involves the N-vortex distribution function μ_{sys}. We shall close the system by assuming that μ_{sys} can be approximated by a product of N one-particle distribution functions in the form

$$\mu_{sys}(\mathbf{r}_1, ..., \mathbf{r}_N, t) \simeq \prod_{k=1}^{N} P(\mathbf{r}_k, t). \tag{234}$$

Inserting (234) in (231), we obtain

$$\frac{\partial P}{\partial t} + \langle \mathbf{V} \rangle \frac{\partial P}{\partial \mathbf{r}} = \frac{\partial}{\partial r^{\mu}} \int \prod_{k=1}^{N} d^2\mathbf{r}_k \int_0^t d\tau \sum_{i=1}^{N} \sum_{j=1}^{N} \mathcal{V}^{\mu}(i \to 0)$$
$$\times \mathcal{G}(t, t-\tau) \left(\mathcal{V}^{\nu}(j \to 0) \frac{\partial}{\partial r^{\nu}} + \mathcal{V}^{\nu}(0 \to j) \frac{\partial}{\partial r_j^{\nu}} \right) P(\mathbf{r}, t-\tau) \prod_{k=1}^{N} P(\mathbf{r}_k, t-\tau). \tag{235}$$

If we assume that between t and $t - \tau$ the trajectories of the particles are determined from the smooth velocity field created by the vorticity distribution $\langle \omega \rangle = N\gamma P(\mathbf{r}, t)$, the foregoing equation simplifies in

$$\frac{\partial P}{\partial t} + \langle \mathbf{V} \rangle \frac{\partial P}{\partial \mathbf{r}} = N \frac{\partial}{\partial r^{\mu}} \int_0^t d\tau \int d^2\mathbf{r}_1 V^{\mu}(1 \to 0)_t$$
$$\times \left\{ V^{\nu}(1 \to 0) P_1 \frac{\partial P}{\partial r^{\nu}} + V^{\nu}(0 \to 1) P \frac{\partial P_1}{\partial r_1^{\nu}} \right\}_{t-\tau}, \tag{236}$$

where $P = P(\mathbf{r}, t)$ and $P_1 = P(\mathbf{r}_1, t)$. Equation (236) is a non Markovian integrodifferential equation since the probability density $P(\mathbf{r}, t)$ in $\mathbf{r}$ at time t depends on the value of the whole distribution of probability $P(\mathbf{r}_1, t - \tau)$ at earlier times through an integration on $\mathbf{r}_1$ and τ. Equation (236) is therefore non local in space and time. It can be shown [36] that this kinetic equation rigorously conserves angular momentum in a circular domain and linear impulse in a channel (or in an infinite domain). However, under this form, it has not been possible to prove the conservation of energy and the H-theorem for the Boltzmann entropy (41).

5.5 The Case of Short Decorrelation Times

If we assume that the decorrelation time τ is short (which does not need to be the case) and implement a strong Markov approximation, we obtain

$$\frac{\partial P}{\partial t} + \langle \mathbf{V} \rangle \frac{\partial P}{\partial \mathbf{r}} = \frac{N\tau}{2} \frac{\partial}{\partial r^{\mu}} \int d^2\mathbf{r}_1 V^{\mu}(1 \to 0)$$
$$\times \left\{ V^{\nu}(1 \to 0) P_1 \frac{\partial P}{\partial r^{\nu}} + V^{\nu}(0 \to 1) P \frac{\partial P_1}{\partial r_1^{\nu}} \right\}. \tag{237}$$

In the case of an infinite domain $\mathbf{V}(0 \to 1) = -\mathbf{V}(0 \to 1)$ and we have the further simplification

$$\frac{\partial P}{\partial t} + \langle \mathbf{V} \rangle \frac{\partial P}{\partial \mathbf{r}} = \frac{N\gamma^2}{8\pi^2}\tau \frac{\partial}{\partial r^\mu} \int d^2\mathbf{r}_1 K'^{\mu\nu}(\boldsymbol{\xi}) \left(P_1 \frac{\partial P}{\partial r^\nu} - P \frac{\partial P_1}{\partial r_1^\nu} \right), \tag{238}$$

where

$$K'^{\mu\nu}(\boldsymbol{\xi}) = \frac{\xi_\perp^\mu \xi_\perp^\nu}{\xi^4} = \frac{\xi^2 \delta^{\mu\nu} - \xi^\mu \xi^\nu}{\xi^4} \tag{239}$$

and $\boldsymbol{\xi} = \mathbf{r}_1 - \mathbf{r}$. To arrive at (239) we have explicitly used the form of the Kernel (14), and to get the second equality we have used the fact that we are in two dimensions. Note that the symmetrical form of (238) is reminiscent of the Landau equation introduced in plasma physics and in stellar dynamics (see, e.g., [73]):

$$\frac{\partial f}{\partial t} = \frac{\partial}{\partial v^\mu} \int d^3\mathbf{v}_1 K^{\mu\nu} \left(f_1 \frac{\partial f}{\partial v^\nu} - f \frac{\partial f_1}{\partial v_1^\nu} \right), \tag{240}$$

with

$$K^{\mu\nu} = 2\pi N G^2 m^2 \ln\left(\frac{L_{max}}{L_{min}} \right) \frac{u^2 \delta^{\mu\nu} - u^\mu u^\nu}{u^3}, \tag{241}$$

and $\mathbf{u} = \mathbf{v} - \mathbf{v_1}$, $f = f(\mathbf{v}, t)$, $f_1 = f(\mathbf{v}_1, t)$. In this analogy, the position $\mathbf{r}$ of the vortices plays the role of the velocity $\mathbf{v}$ of the electric charges or stars and the spatial distribution $P(\mathbf{r}, t)$ the role of the velocity distribution $f(\mathbf{v}, t)$. Therefore, we can directly infer the conservation of linear impulse $\mathbf{P}_\perp = \int \langle \omega \rangle \mathbf{r} d^2\mathbf{r}$ and angular momentum $L = \int \langle \omega \rangle r^2 d^2\mathbf{r}$ which play respectively the role of impulse $\mathbf{P} = \int f \mathbf{v} d^3\mathbf{v}$ and kinetic energy $K = \int f \frac{v^2}{2} d^3\mathbf{v}$ in plasma physics. We can also prove a H-theorem for the Boltzmann entropy (41) exactly like in the case of the Landau equation. Finally, the solutions of (238) converge towards the Gaussian vortex (the equivalent of the Maxwellian distribution in plasma physics with $\mathbf{r}$ in place of $\mathbf{v}$):

$$P(\mathbf{r}) = A e^{-\frac{1}{2}\alpha\gamma(\mathbf{r} - \mathbf{r}_0)^2} \tag{242}$$

which is the maximum entropy state at fixed circulation, angular momentum and impulse. It is in general *different* from the Boltzmann distribution (44) with the relative streamfunction $\psi' = \psi + \frac{\Omega}{2} r^2 - \mathbf{U}_\perp \mathbf{r}$, except in the particular limit $\beta \to 0$, $\Omega \to +\infty$ with fixed $\alpha = \beta\Omega/2$ corresponding to the statistical equilibrium (88). This clearly indicates that (238) does *not* conserve energy.

5.6 A Generalized Kinetic Equation

Now, if we account properly for memory effects in (236), we can obtain a generalized kinetic equation which guaranties the conservation of energy (in addition to the other constraints) and is therefore more satisfactory. If the distribution of vortices is axisymmetric, it is possible to calculate the memory function appearing in (236) explicitly if we assume that the correlation time is smaller than the typical time on which the average vorticity changes appreciably [36]. In this approximation, the point vortices follow, between t and $t-\tau$, circular trajectories with angular velocity $\Omega(r,t) = \langle V_\theta \rangle(r,t)/r$ and (236) simplifies in (see Appendix B):

$$\frac{\partial P}{\partial t} = -\frac{N\gamma^2}{4r}\frac{\partial}{\partial r}\int_0^{+\infty} r_1 dr_1 \delta(\Omega - \Omega_1) \ln\left[1 - \left(\frac{r_<}{r_>}\right)^2\right]\left\{\frac{1}{r}P_1\frac{\partial P}{\partial r} - \frac{1}{r_1}P\frac{\partial P_1}{\partial r_1}\right\}, \tag{243}$$

where $\Omega = \Omega(r,t)$, $\Omega_1 = \Omega(r_1,t)$ and $r_>$ (resp. $r_<$) is the biggest (resp. smallest) of r and r_1. The angular velocity is related to the vorticity by

$$\langle \omega \rangle = \frac{1}{r}\frac{\partial}{\partial r}(\Omega r^2). \tag{244}$$

We can propose an approximation of the general kinetic equation (236) which encompasses the axisymmetric form previously derived. Memory effects are not neglected, unlike in (238), but they are simplified in a way which preserves all the conservation laws of the system (as discussed below). We propose the generalized kinetic equation [36]:

$$\frac{\partial P}{\partial t} + \langle \mathbf{V} \rangle \nabla P = \frac{N\gamma^2}{8}\frac{\partial}{\partial r^\mu}\int d^2\mathbf{r}_1 K^{\mu\nu}\delta(\boldsymbol{\xi}.\mathbf{v})\left(P_1\frac{\partial P}{\partial r^\nu} - P\frac{\partial P_1}{\partial r_1^\nu}\right), \tag{245}$$

with

$$K^{\mu\nu}(\boldsymbol{\xi}) = \frac{\xi_\perp^\mu \xi_\perp^\nu}{\xi^2} = \frac{\xi^2\delta^{\mu\nu} - \xi^\mu\xi^\nu}{\xi^2}, \tag{246}$$

and $\boldsymbol{\xi} = \mathbf{r}_1 - \mathbf{r}$, $\mathbf{v} = \langle \mathbf{V} \rangle(\mathbf{r}_1,t) - \langle \mathbf{V} \rangle(\mathbf{r},t)$. In the thermodynamic limit $N \to +\infty$ and $\gamma \sim 1/N$ (see Sect. 2.7), the kinetic equation (245) reduces to the Vlasov equation (233). However, in practice, N is always finite and the correlations between point vortices must be taken into account. The "collision term" in (245) gives the first order correction $O(1/N)$ to the Vlasov limit.

From (243) and (245), it is clear that the relaxation of the point vortices is due to a phenomenon of resonance. Only the points $\mathbf{r}_1$ satisfying the condition $\boldsymbol{\xi}\cdot\mathbf{v} = 0$ with $\mathbf{r}_1 \neq \mathbf{r}$ contribute to the diffusion current in $\mathbf{r}$. In the axisymmetric case, this condition of resonance reduces to $\Omega(r_1) = \Omega(r)$, which supposes that the angular velocity profile is non monotonic. The occurence of a δ-function in (243) and (245) is the main difference with the Landau equation (240). In the present context, it ensures the conservation of the "potential" energy of the vortices $E = \frac{1}{2}\int \omega\psi d^2\mathbf{r}$ which has no counterpart in Landau's theory applying to spatially uniform plasmas. It can also be noted that, contrary to the Landau equation, the kinetic equations (238) and (245) do *not* suffer the well-known logarithmic divergence appearing in the context of Coulombian plasmas and stellar systems (see, e.g., [73]). This is due essentially to the lower dimension of space ($D = 2$ instead of $D = 3$) and to the different nature of the interactions.

5.7 Conservation Laws and H-theorem

We now derive the conservation laws and the H-theorem satisfied by (245). The conservation of the circulation is straightforward since the right hand side of (245) can be written as the divergence of a current. To prove the conservation of angular momentum, we take the time derivative of $L = N\gamma\int Pr^2 d^2\mathbf{r}$, substitute for (245), permut the dummy variables $\mathbf{r}$ and $\mathbf{r}_1$ and add the resulting expressions. This yields

$$\dot{L} = \frac{N^2\gamma^3}{8}\int d^2\mathbf{r}d^2\mathbf{r}_1 K^{\mu\nu}\xi^\mu\delta(\boldsymbol{\xi}.\mathbf{v})\left(P_1\frac{\partial P}{\partial r^\nu} - P\frac{\partial P_1}{\partial r_1^\nu}\right). \tag{247}$$

From (246), we immediately verify that

$$K^{\mu\nu}\xi^\mu = 0, \tag{248}$$

which proves the conservation of angular momentum. We can prove the conservation of linear impulse in a similar manner [36]. For the conservation of energy, we start from (39) and follow the same procedure. This yields

$$\dot{E} = \frac{N^2\gamma^3}{16}\int d^2\mathbf{r}d^2\mathbf{r}_1 K^{\mu\nu}v_\perp^\mu\delta(\boldsymbol{\xi}.\mathbf{v})\left(P_1\frac{\partial P}{\partial r^\nu} - P\frac{\partial P_1}{\partial r_1^\nu}\right). \tag{249}$$

Considering the form of the tensor (246), we have

$$K^{\mu\nu}v_\perp^\mu = \frac{\xi_\perp^\nu}{\xi^2}(\boldsymbol{\xi}.\mathbf{v}). \tag{250}$$

When substituted in (249), we see that the occurence of the δ-function in the kinetic equation implies $\dot{E} = 0$. Finally, for the rate of entropy production we have, according to (41) and (245):

$$\dot{S} = \frac{N^2\gamma^2}{8}\int d^2\mathbf{r}d^2\mathbf{r}_1\frac{1}{PP_1}P_1\frac{\partial P}{\partial r^\mu}K^{\mu\nu}\delta(\boldsymbol{\xi}.\mathbf{v})\left(P_1\frac{\partial P}{\partial r^\nu} - P\frac{\partial P_1}{\partial r_1^\nu}\right). \tag{251}$$

Permutting the dummy variables $\mathbf{r}$ and $\mathbf{r}_1$ and adding the resulting expression to (251), we obtain

$$\dot{S} = \frac{N^2\gamma^2}{16}\int d^2\mathbf{r}d^2\mathbf{r}_1\frac{1}{PP_1}\delta(\boldsymbol{\xi}.\mathbf{v})\left(P_1\frac{\partial P}{\partial r^\mu} - P\frac{\partial P_1}{\partial r_1^\mu}\right)K^{\mu\nu}\left(P_1\frac{\partial P}{\partial r^\nu} - P\frac{\partial P_1}{\partial r_1^\nu}\right). \tag{252}$$

Now, for any vector, $A^\mu K^{\mu\nu}A^\nu = (\mathbf{A}\cdot\boldsymbol{\xi}_\perp)^2/\xi^2 \geq 0$. This proves a H-theorem ($\dot{S} \geq 0$) for the kinetic equation (245). It should be emphasized that the conservation laws and the H-theorem result essentially from the *symmetry* of the kinetic equation. This is satisfying from a physical point of view.

It is also easy to show that the Boltzmann distribution

$$P = Ae^{-\beta\gamma(\psi + \frac{\Omega}{2}r^2 - \mathbf{U}_\perp\mathbf{r})}, \tag{253}$$

is a stationary solution of (245). Noting that

$$\frac{\partial P}{\partial r^\nu} = -\beta\gamma\left(\frac{\partial\psi}{\partial r^\nu} + \Omega r^\nu - U_\perp^\nu\right)P, \tag{254}$$

we have successively

$$K^{\mu\nu}\left(P_1\frac{\partial P}{\partial r^\nu} - P\frac{\partial P_1}{\partial r_1^\nu}\right) = \beta\gamma P P_1 K^{\mu\nu}(v_\perp^\nu + \Omega\xi^\nu) = \beta\gamma P P_1 \frac{\xi_\perp^\mu}{\xi^2}(\boldsymbol{\xi}.\mathbf{v}), \tag{255}$$

where we have used (248) and (250). When (255) is substituted in (245), we find that the right hand side cancels out due to the δ-function. The advective term is also zero since $P = f(\psi')$. Therefore, the Boltzmann distribution (253) is a stationary solution of (245). Note, however, that this is not the only solution, unlike for ordinary kinetic equations. Any stationary solution of the Euler equation satisfying in addition $\boldsymbol{\xi}.\mathbf{v} \neq 0$ for any couple of points $\mathbf{r}$, $\mathbf{r}_1$ (with $\mathbf{r} \neq \mathbf{r}_1$) is a solution of (245). Physically, this implies that the system needs sufficiently strong resonances to relax towards the maximum entropy state. If this condition is not realized, the system can remain frozen in a sort of "metastable" equilibrium state. Further evolution of the system will require non trivial correlations between point vortices which are not taken into account in the present theory.

5.8 The Thermal Bath Approximation

A direct connexion between the generalized kinetic equation (245) and the Fokker-Planck equation of Sect. 4 can be found. Introducing a diffusion tensor

$$D^{\mu\nu} = \frac{N\gamma^2}{8}\int d^2\mathbf{r}_1 K^{\mu\nu}\delta(\boldsymbol{\xi}.\mathbf{v})P_1, \tag{256}$$

and a drift term

$$\eta^\mu = -\frac{N\gamma^2}{8}\int d^2\mathbf{r}_1 K^{\mu\nu}\delta(\boldsymbol{\xi}.\mathbf{v})\frac{\partial P_1}{\partial r_1^\nu}, \tag{257}$$

we can rewrite (245) in the more illuminating form

$$\frac{\partial P}{\partial t} + \langle\mathbf{V}\rangle\nabla P = \frac{\partial}{\partial r^\mu}\left[D^{\mu\nu}\frac{\partial P}{\partial r^\nu} + P\eta^\mu\right], \tag{258}$$

similar to a general Fokker-Planck equation. Note, however, that (258) is an integrodifferential equation since the density probability $P(\mathbf{r}, t)$ in $\mathbf{r}$ at time t depends on the value of the whole distribution of probability $P(\mathbf{r}_1, t)$ at the same time by an integration over $\mathbf{r}_1$. By contrast, the Fokker-Planck equation (182) is a differential equation. The usual way to transform an integrodifferential equation into a differential equation is to make a guess for the function $P(\mathbf{r}_1)$ appearing under the integral sign and refine the guess by successive iterations. In practice we simply make one sensible guess. Therefore, if we are close to equilibrium, it seems natural to replace the function P_1 appearing in the integrals by the Boltzmann distribution

$$P(\mathbf{r}_1) = Ae^{-\beta\gamma\psi(\mathbf{r}_1)}. \tag{259}$$

This corresponds to the "thermal bath approximation" of Sect. 4: the vortices have not yet relaxed completely, but when we focus on the relaxation of a given point vortex (described by P) we can consider, in a first approximation, that the rest of the system (described by P_1) is at equilibrium. Within this approximation, the diffusion coefficient and the drift simplify in

$$\eta^\mu = \beta\gamma D^{\mu\nu}\frac{\partial\psi}{\partial r^\nu}, \tag{260}$$

$$D^{\mu\nu} = \frac{N\gamma^2}{8}P(\mathbf{r},t)\int K^{\mu\nu}\delta(\boldsymbol{\xi}.\mathbf{v})d^2\boldsymbol{\xi}, \tag{261}$$

where we have made the local approximation. If we assume that the correlation time is short, i.e. if we replace $\xi^2\delta(\boldsymbol{\xi}.\mathbf{v})$ by τ/π^2 (compare (245) and (238)), we obtain

$$\boldsymbol{\eta} = \beta\gamma D\nabla\psi, \tag{262}$$

$$D = \frac{\gamma\tau}{16\pi}\ln N\langle\omega\rangle, \tag{263}$$

and (258) reduces to the Fokker-Planck equation

$$\frac{\partial P}{\partial t} + \langle\mathbf{V}\rangle\nabla P = \nabla(D(\nabla P + \beta\gamma P\nabla\psi)). \tag{264}$$

This approximation is, however, not very satisfactory since the decorrelation time τ appears as a free parameter. In fact, the decorrelation time can be determined self-consistently from the above formulae by evaluating properly the δ-function in (261). Expanding the velocity difference $\mathbf{v} = \langle\mathbf{V}_1\rangle - \langle\mathbf{V}\rangle$ in a Taylor series in $\boldsymbol{\xi} = \mathbf{r}_1 - \mathbf{r}$, we obtain to first order in the expansion

$$\boldsymbol{\xi}.\mathbf{v} = \Sigma^{\mu\nu}\xi^\mu\xi^\nu, \tag{265}$$

where

$$\Sigma^{\mu\nu} = \frac{1}{2}\left(\frac{\partial\langle V\rangle^\mu}{\partial r^\nu} + \frac{\partial\langle V\rangle^\nu}{\partial r^\mu}\right), \tag{266}$$

is the stress tensor. It has the property of symmetry $\Sigma^{\mu\nu} = \Sigma^{\nu\mu}$. Since the flow is divergenceless, we also have $\Sigma^{xx} + \Sigma^{yy} = 0$. In terms of the stress tensor (266), the diffusion tensor (261) can be rewritten

$$D^{\mu\nu} = \frac{NP\gamma^2}{8}\int\frac{\xi^2\delta^{\mu\nu} - \xi^\mu\xi^\nu}{\xi^2}\delta(\Sigma^{\mu\nu}\xi^\mu\xi^\nu)d^2\boldsymbol{\xi}. \tag{267}$$

This integral can be performed easily by working in a basis (ξ_1', ξ_2') in which the tensor $\Sigma^{\mu\nu}$ is anti-diagonal [36]. Then,

$$D^{\mu\nu} = \frac{NP\gamma^2}{8}\int\frac{\xi'^2\delta^{\mu\nu} - \xi'^\mu\xi'^\nu}{\xi'^2}\delta(|\Sigma(\mathbf{r})|\xi_1'\xi_2')\ d\xi_1'd\xi_2', \tag{268}$$

where we have set $|\Sigma(\mathbf{r})| = 2\sqrt{-\det(\Sigma)}$. Clearly, this quantity is invariant by a change of reference frame and it measures the local shear of the flow. It is easy to check that the diffusion is isotropic and that $D^{\mu\nu} = D\delta^{\mu\nu}$ with

$$D = \frac{NP\gamma^2}{8}\frac{1}{|\Sigma(\mathbf{r})|}\int \frac{\xi_2'^2}{\xi_1'^2+\xi_2'^2}\delta(\xi_1'\xi_2')\, d\xi_1' d\xi_2'. \tag{269}$$

Setting $\xi_1' = \xi\cos\theta$ and $\xi_2' = \xi\sin\theta$ where $\xi = \xi' = |\mathbf{r}_1 - \mathbf{r}|$, we obtain

$$D = \frac{NP\gamma^2}{8}\frac{1}{|\Sigma(\mathbf{r})|}\int_0^{+\infty} \xi d\xi \int_0^{2\pi} d\theta \sin^2\theta\delta(\xi^2\cos\theta\sin\theta), \tag{270}$$

or, equivalently,

$$D = \frac{NP\gamma^2}{4}\frac{1}{|\Sigma(\mathbf{r})|}\int_0^{+\infty} \frac{d\xi}{\xi}\int_0^{\pi} d\theta \sin\theta\delta(\cos\theta). \tag{271}$$

As explained previously, we regularize the logarithmic divergence by introducing appropriate cut-offs at small and large scales. With the change of variables $t = \cos\theta$, we finally obtain

$$D = \frac{NP\gamma^2}{8}\frac{1}{|\Sigma(\mathbf{r})|}\ln N \int_{-1}^{+1} dt\ \delta(t) = \frac{NP\gamma^2}{8}\frac{1}{|\Sigma(\mathbf{r})|}\ln N, \tag{272}$$

which establishes (189) in the general case.

5.9 The Collisional Relaxation Time

We can deduce from this kinetic theory the "collisional" relaxation time of the point vortex gas. Considering the Fokker-Planck equation (264), it is easy to check that, for $t \to +\infty$, the distribution function $P(\mathbf{r}, t)$ will converge towards the Boltzmann distribution (44). The relaxation time corresponds typically to the time needed by a vortex to diffuse over a distance R, the system size. Therefore, $t_{relax} \sim R^2/D$, with $D \sim \gamma \ln N$ according to (272). Using $\Gamma = N\gamma$ and introducing the dynamical time $t_D \sim \langle\omega\rangle^{-1} \sim R^2/\Gamma$, we obtain the estimate

$$t_{relax} \sim \frac{N}{\ln N} t_D, \tag{273}$$

as in the case of collisional stellar systems [9]. Since the statistical description is expected to yield relevant results for large N, we conclude that the "collisional" relaxation of point vortices towards the Boltzmann distribution (44) is a *very slow process*. It can certainly not account for most of the numerical simulations and experiments of 2D turbulence and point vortex dynamics in which an equilibrium state is established *extremely rapidly*. This implies that a more violent relaxation mechanism must be at work in the system (see Sect. 6). This is a remark of crucial importance because it means that all the results established previously, including the Boltzmann distribution (44), must be revised.

6 Violent Relaxation of 2d Vortices and Stellar Systems

6.1 The Euler and the Vlasov Equations

In the preceding sections, we have focused our attention to the point vortex model as an idealization of more realistic flows which necessarily involve a continuous vorticity distribution. This approximation is interesting in a first approach because it leads to a system of N particles in interaction (like electric charges or stars) for which the methods of statistical mechanics are directly applicable. In addition, this model keeps the specificity of two-dimensional vorticity flows such as long-range interactions between vortices and structure formation. However, there are many different ways to approximate a continuous vorticity field by a cloud of point vortices and different approximations can lead to different statistical equilibrium states (this difficulty was underlined by Onsager [101]). Therefore, if we want to apply the results of statistical mechanics to realistic situations (e.g., geophysical flows) it is necessary to go beyond the point vortex model and develop a statistical mechanics for continuous vorticity fields.

For flows of geophysical or astrophysical interest, the Reynolds numbers are so high that the molecular viscosity is not expected to play a crucial role in the dynamics. Therefore, these flows are described in the simplest model by the Euler-Poisson system

$$\frac{\partial \omega}{\partial t} + \mathbf{u}\nabla\omega = 0, \tag{274}$$

$$\omega = -\Delta\psi. \tag{275}$$

It can be recalled that these equations also model the early dynamics of a cloud of point vortices before correlations between vortices have developed (in that case, ω is proportional to the one-body distribution function $P(\mathbf{r}, t)$, see (233)). As discussed in Sect. 5.9, this is the regime of physical interest since the "collisional" relaxation of point vortices is in general very slow.

Similarly, for a majority of stellar systems, including the important class of elliptical galaxies, the relaxation time by two-body encounters is $\sim 10^{12}$ times larger than the age of the universe. Therefore, the dynamics of stars in a galaxy is essentially *collisionless* and appropriately described by the Vlasov-Poisson system

$$\frac{\partial f}{\partial t} + \mathbf{v}\frac{\partial f}{\partial \mathbf{r}} + \mathbf{F}\frac{\partial f}{\partial \mathbf{v}} = 0, \tag{276}$$

$$\Delta\Phi = 4\pi G \int f d^3\mathbf{v}. \tag{277}$$

The morphological similarity of the Euler-Poisson and Vlasov-Poisson systems is another manifestation of the close analogy between 2D vortices and stellar systems. If we make the correspondance between the vorticity and the distribution function ($\omega \leftrightarrow f$) and between the stream function and the gravitational potential ($\psi \leftrightarrow \Phi$), these two equations describe the advection of a density by an incompressible flow with which it interacts via a Poisson equation. Then, the density is not advected passively by the flow but is coupled to its motion. This

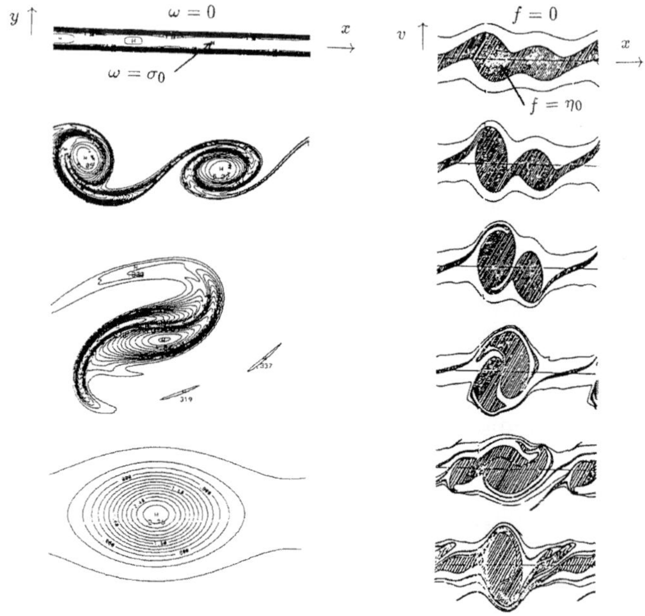

Fig. 13. Violent relaxation of two-dimensional vorticity flows and self-gravitating systems. The left panel corresponds to the nonlinear developement of the Kelvin-Helmholtz instability. The equilibrium state is a large-scale vortex which is well-described by statistical mechanics [119]. The right panel corresponds to a simulation of the Vlasov-Poisson system showing a mixing process and the formation of a coherent structure in phase space with a Fermi-Dirac distribution [7]. This simulation is restricted to a one-dimensional system but the process remains the same in higher dimensions.

coupling is responsible for violent fluctuations of the stream function or gravitational potential. These fluctuations will mix the vorticity or the phase elements at small scales and induce a self-organization and the appearance of structures at larger scales (see Fig. 13). This violently changing potential provides a mechanism analogous to a relaxation in a gas, but the specificity of this relaxation is that it is *collisionless* and due to the long-range nature of the interactions. It is now clear that this "chaotic mixing" is the driving source of relaxation in two-dimensional turbulence and stellar dynamics. The kinetic theory presented in Sect. 5 for point vortices (and the one developed by Chandrasekhar for stars) is only valid in situations in which this chaotic mixing is prevented or has died away.

6.2 The Statistical Equilibrium

During the mixing process, the Vlasov-Poisson and the Euler-Poisson systems generate intermingled filaments at smaller and smaller scales. Therefore, a *deterministic* description of the flow would require a rapidly increasing amount of information as time goes on. For that reason, it is appropriate to undertake a *probabilistic* description in order to smooth out the small scales and concentrate on the locally averaged quantities. This statistical approach, called the theory of "violent relaxation", was introduced by Lynden-Bell [90] in 1967 for collisionless stellar systems and rediscovered independantly by Kuzmin [83], Miller [95,96] and Robert and Sommeria [111] for two-dimensional vorticity flows. The analogy between these two statistical mechanics (including the relaxation towards equilibrium) was discussed in detail by Chavanis [29,54,32,35]. On the other hand, a rigourous justification of this statistical approach has been given by Robert [109] by using the concept of Young measures and large deviations. This theory improves upon previous works based on a spectral representation of the flow [80], which do not respect all the conservation laws of the invisicd dynamics. In the following, we present the statistical mechanics of violent relaxation in two-dimensional vorticity flows. A discussion of the statistical mechanics of violent relaxation in stellar systems, closely following the presentation of this paper, can be found in [42].

In the statistical approach, the exact knowledge of the "fine-grained" or microscopic vorticity field is replaced by the probability density $\rho(\mathbf{r},\sigma)$ of finding the vorticity level σ in $\mathbf{r}$. The normalization condition yields at each point

$$\int \rho(\mathbf{r},\sigma)d\sigma = 1, \tag{278}$$

and the locally averaged (coarse-grained) vorticity is expressed in terms of the density probability in the form

$$\overline{\omega} = \int \rho(\mathbf{r},\sigma)\sigma d\sigma. \tag{279}$$

More generally, the moments of the vorticity are defined by

$$\overline{\omega^n} = \int \rho(\mathbf{r},\sigma)\sigma^n d\sigma. \tag{280}$$

During the evolution, the energy

$$E = \frac{1}{2}\int \overline{\omega}\psi d^2\mathbf{r}, \tag{281}$$

is conserved as well as the total area of each level of vorticity

$$\gamma(\sigma) = \int \rho(\mathbf{r},\sigma)d^2\mathbf{r}. \tag{282}$$

These last constraints are equivalent to the conservation of the Casimir integrals $C_h = \int h(\omega)d^2\mathbf{r}$ for any continuous function h. Such integrals are conserved by the Euler equation because the fluid particles keep their vorticity (on account of the transport equation $d\omega/dt = 0$) and their surface (on account of the incompressibility of the flow).

After a complex evolution the system is expected to be in the most probable, i.e. most mixed state, consistent with all the constraints imposed by the dynamics. We define the mixing entropy as the logarithm of the number of microscopic configurations associated with the same macroscopic state (characterized by the probability density $\rho(\mathbf{r},\sigma)$). We divide the macrocells $(\mathbf{r},\mathbf{r}+d\mathbf{r})$ into ν microcells and denote by n_{ij} the number of microcells occupied by the vorticity level σ_j in the i-th macrocell. A simple combinatorial argument indicates that the number of microstates associated with the macrostate $\{n_{ij}\}$ is

$$W(\{n_{ij}\}) = \prod_i N_j! \prod_i \frac{\nu!}{n_{ij}!}, \tag{283}$$

where $N_j = \sum_i n_{ij}$ is the total number of microcells occupied by σ_j. We have to add the normalization condition $\sum_j n_{ij} = \nu$, equivalent to (278), which prevents overlapping of different vorticity levels. This constraint plays a role similar to the Pauli exclusion principle in quantum mechanics. Morphologically, the statistics (283) corresponds to a 4th type of statistics since the particles are distinguishable but subject to an exclusion principle [90]. There is no such exclusion for point vortices since they are free a priori to approach each other without limitation.

Taking the logarithm of W and passing to the continuum limit with the aid of the Stirling formula, we get

$$S = -\int \rho(\mathbf{r},\sigma)\ln\rho(\mathbf{r},\sigma)d^2\mathbf{r}d\sigma. \tag{284}$$

The most probable macroscopic state is obtained by maximizing the mixing entropy (284) with fixed energy (281), global vorticity distribution (282) and local normalization (278). This problem is treated by introducing Lagrange multipliers, so that the first variations satisfy

$$\delta S - \beta\delta E - \int \alpha(\sigma)\delta\gamma(\sigma)d\sigma - \int \zeta(\mathbf{r})\delta\Big(\int \rho(\mathbf{r},\sigma)d\sigma\Big)d^2\mathbf{r} = 0. \tag{285}$$

The resulting optimal probability is a Gibbs state which can be expressed as

$$\rho(\mathbf{r},\sigma) = \frac{1}{Z(\psi)}g(\sigma)e^{-\beta\sigma\psi}, \tag{286}$$

where $Z(\psi) \equiv e^{\zeta(\mathbf{r})+1}$ and $g(\sigma) \equiv e^{-\alpha(\sigma)}$. The normalization condition (278) leads to a value of the partition function Z of the form

$$Z = \int g(\sigma)e^{-\beta\sigma\psi}d\sigma, \tag{287}$$

and the locally averaged vorticity (279) is expressed as a function of ψ according to

$$\overline{\omega} = \frac{\int g(\sigma)\sigma e^{-\beta\sigma\psi}d\sigma}{\int g(\sigma)e^{-\beta\sigma\psi}d\sigma} = f(\psi). \tag{288}$$

This expression can be rewritten

$$\overline{\omega} = -\frac{1}{\beta}\frac{d\ln Z}{d\psi}. \tag{289}$$

Differentiating (288) with respect to ψ, we check that the variance of the vorticity can be written [53]

$$\omega_2 = \overline{\omega^2} - \overline{\omega}^2 = -\frac{1}{\beta}f'(\psi), \tag{290}$$

or, alternatively,

$$\omega_2 = \frac{1}{\beta^2}\frac{d^2\ln Z}{d\psi^2}. \tag{291}$$

Therefore, the slope of the function $\overline{\omega} = f(\psi)$ is directly related to the variance of the vorticity distribution. Since $\omega_2 > 0$, we find that the function $\overline{\omega} = f(\psi)$ is monotonic; it is decreasing for $\beta > 0$ and increasing for $\beta < 0$ (it is constant for $\beta = 0$). Another proof of this result is given in [111].

Two particular cases are worth considering. If the local distribution of vorticty is Gaussian, then the $\omega - \psi$ relationship (288) is linear [95,96]. On the other hand, in the case of a single level of vorticity σ_0 (in addition to the level $\sigma = 0$), the coarse-grained vorticity $\omega = \rho(\mathbf{r},\sigma_0)\sigma_0$ takes explicitly the form

$$\overline{\omega} = \frac{\sigma_0}{1 + \lambda e^{\beta\sigma_0\psi}}. \tag{292}$$

This is formally similar to the Fermi-Dirac statistics. Here, the exclusion principle $\overline{\omega} \leq \sigma_0$ is due to the Liouville theorem (i.e., the conservation of the fine-grained vorticity) not to quantum mechanics. Because of the averaging procedure, the coarse-grained vorticity can only *decrease* by internal mixing, as irrotational flow is incorporated into the patch σ_0, and this results in an "effective" exclusion principle. In the dilute limit $\overline{\omega} \ll \sigma_0$, (292) reduces to the Boltzmann distribution $\overline{\omega} = Ae^{-\beta\sigma_0\psi}$ as in the point vortex model [70,107].

Similar results are obtained in the context of collisionless stellar system described by the Vlasov-Poisson system [90,42]. The equivalent of (292) is the Lynden-Bell distribution function

$$\overline{f} = \frac{\eta_0}{1 + \lambda e^{\beta\eta_0(\frac{v^2}{2}+\Phi)}}, \tag{293}$$

which formally coincides with the distribution function of the self-gravitating Fermi gas. In the non-degenerate limit $\overline{f} \ll \eta_0$, it reduces to the Maxwell-Boltzmann distribution $\overline{f} = Ae^{-\beta\eta_0(\frac{v^2}{2}+\Phi)}$. Therefore, the theory of violent relaxation naturally explains the observed isothermal cores of elliptical galaxies without recourse to collisions [90,63]

6.3 The Maximum Entropy Production Principle

Basically, a two-dimensional incompressible turbulent flow at high Reynolds numbers is described by the Euler-Poisson system. In principle, these equations determine completely the evolution of the vorticity field $\omega(\mathbf{r},t)$. However, in practice, we are not interested by the finely striated structure of the flow but only by its smoothed-out structure. Indeed, the observations and the numerical simulations are always realized with a finite resolution. Moreover, the coarse-grained vorticity $\overline{\omega}$ is expected to converge towards an equilibrium state, the Gibbs state (286), contrary to the exact vorticity field ω which develops smaller and smaller scales. There is also a technical difficulty to simulate an inviscid dynamics due precisely to the developement of this small-scale motion. Contour dynamics methods need to introduce a "surgery" and spectral codes a "viscosity" in order to prevent numerical instabilities. However, this artificial viscosity breaks the conservation laws of the Euler equations. What we would like to obtain is a set of relaxation equations which smooth out the small scales while conserving all the constraints of the Euler equation. Such a parametrization has been proposed by Robert and Sommeria [112] in terms of a phenomenological Maximum Entropy Production Principle (MEPP).

Let us decompose the vorticity ω and velocity $\mathbf{u}$ into a mean and a fluctuating part, namely $\omega=\overline{\omega}+\tilde{\omega}$, $\mathbf{u}=\overline{\mathbf{u}}+\tilde{\mathbf{u}}$. Taking the local average of the Euler equation (274), we get

$$\frac{\partial\overline{\omega}}{\partial t}+\nabla(\overline{\omega}\,\overline{\mathbf{u}})=-\nabla\mathbf{J}_{\omega}, \tag{294}$$

where the current $\mathbf{J}_{\omega}=\overline{\tilde{\omega}\tilde{\mathbf{u}}}$ represents the correlations of the fine-grained fluctuations. Equation (294) can be viewed as a local conservation law for the circulation $\Gamma=\int\overline{\omega}d^2\mathbf{r}$. To apply the MEPP, we need to consider not only the locally averaged vorticity field $\overline{\omega}$ but the whole probability distribution $\rho(\mathbf{r},\sigma,t)$ now evolving with time t. The conservation of the global vorticity distribution $\gamma(\sigma)=\int\rho d^2\mathbf{r}$ can be written in the local form

$$\frac{\partial\rho}{\partial t}+\nabla(\rho\mathbf{u})=-\nabla\mathbf{J}, \tag{295}$$

where $\mathbf{J}(\mathbf{r},\sigma,\mathbf{t})$ is the (unknown) current associated with the vorticity level σ. Integrating (295) over all the vorticity levels σ, using (278), and comparing with (3), we find the constraint

$$\int\mathbf{J}(\mathbf{r},\sigma,t)d\sigma=\mathbf{0}. \tag{296}$$

Multiplying (295) by σ, integrating over all the vorticity levels, using (279), and comparing with (294), we get $\int\mathbf{J}(\mathbf{r},\sigma,t)\sigma d\sigma=\mathbf{J}_{\omega}$.

We can express the time variation of energy in terms of $\mathbf{J}$, using (281) and (294), leading to the energy conservation constraint

$$\dot{E}=\int\mathbf{J}_{\omega}\nabla\psi\; d^2\mathbf{r}=0. \tag{297}$$

Using (284) and (295), we similarly express the rate of entropy production as

$$\dot{S}=-\int\mathbf{J}\nabla(\ln\rho)d^2\mathbf{r}d\sigma. \tag{298}$$

The Maximum Entropy Production Principle (MEPP) consists in choosing the current $\mathbf{J}$ which maximizes the rate of entropy production $\dot{S}$ respecting the constraints $\dot{E} = 0$, $\int \mathbf{J} d\sigma = \mathbf{0}$ and $\int \frac{J^2}{2\rho} d\sigma \leq C(\mathbf{r}, t)$. The last constraint expresses a bound (unknown) on the value of the diffusion current. Convexity arguments justify that this bound is always reached so that the inequality can be replaced by an equality. The corresponding condition on first variations can be written at each time t:

$$\delta\dot{S} - \beta(t)\delta\dot{E} - \int \boldsymbol{\zeta}(\mathbf{r}, t)\delta\Big(\int \mathbf{J} d\sigma\Big) d^2\mathbf{r} - \int D^{-1}(\mathbf{r}, t)\delta\Big(\int \frac{J^2}{2\rho} d\sigma\Big) d^2\mathbf{r} = 0, \tag{299}$$

and leads to a current of the form

$$\mathbf{J} = -D(\mathbf{r}, t)\Big[\nabla\rho + \beta(t)\rho(\sigma - \overline{\omega})\nabla\psi\Big]. \tag{300}$$

The Lagrange multiplier $\boldsymbol{\zeta}$ has been eliminated, using the condition (296) of local normalization conservation. The conservation of energy (297) at any time determines the evolution of the Lagrange multiplier $\beta(t)$ according to

$$\beta(t) = -\frac{\int D\nabla\overline{\omega}\nabla\psi d^2\mathbf{r}}{\int D\omega_2(\nabla\psi)^2 d^2\mathbf{r}}. \tag{301}$$

The entropy production (298) can be written

$$\dot{S} = -\int \frac{\mathbf{J}}{\rho}(\nabla\rho + \beta\rho(\sigma - \overline{\omega})\nabla\psi) d^2\mathbf{r} d\sigma + \beta\int \mathbf{J}(\sigma - \overline{\omega})\nabla\psi d^2\mathbf{r} d\sigma. \tag{302}$$

Using (296) and (297), the second integral is seen to cancel out. Inserting (300) in the first integral, we find

$$\dot{S} = \int \frac{J^2}{D\rho} d^2\mathbf{r} d\sigma, \tag{303}$$

which is positive provided that $D \geq 0$. A stationary solution $\dot{S} = 0$ is such that $\mathbf{J} = \mathbf{0}$ yielding, together with (300),

$$\nabla(\ln\rho) + \beta(\sigma - \overline{\omega})\nabla\psi = \mathbf{0}. \tag{304}$$

For any reference vorticity level σ_0, it writes

$$\nabla(\ln\rho_0) + \beta(\sigma_0 - \overline{\omega})\nabla\psi = \mathbf{0}. \tag{305}$$

Substracting (304) and (305), we obtain $\nabla\ln(\rho/\rho_0) + \beta(\sigma - \sigma_0)\nabla\psi = \mathbf{0}$, which is immediately integrated into

$$\rho(\mathbf{r}, \sigma) = \frac{1}{Z(\mathbf{r})} g(\sigma) e^{-\beta\sigma\psi}, \tag{306}$$

where $Z^{-1}(\mathbf{r}) \equiv \rho_0(\mathbf{r})e^{\beta\sigma_0\psi(\mathbf{r})}$ and $g(\sigma) \equiv e^{A(\sigma)}$, $A(\sigma)$ being a constant of integration. Therefore, entropy increases until the Gibbs state (286) is reached, with $\beta = \lim_{t\to\infty}\beta(t)$. Furthermore, we can show that a stationary solution of these relaxation equations is linearly stable if, and only if, it is an entropy *maximum* (in preparation). Therefore, this numerical algorithm selects the maxima (and not the minima or the saddle points) among all critical points of entropy. When several entropy maxima subsist for the same values of the constraints, the choice of equilibrium depends on a complicated notion of "basin of attraction" and not simply whether the solution is a local or a global entropy maximum (see [46] in a related context).

The relaxation equations (295), (300) and (301) can be simplified in the single level approximation. In that case, (294) is explicitly given by

$$\frac{\partial\overline{\omega}}{\partial t} + \overline{\mathbf{u}}\nabla\overline{\omega} = \nabla\left[D\left(\nabla\overline{\omega} + \beta(t)\overline{\omega}(\sigma_0 - \overline{\omega})\nabla\psi\right)\right]. \tag{307}$$

In the dilute limit $\overline{\omega} \ll \sigma_0$, it takes a form similar to the Fokker-Planck equation (206) obtained for point vortices. These equations both involve a diffusion and a drift, but these terms have a different physical interpretation in each case.

It is also instructive to apply this thermodynamical approach to the Vlasov-Poisson system. In the single level approximation, it leads to the following equation for the coarse-grained distribution function [54,31]:

$$\frac{\partial\overline{f}}{\partial t} + \mathbf{v}\frac{\partial\overline{f}}{\partial\mathbf{r}} + \mathbf{F}\frac{\partial\overline{f}}{\partial\mathbf{v}} = \frac{\partial}{\partial\mathbf{v}}\left[D\left(\frac{\partial\overline{f}}{\partial\mathbf{v}} + \beta(t)\overline{f}(\eta_0 - \overline{f})\mathbf{v}\right)\right], \tag{308}$$

which is a generalized form of the familiar Fokker-Planck equation (208) recovered in the non degenerate limit $\overline{f} \ll \eta_0$. We can check that (308) returns the Lynden-Bell distribution function (293) at equilibrium.

The diffusion coefficient D is not determined by the MEPP as it depends on the unknown bound C on the current. For the purpose of reaching the Gibbs state (286), the diffusion coefficient can simply be chosen arbitrarily (but with a positive value in order to ensure entropy increase). However, the precise form of the diffusion coefficient is important in order to determine the relaxation time and take into account kinetic confinement and incomplete relaxation (see below). In the context of 2D turbulence, Robert and Rosier [110] have proposed a simple evaluation of D by using an analogy with the diffusion of a passive scalar ω subjected to a turbulent velocity field $\mathbf{u} = \overline{\mathbf{u}} + \tilde{\mathbf{u}}$ (see also [54], Appendix B). In that case, the mean value $\overline{\omega}$ satisfies a convection-diffusion equation

$$\frac{\partial\overline{\omega}}{\partial t} + (\mathbf{u}\nabla)\overline{\omega} = \nabla(D\nabla\overline{\omega}), \tag{309}$$

with a diffusion coefficient given by

$$D = \frac{1}{4}\tau\overline{\tilde{\mathbf{u}}^2}(\mathbf{r}, t), \tag{310}$$

where τ is the decorrelation time of the velocity fluctuations. Equation (307) reduces to (310) when $\beta = 0$, i.e. when the energy constraint is not active. In the Euler-Poisson system, the velocity is produced by the vorticity itself via the Biot and Savart formula

$$\mathbf{u}(\mathbf{r},t) = \int \omega(\mathbf{r}',t)\mathbf{V}(\mathbf{r}'\to\mathbf{r})d^2\mathbf{r}', \tag{311}$$

where $\mathbf{V}(\mathbf{r}'\to\mathbf{r}) = -(1/2\pi)\mathbf{z}\times(\mathbf{r}'-\mathbf{r})/|\mathbf{r}'-\mathbf{r}|^2$. Inserting (311) in (310), we obtain

$$D = \frac{\tau}{4}\int \overline{\tilde{\omega}(\mathbf{r}',t)\tilde{\omega}(\mathbf{r}'',t)}\mathbf{V}(\mathbf{r}'\to\mathbf{r})\mathbf{V}(\mathbf{r}''\to\mathbf{r})d^2\mathbf{r}'d^2\mathbf{r}''. \tag{312}$$

If we neglect the spatial correlations of the vorticity fluctuations on scales larger than ϵ, the resolution scale, we have

$$\overline{\tilde{\omega}(\mathbf{r}',t)\tilde{\omega}(\mathbf{r}'',t)} = \epsilon^2\overline{\tilde{\omega}^2}(\mathbf{r}',t)\delta(\mathbf{r}'-\mathbf{r}''), \tag{313}$$

and (312) reduces to

$$D = \frac{\tau}{4}\epsilon^2\int \overline{\tilde{\omega}^2}(\mathbf{r}',t)\mathbf{V}^2(\mathbf{r}'\to\mathbf{r})d^2\mathbf{r}'. \tag{314}$$

Making a local approximation and introducing an upper cut-off a, we obtain

$$D = \frac{\tau}{4}\epsilon^2\overline{\tilde{\omega}^2}(\mathbf{r},t)\int_{\epsilon}^{a}\left(\frac{1}{2\pi\xi}\right)^2 2\pi\xi d\xi, \tag{315}$$

with $\overline{\tilde{\omega}^2} = \overline{\omega^2} - \overline{\omega}^2$. This leads to the following expression for the diffusion coefficient

$$D = (\overline{\omega^2} - \overline{\omega}^2)\frac{\tau\epsilon^2}{8\pi}\ln\left(\frac{a}{\epsilon}\right). \tag{316}$$

We note that the diffusion coefficient vanishes in regions where there is no fluctuation of the vorticity at small scales, i.e. $\omega_2 = 0$. At the contact with the unmixed flow, the diffusion current also vanishes resulting in a confinement of the vorticity. This leads to the concept of *incomplete relaxation*: in a large-scale turbulent flow, the Gibbs state (286) is satisfied only in restricted regions of space where mixing is sufficiently efficient to justify an ergodic hypothesis. Outside these domains, the relaxation is slowed down or even stopped. This kinetic confinement is illustrated by the numerical simulations of Robert and Rosier [110] and further discussed by Chavanis and Sommeria [51]. In this viewpoint, the vortices of two-dimensional turbulence can be considered as restricted equilibrium states or *maximum entropy bubbles* [51] separated from each other by an almost irrotational background. A similar kinetic confinement can be advocated in the case of stellar systems [54,42] to account for incomplete relaxation [90] and solve the infinite mass problem.

6.4 Recent Developements

An interesting problem in fluid dynamics is to obtain a classification of the "zoology" of vortices (monopoles, translating and rotating dipoles, tripoles...)

met in two-dimensional flows. The statistical mechanics approach presented in Sect. 6.2 provides a general framework to tackle this problem as it selects the most probable structures among all possible solutions of the 2D Euler equations. However, the prediction is not straightforward because, in the general case, we have to take into account an infinite set of constraints, namely the conservation of all the Casimirs C_h, in addition to energy E, angular momentum L and impulse P. A numerical algorithm has been developed by Turkington and Whitaker [124] to solve this problem and several calculations have been performed in rectangular or circular domains for a finite number of vorticity levels and for particular values of the integral constraints. However, many structures are found and it is difficult to have a clear picture of the bifurcation diagram in parameter space. For that reason, Chavanis and Sommeria [49,51] have considered a particular limit of the statistical theory, the so-called "strong mixing limit", in which the study of these bifurcations can be performed analytically. This limit corresponds to $\beta\sigma\psi \ll 1$ so that the equations of the problem can be expanded in terms of this small parameter (this is like the Debye-Hückel approximation in plasma physics). To zeroth order in the expansion, the density probability of each level is uniform which corresponds to a completely mixed state. To first order, the relationship between vorticity and streamfunction is linear and this can justify an inviscid minimum enstrophy principle (for the coarse-grained enstrophy $\Gamma_2^{c.g.} = \int \overline{\omega}^2 d^2\mathbf{r}$) [49]. In that case, the equilibrium flow only depends on E, L, P and the first moment Γ of the vorticity (the fine-grained enstrophy $\Gamma_2^{f.g.} = \int \overline{\omega^2} d^2\mathbf{r}$ serves as a normalization factor). This particular limit of the statistical theory already exhibits a rich bifurcation diagram and often provides a good approximation of more general situations (in particular for weakly energetic flows). It is also possible, in principle, to go to higher orders in the expansion in which case more and more vorticity moments $\Gamma_3, \Gamma_4, ...$ are necessary to describe the structure of equilibrium. This limit makes therefore a hierarchy between the constraints as it shows that, in many situations, only the lowest moments of the vorticity are important to characterize the equilibrium state. Because of this hierarchy, we can make some relevant predictions without the complete knowledge of the initial condition.

The statistical mechanics approach can be extended immediately to the quasi-geostrophic (QG) equations [94] by simply replacing the vorticity by the potential vorticity (PV). The formalism has also been generalized by Chavanis and Sommeria [53] to the shallow-water (SW) equations. These equations are more relevant to describe geophysical flows than the 2D Euler equations [104]. In the QG approximation, Bouchet and Sommeria [14] have explained the formation of jets and vortices in planetary atmospheres in terms of statistical mechanics, as initiated in [118]. In particular, the annular jet structure of Jupiter's Great Red Spot (see Figs. 14-15) is reproduced and explained as the coexistence of two thermodynamical phases in contact (a picture which is rigorously valid in a small Rossby radius expansion). These results can be extended to the more general case of shallow water equations [13]. In this geophysical context, the deformation of the fluid surface tends to reduce the range of interaction between

Fig. 14. Jupiter's great red spot

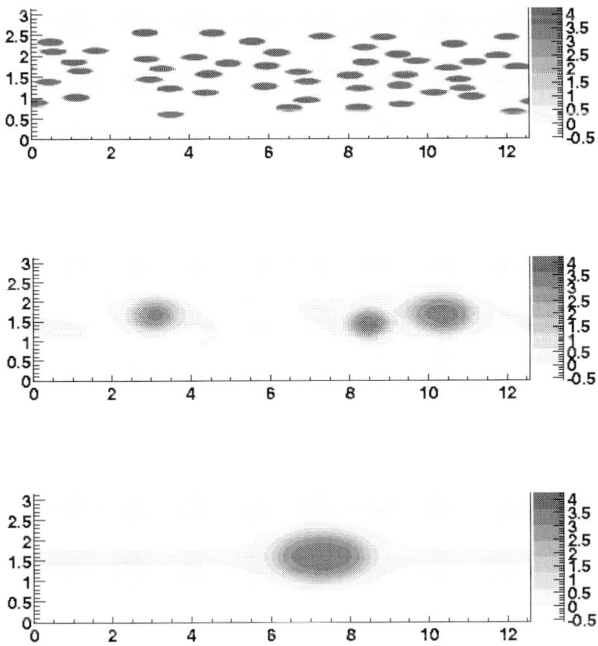

Fig. 15. Relaxation towards statistical equilibrium in a QG model of Jupiter's great red spot (from [12]). Three successive potential vorticity (PV) fields are represented as grey levels. The initial condition (top) is made of small PV patches. These patches organize into vortices (middle) that eventually merge into a single one (bottom). This sequence is obtained with the relaxation equations described in Sect. 6.3, applied to the Q.G. situation: entropy increases with time while energy is exactly conserved. At equilibrium, the vortex is an oval spot of quasiuniform PV surrounded by strong gradients, corresponding to an annular jet.

vortices. This situation is comparable to what happens in a neutral plasma: the interaction between vortices is shielded on a distance of the order of the Rossby length, the analogous of the Debye length in plasma physics.

The relaxation equations presented in Sect. 6.3 provide a convenient parameterization of sub-grid scale eddies which drives the system toward statistical equilibrium by a continuous time evolution. Such relaxation equations can be used both as a realistic coarse resolution model of the turbulent evolution, and as a method of determination of the statistical equilibrium resulting from this evolution. However, these equations do not preserve the invariance properties of the Euler equations by a change of reference frame. In addition, the conservation of energy is enforced by a formal Lagrange multiplier $\beta(t)$ which is *uniform* in space. This may be a limitation to describe large-scale turbulent flows which organize locally in several types of structures with a different temperature. A generalization of the MEPP has been proposed by Chavanis and Sommeria [50] as an attempt to solve these difficulties. More general, but also more complex, relaxation equations are obtained. They involve a space dependant temperature which tends to be uniform in each vortex but with an a priori a different value from one vortex to the other. A simplified version of these relaxation equations has been solved numerically by Kazantzev et al. [78] in an oceanic context.

Despite its practical interest, the main drawback of the MEPP is its *ad hoc* nature. In [34], we have attempted to justify the relaxation equation (307) from first principles starting directly from the 2D Euler equation. A systematic derivation can be obtained in the so-called quasilinear approximation. This approximation is well-known in plasma physics and stellar dynamics [71,114,42] for the Vlasov-Poisson system and we tried to extend it to the Euler-Poisson system. Substracting (274) and (294) and neglecting the non linear terms in the equation for $\tilde{\omega}$, the basic equations of the quasilinear theory are

$$\frac{\partial\overline{\omega}}{\partial t}+\overline{\mathbf{u}}\nabla\overline{\omega}=-\nabla\overline{\tilde{\omega}\tilde{\mathbf{u}}}, \tag{317}$$

$$\frac{\partial\tilde{\omega}}{\partial t}+\overline{\mathbf{u}}\nabla\tilde{\omega}=-\tilde{\mathbf{u}}\nabla\overline{\omega}. \tag{318}$$

It is possible to solve (318) formally with the aid of Green's functions and substitute the resulting expression for $\tilde{\omega}(\mathbf{r},t)$ back into (317). Implementing a closure relation of the form (313), a kinetic equation can be obtained for $\overline{\omega}$. In the single level approximation, it reads [34]:

$$\frac{\partial\overline{\omega}}{\partial t}+\overline{\mathbf{u}}\nabla\overline{\omega}=\epsilon^2\frac{\partial}{\partial r^\mu}\int_0^t ds\int d^2\mathbf{r}' V^\mu(\mathbf{r}'\to\mathbf{r})_t$$
$$\times\left\{V^\nu(\mathbf{r}'\to\mathbf{r})\overline{\omega}'(\sigma_0-\overline{\omega}')\frac{\partial\overline{\omega}}{\partial r^\nu}+V^\nu(\mathbf{r}\to\mathbf{r}')\overline{\omega}(\sigma_0-\overline{\omega})\frac{\partial\overline{\omega}'}{\partial r'^\nu}\right\}_{t-s}. \tag{319}$$

Assuming in addition that the decorrelation time τ is short, as in the stochastic model of Robert and Rosier [110], we are led to the following equation

$$\frac{\partial\overline{\omega}}{\partial t}+\overline{\mathbf{u}}\nabla\overline{\omega}=\frac{\epsilon^2\tau}{8\pi^2}\frac{\partial}{\partial r^\mu}\int d^2\mathbf{r}' K^{\mu\nu}(\mathbf{r}'-\mathbf{r})\left\{\overline{\omega}'(\sigma_0-\overline{\omega}')\frac{\partial\overline{\omega}}{\partial r^\nu}-\overline{\omega}(\sigma_0-\overline{\omega})\frac{\partial\overline{\omega}'}{\partial r'^\nu}\right\}, \tag{320}$$

$$K^{\mu\nu}(\mathbf{r}'-\mathbf{r})=\frac{\xi^2\delta^{\mu\nu}-\xi^\mu\xi^\nu}{\xi^4}, \tag{321}$$

with $\boldsymbol{\xi}=\mathbf{r}'-\mathbf{r}$. This equation includes a diffusion and a drift, as in the MEPP, but these terms are obtained here directly from a local average of the Euler equation. On the other hand, the conservation of angular momentum results from the symmetry of the diffusion current instead of an ad hoc Lagrange multiplier. This symmetric structure respects in addition the invariance properties of the Euler equation. Finally, a H-theorem for the Fermi-Dirac entropy

$$S=-\int\left\{\frac{\overline{\omega}}{\sigma_0}\ln\frac{\overline{\omega}}{\sigma_0}+\left(1-\frac{\overline{\omega}}{\sigma_0}\right)\ln\left(1-\frac{\overline{\omega}}{\sigma_0}\right)\right\}d^2\mathbf{r}, \tag{322}$$

can be *derived* from this kinetic equation instead of being *postulated* as in the MEPP. Our approach provides therefore an alternative, dynamical, justification of the mixing entropy introduced by Miller [95] and Robert and Sommeria [112] at statistical equilibrium.

Equations (320) and (321) are similar to the kinetic equations (238) and (239) obtained for point vortices. There are, however, two important differences: (i) the drift and the diffusion involve the product $\overline{\omega}\times(\sigma_0-\overline{\omega})$ instead of $\langle\omega\rangle$. This nonlinearity ensures that the constraint $\overline{\omega}\leq\sigma_0$ is satisfied at any time. (ii) The diffusion coefficient is proportional to the circulation $\sigma_0\epsilon^2$ of a completely filled macrocell, instead of the circulation γ of a point vortex. In general, $\sigma_0\epsilon^2\gg\gamma$ so that the relaxation by collisionless mixing is much more rapid than the collisional relaxation. From the above theory, we find that the time scale of the violent relaxation is of order t_D, the dynamical time, whereas the collisional relaxation of point vortices operates on a time scale $\sim(N/\ln N)t_D$.

If we are close to equilibrium, we can implement a "thermal bath approximation" and replace the vorticity $\omega'\equiv\omega(\mathbf{r}',t)$ by its equilibrium form (292)[34]. Then, (320) reduces to (307) derived from the MEPP and the diffusion coefficient coincides with the estimate (316) based on the passive scalar model ($\omega_2=\overline{\omega}(\sigma_0-\overline{\omega})$ in the single level approximation). All these results are satisfactory. However, (320) does *not* conserve energy and this marks a flaw in the previous description. It is probable that this constraint is broken by the small correlation time hypothesis, but we do not see how to simplify (319) further without this assumption. We could try to satisfy the energy constraint by introducing a δ-function term in (320) as in (245) but it is not clear how one can justify this procedure in the present context. The conservation of energy (which is intimately related to the form of the drift term) is the most serious problem that we have encountered in trying to develop a kinetic theory of 2D turbulence. Up to date, it has not been answered satisfactory. Further scrutinity should be given to the non markovian kinetic equation (319) which may conserve energy although we were not able to prove it.

It has to be emphasized that the quasilinear theory is *not* a theory of violent relaxation in the usual sense since it only applies to the late quiescent stages of the relaxation ("gentle relaxation"), when the fluctuations have weaken and a linearization procedure can be implemented. In order to give a more relevant description of violent relaxation, it will be important in future works to make a link between statistical mechanics and chaotic dynamics, in particular in the point vortex model. This may give a new estimate of the diffusion coefficient and of the decorrelation time τ, which in this context could be related to a Lyapunov exponent. We feel that this track is an important one to make progress in the understanding of 2D turbulence and point vortex dynamics.

In the quasilinear theory, it is implicitly assumed that the decomposition $\omega = \overline{\omega} + \tilde{\omega}$ is obvious and that $\overline{\omega}$ should be regarded as a statistical average of $\omega(\mathbf{r}, t, \zeta)$ over different realizations ζ of the flow. This implies in particular that $\overline{\overline{\omega}} = \overline{\omega}$. However, a different approach is considered by Laval et al. [87] and Bouchet [12] who define $\overline{\omega}$ as the convolution of $\omega(\mathbf{r}, t)$ with a Gaussian window of size ϵ (or by a truncation in Fourier space). In that case $\overline{\overline{\omega}} \neq \overline{\omega}$ and new terms arise in (317), in particular a term $\mathbf{J}_d = \overline{\overline{\omega}\overline{\mathbf{u}}} - \overline{\omega}\ \overline{\mathbf{u}}$ which dominates over the others [12]. When $\epsilon \to 0$, this term becomes equivalent to an anisotropic diffusion $\epsilon^2 \mathbf{\Sigma}' \nabla \overline{\omega}$ with a diffusion coefficient (or turbulent viscosity) related to the stress tensor $\Sigma'_{ij} = \partial_i \overline{u}_j$. This term alone conserves energy but there exists directions in which the viscosity is negative leading to instabilities. In order to circumvent this difficulty, Bouchet [12] proposes to project the diffusion current on directions in which the viscosity is positive and to introduce a drift term, as in the MEPP, in order to recover the conservation of energy lost by this procedure. This leads to an operational subgridscale model of 2D turbulence which appears to be more efficient than other parametrizations. As for the MEPP, its drawback is its *ad hoc* nature but it is not clear at present if it will be possible one day to do much better and derive a parametrization of 2D turbulence from first principles as attempted in the quasilinear theory.

6.5 The Limits $t \to +\infty$ and $N \to +\infty$s

We have indicated previously that a system of point vortices or point mass stars can achieve two successive equilibrium states. On a short time scale, the correlations between particles have not yet developed and the system is described by the Vlasov (or Euler) equation. In this regime, the dynamics is collisionless. Yet, for systems with long-range interactions, the collective nature of the evolution is responsible for an effective relaxation process, called violent relaxation, which leads to a metaequilibrium state (292)(293) on a very short time scale. On a longer time scale, the fluctuations of the potential have died away and the developement of correlations between stars or between point vortices leads to another, slower, relaxation process. This corresponds to the "collisional" regime. This second process is more standard and leads to a true equilibrium state (44)(101).

In the case of continuous vorticity fields (instead of point vortices), the second stage is replaced by a viscous decay of the vorticity due to inherent viscosity.

We can discuss these two successive equilibrium states in a slightly different manner. Let us consider a fixed interval of time and let the number of particles $N \to +\infty$. In that case, the system is rigorously described by the Vlasov (or Euler) equation, and a metaequilibrium state is achieved on a timescale independant on N. Alternatively, if we fix N and let $t \to +\infty$, the system will relax to the true equilibrium state resulting from a collisional evolution. These two equilibrium states are of course physically distinct. This implies that the order of the limits $N \to +\infty$ and $t \to +\infty$ is not interchangeable.

In the process of violent relaxation, the statistical mechanics is not as firmly established as in the collisional regime, although it is often the process of most interest. Indeed, the mixing required for the validity of the ergodic hypothesis is fed by the fluctuations of the potential. As these fluctuations decay as we approach equilibrium (by definition!), the mixing becomes less and less efficient and this can lead to an *incomplete relaxation*. Since the Boltzmann-Gibbs entropy does not always give a good description of the equilibrium state, it has been proposed sometimes to use a wider class of functionals to describe the process of violent relaxation [122]. Among them, the q-entropies introduced by Tsallis [123] have been shown to give in some cases a good fit of the equilibrium state. However, since the value of q needs to be adjusted in each case, it is not clear whether this agreement is the signal of a generalized thermodynamics or just a coincidence [2]. It is clear that the Boltzmann entropy does not always give a perfect description of the equilibrium state but there is no convincing reason, up to date, why the system would select another "universal" form of entropy in the context of violent relaxation [44]. In any case, the formalism developed by Tsallis and co-workers is nice to generalize to a wider class of functionals the results obtained with the Boltzmann entropy. Since this generalization is often analytically tractable (as it leads to power laws), this may explain the interest and the attractive nature of this approach.

7 Conclusion

The statistical mechanics of 2D vortices and stellar systems appear to be remarkably similar despite the different nature of these systems. We have tried

[2] In 2D turbulence, Boghosian [10] justifies a form of minimum enstrophy principle from Tsallis thermodynamics in order to interpret the experimenal results of Huang and Driscoll [66] in a magnetized plasma. This is because the enstrophy $\Gamma_2 = \int \omega^2 d^2\mathbf{r}$ is a particular q-entropy. However, this is essentially coincidental and the minimum enstrophy principle can lead to inconsistencies as discussed in [16]. In astrophysics, Tsallis entropies lead to pure polytropes [106]. These distribution functions are known for a long time but they do not give a particularly good description of elliptical galaxies or other stellar systems. Therefore, the relevance of Tsallis generalized thermodynamics in 2D turbulence and stellar dynamics remains questionable [16,39].

to develop this analogy in different directions. First of all, the N-star and N-vortex problems both involve an unshielded long-range potential generated by the density of particles themselves. At statistical equilibrium, these systems are described by a Boltzmann-Poisson equation whose solutions characterize organized states (at negative temperatures for vortices). In the case of stars, the relaxation towards equilibrium can be viewed as a competition between a diffusion and a friction. We have proposed to describe the relaxation of point vortices similarly in terms of a diffusion and a drift. The diffusion is due to the fluctuations of the force experienced by a star or to the fluctuations of the velocity field moving a vortex. The statistics of these fluctuations can be studied by similar mathematical methods. The fluctuations of the gravitational field are described by a particular Lévy law, called the Holtzmark distribution, and the fluctuations of the velocity of vortices are described by a marginal Gaussian distribution, intermediate between Gaussian and Lévy laws. The friction experienced by a star is due fundamentally to the inhomogeneity of the velocity distribution of the stellar cloud. Analogously, the drift experienced by a vortex results from the spatial inhomogeneity of the vortex cloud. The friction and the drift can be understood similarly in terms of a polarization process and a back reaction of the system. In the thermal bath approximation, the coefficients of friction and drift are given by an Einstein relation and the one-body distribution function satisfies a Fokker-Planck equation. Further away from equilibrium, the collisional dynamics of stars is described by the gravitational Landau equation. We have derived a new kinetic equation that should be appropriate to the "collisional" relaxation of point vortices.

A system of stars or vortices can also undergo a form of violent relaxation. This is essentially a collisionless process driven by the rapid fluctuations of the potential as a result of collective effects (chaotic mixing). In this collisionless regime, the stars and the vortices are described by the Vlasov-Poisson and the Euler-Poisson systems. These equations present a similar structure and a statistical mechanics can be developed to predict the "most probable state" resulting from a complex evolution driven by a mixing process. The relaxation towards equilibrium can be incomplete and an out-of-equilibrium study is necessary to understand what limits relaxation and causes a kinetic confinement of the system in a "maximum entropy bubble". Relaxation equations have been derived either from a heuristic Maximum Entropy Production Principle or from a more controllable kinetic theory, in an asymptotic regime of the dynamics (gentle relaxation) in which a quasilinear approximation can be implemented.

It should be noted that many results presented in this paper, in particular those corresponding to the kinetic theory of 2D vortices presented in Sects. 3-5, are very recent and need to be completed and further discussed. In particular, it appears indispensable to carry out extensive numerical simulations to test their relevance and determine their domains of applicability. It is plausible that the true dynamics of stars and vortices is more complex than the picture that has been given here. In addition, the description of chaos in these systems has not been addressed at all in this paper although it is presumably an essential ingredient to understand their dynamics. We are thus far from reaching a complete understanding of these systems with long-range interactions. We feel, however, that the analogy between the statistical mechanics of stars and vortices that we have investigated is correct in its mains lines and should lead again to fruitful developements.

Appendix A: The Calculation of the Diffusion Coefficient

In this Appendix, we calculate the diffusion coefficient of a point vortex evolving in an inhomogeneous vortex cloud, using the Kubo formula. To evaluate the velocity autocorrelation function $C(\tau) = \langle V(t)V(t-\tau)\rangle$, we shall assume that between t and $t-\tau$, the point vortices follow the streamlines of the equilibrium flow. This is a reasonable approximation in the case of strong shears.

A.1 Unidirectional Flow

The trajectory of a point vortex advected by a unidirectional equilibrium flow is simply:

$$y(t-\tau) = y(t), \tag{323}$$

$$x(t-\tau) = x(t) - \langle V\rangle_{eq}(y)\tau. \tag{324}$$

The velocity auto-correlation function appearing in (185) can be written explicitly

$$C(\tau) = \frac{N\gamma^2}{4\pi^2}\int dx_1 dy_1 \frac{x_1 - x}{(x_1-x)^2 + (y_1-y)^2}(t) \times \frac{x_1 - x}{(x_1-x)^2+(y_1-y)^2}(t-\tau)P_{eq}(y), \tag{325}$$

where we have used (14). The second term involves the quantity

$$(x_1 - x)(t-\tau) = x_1 - x + (\langle V\rangle_{eq}(y_1) - \langle V\rangle_{eq}(y))\tau. \tag{326}$$

Since the integral in (325) diverges as $\mathbf{r}_1 \to \mathbf{r}$, we can make a local approximation and expand the velocity difference in a Taylor series in $y_1 - y$. To first order, we have

$$\langle V\rangle_{eq}(y_1) - \langle V\rangle_{eq}(y) \simeq -\Sigma(y)(y_1 - y), \tag{327}$$

where $\Sigma(y)$ is the local shear of the flow. Introducing the variables $X \equiv x_1 - x$ and $Y \equiv y_1 - y$, we obtain

$$C(\tau) = \frac{N\gamma^2}{4\pi^2}P_{eq}(y)\int dXdY \frac{X}{X^2+Y^2}\,\frac{X+\Sigma(y)Y\tau}{(X+\Sigma(y)Y\tau)^2+Y^2}. \tag{328}$$

The integration over X can be performed easily since the integrand is just a rational function of polynomials. After straightforward calculations, we find

$$C(\tau) = \frac{N\gamma^2}{4\pi} P_{eq}(y) \frac{1}{1 + \frac{1}{4}\Sigma^2(y)\tau^2} \int_0^{+\infty} \frac{dY}{Y}. \tag{329}$$

The integral over Y diverges logarithmically for both small and large Y. The reason for this divergence has been explained in Sect. 3. Introducing two cut-offs at scales d (the inter-vortex distance) and R (the domain size) and noting that $\ln(R/d) \sim \frac{1}{2}\ln N$, we obtain

$$C(\tau) = \frac{N\gamma^2}{8\pi} \ln N \frac{1}{1 + \frac{1}{4}\Sigma^2(y)\tau^2} P_{eq}(y). \tag{330}$$

For $\tau \to +\infty$, the correlation function decreases like τ^{-2}. This is a slow decay but it is sufficient to ensure the convergence of the diffusion coefficient (185). Using

$$\int_0^t C(\tau)d\tau = \frac{N\gamma^2}{4\pi} \frac{\ln N}{|\Sigma(y)|} \arctan\left(\frac{1}{2}|\Sigma(y)|t\right) P_{eq}(y), \tag{331}$$

and taking the limit $t \to +\infty$, we establish (189).

A.2 Axisymmetric Flow

In an axisymmetric flow, the trajectory of a point vortex takes the simple form:

$$r(t-\tau) = r(t), \tag{332}$$

$$\theta(t-\tau) = \theta(t) - \frac{\langle V \rangle_{eq}(r)}{r}\tau. \tag{333}$$

As indicated in Sect. 4.2, we are particularly interested by the $r(t)r(t-\tau)$ component of the correlation tensor. Let us introduce the separation $\delta\mathbf{r} \equiv \mathbf{r}_1 - \mathbf{r}$ between the field vortex 1 and the test vortex. In the local approximation, $\delta\mathbf{r}$ can be considered as a small quantity. Therefore, we can write

$$\delta\mathbf{r} = r\delta\theta\mathbf{e}_\theta + \delta r\mathbf{e}_r \equiv X\mathbf{e}_\theta + Y\mathbf{e}_r, \tag{334}$$

$$d^2\mathbf{r}_1 = d^2(\delta\mathbf{r}) = dXdY. \tag{335}$$

With these notations, the correlation function appearing in (188) can be rewritten

$$C(\tau) = \frac{N\gamma^2}{4\pi^2} P_{eq}(r) \int dXdY \frac{X}{X^2+Y^2}(t) \frac{X}{X^2+Y^2}(t-\tau). \tag{336}$$

Now,

$$\begin{aligned}Y(t-\tau) &= \delta r(t-\tau) = r_1(t-\tau) - r(t-\tau) \\ &= r_1(t) - r(t) = Y(t) = Y,\end{aligned} \tag{337}$$

and

$$\begin{aligned}X(t-\tau) &= r(t-\tau)\delta\theta(t-\tau) = r(t-\tau)(\theta_1(t-\tau) - \theta(t-\tau)) \\ &= r\left(\theta_1(t) - \theta(t) - \left(\frac{\langle V\rangle_{eq}(r_1)}{r_1} - \frac{\langle V\rangle_{eq}(r)}{r}\right)\tau\right).\end{aligned} \tag{338}$$

In the local approximation, we can expand the last term in (338) in a Taylor series in $r_1 - r$. This yields

$$X(t-\tau) = r(\theta_1(t) - \theta(t)) - r\frac{d}{dr}\left(\frac{\langle V\rangle_{eq}(r)}{r}\right)(r_1 - r)\tau = X - \Sigma(r)Y\tau, \tag{339}$$

where $\Sigma(r)$ is the local shear of the flow. Substituting (337) and (339) in (336), we get

$$C(\tau) = \frac{N\gamma^2}{4\pi^2}P_{eq}(r)\int dXdY\frac{X}{X^2+Y^2}\frac{X-\Sigma(r)Y\tau}{(X-\Sigma(r)Y\tau)^2+Y^2}. \tag{340}$$

This integral is similar to (328), so we again obtain (189).

Appendix B: Calculation of the Memory Function

In this Appendix, we calculate the memory function that occurs in (236). If we assume that $P = P(r,t)$, then (236) simplifies to

$$\frac{\partial P}{\partial t} = -\frac{1}{r}\frac{\partial}{\partial r}(rJ_r), \tag{341}$$

where

$$\begin{aligned}J_r = &-N\int_0^t d\tau \int d^2\mathbf{r}_1 V_{r(t)}(1\to 0)_t \\ &\times\left\{V_{r(t-\tau)}(1\to 0)P_1\frac{\partial P}{\partial r} - V_{r_1(t-\tau)}(1\to 0)P\frac{\partial P_1}{\partial r_1}\right\}_{t-\tau},\end{aligned} \tag{342}$$

and where $V_{r(t)}(1\to 0)$ is the component of the vector $\mathbf{V}(1\to 0)$ in the direction of $\mathbf{r}(t)$. If we denote by $(r(t),\theta(t))$ and $(r_1(t),\theta_1(t))$ the polar coordinates that specify the position of the point vortices 0 and 1 at time t, we easily find that

$$V_{r(t)}(1\to 0) = -\frac{\gamma}{2\pi}\frac{r_1\sin(\theta-\theta_1)}{r_1^2+r^2-2rr_1\cos(\theta-\theta_1)}. \tag{343}$$

We shall assume that between t and $t-\tau$, the point vortices follow circular trajectories with angular velocity $\Omega(r,t)$. In that case, $r(t-\tau)=r$ and $\theta(t-\tau)=\theta-\Omega(r,t)\tau$. Then, we obtain

$$V_{r(t-\tau)}(1\to 0)=-\frac{\gamma}{2\pi}\frac{r_1\sin(\theta-\theta_1-\Delta\Omega\tau)}{r_1^2+r^2-2rr_1\cos(\theta-\theta_1-\Delta\Omega\tau)} \tag{344}$$

with

$$\Delta\Omega=\Omega(r,t)-\Omega(r_1,t). \tag{345}$$

We find similarly that $V_{r_1(t-\tau)}(1\to 0)=\frac{r}{r_1}V_{r(t-\tau)}(1\to 0)$. Our previous assumptions also imply that $P(r(t-\tau),t-\tau)\simeq P(r,t)$ between t and $t-\tau$. In words, this means that the correlation time is smaller than the time scale on which the average vorticity changes appreciably. We do not assume that it is *much* smaller as in Sect. 5.5, so this approximation is not over restrictive. In that case, the diffusion current becomes

$$J_r=-N\int_0^{+\infty}d\tau\int_0^{2\pi}d\theta_1\int_0^{+\infty}rr_1dr_1V_{r(t)}(1\to 0)$$
$$\times V_{r(t-\tau)}(1\to 0)\left[\frac{1}{r}P_1\frac{\partial P}{\partial r}-\frac{1}{r_1}P\frac{\partial P_1}{\partial r_1}\right], \tag{346}$$

where the time integral has been extended to $+\infty$. We now need to evaluate the memory function

$$M=\int_0^{+\infty}d\tau\int_0^{2\pi}d\theta_1V_{r(t)}(1\to 0)V_{r(t-\tau)}(1\to 0). \tag{347}$$

Introducing the notations $\phi=\theta_1-\theta$ and

$$\lambda=\frac{2rr_1}{r_1^2+r^2}<1, \tag{348}$$

we have explicitly

$$M=\left(\frac{\gamma\lambda}{4\pi r}\right)^2\int_0^{+\infty}d\tau\int_0^{2\pi}d\phi\frac{\sin\phi}{1-\lambda\cos\phi}\,\frac{\sin(\phi+\Delta\Omega\tau)}{1-\lambda\cos(\phi+\Delta\Omega\tau)}. \tag{349}$$

This can also be written

$$M=\left(\frac{\gamma}{4\pi r}\right)^2\int_0^{+\infty}d\tau\int_0^{2\pi}d\phi V'(\phi)V'(\phi+\Delta\Omega\tau), \tag{350}$$

where

$$V(\phi)=\ln(1-\lambda\cos\phi). \tag{351}$$

We now write the function $V(\phi)$ in the form of a Fourier series,

$$V(\phi)=\sum_{n=-\infty}^{+\infty}a_ne^{in\phi}\qquad\text{with}\qquad a_n=\frac{1}{2\pi}\int_{-\pi}^{\pi}V(\phi)e^{-in\phi}d\phi. \tag{352}$$

The memory function becomes

$$M = -\frac{1}{2}\left(\frac{\gamma}{4\pi r}\right)^2 \int_{-\infty}^{+\infty} d\tau \int_0^{2\pi} d\phi \sum_{n,m=-\infty}^{+\infty} nma_n a_m e^{i(n+m)\phi} e^{im\Delta\Omega\tau}. \quad (353)$$

Carrying out the integrations on ϕ and τ using the integral representation of the delta function

$$\delta(x) = \frac{1}{2\pi}\int_{-\infty}^{+\infty} e^{-i\rho x} d\rho, \quad (354)$$

we are left with

$$M = -\frac{\gamma^2}{8r^2}\sum_{n,m=-\infty}^{+\infty} nma_n a_m \delta_{n,-m}\delta(m\Delta\Omega) = \frac{\gamma^2}{8r^2}\delta(\Delta\Omega)\sum_{n=-\infty}^{+\infty}|n|a_n^2. \quad (355)$$

It remains for us to evaluate the series that appears in the last expression of the memory function. Using the identities

$$\int_0^{\pi} \ln(1-\lambda\cos\phi)\cos(n\phi)d\phi = -\frac{\pi}{n}\left(\frac{1}{\lambda} - \sqrt{\frac{1}{\lambda^2}-1}\right)^n \qquad (n>0), \quad (356)$$

$$\int_0^{\pi} \ln(1-\lambda\cos\phi)d\phi = \pi\ln\left(\frac{1}{2}+\frac{\sqrt{1-\lambda^2}}{2}\right), \quad (357)$$

and the definition (348) of λ, we find that $a_0 < \infty$ and, for $n > 0$,

$$a_n = -\frac{1}{n}\left(\frac{(r_1^2+r^2)-|r_1^2-r^2|}{2rr_1}\right)^n = -\frac{1}{n}\left(\frac{r_<}{r_>}\right)^n, \quad (358)$$

where $r_>$ (resp. $r_<$) is the biggest (resp. smallest) of r and r_1. Therefore, the value of the series is

$$\sum_{n=-\infty}^{+\infty}|n|a_n^2 = 2\sum_{n=1}^{+\infty} na_n^2 = 2\sum_{n=1}^{+\infty}\frac{1}{n}\left(\frac{r_<}{r_>}\right)^{2n} = -2\ln\left[1-\left(\frac{r_<}{r_>}\right)^2\right]. \quad (359)$$

The memory function takes the form

$$M = -\frac{\gamma^2}{4r^2}\delta(\Delta\Omega)\ln\left[1-\left(\frac{r_<}{r_>}\right)^2\right] \quad (360)$$

and the diffusion current in the axisymmetric case can be written

$$J_r = \frac{N\gamma^2}{4r}\int_0^{+\infty} r_1 dr_1 \delta(\Omega-\Omega_1)\ln\left[1-\left(\frac{r_<}{r_>}\right)^2\right]\left\{\frac{1}{r}P_1\frac{\partial P}{\partial r} - \frac{1}{r_1}P\frac{\partial P_1}{\partial r_1}\right\}. \quad (361)$$

This leads to the kinetic equation (243).

References

1. T. Agekyan, Sov. Astron. **5**, 809 (1962)
2. V.A. Antonov, Vest. Leningr. Gos. Univ. **7**, 135 (1962)
3. E.B. Aronson, C.J. Hansen, Astrophys. J. **177**, 145 (1972)
4. R. Balescu, Statistical Mechanics of Charged Particles (Interscience, New York, 1963)
5. P. Barge, J. Sommeria, Astron. Astrophys. **295**, L1 (1995)
6. J. Barré, D. Mukamel, S. Ruffo, Phys. Rev. Lett. **87**, 030601 (2001)
7. P. Bertrand, Contribution à l'étude de modèles mathématiques de plasmas non collisionnels, PhD thesis, Université de Nancy I (1972)
8. N. Bilic, R.D. Viollier, Phys. Lett. B **408**, 75 (1997)
9. J. Binney, S. Tremaine, Galactic Dynamics (Princeton Series in Astrophysics, 1987)
10. B.M. Boghosian, Phys. Rev. E **53**, 4754 (1996)
11. J.P. Bouchaud, A. Georges, Phys. Rep. **195**, 127 (1990)
12. F. Bouchet, Mécanique statistique pour des écoulements géophysiques, PhD thesis, Université J. Fourier (2001)
13. F. Bouchet, P.H. Chavanis, J. Sommeria, "Statistical mechanics of Jupiter's Great Red Spot in the shallow water model", preprint.
14. F. Bouchet, J. Sommeria, J. Fluid. Mech. **464**, 165 (2002)
15. A. Bracco, P.H. Chavanis, A. Provenzale, E. Spiegel, Phys. Fluids **11**, 2280 (1999)
16. H. Brands, P.H. Chavanis, R. Pasmanter, J. Sommeria, Phys. Fluids **11**, 3465 (1999)
17. E. Caglioti, P.L. Lions, C. Marchioro, M. Pulvirenti, Commun. Math. Phys. **143**, 501 (1992)
18. G.F. Carnevale, J.C. McWilliams, Y. Pomeau, J.B. Weiss, W.R. Young, Phys. Rev. Lett. **66**, 2735 (1991)
19. S. Chandrasekhar, An Introduction to the Theory of Stellar Structure (Dover 1939)
20. S. Chandrasekhar, Astrophys. J. **94**, 511 (1941)
21. S. Chandrasekhar, Principles of stellar dynamics (Dover 1942)
22. S. Chandrasekhar, Rev. Mod. Phys. **15**, 1 (1943)
23. S. Chandrasekhar, Astrophys. J. **97**, 255 (1943)
24. S. Chandrasekhar, Astrophys. J. **99**, 25 (1944)
25. S. Chandrasekhar, Astrophys. J. **99**, 47 (1944)
26. S. Chandrasekhar, Rev. Mod. Phys. **21**, 383 (1949)
27. S. Chandrasekhar & J. von Neumann: Astrophys. J. **95**, 489 (1942)
28. S. Chandrasekhar & J. von Neumann: Astrophys. J. **97**, 1 (1943)
29. P.H. Chavanis, Contribution à la mécanique statistique des tourbillons bidimensionnels. Analogie avec la relaxation violente des systèmes stellaires, Thèse de doctorat, Ecole Normale Supérieure de Lyon (1996)
30. P.H. Chavanis: Phys. Rev. E **58**, R1199 (1998)
31. P.H. Chavanis: Mon. Not. R. astr. Soc. **300**, 981 (1998)
32. P.H. Chavanis: Annals of the New York Academy of Sciences **867**, 120 (1998)
33. P.H. Chavanis: Astron. Astrophys. **356**, 1089 (2000)
34. P.H. Chavanis: Phys. Rev. Lett. **84**, 5512 (2000)
35. P.H. Chavanis, "On the analogy between two-dimensional vortices and stellar systems", Proceedings of the IUTAM Symposium on Geometry and Statistics of Turbulence (2001), T. Kambe, T. Nakano, T. Miyauchi Eds. (Kluwer Academic Publishers)

36. P.H. Chavanis: Phys. Rev. E **64**, 026309 (2001)
37. P.H. Chavanis: Astron. Astrophys. **381**, 340 (2002)
38. P.H. Chavanis: Astron. Astrophys. **381**, 709 (2002)
39. P.H. Chavanis: Astron. Astrophys. **386**, 732 (2002)
40. P.H. Chavanis: Phys. Rev. E **65**, 056302 (2002)
41. P.H. Chavanis: Phys. Rev. E **65**, 056123 (2002)
42. P.H. Chavanis, "Statistical mechanics of violent relaxation in stellar systems", Proceedings of the Conference on Multiscale Problems in Science and Technology (Springer 2002); also available on astro-ph.
43. P.H. Chavanis, "The self-gravitating Fermi gas", Proceedings of the Conference Dark2002: 4th International Heidelberg Conference on Dark Matter in Astro and Particle Physics, 4-9 Feb 2002, Cape Town, South African Astroparticles (Springer); also available on astro-ph/0205426.
44. P.H. Chavanis, "Gravitational instability of isothermal and polytropic spheres", submitted to Astron. Astrophys. [astro-ph/0207080].
45. P.H. Chavanis, I. Ispolatov: Phys. Rev. E **66**, 036109 (2002)
46. P.H. Chavanis, C. Rosier, C. Sire "Thermodynamics of self-gravitating systems", Phys. Rev. E **66**, 036105 (2002)
47. P.H. Chavanis, C. Sire: Phys. Rev. E **62**, 490 (2000)
48. P.H. Chavanis, C. Sire: Phys. Fluids **13**, 1904 (2001)
49. P.H. Chavanis, J. Sommeria: J. Fluid Mech. **314**, 267 (1996)
50. P.H. Chavanis, J. Sommeria: Phys. Rev. Lett. **78**, 3302 (1997)
51. P.H. Chavanis, J. Sommeria: J. Fluid Mech. **356**, 259 (1998)
52. P.H. Chavanis, J. Sommeria: Mon. Not. R. astr. Soc. **296**, 569 (1998)
53. P.H. Chavanis, J. Sommeria: Phys. Rev. E **65**, 026302 (2002)
54. P.H. Chavanis, J. Sommeria, R. Robert: Astrophys. J. **471**, 385 (1996)
55. K.V. Chukbar: Plasma Physics Reports. **25**, 77 (1999)
56. H. Cohn: Astrophys. J. **242**, 765 (1980)
57. H.J. de Vega, N. Sanchez: Nucl. Phys. B **625**, 409 (2002)
58. G.L. Eyink, H. Spohn: J. Stat. Phys. **70**, 833 (1993)
59. E. Follana, V. Laliena: Phys. Rev. E **61**, 6270 (2000)
60. P. Godon, M. Livio: Astrophys. J. **537**, 396 (2000)
61. A.E. Hansen, D. Marteau, P. Tabeling: Phys. Rev. E **58**, 7261 (1998)
62. P. Hertel, W. Thirring, "Thermodynamic instability of a system of gravitating fermions" in Quanten und Felder, edited by H.P. Dürr (Vieweg, Braunschweig, 1971)
63. J. Hjorth, J. Madsen: Mon. Not. R. astr. Soc. **265**, 237 (1993)
64. J. Holtsmark, Ann. Phys. (Leipzig) **58**, 577 (1919)
65. G. Horwitz, J. Katz: Astrophys. J. **222**, 941 (1978)
66. X.P. Huang, C.F. Driscoll: Phys. Rev. Lett. **72**, 2187 (1994)
67. I.A. Ibragimov, Yu. V. Linnik, Independant and Stationary Sequences of Random Variables (Wolters-Noordhoff, Groningen, 1971)
68. S. Inagaki, D. Lynden-Bell: Mon. Not. R. astr. Soc. **205**, 913 (1983)
69. J. Jimenez: J. Fluid Mech. **313**, 223 (1996)
70. G. Joyce, D. Montgomory: J. Plasma Phys. **10**, 107 (1973)
71. B.B. Kadomtsev, O.P. Pogutse: Phys. Rev. Lett. **25**, 1155 (1970)
72. H.E. Kandrup: Physics Reports **63**, 1 (1980)
73. H.E. Kandrup: Astrophys. J. **244**, 316 (1981)
74. H.E. Kandrup: Astrophys. & Space Sci. **97**, 435 (1983)
75. J. Katz: Mon. Not. R. astr. Soc. **183**, 765 (1978)

76. J. Katz, D. Lynden-Bell: Mon. Not. R. astr. Soc. **184**, 709 (1978)
77. J. Katz, I. Okamoto: Mon. Not. R. astr. Soc. **317**, 163 (2000)
78. E. Kazantzev, J. Sommeria, J. Verron: J. Phys. Ocean. **28**, 1017 (1998)
79. G. Kirchhoff, in Lectures in Mathematical Physics, Mechanics (Teubner, Leipzig, 1877)
80. R.H. Kraichnan: J. Fluid Mech. **67**, 155 (1975)
81. H.A. Kramers, Physica **7**, 284 (1940)
82. B.N. Kuvshinov, T.J. Schep: Phys. Rev. Lett. **84**, 650 (2000)
83. G.A. Kuzmin, "Statistical mechanics of the organization into two-dimensional coherent structures", in Structural Turbulence, edited by M.A. Goldshtik (Acad. Naouk CCCP Novosibirsk, Institute of ThermoPhysics, 1982), pp. 103-114.
84. C. Lancellotti, M. Kiessling: Astrophys. J. **549**, L93 (2001)
85. R.B. Larson: Mon. Not. R. astr. Soc. **147**, 323 (1970)
86. J.P. Laval, P.H. Chavanis, B. Dubrulle, C. Sire: Phys. Rev E **63**, 065301(R) (2001)
87. J.P. Laval, B. Dubrulle, S. Nazarenko: Phys. Rev. Lett. **83**, 4061 (1999)
88. E.P. Lee: Astrophys. J. **151**, 687 (1968)
89. T.S. Lundgren, Y.B. Pointin: J. Stat. Phys. **17**, 323 (1977)
90. D. Lynden-Bell: Mon. Not. R. astr. Soc. **136**, 101 (1967)
91. D. Lynden-Bell, P.P. Eggleton: Mon. Not. R. astr. Soc. **191**, 483 (1980)
92. D. Lynden-Bell, R. Wood: Mon. Not. R. astr. Soc. **138**, 495 (1968)
93. J.C. McWilliams: J. Fluid Mech. **146**, 21 (1984)
94. J. Michel, R. Robert: J. Phys. Stat. **77**, 645 (1994)
95. J. Miller: Phys. Rev. Lett. **65**, 2137 (1990)
96. J. Miller, P.B. Weichman, M.C. Cross: Phys. Rev. A **45**, 2328 (1992)
97. I.A. Min, I. Mezic, A. Leonard: Phys. Fluids **8**, 1169 (1996)
98. P.K. Newton, The N-Vortex Problem: Analytical Techniques, Springer-Verlag, Applied Mathematical Sciences Vol. 145, May 2001.
99. M.V. Nezlin, E.N. Snezhkin, Rossby vortices, spiral structures, solitons (Springer-Verlag 1993)
100. E.A. Novikov: Sov. Phys. JETP **41**, 937 (1975)
101. L. Onsager: Nuovo Cimento Suppl. **6**, 279 (1949)
102. T. Padmanabhan: Astrophys. J. Supp. **71**, 651 (1989)
103. T. Padmanabhan: Phys. Rep. **188**, 285 (1990)
104. J. Pedlosky, Geophysical fluid dynamics (Springer-Verlag, 1996)
105. M.V. Penston: Mon. Not. R. astr. Soc. **144**, 425 (1969)
106. A. Plastino, A.R. Plastino: Phys. Lett. A **226**, 257 (1997)
107. Y.B. Pointin, T.S. Lundgren: Phys. Fluids. **19**, 1459 (1976)
108. H. Risken, The Fokker-Planck equation (Springer, 1989)
109. R. Robert: J. Stat. Phys. **65**, 531 (1991)
110. R. Robert, C. Rosier: J. Stat. Phys. **86**, 481 (1997)
111. R. Robert, J. Sommeria: J. Fluid Mech. **229**, 291 (1991)
112. R. Robert, J. Sommeria: Phys. Rev. Lett. **60**, 2776 (1992)
113. A. Salzberg, S. Prager, J. Chem. Phys. **38**, 2587 (1963)
114. G. Severne, M. Luwel: Astrophys. & Space Sci. **72**, 293 (1980)
115. C. Sire, P.H. Chavanis: Phys. Rev. E **61**, 6644 (2000)
116. C. Sire, P.H. Chavanis: Phys. Rev. E, in press [cond-mat/0204303].
117. A.R. Smith, T.M. O'Neil: Phys. Fluids B **2**, 2961 (1990)
118. J. Sommeria, C. Nore, T. Dumont, R. Robert: C.R. Acad. Sci. II **312**, 999 (1991)
119. J. Sommeria, C. Staquet, R. Robert: J. Fluid Mech. **233**, 661 (1991)
120. B. Stahl, M.K.H. Kiessling, K. Schindler: Planet. Space Sci. **43**, 271 (1994)

121. P. Tanga, A. Babiano, B. Dubrulle, A. Provenzale: Icarus **121**, 158 (1996)
122. S. Tremaine, M. Hénon, D. Lynden-Bell: Mon. Not. R. astr. Soc. **219** 285 (1986)
123. C. Tsallis: J. Stat. Phys. **52** 479 (1988)
124. B. Turkington, N. Whitaker: SIAM J. Sci. Comput. **17**, 1414 (1996)
125. J.B. Weiss, J.C. McWilliams: Phys. Fluids A **5**, 608 (1993)
126. J.B. Weiss, A. Provenzale and J.C. McWilliams: Phys. Fluids **10**, 1929 (1998)
127. C.R. Willis, R.H. Picard: Phys. Rev. A **9**, 1343 (1974)
128. V.P. Youngkins, B.N. Miller: Phys. Rev. E **62**, 4582 (2000)

Part III

Bose–Einstein Condensation

Coherence and Superfluidity of Gaseous Bose–Einstein Condensates

Jean Dalibard

Laboratoire Kastler Brossel*, 24 rue Lhomond, 75005 Paris, France

Abstract. This paper gives a summary of the experimental production and the theoretical description of a gaseous Bose–Einstein condensate. We present the standard procedure to obtain a condensate: laser cooling, followed by evaporative cooling in a magnetic trap. We then turn to the theoretical treatment of an interacting Bose–Einstein condensate. We justify the use of a contact potential to describe atom interactions and we outline the derivation of the Gross-Pitaevskii equation. We then discuss experiments revealing the coherence properties of a single condensate, and the relative coherence between two independent condensates. Finally, we address the problem of the superfluidity of a condensate, by focussing on its rotational properties and the nucleation of quantized vortices.

1 Introduction

The story of Bose–Einstein condensation starts in 1924 when a young Bengali physicist, Satyendra Bose, sent to Einstein a manuscript about his recent work [1]. In this paper he showed that one can recover Planck's law for black body radiation by treating the photons as a gas of indiscernable particles [2]. Einstein read Bose's work with great interest and subsequently wrote two papers in which he generalized Bose's argument to the case of material particles [3,4]. In particular, in his second paper, he presented the condensation phenomenon. When an ideal gas obeying Bose statistics is cooled below a critical temperature T_c given by:

$$k_B T_c = 3.31 \frac{n^{2/3}\hbar^2}{m} , \tag{1}$$

where n is the density of the gas and m the mass of a particle, a macroscopic fraction of the particles accumulates in a single quantum state, the ground state of the box. In other words, a phase transition is predicted to occur when the distance between particles $n^{-1/3}$ becomes smaller than 0.73 λ, where λ is the thermal wavelength of the particles $\lambda = \hbar\sqrt{2\pi/mk_BT}$.

"It is a nice theory, but does it contain any truth?" By these words Einstein described his work to his friend Ehrenfest, before leaving this field of research. The phenomenon predicted by Einstein remained controversial and somewhat

* Unité de Recherche de l'Ecole normale supérieure et de l'Université Pierre et Marie Curie, associée au CNRS.

mysterious until 1937, when the superfluidity of liquid helium was discovered by Kapitza [5], Allen and Misener [6]. London then noticed that the temperature of the superfluid transition, $T_s = 2.2$ K, is remarkably close to the temperature of the Bose–Einstein condensation for an ideal gas with the same density as liquid helium, $T_c = 3.2$ K, and he made the connection between the two phenomena [7].

London's key idea is the starting point of all modern theoretical models of liquid helium. However, the connection between Bose–Einstein condensation and superfluidity is not obvious. Superfluidity occurs because of the interaction between particles, while Einstein was dealing with an ideal gas. More quantitatively, the condensate in liquid helium exists and can be detected experimentally, but it never contains more than 10 % of the particles of the fluid, while all particles should be condensed at $T = 0$ for an ideal gas.

Since 1970, the research of systems closer to Einstein's theory has been very active. Two classes of systems are now available which correspond to the initial model of an ideal, or weakly interacting, Bose gas: atomic gases [8] and excitons in solid [9,10]. In the present paper, we focus on the case of atomic assemblies. The Bose–Einstein condensation has already been found experimentally for many atomic species; by order of increasing mass, one finds atomic hydrogen [11], metastable helium [12,13], lithium [14], sodium [15], potassium [16], and rubidium [17]. In Sect. 2, we first describe briefly the principle of production of these condensates. We then discuss the notion of a macroscopic wave function (Sect. 3) and show how it is related to measurable coherence properties of the gas (Sect. 4). In Sect. 5, we discuss the superfluid properties of these gases, focusing in particular on the rotation of a condensate and the nucleation of quantized vortices. Finally we draw some conclusions and perspectives (Sect. 6).

2 How to Prepare a Gaseous Bose–Einstein Condensate

Except for atomic hydrogen [11], all experiments with gaseous condensates start with laser cooling. By taking advantage of the exchange of momentum between atoms and resonant photons, it is possible to bring a gas of atoms from room temperature down to sub-millikelvin temperatures [18–20]. This is down in a magneto-optical trap or an optical molasses. Laser cooled atoms do not yet form a condensate because their density is too low. Indeed inelastic light-assisted collisions limit the achievable density to $n \sim 10^{11}$ cm^{-3}, and the phase space density is only $n\lambda^3 \sim 10^{-6}$, while it should be larger than 1 for entering the regime of quantum degeneracy.

The second step in the experiment consists in transferring the atoms from the optical molasses into a non dissipative trap, in most cases a magnetic trap. In this device, one takes advantage of the magnetic interaction energy $E_{\text{mag}} = -\boldsymbol{\mu} \cdot \boldsymbol{B}$ between the atomic magnetic moment $\boldsymbol{\mu}$ and the magnetic field $\boldsymbol{B}$ to achieve a minimum of potential energy. More precisely, one usually chooses an appropriate configuration of coils or of permanent magnets to produce a static magnetic field which has a local minimum for $|\boldsymbol{B}|$ in a given point O (Ioffe-Pritchard trap). The atoms are prepared in a low-field seeking state, i.e. $\boldsymbol{\mu}$ and $\boldsymbol{B}$ are opposite.

The energy $E_{\text{mag}} = +|\boldsymbol{\mu}|\ |\boldsymbol{B}|$ is then also minimum in O. It plays the role of a potential energy for the atomic center-of-mass motion with a harmonic confinement potential. Note that we assume here that the direction of the magnetic moment follows adiabatically the direction of the magnetic field when the atom moves. This approximation is valid since the characteristic Larmor frequency $\mu B/h$ ($\sim$ 1 MHz) is much larger than the oscillation frequency of the atom in the magnetic trap ($\sim$ 100 Hz).

At this stage, the phase space density has not yet been increased with respect to the phase space density in the optical molasses. The gain in phase space is provided by evaporative cooling [21]. It consists in eliminating the fastest atoms of the trapped gas, and letting the remaining atoms thermalize to a lower temperature. The filtering of the most energetic atoms is done spatially. One generally uses a radio-frequency ν which is resonant with atoms located at a certain distance from the center. The magnetic moment of these atoms flips and the atoms then see an expulsive potential $-|\boldsymbol{\mu}|\ |\boldsymbol{B}|$ instead of the trapping one. The atoms with low energy stay close to the center of the trap, where the magnetic field $|\boldsymbol{B}|$ is not large enough to put them in resonance with the radio-frequency.

For efficient cooling, one usually sweeps the radio-frequency to maintain a constant ratio between the effective depth of the potential U and the thermal energy $k_{\text{B}}T$. One tries to keep the collision rate at the center of the trap constant during the evaporation process. In a harmonic trap, the collision rate scales as N/T, where N is the number of atoms. On the other hand, the phase space density at the center of the trap $n\lambda^3$ scales as N/T^3 and it must be multiplied by $\sim 10^6$ to reach the Bose–Einstein condensation threshold. Therefore a typical evaporation process consists in dividing the number of atoms and the temperature by 10^3, which leads to the desired gain. Starting with 10^9 atoms at 100 μK, one ends up with 10^6 atoms at 100 nK. The duration of the evaporation sequence varies from a few seconds to minutes, depending on the initial collision rate.

The last step of the process is the visualization of the atom cloud. This is usually performed by shining the remaining atom cloud with a flash of resonant light, and imaging the shadow of the atom cloud onto a CCD camera. One has access in this way to the spatial distribution of the atoms. One can also release the atoms from the magnetic trap and wait for an ajustable time of flight (typically from 10 ms to 50 ms) before sending the flash of light. In this way, one measures the velocity distribution of the cloud after the release from the trap. A typical image is shown in Fig. 1. Note that a non destructive *in situ* imaging is also possible by using an off-resonant pulse of light and by measuring the index of refraction of the cloud [22].

It is important to note that the atom clouds prepared in this way are metastable gases. The thermodynamic equilibrium for sodium or rubidium atoms at a temperature of 1 microkelvin or below is a solid phase, with a completely negligible vapor pressure. In the present case, we are able to produce a vapor phase thanks to laser and evaporative cooling. This allows to produce a cold atomic gas without putting the atoms in contact with a material heat reservoir

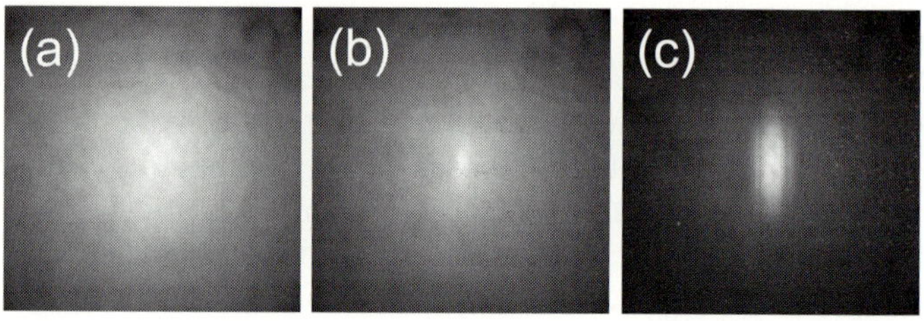

Fig. 1. Bose–Einstein condensation of a cold rubidium ^{87}Rb cloud. The pictures are taken after a 25 ms time of flight. The density profile of the cloud gives access to the velocity distribution, just after the release from the trap. (a) The temperature T is notably larger than the critical temperature T_c. The velocity distribution is isotropic and close to a Maxwell–Boltzmann distribution: $mv_i^2 \sim k_B T$, with $i = x, y, z$. (b) $T \leq T_c$: atoms start to accumulate in the ground state of the trap. The ellipticity of the condensate is a consequence of the anisotropy of the magnetic trap and it constitutes a clear signature of the condensate (c) $T \ll T_c$: the major fraction of the $\sim 4 \times 10^5$ atoms occupy a single quantum state (pictures by F. Chevy and K. Madison, LKB Paris)

at microkelvin temperature. The metastability of these gases manifests itself through losses due 3-body collisions. In such a collision, two atoms can form a dimer, and the third atom carries away the released momentum and energy. Due to these losses, the lifetime of a gaseous condensate is limited at most to a few tens of seconds.

3 Theoretical Description of a Condensate

In his initial model, Einstein was dealing with an ideal gas. In reality, atom-atom interactions cannot be neglected in a Bose–Einstein condensate, since the density at the center of the cloud is relatively high (from 10^{13} cm^{-3} to 10^{16} cm^{-3}). These interactions have important consequences on the shape and the size of the condensate, as well as its dynamical properties: spectrum and shape of the excitations, superfluidity.

We discuss in this section a simple approach to describe a pure Bose–Einstein condensate ($T = 0$), taking into account the atom-atom interactions in a mean-field approach (for more details, see e.g. [23] and refs. in). We start with the N-body hamiltonian:

$$H = \sum_{i=1}^{N} \left(\frac{\boldsymbol{p}_i^2}{2m} + V(\boldsymbol{r}_i) \right) + \sum_i \sum_{j<i} W(\boldsymbol{r}_i - \boldsymbol{r}_j) \, , \tag{2}$$

where V stands for the trapping potential and W for the two-body interaction potential. We look for the ground state of the system within the Hartree approximation. In other words, we use the variational method with the class of functions:

$$\Psi(\boldsymbol{r}_1, \boldsymbol{r}_2, \ldots, \boldsymbol{r}_N) = \psi(\boldsymbol{r}_1)\, \psi(\boldsymbol{r}_2) \, \ldots \, \psi(\boldsymbol{r}_N) \, . \tag{3}$$

This approximation reproduces the basic ingredient of Einstein's model, assuming that all atoms occupy the same quantum state. The important modification with respect to Einstein's result is that the state $\psi(\boldsymbol{r})$ is now determined using a self-consistent method, instead of being simply the ground state of the 1-body hamiltonian $\boldsymbol{p}^2/2m + V(\boldsymbol{r})$ for an ideal gas. It is worth noting that the description of the N-body system by a factorized wave function like (3) amounts to neglect all correlations between particles. As we shall see, this is quite a good approximation for the gaseous Bose–Einstein condensates. On the other hand, it would not be valid for a dense system such as liquid helium.

3.1 The Scattering Length

In addition to the restriction to Hartree functions, another ingredient is needed to proceed further. It concerns the 2-body potential $W(\boldsymbol{r})$ and it takes advantage of the very low temperature of our problem.

We first recall that the problem of a collision between two particles with mass m_1 and m_2, interacting with a potential $W(\boldsymbol{r}_1 - \boldsymbol{r}_2)$, is equivalent to the problem of the scattering of a particle with the reduced mass $\mu = m_1 m_2/(m_1+m_2)$ by the potential $W(\boldsymbol{r})$ (see e.g. [24,25]). The solution of the scattering problem amounts to look for the eigenfunctions of the 1 body hamiltonian $H_1 = p^2/(2\mu) + W(\boldsymbol{r})$ with a positive energy $E_k = \hbar^2 k^2/(2\mu)$. These eigenfunctions are called *scattering states*. The convenient asymptotic expression for a scattering state is:

$$\psi_{\boldsymbol{k}}(\boldsymbol{r}) \propto e^{i\boldsymbol{k}\cdot\boldsymbol{r}} + f(k, \boldsymbol{n}, \boldsymbol{n}')\frac{e^{ikr}}{r} , \tag{4}$$

which is valid for r large compared with the range of W. It correspond to the scattering of an incident plane wave of wave vector $\boldsymbol{k} = k\boldsymbol{n}$, where $\boldsymbol{n}$ is a unit vector. The quantity $f(k, \boldsymbol{n}, \boldsymbol{n}')$, where we introduce the unit vector $\boldsymbol{n}' = \boldsymbol{r}/r$, gives the scattering amplitude from $\boldsymbol{n}$ to $\boldsymbol{n}'$ for the energy E_k.

At low energy, for a potential $W(\boldsymbol{r})$ decreasing faster than r^{-3}, one can show that the scattering becomes isotropic, i.e. $f(k, \boldsymbol{n}, \boldsymbol{n}')$ does not depend on $\boldsymbol{n}$ and $\boldsymbol{n}'$. In addition, when k tends to zero, the scattering amplitude usually tends to a finite value and one sets:

$$\lim_{k\to 0} f(k, \boldsymbol{n}, \boldsymbol{n}') = -a \tag{5}$$

The quantity a, which has the dimension of a length, is called the *scattering length* (in the context of cold atoms, see e.g. [26]). Assuming for simplicity that W is a spherically symmetric potential, the calculation of a requires only the knowledge of the zero-energy eigenstates of the hamiltonian H_1. More precisely, one can consider the Schrödinger equation at zero energy for the radial function $u(r) = r\,\psi(r)$:

$$\frac{d^2u}{dr^2} - \frac{2\mu W(r)}{\hbar^2}u = 0 \tag{6}$$

and simply extract the scattering length from the asymptotic behavior of its solutions: $u(r) \propto r - a$.

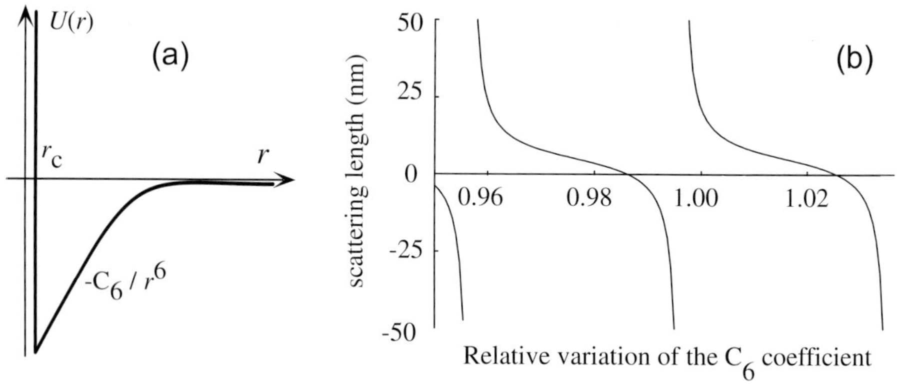

Fig. 2. (a) Van der Waals potential C_6/r^6 between two neutral atoms; the potential is truncated in r_c in order to account for the repulsion effect between the electronic cores. (b) Variations of the scattering length a (in nanometers) for a small relative change of the C_6 coefficient (numerical data for the cesium atom)

If the Born approximation is valid, a is simply given by

$$a = \frac{\mu}{2\pi\hbar^2} \int W(\boldsymbol{r})\, d^3r \ . \tag{7}$$

Unfortunately, for usual potentials between alkali atoms with a long-range attraction in $-C_6/r^6$ (Van der Waals interaction), this Born approximation is not valid since there are many bound states in the potential well $W(\boldsymbol{r})$. One has to know precisely the potential W and solve (6) to determine a. One finds that a depends strongly on the parameters of the problem, such as the C_6 coefficient (see Fig. 2). When one increases continuously C_6, one introduces new bound states in the problem. Each time a new bound state is about to appear, the scattering length tends to $-\infty$ (Levinson theorem). Once C_6 is set just above the critical strength for the appearance of the new bound state, a is now large and positive. It then evolves slowly with C_6 until another bound state is about to appear.

In the context of the present book, it is worth noting that this description in terms of a scattering length is valid only if the potential $W(\boldsymbol{r})$ decreases faster than r^{-3} at infinity. If $W(\boldsymbol{r})$ decreases as r^{-3} or slower, the differential scattering cross-section is not isotropic anymore even in the limit of a zero-energy: all partial waves contribute even for $k = 0$. This is not just an academic counterexample. For instance atoms have a magnetic moment and the magnetic dipolar interaction varies as r^{-3}. This is usually negligible for "standard" atoms, but it could play a role for atoms (or molecules) with a large magnetic moment (or with a permanent electric dipole moment). One can also use laser light to induce a long-range interaction between atoms, which cannot be described by a scattering length [27].

3.2 The Gross-Pitaevski Equation

Once the scattering length has been calculated or measured experimentally, we can now take advantage of the fact that two potentials having the same scattering length a will lead to the same properties for the cold N-body system. Consequently, one usually replaces the real potential W by the much simpler contact potential[1]:

$$W(\boldsymbol{r}_i - \boldsymbol{r}_j) \longrightarrow \frac{4\pi\hbar^2 a}{m}\delta(\boldsymbol{r}_i - \boldsymbol{r}_j) . \tag{8}$$

The variational method can then be applied and the minimization of the energy functional leads to the following equation for the normalized *condensate wave function* $\psi(\boldsymbol{r})$:

$$-\frac{\hbar^2}{2m}\Delta\psi(\boldsymbol{r}) + V(\boldsymbol{r})\,\psi(\boldsymbol{r}) + \frac{4\pi\hbar^2 a}{m}(N-1)|\psi(\boldsymbol{r})|^2\,\psi(\boldsymbol{r}) = \mu\,\psi(\boldsymbol{r}) . \tag{9}$$

This is the Gross-Pitaevski equation and the energy eigenvalue μ is the chemical potential. The physical interpretation of this equation is simple. A particle with wave function $\psi(\boldsymbol{r})$ evolves in the external potential $V(\boldsymbol{r})$ plus the mean field potential created by the $N-1$ remaining particles; this mean field is proportional to the scattering length a and to the local density $(N-1)|\psi(\boldsymbol{r})|^2$. Of course, this equation is of interest for $N \gg 1$, so that one takes $N-1 \simeq N$.

This equation is used to describe the ground state of a condensate as well as its excitations. One uses in the latter case the time-dependent version of (9):

$$i\hbar\frac{\partial\psi(\boldsymbol{r},t)}{\partial t} = -\frac{\hbar^2}{2m}\Delta\psi(\boldsymbol{r},t)+V(\boldsymbol{r})\,\psi(\boldsymbol{r},t)+\frac{4\pi\hbar^2 a}{m}(N-1)|\psi(\boldsymbol{r},t)|^2\,\psi(\boldsymbol{r},t) . \tag{10}$$

By a linear expansion $\psi(\boldsymbol{r},t) = \psi_0(\boldsymbol{r}) + \delta\psi(\boldsymbol{r},t)$ around the steady-state solution $\psi_0(\boldsymbol{r})$, one obtains the frequency and the spatial structure of the main vibration modes of the condensate (for example phonons for an homogeneous condensate) . This procedure yields the same results as the analysis based upon the Bogoliubov transformation and discussed by M. Holthaus in his contribution to this volume (see also [29]).

The predictions derived from this equation are in excellent agreement with experiments dealing with a quasi-pure condensate. At $T = 0$, the validity criterion is $na^3 \ll 1$, which expresses the fact that the fluid must be dilute. For typical gaseous condensates, one has $na^3 \sim 10^{-5}$, so that the mean-field approximation is excellent. This would not be the case for liquid helium, in which case $na^3 > 1$.

The critical dependence of the scattering length with the parameters of the 2-body interaction potential $W(\boldsymbol{r})$ opens the way to a manipulation in real time of the effective interaction between the atoms in the condensate. By varying

[1] Rather than a pure $\delta(\boldsymbol{r})$ potential which is singular with respect to scattering, one rather uses the well behaved pseudo-potential W_{p} defined by $W_{\mathrm{p}}(\boldsymbol{r})\psi(\boldsymbol{r}) = (4\pi\hbar^2 a/m)\,\delta(\boldsymbol{r})\,\partial(r\psi(\boldsymbol{r}))/\partial r$ [28,26].

for example the external magnetic field, and taking advantage of a *Feschbach resonance*, one can tune the scattering length all the way from a large and positive scattering length (strongly repulsive effective interactions) to a large and negative scattering length (strongly attractive effective interactions) [30].

4 Coherence of a Bose–Einstein Condensate

The experimental study of the coherence of a Bose–Einstein condensate addresses the question of the validity of the description of this N-body system in terms of a macroscopic wave function $\psi(\boldsymbol{r})$. Another related question deals with the relative coherence between two independent condensates. In this section, we briefly review some of the experimental answers which have been given to these two problems.

4.1 Phase Coherence of a Single Condensate

In order to study the transverse coherence of a light beam, the standard procedure consists in sending the beam onto a screen pierced with two small holes and measuring the contrast of the interferences between the light waves diffracted by these holes. This scheme has been transposed to trapped Bose–Einstein condensates by the Munich group [31]. The equivalent of a hole pierced in a screen is a radio-frequency field, which extracts the atoms at a given location in the magnetic trap [32]. When a single radio-frequency is applied, one obtains a beam of atoms leaking out of the condensate, which simply falls in the earth gravity field. By applying two different radio-frequency fields, one achieves the desired situation, with two beams emerging from two different locations of the atom cloud.

The results of the experiment fully confirm the validity of the description of the condensate by a Hartree function (see Fig. 3). When the temperature T is much lower than the critical temperature T_{c}, one observe an interference with a high contrast in the region where the two extracted beams overlap (Fig. 3a). This holds even for a large difference between the two radio-frequencies, in which case the separation between the two "holes" is of the order of the size of the condensate. On the contrary, when $T > T_{\mathrm{c}}$, no detectable interference is seen in the region where the two extracted beams overlap, unless the distance between the two holes is smaller than 200 nm, i.e. the coherence length of the gas for these experimental conditions (Fig. 3b).

4.2 Coherence Between Two Condensates

We consider now two Bose–Einstein condensates of the same atomic species. Each condensate contains in average N atoms, and the condensates have been prepared independently in different traps. We release these condensates from the traps, we let them overlap, and we measure the atom density distribution. Is it

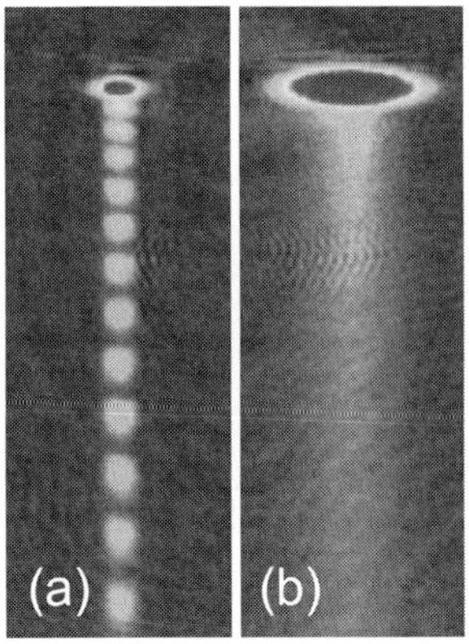

Fig. 3. Interferences between two beams extracted from a cold rubidium cloud, observed in Munich. (a) The cloud consists in a quasi-pure condensate and a high fringe contrast is observed, revealing a global phase coherence of the gas. (b) For a cloud above the critical temperature, no phase coherence is observed between the two beams if the extraction points are distant by more than 200 nm (courtesy of I. Bloch)

possible to observe an interference pattern revealing a coherence between the two condensates?

The problem that we are facing here is analogous to the question raised by P.W. Anderson [33]: Do two superfluids which have never "seen" one another possess a definite relative phase? As pointed out in [34], the question is meaningless as long as no measurement is performed onto the system. A detailed analysis of the measurement process has been performed in [35–37]. The main conclusion is that two equivalent points of view on the system can be considered.

1. In the first point of view, each condensate can be described initially by a coherent state, respectively $|\psi|e^{i\phi_1}$ and $|\psi|e^{i\phi_2}$, with $|\psi|^2 = N$. Here the phases ϕ_i are random variables, which take different values for each new preparation of the two condensates. In this case, interferences are clearly expected, and their measurement "reveals" the relative phase $\phi_1 - \phi_2$.
2. Alternatively, one can consider that the initial state of the two condensates is a factorized number (or Fock) state $|N_1, N_2\rangle$, with N_1 (resp. N_2) atoms in condensate 1 (resp. 2) . The random variables N_1 and N_2 have a Poisson distribution. In this second point of view, the succession of detection events (an atom in $\boldsymbol{r}_1$, an atom in $\boldsymbol{r}_2$, ...) "builds up" the relative phase, by producing a quantum entanglement between the states of the two condensates

There is no possibility to favor one of these two points of view from experimental results. The probability for a given result (i.e. one atom in $\boldsymbol{r}_1$, one atom in $\boldsymbol{r}_2$, ..., one atom in $\boldsymbol{r}_{2N}$) is the same in the two points of view.

The experiment has been performed at MIT in 1997, with two condensates prepared in a double well potential [38]. This potential was realized using an

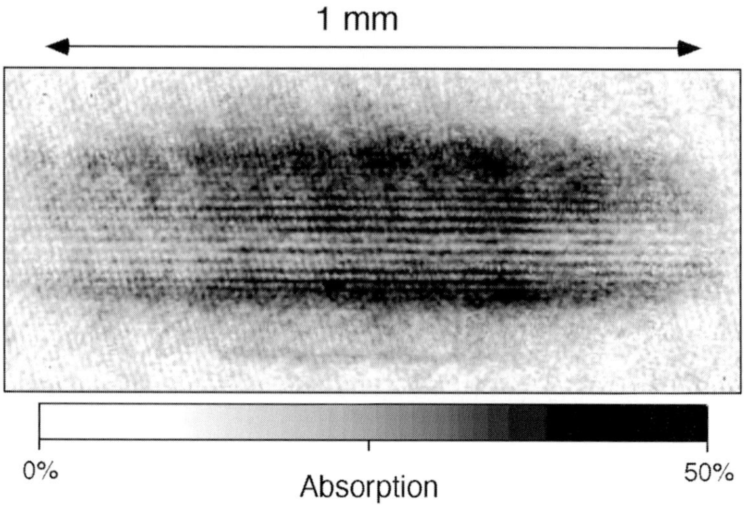

Fig. 4. Interference pattern obtained at MIT by measuring the density profile of the superposition of two sodium condensates. The two condensates have been prepared in two traps, centered in different locations. They have subsequently been released from the traps and expanded during a 40 ms free flight period, which allowed them to overlap (courtesy of W. Ketterle)

elongated magnetic trap, separated in two parts by a thin sheet of light creating a impenetrable barrier between the two sides of the trap. The experimental results show a spectacular phase coherence between the two independent condensates (see Fig. 4).

5 Superfluidity of a Bose–Einstein Condensate

Superfluidity, originally discovered and studied in the context of superconductors and later in the system of superfluid liquid Helium, is a hallmark property of interacting quantum fluids and encompasses a whole class of fundamental phenomena [29,39]. As shown by Landau, superfluidity is a direct consequence of repulsive interactions, which give rise to a phonon-like dispersion relation for the lowest lying excitations [29]. In addition to the observation of specific oscillation modes of the condensate (see e.g. [40,41]), the study of superfluidity in a dilute BEC has been the subject of two classes of study. The first one has been performed at MIT. It consists in measuring the energy deposited in the condensate by a large moving "object" (a hole created by a repulsive laser) or the friction force acting on a moving atom, discernable from the condensate [42–44]. The second type of study is related to the rotation properties of the condensate and the possible nucleation of quantized vortices [45]. This is the subject that we address in the following.

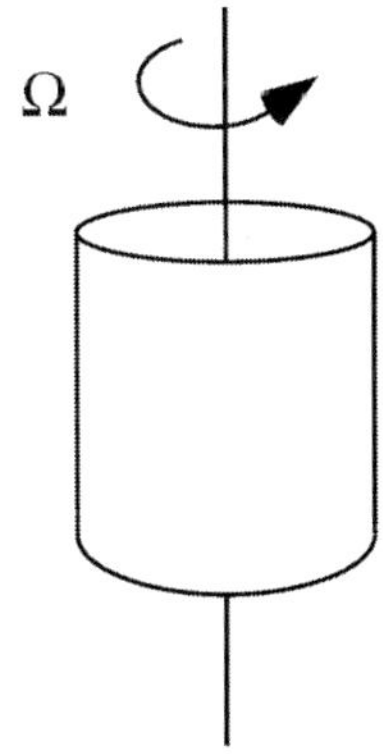

Fig. 5. The rotating bucket experiment

5.1 The Rotating Bucket Experiment

Consider a superfluid placed in a bucket rotating at angular frequency Ω (Fig. 5). If Ω is smaller than a critical value Ω_c, the superfluid will not circulate, which is is a direct manifestation of superfluidity. The existence of this critical angular frequency is analogous to that of a critical linear flow velocity below which the condensate exhibits viscous free behaviour. Here the slow motion of the rough walls of the bucket does not rotate the superfluid.

When Ω is increased beyond the critical frequency Ω_c, the superfluid is set into motion. As pointed out by Onsager [46] and Feynman [47], the corresponding velocity field is subject to very strong constraints due to its quantum nature. Consider the macroscopic wave function $\psi(\boldsymbol{r})$ describing the state of the superfluid. It can be written:

$$\psi(\boldsymbol{r}) = \sqrt{\rho(\boldsymbol{r})}\ \exp i\theta(\boldsymbol{r}) \ , \tag{11}$$

where $\rho(\boldsymbol{r})$ is the superfluid density. In a place where the density is not zero, the corresponding velocity field is given by:

$$\boldsymbol{v}(\boldsymbol{r}) = \frac{\hbar}{m} \boldsymbol{\nabla}\theta(\boldsymbol{r}) \ . \tag{12}$$

From (12) it is clear that the circulation of the velocity along any closed contour is quantized as a multiple of h/m:

$$\oint \boldsymbol{v} \cdot \mathrm{d}\boldsymbol{r} = n\frac{h}{m} \qquad \text{where } n \text{ is an integer.} \tag{13}$$

Just above Ω_c, the superfluid wave function has a singular line or *vortex line*, along which the density is zero. On any closed path going around this line, the phase of the wave function varies continuously from 0 to 2π.

It is worth emphasizing that these vortices are universal structures associated with a circulating quantum flow. Besides superfluid liquid helium, other large quantum systems such as neutron stars and superconductors support quantized vortices. In the latter case, the rotation vector is induced by an applied magnetic field which modifies the motion of the charges, and the quantization of the circulation of the velocity field results in magnetic flux quantization.

For a rotating frequency Ω notably larger than Ω_c, several vortex lines can be generated, and they form a regular triangular lattice. Such regular lattices were originally predicted to occur in type-II superconductors by Abrikosov [48] and were subsequently observed experimentally by imaging with an electron microscope small ferromagnetic particles which, when scattered on the surface of a superconductor in a magnetic field, accumulate at the flux line exit points [49,50]. Evidence of vortex lattice arrangements in liquid helium was demonstrated by trapping electrons at the core of each vortex and then accelerating the electrons along the vortex lines to a phosphorus screen [51].

5.2 Experimental Observation of Vortices in a Gaseous Condensate

The first method which has been successful in producing a vortex in a gaseous condensate uses a "phase printing method". It has been used by the Boulder group to create a vortex in a double component condensate with one component standing still at the center of a magnetic trap and the other component in quantized rotation around the first one [52]. In subsequent experiments, this group also succeeded in emptying the core of the double component vortex [53], and showed that vortex rings can form as decay products of a dark soliton [54].

The second method, which is the most commonly used nowadays, consists in transposing the rotating bucket concept to a trapped gas [55–59]. One starts with a gaseous condensate confined in a magnetic trap, which creates an axisymmetric harmonic potential:

$$U(\boldsymbol{r}) = \frac{1}{2} m\omega_\perp^2 (x^2 + y^2) + \frac{1}{2} m\omega_z^2 z^2 \ . \tag{14}$$

In our experiment in Paris, where we developed this method, the transverse frequency $\omega_\perp/2\pi$ varies from 90 to 220 Hz, and the axial frequency is $\omega_z/2\pi =$ 12 Hz. In this case, the condensate is cigar-shaped, with a length of $\sim$ 110 μm and a diameter of 7 μm for 2×10^5 atoms and a trapping frequency of 170 Hz.

The condensate is then stirred by a additional potential in the xy plane. This potential is harmonic and anisotropic:

$$\delta U(\boldsymbol{r}) = \frac{\epsilon}{2} m\omega_\perp^2 (X^2 - Y^2) \ , \tag{15}$$

with ϵ of the order of a few percent. Its eigenaxes X and Y rotate with an angular frequency Ω in the laboratory frame.

$$X = x\cos(\Omega t) + y\sin(\Omega t) \qquad Y = -x\sin(\Omega t) + y\cos(\Omega t) \ . \tag{16}$$

The stirring frequency Ω is usually chosen in the interval $(0, \omega_\perp)$. At the upper value of this interval, the centrifugal force equals the transverse restoring force of

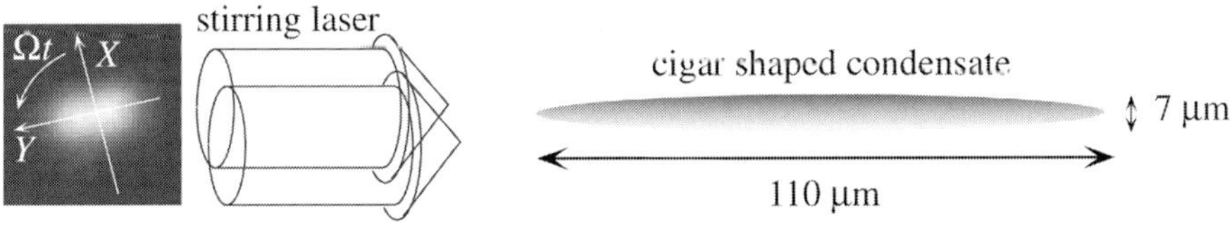

Fig. 6. The cigar-shaped condensate is stirred using the dipole potential created by a laser "spoon". The laser waist is 20 µm; its axis is toggled between two symmetric positions about the trap axis, separated by 16 µm. The laser intensity profile averaged after this toggling is displayed on the left of the figure. The resulting anisotropic potential is then rotated at the stirring frequency Ω

the trap. A dynamical instability of the center of mass motion of the condensate occurs when Ω lies in the interval $[\sqrt{1-\epsilon}\,\omega_\perp, \sqrt{1+\epsilon}\,\omega_\perp]$ [60].

In our setup, the stirring of the condensate is provided by a focused 500 µW laser beam of wavelength 852 nm and waist $w_0 = 20$ µm, whose motion is controlled using acousto-optic deflectors (Fig. 6) [55]. This laser propagates along the axis of the cigar, and it toggles back and forth very rapidly between two symmetric positions located at 8 µm from the center of the condensate. The toggling frequency is 100 kHz, which is much larger than both ω_z and $\omega_\perp$. Other stirring mechanisms, based upon a strongly focused laser [56,57] or a rotating magnetic field [58,59], have also been developed.

Our experimental procedure begins with the production of a quasi-pure condensate of 10^5 atoms. At this point, the stirring laser is switched on and the condensate is allowed to evolve in the combined magnetic and optical potential for a controlled duration ($\sim$ 200 ms). Finally, the stirring potential is extinguished and the condensate density profile along the stirring axis is measured to detect the presence of vortices.

We now address the question of the vortex visibility. In a fluid of density ρ, the radius of the vortex core is determined by the healing length $\xi = (8\pi\rho a)^{-1/2}$, where a is the scattering length [29]. For our experimental conditions, $\xi \sim 0.2$ µm, which is too small to be observed optically with resonant light at 0.8 µm. Fortunately this size can be expanded using the time-of-flight technique [61–64]. When we release the atom cloud from the magnetic trap and let it expand for a duration T, the transverse dimensions of the condensate and of the vortex core are increased by a factor $\omega_\perp T \sim 35$ for $T = 27$ ms and a transverse trapping frequency $\omega_\perp/2\pi = 170$ Hz. The detection of the expanded condensate density profile is then performed by imaging the absorption of a resonant laser beam propagating along the z axis, as explained in Sect. 2.

5.3 Single and Multiple Vortices

We now discuss the results of this experiment. When the stirring frequency is below a threshold frequency Ω_c depending on the stirring ellipticity, no modification of the condensate is observed (Fig. 7a). Just above this critical frequency (within 1 or 2 Hz), a density dip appears at the center of the cloud, with a

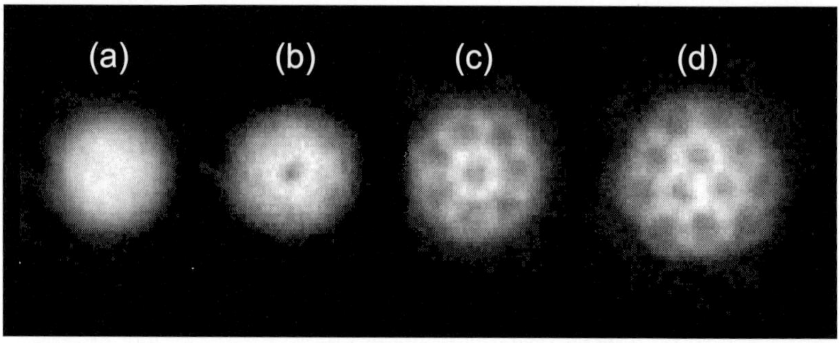

Fig. 7. Density profile after a time-of-flight expansion of a Bose–Einstein condensate stirred below (a), just above (b), and notably above (c,d) the threshold frequency for nucleating a vortex. For large number of vortices, one usually observes an Abrikosov triangular lattice.

reduction of the optical thickness at this location which reaches 50% (Fig. 7b). When we stir the condensate at a frequency higher than Ω_c, more vortices are nucleated (Fig. 7cd). The maximum number of vortices generated in our experiment is 14. Similar observations have been made recently at MIT, in Boulder an in Oxford [56–59,65]. In particular for condensates bigger than ours, large arrays of vortices (up to 150) have been nucleated [56,65].

The existence of regular vortex lattices is a consequence of the balance between the long-range repulsive interaction between two vortex lines and the restoring force that acts on a vortex line centering it on the condensate. In the large vortex number limit, the density of vortices can be deduced from the correspondence principle. In this limit, the coarse grain average (on a scale larger than the distance between two vortices) of the quantum velocity field should be the same as that for classical, rigid-body rotation $\boldsymbol{v} = \boldsymbol{\Omega} \times \boldsymbol{r}$ [47].

In order to recover this linear variation of the velocity field with $\boldsymbol{r}$, the distance from the rotation axis, the surface density of vortices ρ_v must be uniform. The circulation of the velocity field $\oint \boldsymbol{v} \cdot \mathrm{d}\boldsymbol{r}$ on a circle of radius R centered on the condensate is then $\mathcal{N}(R)\ h/m$, where $\mathcal{N}(R) = \rho_\mathrm{v} \pi R^2$ is the number of vortices contained in the circle (see (13)). This is equivalent to the rigid body circulation $2\pi R^2 \Omega$ if

$$\rho_\mathrm{v} = \frac{2m\Omega}{h} \ . \tag{17}$$

Consider for instance the vortex lattice shown in Fig. 8, which contains $\mathcal{N} = 12$ visible vortices. The circle drawn on the edge of the condensate has a radius 80 µm after the time-of-flight expansion, i.e. $R = 5$ µm before expansion (using the expansion factor $\omega_\perp T$ given above). The stirring frequency in this experiment was $\Omega/2\pi = 77$ Hz, which yields the ratio between the average velocity $\bar{v}$ on the circle and the velocity v_r corresponding rigid body rotation:

$$\frac{\bar{v}}{v_\mathrm{r}} = \frac{\mathcal{N}\hbar}{m\Omega R^2} \simeq 0.7 \ . \tag{18}$$

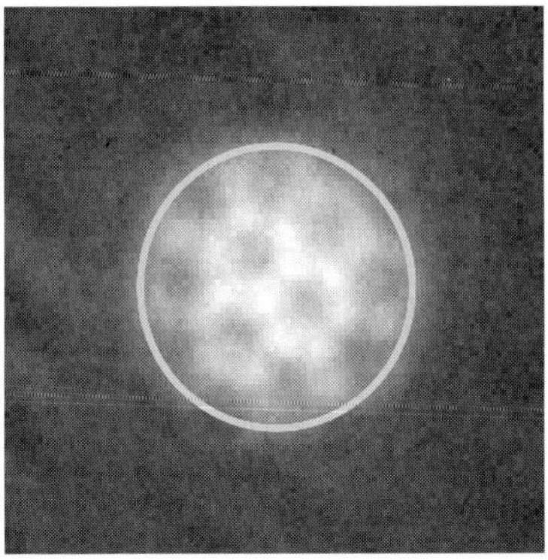

Fig. 8. Array of $\mathcal{N} = 12$ vortices obtained in a magnetic trap with $\bar{\omega}_\perp/2\pi = 103$ Hz and $\Omega/2\pi = 77$ Hz. The circulation of the velocity field on the circle is given by $\mathcal{N}h/m$, and is comparable with the one expected from rigid body rotation (see text).

This shows that the coarse grain average velocity field of the condensate shown in Fig. 8 is close that of a rigid body rotating at the stirring frequency Ω (see also [57]).

6 Conclusions and Perspectives

We have presented in this paper some aspects of the very rich physics of Bose–Einstein condensates. The choice of the themes is a result of our own research in Paris, and it is clear than only a very small fraction of the work currently pursued has been evoked here.

We have chosen to emphasize first the coherence properties of the condensate. From a theoretical point of view, this coherence emerges from the description of the condensate in terms of a Hartree wave function, in which all atoms occupy the same state. This coherence has been tested in spectacular experiments, performed either with a single condensate or with a pair of independent condensates. The coherence of these condensates opens the way to *atom lasers*. By extracting atoms from a trapped condensate, using e.g. a radiofrequency wave which flips the magnetic moment of the atoms at the center of the condensate, one can form a quasi-continuous beam of atoms [32]. The expected coherence properties of this atomic beam are similar to those of a monochromatic laser beam. The atom lasers realized up to now are not fully continuous, since one has to reform the condensate once it has been depleted. Several ways can be considered to form a truly continuous atom laser. One can transpose to the space domain the temporal procedure that we outlined at the beginning of this paper. In such a scheme, a slow atomic beam is injected at one end of a magnetic guide, and transverse evaporative cooling takes place as the atoms progress along the guide [66]. After some distance, a degenerate beam of atoms can be expected if the initial collision rate is large enough to initiate an efficient evaporative cooling. Recent progress along this line is presented in [67]. Another option is to merge two traps each containing a condensate, so that the trap feeding the atom laser is refilled before it is completely emptied [68].

We have also discussed the question of the superfluidity of a 3D gaseous Bose–Einstein condensate. We have shown that this question has now been answered in a positive way. All standard criteria for superfluidity, existence of a critical linear velocity, nucleation of quantized vortices for a large enough stirring frequency, have been observed. A question that we still intend to explore concerns the shape of a single vortex line and the nature of its dynamics. For example, recent theoretical studies predict a bend vortex line, even in a perfectly axisymmetric trap [69–72,64]. The question of the decay of a single vortex line [73] or of a vortex lattice [74] also needs some further experimental investigations.

The possibility to measure a phase coherence over the whole extension of the gas and the observation of large regular vortex arrays show that these quantum degenerate gases indeed belong to the class of systems with long-range interactions considered in the present book, although the binary interactions between atoms is treated in a contact approximation. We note as a concluding remark that the experimental techniques that we have described in the present paper can be extended to Fermi gases. Atomic samples with a temperature well below the Fermi temperature have now been achieved in various laboratories [75–77] and this opens the way to new physical phenomena, such as the observation of a BCS type transition in these dilute gases.

Acknowledgements

I thank V. Bretin, F. Chevy, K. Madison and P. Rosenbuch for their participation to the work reported here and for many useful discussions. This work was partially supported by CNRS, Collège de France, Région Ile de France, DGA, DRED and EC (TMR network ERB FMRX-CT96-0002).

References

1. A. Pais: *Subtle is the Lord* (Oxford University Press, Oxford 1982)
2. S.N. Bose: Z. Phys. **26**, 178 (1924)
3. A. Einstein: Sitzber. Klg. Preuss. Akad. Wiss. 261 (1924)
4. A. Einstein: Sitzber. Klg. Preuss. Akad. Wiss. 3 (1925)
5. P. Kapitza, Nature **141**, 74 (1938)
6. J.F. Allen, A.D. Misener, Nature **141**, 75 (1938)
7. F. London: Nature **141**, 643 (1938)
8. M. Inguscio, S. Stringari, C.E. Wieman: *Proceedings of the International School of Physics Enrico Fermi*, Course CXL, *Bose–Einstein Condensation in Atomic Gases* (IOS Press, Amsterdam 1999)
9. J.P. Wolfe, J.L. Lin, D.W. Snoke: 'Bose–Einstein Condensation of a Nearly Ideal Gas: Excitons in Cu_20'. In *Bose–Einstein Condensation*, ed. by A. Griffin, D.W. Snoke, S. Stringari (Cambridge University Press, Cambridge 1995) pp. 281–329
10. A. Mysyrowicz: 'Bose–Einstein Condensation of Excitonic Particles in Semiconductors'. In *Bose–Einstein Condensation*, ed. by A. Griffin, D.W. Snoke, S. Stringari (Cambridge University Press, Cambridge 1995) pp. 330–354

11. D.G. Fried, T. C. Killian, L. Willmann, D. Landhuis, S.C. Moss, D. Kleppner, T.J. Greytak: Phys. Rev. Lett. **81**, 3811 (1998)
12. A. Robert, O. Sirjean, A. Browaeys, J. Poupard, S. Nowak, D. Boiron, C.I. Westbrook, A. Aspect: Science **292**, 461 (2001)
13. F. Pereira Dos Santos, J. Léonard, Junmin Wang, C. J. Barrelet, F. Perales, E. Rasel, C. S. Unnikrishnan, M. Leduc, C. Cohen-Tannoudji: Phys. Rev. Lett. **86**, 3459 (2001)
14. C.C. Bradley, C.A. Sackett, R.G. Hulet: Phys. Rev. Lett. **78**, 985 (1997) See also C.C. Bradley, C.A. Sackett, J.J. Tollett, R. G. Hulet: Phys. Rev. Lett. **75**, 1687 (1995)
15. K.B. Davis, M.O. Mewes, N. Van Druten, D. Durfee, D. Kurn, W. Ketterle: Phys. Rev. Lett. **75**, 3969 (1995)
16. G. Modugno, G. Ferrari, G. Roati, R. J. Brecha, A. Simoni, M. Inguscio: Science **294**,1320-1322 (2001)
17. M.H. Anderson, J. Ensher, M. Matthews, C. Wieman, E. Cornell: Science **269**, 198 (1995)
18. S. Chu: Rev. Mod. Phys. **70**, 685 (1998)
19. C. Cohen-Tannoudji: Rev. Mod. Phys. **70**, 707 (1998)
20. W.D. Phillips: Rev. Mod. Phys. **70**, 721 (1998)
21. W. Ketterle, N.J. Van Druten: Advances in Atomic, Molecular, and Optical Physics, **37**, 181 (1996)
22. M.R. Andrews, D.M. Kurn, H.-J. Miesner, D.S. Durfee, C.G. Townsend, S. Inouye, W. Ketterle: Phys. Rev. Lett. **79**, 553 (1997)
23. F. Dalfovo, S. Giorgini, l.P. Pitaevskii, S. Stringari: Rev. Mod. Phys. **71**, 463 (1999)
24. L.D. Landau, E. M. Lifshitz: *Quantum Mechanics*, § 131 (Pergamon Press, Oxford 1977)
25. C. J. Joachain: *Quantum collision theory* (North-Holland, Amsterdam 1983)
26. J. Dalibard: 'Collisional dynamics of ultra-cold atomic gases'. In: *Proceedings of the International School of Physics Enrico Fermi, Course CXL, Bose–Einstein Condensation in Atomic Gases*, ed. by M. Inguscio, S. Stringari, C.E. Wieman (IOS Press, Amsterdam 1999) pp. 321–350
27. See the contribution of G. Kurizki to the present book and refs. in
28. K. Huang: *Statistical Mechanics* (Wiley, New York 1987)
29. E.M. Lifshitz, L.P. Pitaevskii: *Statistical Physics, Part 2* (Butterworth-Heinemann,Oxford 1980)
30. S. Inouye, M.R. Andrews, J. Stenger, H.-J. Miesner, D.M. Stamper-Kurn, W. Ketterle: Nature **392**, 151 (1998)
31. I. Bloch, T. W. Hänsch, T. Esslinger: Nature **403**, 166 (2000)
32. I. Bloch, T.W. Hänsch, T. Esslinger: Phys. Rev. Lett. **82**, 3008 (1999).
33. P.W. Anderson. 'Measurement in Quantum Theory and the Problem of Complex Systems'. In *The Lesson of Quantum Theory*, ed. by J. de Boer, E. Dal, O. Ulfbeck (Elsevier , Amsterdam 1986) pp. 23 34
34. A. J. Leggett, F. Sols: Found. of Physics **21**, 353 (1991)
35. J. Javanainen, S.M. Yoo: Phys. Rev. Lett. **76**, 161 (1996)
36. J.I. Cirac, C.W. Gardiner, M. Narschewski, P. Zoller: Phys. Rev. A **54**, R3714 (1996)
37. Y. Castin, J. Dalibard: Phys. Rev. A **55**, 4330 (1997)
38. M.R. Andrews, C.G. Townsend, H.-J. Miesner, D.S. Durfee, D.M. Kurn, W. Ketterle: Science **275**, 637 (1997)
39. R.J. Donnelly: *Quantized Vortices in Helium II* (Cambridge University Press, Cambridge 1991)

40. D. Guéry-Odelin, S. Stringari: Phys. Rev. Lett. **83**, 4452 (1999)
41. O.M. Maragò, S.A. Hopkins, J. Arlt, E. Hodby, G. Hechenblaikner, C. J. Foot: Phys. Rev. Lett. **84**, 2056 (2000)
42. C. Raman, M. Köhl, R. Onofrio, D. S. Durfee, C.E. Kuklewicz, Z. Hadzibabic, W. Ketterle: Phys. Rev. Lett. **83**, 2502 (1999)
43. A.P. Chikkatur, A. Gorlitz, D.M. Stamper-Kurn, S. Inouye, S. Gupta, W. Ketterle: Phys. Rev. Lett. **85**, 483 (2000)
44. R. Onofrio, C. Raman, J.M. Vogels, J. Abo-Shaeer, A.P. Chikkatur, W. Ketterle: Phys. Rev. Lett. **85**, 2228 (2000)
45. A. Fetter, A Svidzinsky: J. Phys. Condens. Matter **13**, 135 (2001)
46. L. Onsager: Nuovo Cimento **6**, suppl. 2, 249 (1949)
47. R.P. Feynman: 'Application of Quantum Mechanics to Liquid Helium'. In *Progress in Low Temperature Physics*, vol. 1, ed. by C. J. Gorter (North-Holland, Amsterdam, 1955) pp. 17–53
48. A.A. Abrikosov: JETP **32**, 1442 (1957)
49. H. Träuble, U. Essmann: J. Appl. Phys. **25**, 273 (1968)
50. N.V. Sarma: Philos. Mag. **17**, 1233 (1968)
51. E.J. Yarmchuk, M.J.V. Gordon, R.E. Packard: Phys. Rev. Lett. **43**, 214 (1979)
52. M.R. Matthews, B.P. Anderson, P.C. Haljan, D.S. Hall, C.E. Wieman, E.A. Cornell: Phys. Rev. Lett. **83**, 2498 (1999)
53. B.P. Anderson, P.C. Haljan, C.E. Wieman, E.A. Cornell: Phys. Rev. Lett. **85**, 2857 (2000)
54. B.P. Anderson, P.C. Haljan, C.A. Regal, D.L. Feder, L.A. Collins, C.W. Clark, E.A. Cornell: Phys. Rev. Lett. **86**, 2926 (2001)
55. K.W. Madison, F. Chevy, W. Wohlleben, J. Dalibard: Phys. Rev. Lett. **84**, 806 (2000)
56. J.R. Abo-Shaeer, C. Raman, J.M. Vogels, W. Ketterle: Science **292**, 476 (2001)
57. C. Raman, J.R. Abo-Shaeer, J.M. Vogels, K. Xu, W. Ketterle: Phys. Rev. Lett. **87**, 210402 (2001)
58. P.C. Haljan, I. Coddington, P. Engels, E.A. Cornell: Phys. Rev. Lett. **87**, 210403 (2001)
59. E. Hodby, G. Hechenblaikner, S.A. Hopkins, O.M. Maragò, C.J. Foot: Phys. Rev. Lett. **88**, 010405 (2002)
60. P. Rosenbusch, D.S. Petrov, S. Sinha, F. Chevy, V. Bretin, Y. Castin, G. Shlyapnikov, J. Dalibard: Phys. Rev. Lett. **88**, 250403 (2002)
61. Y. Castin, R. Dum: Phys. Rev. Lett. **77**, 5315 (1996)
62. E. Lundh, C.J. Pethick, H. Smith: Phys. Rev. A **58**, 4816 (1998)
63. F. Dalfovo, M. Modugno: Phys. Rev. A **61**, 023605 (2000)
64. M. Modugno, L. Pricoupenko, Y. Castin: cond-mat/02033597.
65. P. Engels, I. Coddington, P. C. Haljan, E. A. Cornell: cond-mat/0204449
66. E. Mandonnet, A. Minguzzi, R. Dum, I. Carusotto, Y.Castin, J. Dalibard: Eur. Phys. J. D **10**, 9 (2000)
67. P. Cren, C.F. Roos, A. Aclan, J. Dalibard, D. Guéry-Odelin: to be published in Eur. Phys. J. D (2002)
68. A.N. Chikkatur (MIT group), private communication
69. J.J. García-Ripoll, V.M. Pérez-García: Phys. Rev. A **63**, 041603 (2001)
70. J.J. García-Ripoll, V.M. Pérez-García: Phys. Rev. A **64**, 053611 (2001)
71. A. Aftalion, T. Riviere: Phys. Rev. A **64**, 04s3611 (2001)
72. A. Aftalion, R. L. Jerrard: cond-mat/0204475
73. P.O. Fedichev, G.V. Shlyapnikov: Phys. Rev. A **60**, R1779 (1999)

74. P.O. Fedichev, A.E. Muryshev: cond-mat/0004264
75. B. DeMarco, D.S. Jin: Science **285**, 1703 (1999)
76. A.G. Truscott, K.E. Strecker, W.I. McAlexander, G.B. Partridge, R.G. Hulet: Science **291**, 2570 (2001)
77. F. Schreck, L. Khaykovich, K.L. Corwin, G. Ferrari, T. Bourdel, J. Cubizolles, C. Salomon: Phys. Rev. Lett. **87**, 080403 (2001)

Ultracold Atoms and Bose–Einstein Condensates in Optical Lattices

Oliver Morsch and Ennio Arimondo

INFM, Dipartimento di Fisica, Università di Pisa,
Via Buonarroti 2, I-56127 Pisa, Italy

Abstract. For ultracold and Bose-condensed atoms contained in periodic optical potential wells the quantized nature of their motion is clearly visible. The motion of the atomic wavepacket can also be accurately controlled. For those systems the long-range character of the atomic interaction and of the external potential play a key role in the quantum mechanical evolution. The basic facets of the experimental and theoretical research for atoms within optical lattice structures will be reviewed.

1 Introduction

Crystalline samples of cold atoms, now known as optical lattices , were initially investigated in the dissipative regime, as a tool to provide velocity damping, and hence a reduction of the kinetic energy of the atomic samples (for reviews see [1–4]). In fact the sub-Doppler cooling regime relies on the action of laser beams on the atomic motion in a standing wave configuration. The study of that regime also implied the possibility of trapping atoms in the sub-wavelength sized potential wells created by the laser beams. As soon it was clear that the standing wave patterns created by several intersecting laser beams provided by low-power diode lasers could be used to trap atoms in periodic structures, the field exploded. For instance a large experimental effort was made to probe directly the bound states of the atoms within the optical potential. After the initial experiments using one-dimensional (1D) lattices, several schemes were developed allowing an extension to two- and three-dimensions (2D and 3D). Very soon optical lattices were used as a flexible tool to modify the spatial periodicity of the cold atomic samples, and in some cases fancy spatial structures could be produced which the solid state physics community could only dream of. Moreover, important applications for atomic nanolithography have been realized.

Later, the research effort moved into the nondissipative, or conservative, regime, with the aim of reducing the scattering rate in the optical potential wells which ruled out coherent control over the wave-packet atomic motion. In fact, the interest in optical lattices shifted to using them as a test-bed for quantum mechanics. Such a shift in interest was enhanced when ultracold atomic samples represented by quantum degenerate gases were available for loading into the optical periodic potential. Bose–Einstein condensates (BEC's) represent flexible sources whose spatial dimensions and velocity spread can be controlled with large freedom, so that a condensate may be loaded with great accuracy into the periodic potential created by intersecting laser beams. Thus the study of conservative optical lattices used to modify the spatial macroscopic wavefunction of BEC's has greatly expanded in the last few years.

In this work we will report on the most important aspects of the interaction between ultracold atoms, above and below the BEC temperature, within optical lattices. We will concentrate on those features more directly connected with the long-range interactions within the optical lattice. We will not, therefore, discuss some exciting investigations on the dynamical tunneling and chaotic behavior for atoms located within an optical periodic potential whose amplitude is periodically or randomly modulated [5,6]. Furthermore, we will not discuss the use of optical lattices in quantum computation schemes [7]. Moreover the subject of self generated periodic spatial structures will not be covered because separately treated in this book [8].

Section 2 will introduce the basic notions on the creation of optical lattices and the atomic response within the lattice, discussing first the near-resonant optical lattices that evolved from laser cooling schemes and then the non-dissipative, far-detuned optical lattices. Section 3 briefly reviews some of the experiments on ultracold atoms in optical lattices carried out since 1992, whilst Sect. 4 deals with the more recent experiments in which Bose–Einstein condensates within optical lattices have been explored. The experimental results obtained by the Pisa group on Bloch oscillations, Landau-Zener tunneling and optical potential renormalization are there reported. Section 5 concludes the presentation.

2 Basic Notions

2.1 Laser Cooling

The simplest possibility to create a periodic potential for neutral atoms is to exploit the light-shift experienced by the atoms in a spatially modulated light field. In one dimension, this can be achieved by superposing two linearly polarized, counter-propagating laser beams with parallel or perpendicular polarizations (see Fig. 1(a)).

If the polarizations of the two laser beams are perpendicular to each other (see Fig. 2), an atom with two magnetic sub-levels in the ground state will see two interleaved standing waves of σ^+ and σ^- circularly polarized light. This so-called *lin* $\perp$ *lin* configuration was typical of the early experiments on optical lattices, as it provided both localization of the atoms at the troughs of the potential wells and a sub-Doppler cooling mechanism ("Sisyphus-cooling", in which atoms are preferentially pumped from a sub-level with locally high potential energy to the other sub-level with a potential minimum at that point, thus reducing the kinetic energy of the atoms as shown in Fig. 3.). For this combination of effects to work, the laser beams creating the optical lattice were detuned by a few natural linewidths from the atomic resonance (near-resonant optical lattices). In most configurations, the detuning was to the red side of the resonance, resulting in the atoms being trapped at the antinodes of the standing wave creating the lattice.

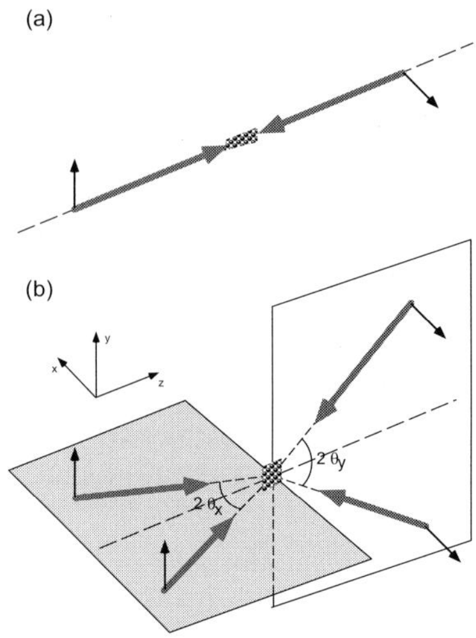

Fig. 1. Optical lattices in one and three dimensions. In both geometries, a frequency difference δ between the lattice beams can be used to create a moving or uniformly accelerated lattice.

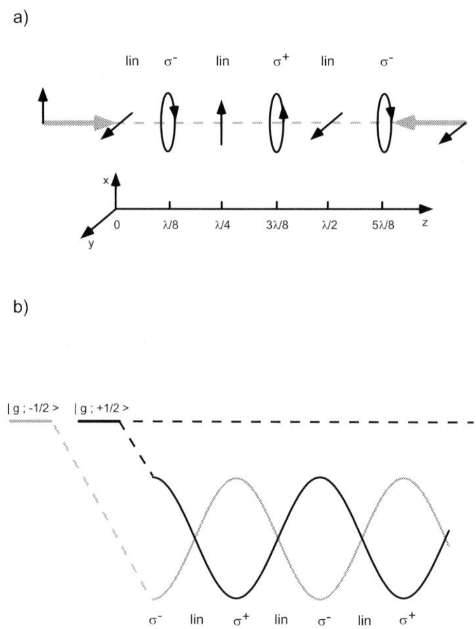

Fig. 2. A one-dimensional optical lattice in the $lin \perp lin$ configuration.

The 1D periodic structure created by two-counter-propagating laser beams can be generalized to two and three dimensions in different ways. Theodor Hänsch and his group in Munich used orthogonal pairs of beams whose relative phases were stabilized to create a constant lattice geometry [9]. A simpler

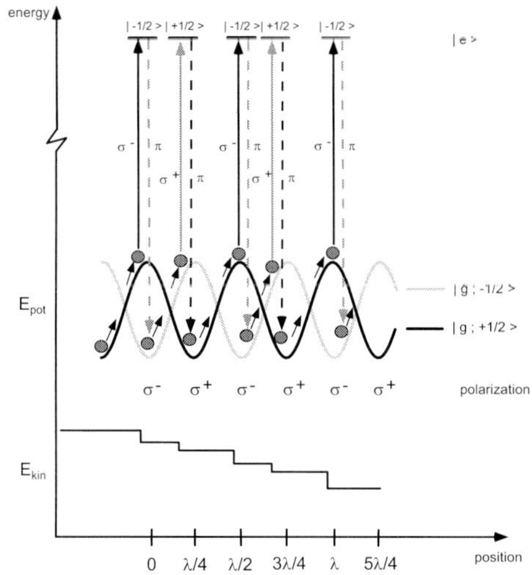

Fig. 3. The principle of Sisyphus cooling in a 1D optical lattice.

approach was pioneered by Gilbert Grynberg's team at the ENS in Paris [10]. Figure 1(b) shows their beam geometry used to create a 3D lattice. In this setup, no active phase-control is necessary as a variation in phase of the beams will only lead to a spatial translation of the lattice without changing its intrinsic geometry. In 2D and 3D lattices, the arrangement of lattice wells can either be anti-ferromagnetic, with adjacent wells having orthogonal circular polarizations (as in the 1D example shown above), or ferromagnetic, in which case adjacent wells have the same circular polarization.

By changing the splitting angle θ between the two pairs of beams, the distance between neighbouring lattice wells in the ENS setup could be varied. This possibility has later also been used for experiments on BECs. In a 1D optical lattice with angle θ, the distance between neighboring wells (lattice constant) d can be varied through the angle θ between between the two laser beams creating a lattice with

$$d = \frac{\pi \sin(\theta/2)}{k_L}, \tag{1}$$

where k_L is the laser wavenumber.

2.2 Conservative Potential

After the exploration of the properties of near-resonant lattices, the research effort was concentrated on far-detuned lattices with detunings ranging from hundreds to thousands of linewidths, using both ultra-cold atoms and BECs. In these optical lattices the dissipative cooling mechanisms are not active and the optical lattice is described by a conservative potential. For a 1D lattice configuration as in Fig. 1 with the two counterpropagating laser beams having the same polarization, the ac-Stark shift created by an off-resonant interaction between the electric field of the laser and the atomic dipole results a potential of the form

$$U(x) = U_0 \sin^2(\pi x/d), \tag{2}$$

where U_0 is the depth of the potential and d the lattice constant. In a lattice configuration in which the two laser beams with wavevector k_L are counter-propagating, the usual choice of units are the recoil momentum $p_{rec} = \hbar k_L = M v_{rec}$ and the recoil energy $E_{rec} = \hbar^2 k_L^2/2M$. In the case of an angle-geometry, it is more intuitive to base the natural units on the lattice spacing d and the projection $k = \pi/d$ of the laser wavevector k_L onto the lattice direction. However, the spatially periodic external potential leads naturally to a solid state physics approach. One can then define a Bloch momentum

$$p_B = \frac{2\hbar\pi}{d} = M v_B, \tag{3}$$

corresponding to the full extent of the first Brillouin zone [11] or, alternatively, to the net momentum exchange in the lattice direction between the atoms and the two laser beams. In that frame of reference a possible choice for the energy unit is the Bloch energy defined as $E_{\mathrm{B}} = \hbar^2(2\pi)^2/Md^2$. These units can also be used for the case of the angle-geometry, making use of (1) for the connection between d and θ.

By introducing a frequency difference δ between the two beams, the lattice potential of (2) can be moved at a constant velocity v_{lat} given by

$$v_{lat} = d\delta \tag{4}$$

or accelerated with an acceleration a given by

$$a = d\frac{d\delta}{dt}. \tag{5}$$

3 Ultracold Atoms

3.1 Early Experiments: Local Properties

The first experiments on optical lattices were aimed at local properties of the atoms trapped in the wells. Making a harmonic approximation to the potential at the centre of each well, it is straightforward to calculate the harmonic trapping frequency for a one-dimensional lattice as

$$\omega_{harm} = 2\frac{E_{rec}}{\hbar}\sqrt{\frac{U_0}{E_{rec}}}. \tag{6}$$

In a typical optical lattice experiment, atoms were first trapped and laser-cooled in a magneto-optical trap (MOT). After further cooling using optical molasses (essentially by switching off the magnetic fields of the MOT and increasing the detuning of the trap beams), the lattice laser beams were switched

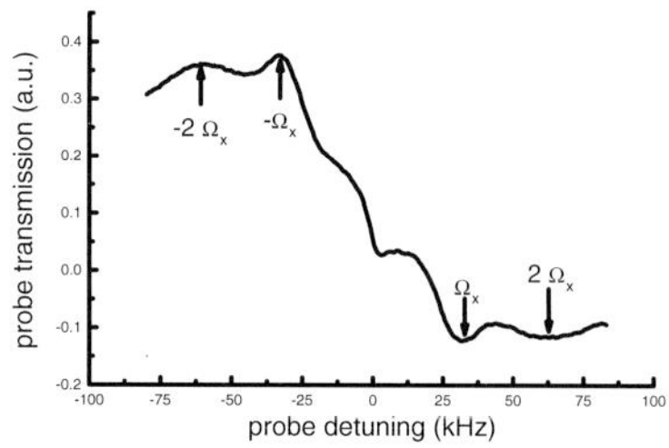

Fig. 4. Raman resonances in pump-probe spectroscopy along the x axis.

on. In order to demonstrate that the atoms were truly localized in the potential wells of the lattice, a series of experiments was carried out using pump-probe techniques [12]. By propagating a probe beam through the optical lattice that had a variable detuning from the lattice beams, Raman resonances could be identified (see Fig. 4). These corresponded to an atom absorbing a photon from a lattice beam and emitting another into the probe beam (or vice versa) whilst changing its vibrational quantum number by unity. The measured positions of the Raman frequencies agreed with the calculated vibrational frequencies. These experiments were helped by the fact that due to Lamb-Dicke suppression of inelastic scattering, the Raman resonances were narrow enough to be individually resolvable.

Other experiments investigating the motion of the atoms in the lattice wells included the observation of breathing oscillations of the atomic wavepackets when then the well-depth was suddenly changed [13] as well as the collapse and revival of coherent wavepacket oscillations induced by rapidly shifting the lattice in space [14]. In both experiments, the wavepacket motion was inferred from the fluorescence light as the re-distribution of photons between the lattice beams due to the motion of the atoms led to intensity fluctuations that could be experimentally detected. Stimulated revivals were observed in 2000 by the lattice group in Hannover [15].

Information about coherence times of the atomic motion in optical lattices can also be obtained by creating coherent transients [16]. In this method, coherences between vibrational states are first created by a short probe pulse. Subsequently, during the lifetime of the coherence lattice photons will be preferentially scattered in the direction of the probe beam (now switched off). From the spectrum of these photons, vibrational frequencies and coherence times can be extracted [17].

3.2 Global Properties

The nick-name "crystals bound by light" that was invented for optical lattices soon after their first experimental realization highlighted a property of these physical systems that was not visible in the early experiments, namely their periodic spatial structure. If optical lattices were, as predicted, "egg cartons" for atoms, then some direct evidence to that effect was desirable. An obvious

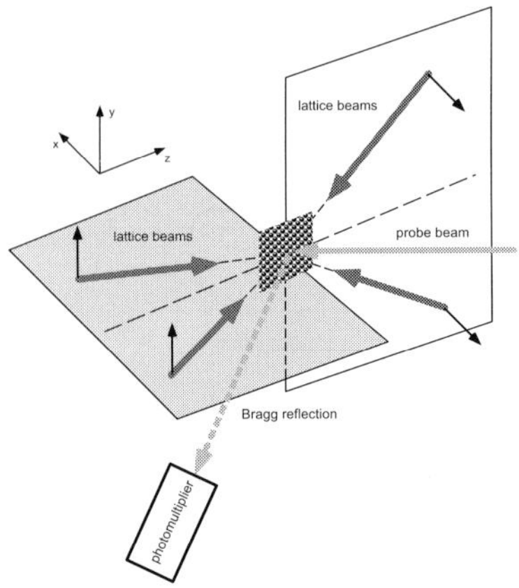

Fig. 5. Bragg scattering experiment on an optical lattice. From the Bragg-reflected probe beam, information about the spatial order and the localization of the atoms can be obtained.

experiment to carry out was Bragg-scattering of a probe beam off the planes of the "crystal", and in 1995 the groups in Munich and Gaithersburg [18,19] managed to do just that (see Fig. 5). They were able to show that when the lattice beams were switched on, long-range spatial order was built up as the atoms were further cooled and localized in the periodically arranged potential wells of the lattice. When the lattice was switched off, the spatial order was lost on a time-scale consistent with the thermal motion of the atoms. By measuring the structure factor associated with the Bragg reflection, it was also possible to obtain information about the spatial localization of the atoms in the lattice wells [20].

Another global property of near-resonant optical lattices that was accessible to experimental verification was their response to external magnetic fields. Using an optical lattice with anti-ferromagnetic order of the atomic spins, the ENS group showed that by applying a magnetic field that differentially shifted the potential energies of the σ^+ and σ^- wells, a redistribution of population between the two sets of wells resulted in a macroscopic magnetization of the lattice [21]. The entire lattice, therefore, exhibited paramagnetic behaviour.

In near-resonant optical lattices, only a few percent of all the lattice sites actually contain an atom. Despite the consequent lack of atom-atom interactions, it was shown in 1996 that propagating excitations similar to sound waves can be created in optical lattices [22]. Through cycles of half-oscillations in the potential wells and subsequent optical pumping into the other magnetic sub-level of the ground state, propagating density modulations were created that could be detected via pump-probe spectroscopy. Moreover the first direct observation of Brillouin-like propagation modes in a dissipative periodic optical lattice was recently reported [23].

3.3 Quasi-periodic Lattices

Whilst so far we have stressed the regular, periodic structure of optical lattices, a number of interesting experiments have also been carried out in a quasi-periodic geometry. By adding an additional laser beam to the four-beam geometry described above, a quasi-periodic lattice structure can be created. The diffusion properties in such a lattice were studied by the group of Philippe Verkerk [24].

3.4 Far-Detuned Lattices

The fact that in a red-detuned, near resonant lattice the atoms are trapped at local maxima of the light intensity means that the coherent motion of the atomic wavepackets is frequently interrupted by spontaneous emission events. Early on in the development of optical lattices, schemes were developed that would provide trapping in locally "gray" or "dark" states in which the atoms absorb very few photons or no photons at all. These schemes were, appropriately, called "gray" or "dark" optical lattices [25].

In order to get rid completely of dissipative effects, it was, however, necessary to increase the detuning of the lattice beams from the atomic resonance and, at the same time, the beam intensity in order to keep a fixed lattice depth. As the spontaneous scattering rate decreases more rapidly with detuning than the light-shift or dipole force providing localization, in this manner optical lattices could be realized in which for the duration of the experiment (usually on the order of tens to hundreds of milliseconds) virtually no photons were scattered. The lack of photon scattering, on the other hand, also ruled out the Sisyphus cooling mechanism present in near-resonant lattices. In order further to cool atoms transferred into a far-resonant lattice, resolved-sideband Raman cooling was successfully employed by several groups [26–28]. As in trapped ion experiments, transitions towards lower vibrational levels were induced using a combination of laser beams and magnetic fields. In this way, a large fraction of the atoms could be cooled to the ground vibrational state of the lattice.

Another possibility to achieve the loading of atoms into the ground vibrational state (or, for shallow lattices, the ground state Bloch band) is to start from a Bose-condensed cloud of atoms, as will be discussed in the second part of this chapter.

3.5 A Lego-Kit for Quantum Systems

Apart from the intrinsic interest in the characteristics of optical lattices, a number of experiments have used optical lattices as a tool for creating a specific quantum system whose properties could then be studied. Especially the precise experimental control of the lattice parameters (well depth, geometry) means that optical lattices can be used as a construction kit for quantum systems.

In a beautiful experiment [33,34], the group of Christophe Salomon at the ENS in Paris demonstrated Bloch oscillations of atomic matter waves delocalized over many lattice sites of a shallow optical lattice by accelerating the lattice and measuring the resulting velocity of the atoms. Related experiments by Mark Raizen's team in Austin investigated Landau-Zener tunneling and Wannier-Stark ladders [29]. In an extension to this work, they were able to experimentally verify the non-exponential nature of the initial decay of a quantum system [30] and the associated Zeno and anti-Zeno effects exhibited by such systems when they are subject to frequent observations [31].

In a near-resonant lattice, Gilbert Grynberg's group at the ENS was able to create an asymmetric optical lattice with a ratchet-like potential that converted the random thermal motion of the atoms into directed motion [32].

4 Bose-Condensed Atoms

While in most of the original optical lattice experiments the atomic clouds had temperatures in the the micro-Kelvin range, corresponding to a few recoil energies of the atoms, atomic samples with sub-recoil energies are now routinely produced in Bose–Einstein condensation experiments. Since the first experimental realizations in 1995, many aspects of Bose–Einstein condensed atomic clouds (BECs) have been studied [39], ranging from collective excitations to superfluid properties and quantized vortices. The properties of BECs in periodic potentials constitute a vast new field of research initially explored in [40,41]. Several experiments have made use of the periodic optical potential produced by a pulsed standing wave to manipulate the condensate or to explore its properties [42–49]. In the following, we will concentrate on studies of the condensate within the periodic optical lattice. The first step in that direction was taken by the investigation of the tunneling of BECs out of the one-dimensional potential wells of a shallow optical lattice in the presence of gravity [50]. More recently the phase properties of the condensate wavefunction occupying the whole optical lattice have involved such intriguing concepts as number squeezing [51] and the Mott insulator transition [52]. The tunneling of the condensate between neighboring wells, controlled by varying the optical lattice potential depth, determine the overall properties of the macroscopic wavefunction. Thus the condensate response within a 1D optical lattice can be described as an array of tunneling junctions, as pointed out in [50] and later explored in [53,54] in connection with the superfluid properties of the condensate wavefunction. Coherent acceleration of BECs adiabatically loaded into optical lattices was demonstrated in [55,56], with Bloch oscillations observed for small values of the lattice depth, while a Landau-Zener breakdown occurred when the lattice depth was further reduced and/or the acceleration increased. The expansion of the condensate array was explored initially in [57] and later in [58]. The high level coherent control over this artificial solid state system was demonstrated in [59], where the BEC was carefully loaded into the lattice ground state by adiabatically turning on the optical lattice. The different dependence of the condensate population on the temperature for the 1D optical lattice was pointed out by [60].

Most experiments were based on the production of the condensate through the standard technique, followed by an adiabatic load into the optical lattice.

Often the magnetic trap was switched off when the optical lattice was switched on. However, a larger condensate density is realized when the interaction between the condensate and the lattice takes place inside the magnetic trap, and both of them are subsequently switched off to allow time-of-flight imaging. The Florence group [53] has developed a different approach by producing the condensate directly inside the optical lattice which is adiabatically loaded during the evaporating cooling stage. The main advantage of this approach is that the phase coherence of the condensate over the whole optical lattice structure is built up during the condensate formation process. The main disadvantage is that the optical lattice is on during the whole evaporation process, and the spontaneous losses produced by the optical lattice should be heavily reduced by further increasing the detuning of the optical lattice lasers. While a large majority of the experiments have loaded the condensate into 1D optical lattices, experiments on 2D lattices were performed by Greiner et al. [61], and the Mott-Hubbard transition of [52] was realized in a 3D optical lattice.

If the momentum spread of the atoms loaded into an optical lattice structure is small compared to the characteristic lattice momentum p_{B}, then their thermal de Broglie wavelength will be large compared to the lattice spacing d and will, therefore, extend over many lattice sites. A description in terms of a coherent delocalized wavepacket within a periodic structure is then appropriate and leads us directly to the Bloch formalism first developed in condensed matter physics. In the tight-binding limit ($U_0 \gg 10\, E_{\mathrm{rec}}$), the condensate in the lattice can be approximated by wavepackets localized at the individual lattice sites (Wannier states). This description is more intuitive than the Bloch picture in the case of experiments in which the condensate is released from a (deep) optical lattice into which it has previously been loaded adiabatically.

4.1 BEC Theory in Optical Lattices

In Bose–Einstein condensates, interactions between the constituent atoms are responsible for the non-linear behavior of the BEC and can lead to interesting phenomena such as solitons [62] and four-wave mixing with matter waves [63]. As the atoms are extremely cold, collisions between them can be treated by considering only s-wave scattering, which is described by the scattering length a_s. Modeling the interatomic interaction as hard-core collisions, one can simplify the treatment using a mean-field description which leads to the famous Gross-Pitaevskii equation [64], the validity of which has been demonstrated in numerous experiments. For a BEC in an optical lattice, one expects an effect due to the mean-field interaction similar to the one responsible for determining the shape of a condensate in the Thomas-Fermi limit: The interplay between the confining potential and the density-dependent mean-field energy leads to a modified ground state that reflects the strength of the mean-field interaction. Applied to a BEC in a periodic potential, one expects the density modulation imposed on the condensate by the potential (higher density in potential troughs, lower density where the potential energy is high) to be modified in the presence of mean-field interactions. In particular, the tendency of the periodic potential to create a locally higher density where the potential energy of the lattice is low will be counteracted by the (repulsive) interaction energy that rises as the local density increases.

The description of a condensate in a 1D array of coupled potentials wells is based on the total Hamiltonian

$$H_{tot} = \frac{-\hbar^2}{2M}\nabla^2 + U_0 \sin^2(\pi\frac{x}{d}) + g|\Psi(\boldsymbol{r})|^2, \tag{7}$$

with the interaction parameter g given by

$$g = \frac{4\pi\hbar^2 a_s}{M}, \tag{8}$$

and $\Psi(\boldsymbol{r})$ the condensate wavefunction at position $\boldsymbol{r}$.

As the interaction term is expected to distort the band structure of the condensate in the lattice [40], it should affect all measurable quantities (Rabi frequency, amplitude of Bloch oscillations, and tunneling probability). In [65], the authors derived an analytical expression in the perturbative limit (assuming $U_0 \ll E_B$) for the effect of the mean-field interaction on the ground state of the condensate in the lattice. Starting from the Gross-Pitaevskii equation for the condensate wavefunction in a one-dimensional optical lattice (*i.e.* a one dimensional Hamiltonian equivalent to that of (7)), they found that the effect of the mean-field interaction could be approximately accounted for by substituting the potential depth U_0 with an effective potential

$$U_{eff} = \frac{U_0}{1+4C}, \tag{9}$$

with the dimensionless parameter C given by [65]

$$C = \frac{\pi n_0 a_s}{k_L^2 \sin^2(\theta/2)} = \frac{n_0 g}{E_{\mathrm{B}}}, \tag{10}$$

corresponding to the ratio of the nonlinear interaction term and the Bloch energy. The C parameter contains the peak condensate density n_0, the scattering length a_s, and the atomic mass M. From the dependence of C on the lattice angle θ it follows that a small angle θ (meaning a large lattice constant d) should result in a large interaction term C. The reduction of the effective potential given by (9) agrees with the intuitive picture of the back-action on the periodic potential of the density modulation of the condensate imposed on it by the lattice potential. For repulsive interactions, this results in the effective potential being lowered with respect to the actual optical potential created by the lattice beams.

4.2 Theoretical Advances

The properties of Bose–Einstein condensates within optical lattices have been examined in a large number of theoretical papers, predicting a variety of phenomena, often making use of the strict analogies with cases previously studied within the context of solid state physics and nonlinear dynamics. We will list here the principal research lines that have characterized this research so far.

The collective excitations of the condensate within optical lattices, and their probe, have been determined [66–68], and those analyses have stimulated the experimental investigations in [47–49]. A strong deformation of the Bloch energy bands of the condensate produced by the nonlinear atomic interactions has been predicted in [69–71]. Different mechanisms of breaking down the Bloch oscillation, all of them connected to the interatomic interactions, have been discussed in [72–75].

The thermal and quantum decoherence for an array of multiple condensates within an optical lattice, introduced in [76,77], are an important issue requiring more detailed studies, both theoretically and experimentally. The Bose-Hubbard-Hamiltonian for atoms in an optical lattice, introduced by Jaksch et al. [41], was analyzed in [78] through a mean-field approximation generalization of the Bogoliubov approach; later it was applied to determine the conditions for the number squeezing in that transition [79].

A number of papers have pointed out the existence of additional solutions for the evolution of the condensates within optical lattices: instabilities, solitons (shape preserving excitations), breathers (excitations characterized by internal oscillations), and self-trapping states or intrinsic localized modes (wavepacket localized around few lattice sites) [74,80–82]. Those predictions have often made use of theoretical analogies with other nonlinear classical and quantum problems, involving the sine-Gordon equation, the discrete nonlinear Schrödinger equation and other nonlinear physics problems. Spatial instabilities of the condensate within the optical lattice, with a spontaneous breaking in the spatial periodicity, have been predicted by Wu and Niu [75].

4.3 Experimental Results

Bloch oscillations. For 1D optical lattice, a linear increase of the detuning δ between the two laser beams forming the lattice provides to the optical lattice the constant acceleration given by (5). As the lattice can only transfer momentum to the condensate in units of $2\hbar k$, the acceleration of the condensate leads to higher momentum classes as the acceleration time increases. With an initial momentum spread of the condensate much less than the Bloch momentum $p_{\rm B}$) and since the adiabatic switching transfers the momentum spread into lattice quasimomentum, the different momentum classes $p = \pm n p_{\rm B}$ (where $n = 0, 1, 2, ...$) occupied by the condensate wavefunction can be resolved directly after the time-of-flight. In the experiments of [55,56] with the optical lattice in the horizontal direction and the atomic wave diffraction monitored in the time-of-flight detection, the accelerated momentum classes showed up as diffraction peaks in the time-of flight absorption images as in Fig. 6. Up to $6\,p_{\rm B}$ momentum could be transferred to the condensate preserving the phase-space density of the condensate during the acceleration, a result indicating that no heating or reduction of the condensate fraction occurred. Measuring the average velocity of the condensate from the occupations

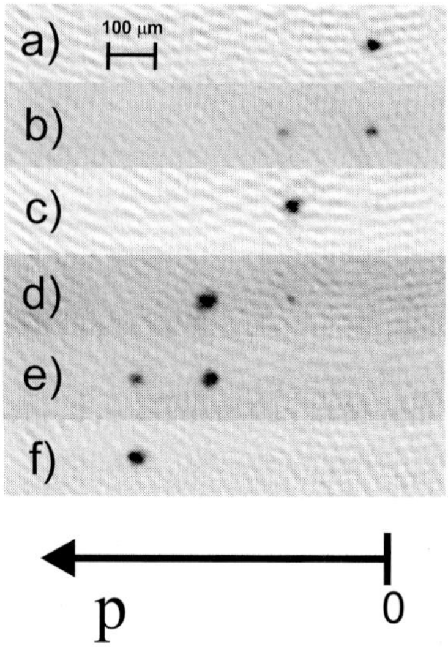

Fig. 6. Coherent acceleration of a Bose–Einstein condensate. In (a)-(f) $U_0 = 2.3\,\mathrm{E_{rec}}$ $a = 9.81\,\mathrm{m\,s^{-2}}$, the condensate being accelerated for $0.1, 0.6, 1.1, 2.1, 3.0$ and $3.9\,\mathrm{ms}$, respectively. The separations between the different spots vary because the detection occurred after different time delays.

of the different momentum states, the Bloch oscillations of the condensate velocity corresponding to a Bloch-period $\tau_\mathrm{B} = h/(Ma_s d)$ could be detected. Note in Fig. 6 that while the Bloch oscillation takes place, the condensate wavefunction coherently occupies two neighboring velocity classes. In a configuration with the optical lattice oriented along the vertical direction [50,56], the different momentum classes emitted from the condensate travel in space separately because of the acceleration due to gravity. Thus the condensate absorption images after the time of flight show one or two atom laser pulses corresponding to the single or double momentum class occupied by the condensate at the time of the release.

Landau-Zener tunneling. At large acceleration of the lattice or at a decreased lattice depth, not all of the condensate could be coherently accelerated up to the final velocity of the lattice. Such condensate loss can be interpreted in terms of Landau-Zener tunneling of the condensate out of the lowest band when the edge of the Brillouin zone is reached. Each time the condensate is accelerated across this edge, the fraction undergoing tunneling into the first excited band is given by the Landau-Zener probability P_LZ [33,55,59]:

$$P_\mathrm{LZ} = e^{-a_c/a} \tag{11}$$

with the critical acceleration a_c given by

$$a_c = \frac{\pi U_0^2}{16\hbar^2 k}. \tag{12}$$

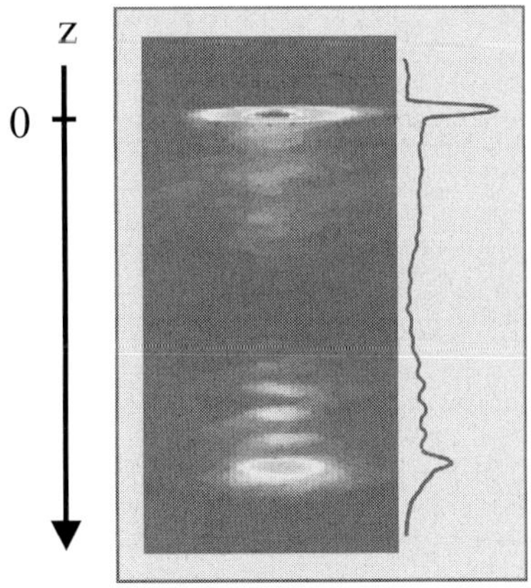

Fig. 7. Time-of-flight image and, on the left, transverse vertical profile for the Landau-Zener tunnelling of a condensate from a 1-D optical lattice oriented vertically. The $p = +|n|p_{\mathrm{B}}$ atomic momentum classes are separated by the gravity, pointing downwards. The large condensate amplitude near $z = 0$ corresponds to the population confined in the lattice. The atomic momentum classes generate the atom laser pulses detected after a 10.1 ms time of flight, with $U_0 = 10\,\mathrm{E_{rec}}$ and condensate acceleration for 10 ms.

Such tunneling produces the following mean velocity v_m of the condensate at the end of the acceleration process for a final velocity v_{B} of the lattice:

$$v_m = (1 - P_{\mathrm{LZ}})v_{\mathrm{B}}. \tag{13}$$

In [55] it was verified that this formula correctly described the tunneling of the condensate at low values of the condensate density by varying both the potential depth and the lattice acceleration. In the configuration with the optical lattice oriented along the vertical direction [50,56], the different momentum classes emitted from the condensate travel in space separately because of the acceleration due to gravity. Thus the condensate absorption images after the time of flight (Fig. 7) showed a part of the population confined in the optical lattice, and several laser pulses corresponding to the different momentum class occupied by the condensate.

Optical potential renormalization. In order to measure accurately the variation of the effective potential U_{eff} with the interaction parameter C as given by (9), the Landau-Zener tunneling out of the lowest Bloch band for small lattice depth was studied in a regime where the parameter C modified the optical potential experienced by the condensate [55,56]. Therefore, the variation of the final mean velocity v_m was studied as a function of the condensate density. The density was varied by changing the mean harmonic frequency of the magnetic trap. From the mean velocity the effective potential was then calculated using the

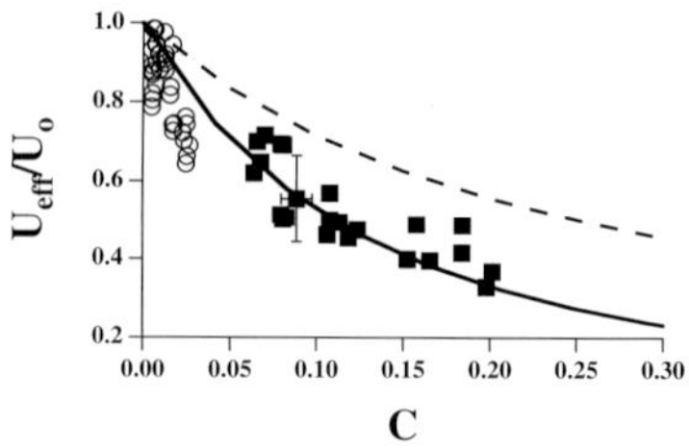

Fig. 8. Dependence on the C parameter for the effective potential U_{eff}, normalized to the applied optical potential U_0, for the two lattice geometries of [55]. The experimental results for the counter-propagating geomotry (open circles) and angle geometry with $\theta = 29$ deg (filled squares) are plotted together with the theoretical prediction of [65](dashed line) and of [83](continuous line). Parameters in these experiments were $a = 23.4\,\text{m}\,\text{s}^{-2}$ and $U_0 = 2.2\,E_{\text{rec}}$ for the counter-propagating lattice and $a = 3.23\,\text{m}\,\text{s}^{-2}$ and $U_0 = 5.7\,E_{\text{rec}}$ for the angle geometry.

Landau-Zener probability given above, with the critical acceleration determined by the effective potential

$$a_c = \frac{\pi U_{\text{eff}}^2}{16\hbar^2 k}. \tag{14}$$

Figure 8 shows the ratio U_{eff}/U_0 as a function of the parameter C for the experimental investigation in two different geometries of the optical lattice, counter-propagating and in an angle-geometry with $\theta = 29$ deg. As predicted by (10), the reduction of the effective potential is much larger in the angle geometry. The theoretical predictions of (9) given by the theory of [65] are also shown in the figure. An effective potential may also be derived within the framework of a tight-binding approximation using Wannier states, i.e. describing the condensate within each potential minimum potential through the solution of the Gross-Pitaevskii equation while neglecting the overlap with neighboring potential minima [83], a very weak tunneling of the condensate preserving the overall phase relation of the condensate wavefunction. The tight-binding predictions presented in Fig. 8 provide a better agreement with experimental results than those of (9).

Squeezed states and Mott insulator. Because the condensate is a superfluid, described by a wavefunction exhibiting long-range phase coherence, if the lattice potential is turned on smoothly, the condensate remains in the superfluid phase. In this regime a delocalized condensate wavefunction minimizes the total energy of the many-body system, the phase of the atomic wavefunction being well determined, with a fluctuating atom number at each lattice site. This applies as long as the atom-atom interactions are small compared to the tunnel coupling. Instead, when the repulsive atom-atom interactions are large compared to the tunnel coupling, the total energy is minimized when each site of the optical lattice is filled with the same number of atoms. Approaching this Mott-isulator quantum phase transition, with the lattice site number commensurate to the atom number, the wavefunction phase coherence is lost. Meanwhile, the fluctuations in the number of atoms per site are reduced and finally go to zero. In addition, in the superfluid regime the excitation spectrum is gapless, whereas the Mott insulator phase exhibits a gap in the excitation spectrum.

The first important step in modifying a condensate from a coherent state to a Fock state was taken by Orzel et al. [51] who realized squeezed states in a 1D optical lattice. By properly choosing the depth of the optical potential and the amplitude of the mean field interaction term in (7), the authors were able to reach a regime where deviation from the coherent state wavefunction for the condensate became significant. An increase in the phase fluctuations of the condensate wavefunction was measured in the interference of atom waves released from the optical lattice. From this, a squeezing in the number of atoms occupying the lattice sites was derived by assuming a minimum uncertainty state.

The experimental realization of the Mott isulator in a 3D optical lattice in [52] required a cubic optical lattice where a volume with 65 sites in each single direction was occupied, with potential depth around $U_0 = 12\,E_{\rm rec}$. Adiabatic loading of a condensate of 2×10^5 atoms into such a lattice produced a state whose coherence properties were tested as usual by the interference pattern following a time-of-flight. At a potential depth larger than $10\,E_{\rm rec}$ the interference pattern presented an incoherent background, gaining more strength increasing the potential depth until that background was the only visible feature in the interference. That loss of interference was a first sign of the Mott-isulator regime, with a Fock state without macroscopic phase coherence characterizing the occupation of the single wells. Evidence of the phase transition was gained by monitoring the quick restoration of the phase coherence, quick as compared to the restoration produced by the inhomogeneous dephasing of the superfluid condensate wavefunction under the application of a magnetic energy gradient. Additional strong evidence was provided by the direct measurements of the gap in the excitation spectrum of the Mott insulator, in contrast to the superfluid phase where the excitation spectrum is gapless.

5 Conclusion and Outlook

We have provided a brief overview of the research work performed on cold atomic samples (laser cooled or evaporatively cooled to the quantum degenerate regime) located within the periodic potential created by an optical interaction, concentrating our attention on the phenomena where the long-range order of the external potential is imprinted onto the atomic wavefunction. As far as quantum degenerate gases within optical lattices are concerned, recently the field has greatly flourished with new experimental groups joining the crowded space and theory groups proposing new exciting schemes and pointing out the occurrence of interesting phenomena. It seems that Bose–Einstein condensates in optical lattices could be the test-bed for a large variety of theoretical models. It is obvious that this flourishing of ideas and of measurements will continue still for some time, because additional configurations could be explored, the control on the atoms and the optical lattices will progress, and finally because other phenomena may occur when Fermi degenerate gases will be loaded in optical lattices. The detailed comparison between theory and experiments dealing with Bose–Einstein condensates within optical lattices requires heavy numerical solutions of the 3D Gross-Pitaevskii equation. However, the recent realization of nearly 1D Bose–Einstein condensates could simplify the theoretical analysis.

But even with non-condensed, ultra-cold gases the versatility of optical lattices in realizing and studying quantum systems will certainly provide interesting challenges to physicists in the future. Studies of cold collisions and diffusion properties in optical lattices are but two of the areas in which a lot of work still needs to be done. Moreover, experiments will test the schemes in which the coherent motion of the atoms in far-detuned optical lattices could be exploited to realize quantum logic gates.

Acknowledgments

The collaboration of D. Ciampini, M. Cristiani and J.H. Müller to the experimental results reported in the present work is gratefully acknowledged. This work was supported by the MURST through the PRIN2000 Initiative, by the INFM through the Progetto di Ricerca Avanzata 'Photon matter', and by the the EU through the Cold Quantum Gases Network, Contract No. HPRN-CT-2000-00125. O.M. gratefully acknowledges a Marie Curie Fellowship from the EU within the IHP Programme.

References

1. P.S. Jessen and I.H. Deutsch, Adv. At. Mol. Opt. Phys. **37**, 95 (1996).
2. G. Grynberg and C. Triché, in *Coherent and Collective Interactions of Particles and Radiation Beams* edited by A. Aspect, W. Barletta, and R. Bonifacio (IOS Press, Amsterdam) (1996).
3. D.R. Meacher, Cont. Phys. **39**, 329 (1998).
4. G. Grynberg and C. Robilliard, Phys. Rep. **355**, 335 (2001).
5. D.A. Steck, W.H. Oskay and M.G. Raizen, Science **293**, 274 (2001) and references therein.
6. W.K. Hensinger, H. Häffner, A. Browaeys, N.R. Heckenberg, K. Helmerson, C. McKenzie, G.J. Milburn, W.D. Phillips, S.L. Rolston, H. Rubinstein-Dunlop, and B. Upcroft, Nature **412**, 52 (2001) and references therein.
7. See, e.g., I.H. Deutsch and P.S. Jessen, Phys. Rev. A **57**, 1972 (1998), and D. Jaksch, H.-J. Briegel, J.I. Cirac, C.W. Gardiner, and P. Zoller, Phys. Rev. Lett. **82**, 1975 (1998).
8. G. Kurizki, S. Giovanazzi, D. O'Dell, A. I. Artemiev*New Regimes in Cold Gases Via Laser-Induced Long-Range Interactions*, in "Dynamics and Thermodynamics of Systems with Long-Range Interactions", T. Dauxois, S. Ruffo, E. Arimondo, M. Wilkens Eds., Lecture Notes in Physics Vol. 602, Springer (2002), (in this volume)
9. A. Hemmerich and T.W. Hänsch, Phys. Rev. Lett. **70**, 410 (1993).

10. G. Grynberg, B. Lounis, P. Verkerk, J.Y. Courtois, and C. Salomon, Phys. Rev. Lett. **70**, 2249 (1993).
11. see, e.g., C. Kittel, *Introduction to Solid State Physics* (Wiley, New York, 1996).
12. P. Verkerk, B. Lounis, C. Salomon, C. Cohen-Tannoudji, J.Y. Courtois, and G. Grynberg, Phys. Rev. Lett. **68**, 3861 (1992).
13. P. Rudy, R. Ejnisman, and N.P. Bigelow, Phys. Rev. Lett. **78**, 4906 (1997).
14. G. Raithel, W.D. Phillips, and S.L. Rolston, Phys. Rev. Lett. **81**, 3615 (1998).
15. F.B.J. Buchkremer, R. Dumke, H. Levsen, G. Birkl, and W. Ertmer, Phys. Rev. Lett. **85**, 3121 (2000).
16. C. Triché, L. Guidoni, P. Verkerk, and G. Grynberg, in *OSA TOPS on ultra-cold atoms and BEC*, edited by K. Burnett, vol. 7, p. 82,(OSA, Washington D.C., 1996).
17. O. Morsch, P.H. Jones, and D.R. Meacher, Phys. Rev. A **61**, 023410 (2000).
18. M. Weidemüller, A. Hemmerich, A. Görlitz, T. Esslinger, and T.W. Hänsch, Phys. Rev. Lett. **75**, 4583 (1995).
19. G. Birkl, M. Gatzke, I.H. Deutsch, S.L. Rolston, and W.D. Phillips, Phys. Rev. Lett. **75**, 2823 (1995).
20. A. Görlitz, M. Weidemüller, T.W. Hänsch, and A. Hemmerich, Phys. Rev. Lett. **78**, 2096 (1997).
21. D.R. Meacher, S. Guibal, C. Mennerat, J.Y. Courtois, K.I. Petsas, and G. Grynberg, Phys. Rev. Lett. **74**, 1958 (1995).
22. J.Y. Courtois, S. Guibal, D.R. Meacher, P. Verkerk, and G. Grynberg, Phys. Rev. Lett. **77**, 40 (1996).
23. L. Sanchez-Palencia, F.-R. Carminati, M. Schiavoni, F. Renzoni, and G. Grynberg , Phys. Rev. Lett. **88**, 133903 (2002).
24. L. Guidoni, B. Dépret, A. di Stefano, and P. Verkerk, Phys. Rev. A **60**, R4233 (1999).
25. A. Hemmerich, M. Weidemüller, T. Esslinger, C. Zimmermann, and T.W. Hänsch, Phys. Rev. Lett. **75**, 37 (1995).
26. S.E. Hamann, D.L. Haycock, G. Klose, P.H. Pax, I.H. Deutsch, and P.S. Jessen, Phys. Rev. Lett. **80**, 4149 (1998).
27. H. Perrin, A. Kuhn, I. Bouchoule, and C. Salomon, Europhys. Lett. **42**, 395 (1998).
28. V. Vuletic, C. Chin, A.J. Kerman, and S. Chu, Phys. Rev. Lett. **81**, 5768 (1998).
29. K.W. Madison, M.C. Fischer, and M.G. Raizen, Phys. Rev. A **60**, R1767 (1999).
30. S.R. Wilkinson, C.F. Bharucha, M.C. Fischer, K.W. Madison, P.R. Morrow, Qian Niu, B. Sundaram, and M.G. Raizen, Nature **387**, 575 (1997).
31. M.C. Fischer, B. Guttierez-Medina, and M.G. Raizen, Phys. Rev. Lett. **87**, 040402 (2001).
32. C. Mennerat-Robilliard, D. Lucas, S. Guibal, J. Tabosa, C. Jurczak, J.Y. Courtois, and G. Grynberg, Phys. Rev. Lett **82**, 851 (1999).
33. M. Ben-Dahan, E. Peik, J. Reichel, Y. Castin, and C. Salomon, Phys. Rev. Lett **76**, 4508 (1996);
34. E. Peik, M. Ben Dahan, I. Bouchoule, Y. Castin, and C. Salomon, Phys. Rev. A **55**, 2989 (1997).
35. S.R. Wilkinson, C.F. Bharucha, K.W. Madison, Qian Niu, and M.G. Raizen, Phys. Rev. Lett. **76**, 4512 (1996).
36. Qian Niu, Xian Gong Zhao, G.A. Georgakis, and M.G. Raizen, Phys. Rev. Lett. **76**, 4504 (1996).
37. M. Raizen,C. Salomon, and Qian Niu, Phys. Today **50** (7), 30 (1997).
38. C.F. Bharucha, K.W. Madison, P.R. Morrow, S.R. Wilkinson, B. Sundaram, and M.G. Raizen, Phys. Rev. A **55**, R857 (1997).

39. See reviews by W. Ketterle, D.S. Durfee and D.M. Stamper-Kurn, and by E. Cornell et al. in *Bose–Einstein condensation in atomic gases*, edited by M. Inguscio, S. Stringari and C. Wieman (IOS Press, Amsterdam) (1999).
40. K. Berg-Sørensen and K. Mølmer, Phys. Rev. A **58**, 1480 (1998).
41. D. Jaksch, C. Bruder, J.I. Cirac, C.W. Gardiner, and P. Zoller, Phys. Rev. Lett. **81**, 3108 (1998).
42. M. Kozuma, L. Deng, E.W. Hagley, J. Wen, R. Lutwak, K. Helmerson, S.L. Rolston, and W.D. Phillips, Phys. Rev. Lett. **82**, 871 (1999).
43. Yu.B. Ovchinnikov, J.H. Müller, M.R. Doery, E.J.D. Vredenbregt, K. Helmerson, S.L. Rolston, and W.D. Phillips, Phys. Rev. Lett. **83**, 284 (1999).
44. E.W. Hagley, L. Deng, M. Kozuma, M. Trippenbach, Y. B. Band, M. Edwards, M. Doery, P. S. Julienne, K. Helmerson, S. L. Rolston, and W. D. Phillips, Phys. Rev. Lett. **83**, 3112 (1999).
45. J. Stenger, S. Inouye, A. P. Chikkatur, D. M. Stamper-Kurn, D. E. Pritchard, and W. Ketterle, Phys. Rev. Lett. **82**, 4569 (1999); and erratum: Phys. Rev. Lett. **84**, 2283 (2000).
46. J.E. Simsarian, J. Denschlag, M. Edwards, C.W. Clark, L. Deng, E.W. Hagley, K. Helmerson, S.L. Rolston, and W.D. Phillips, Phys. Rev. Lett. **85**, 2040 (2000).
47. J. M. Vogels, K. Xu, C. Raman, J. R. Abo-Shaeer, and W. Ketterle, Phys. Rev. Lett. 88, 060402 (2002)
48. J. Steinhauer, R. Ozeri, N. Katz, and N. Davidson, Phys. Rev. Lett. 88, 120407 (2002).
49. R. Ozeri, J. Steinhauer, N. Katz, and N. Davidson, Phys. Rev. Lett. 88, 220401 (2002).
50. B.P. Anderson and M. Kasevich, Science **282**, 1686 (1998).
51. C. Orzel, A.K. Tuchman, M.L. Fenselau, M. Yasuda, and M.A. Kasevich, Science **231**, 2386 (2001).
52. M. Greiner, O. Mandel, T. Esslinger, T.W. Hänsch, and I. Bloch, Nature **415**, 6867 (2002).
53. S. Burger, F. Cataliotti, C. Fort, F. Minardi, M. Inguscio, M.L. Chiofalo, and M.P. Tosi, Phys. Rev. Lett. **86**, 4447 (2001).
54. F.S. Cataliotti, S. Burger, C. Fort, P. Maddaloni, F. Minardi, A. Trombettoni, A. Smerzi, and M. Inguscio, Science **293**, 843 (2001).
55. O. Morsch, J.H. Müller, M. Cristiani, D. Ciampini, and E. Arimondo, Phys. Rev. Lett. **87**, 140402 (2001).
56. M. Cristiani, O. Morsch, J. H. MTller, D. Ciampini, and E. Arimondo, Phys. Rev. A **65**, 063612 (2002).
57. P. Pedri, L. Pitaevskii, S. Stringari, C. Fort, S. Burger, F.S. Cataliotti, P. Maddaloni, F. Minardi, and M. Inguscio, Phys. Rev. Lett. **87**, 220401 (2001).
58. O. Morsch, M. Cristiani, J.H. Müller, D. Ciampini, and E. Arimondo,cond-mat 0204528.
59. J. Hecker-Denschlag, J. E. Simsarian, H. Häffner, C. McKenzie, A. Browaeys, D. Cho, K. Helmerson, S. L. Rolston, and W. D. Phillips, J. Phys. B: At. Mol. Opt. Phys in press (2002).
60. S. Burger, F. S. Cataliotti, C. Fort, P. Maddaloni, F. Minardi and M. Inguscio, Europhys. Lett. **57** 1 (2002).
61. M. Greiner, I. Bloch, O. Mandel, T.W. Hänsch, and T. Esslinger, Phys. Rev. Lett. **87**, 160405 (2001).
62. S. Burger, K. Bongs, S. Dettmer, W. Ertmer, K. Sengstock, A. Sanpera, G.V. Shlyapnikov, and M. Lewenstein, Phys. Rev. Lett. **83**, 5198 (1999).

63. L. Deng, E.W. Hagley, J. Wen, M. Trippenbach, Y. Band, P.S. Julienne, J.E. Simsarian, K. Helmerson, S.L. Rolston, and W.D. Phillips, Nature **398**, 6724 (1999).
64. F. Dalfovo, S. Giorgini, L.P. Pitaevskii, and S. Stringari, Rev. Mod. Phys. **71**, 463 (1999).
65. D.I. Choi and Qian Niu, Phys. Rev. Lett. **82**, 2022 (1999).
66. A. Brunello, F. Dalfovo, L. Pitaevskii, and S. Stringari, Phys. Rev. Lett. **85**, 4422 (2000).
67. M.L. Chiofalo, S. Succi, and M.P. Tosi, Phys. Rev. A **63**, 063613 (2001).
68. M.Krämer, L. Pitaevskii, and S. Stringari, Phys. Rev. Lett. **88**, 180404 (2002).
69. J.C. Bronski, L.D. Carr, B. Deconink, and J.N. Kutz, Phys. Rev. lett. **86**, 1402 (2001).
70. B. Wu, R.B. Diener, and Qian Niu, Phys. Rev. A **65**, 025601 (2002).
71. D.Diakonov, L.M. Jensen, C.J. Pethick, and H. Smith, e-print: cond-mat/0111303.
72. B. Wu and Qian Niu, Phys. Rev. A **61**, 023401 (2000).
73. O.Zobay and B.M. Garraway, Phys. Rev. **61** 033603 (2000).
74. A. Trombettoni and A. Smerzi, Phys. Rev. Lett. **86**, 2353 (2001).
75. B. Wu and Qian Niu, Phys. Rev. A **64**, 061603 (2001).
76. L. Pitaevskii, and S. Stringari, Phys. Rev. Lett. **88**, 180402 (2001).
77. A. Cuccoli, A. Fubini, V. Tognetti, and R. Vaia, Phys. Rev. A **64**, 061601 (2001).
78. D. van Oosten, P. van der Straten, and H.T.C. Stoof, Phys. Rev. A **63** 053601 (2001).
79. K.Burnett, M. Edwards, C.W. Clark, and M. Shotter, J. Phys. B: At. Mol. Opt. Phys. **35**, 1671 (2002).
80. F. Kh. Abdullaev, B. B. Baizakov, S. A. Darmanyan, V. V. Konotop, and M. Salerno, Phys. Rev. A **64**, 043606 (2001).
81. V.V. Konotop and M.Salerno, Phys. Rev. A **65**, 021602(R) (2002).
82. N. Tuskada, Phys. Rev. A **65**, 063608 (2002).
83. M. Kalinski, G. Metakis, and W.P. Schleich, private communication.

Canonical Statistics of Occupation Numbers for Ideal and Weakly Interacting Bose–Einstein Condensates

Dave Boers and Martin Holthaus

Fachbereich Physik, Carl von Ossietzky Universität, D-26111 Oldenburg, Germany

Abstract. We give a self-contained, tutorial review of recent works on the statistics of the number of particles contained in a Bose–Einstein condensate within the canonical ensemble, both for ideal and weakly interacting Bose gases. While in the case of the ideal gas there exists a general mathematical framework for computing the fluctuation of the condensate particles in any trap, the analysis of the weakly interacting case is restricted to a homogeneous gas in the framework of the Bogoliubov theory. In particular, we present a simplified derivation of the pair characteristic function which governs the condensate statistics, first obtained by Kocharovsky, Kocharovsky, and Scully [Physical Review A **61**, 053606 (2000)].

1 Introduction

The enormous experimental progress in the preparation and manipulation of Bose–Einstein condensates [1] has brought back into focus some apparently simple questions which still await a final answer. For instance, the precise dependence of the Bose–Einstein condensation temperature on the interaction strength presently is a subject of lively debate [2]. Even the ideal Bose gas, although thoroughly reviewed in 1977 by Ziff, Uhlenbeck, and Kac [3], has given rise to further questions; the fluctuation of the number of condensate particles in inhomogeneous, ideal Bose gases, together with certain differences between the canonical and the microcanonical ensemble, have been understood only recently [4–10]. Condensate fluctuations have also been investigated for weakly interacting Bose gases, both within the canonical [11–14] and the microcanonical [15,16] ensemble; although the situation appears to be less clear in the latter case.

With respect to condensate fluctuations of the ideal Bose gas, the three standard statistical ensembles are not equivalent: If $\langle n_0 \rangle_{\rm gc}$ denotes the ground-state occupation number for a gas within the grand canonical ensemble, the corresponding mean-square fluctuation is given, without further approximation, by

$$\left\langle (n_0 - \langle n_0 \rangle_{\rm gc})^2 \right\rangle_{\rm gc} = \langle n_0 \rangle_{\rm gc} \left(\langle n_0 \rangle_{\rm gc} + 1 \right) \; .$$

Hence, writing the grand canonical expectation value of the total particle number as $\langle N \rangle_{\rm gc}$, one finds

$$\left\langle (n_0 - \langle n_0 \rangle_{\rm gc})^2 \right\rangle_{\rm gc} \longrightarrow \langle N \rangle_{\rm gc} \left(\langle N \rangle_{\rm gc} + 1 \right)$$

for $T \to 0$, so that the grand canonical r.m.s.-fluctuation of the number of condensed particles is about $\langle N \rangle_{\rm gc}$ at zero temperature. Such huge fluctuations clearly cannot occur when the total particle number is fixed, as in the canonical or in the microcanonical ensemble. In these latter two ensembles the condensate fluctuation of the ideal Bose gas vanishes for $T \to 0$, when all particles settle into the ground state. Still, the predictions for the magnitude of the condensate fluctuations derived from these two ensembles differ at finite temperatures; the canonical fluctuations generally exceed the microcanonical ones [7,8]. Thus, when studying condensate fluctuations for a finite number of Bose particles, it is necessary to discard the usually employed grand canonical ensemble, and to investigate whether some given experimental set-up requires a canonical or a microcanonical description.

In these notes we intend to give a self-contained tutorial review on what is known on condensate fluctuations within the canonical ensemble, both for the ideal and the weakly interacting Bose gas. While there exists a fairly complete analytical treatment for ideal Bose gases in arbitrary trapping potentials, as outlined in Sect. 2, our investigation of weakly interacting Bose gases in Sect. 4 is restricted to a homogeneous gas under periodic boundary conditions, for which rather concise results have been obtained. In order to provide the necessary background, Sect. 3 gives a short, but detailed introduction to the Bogoliubov theory.

Among the main sources of the present review are [8] and [17] for the ideal gas, [18] for the Bogoliubov transformation, and, in particular, [14] for the canonical statistics of weakly interacting Bose–Einstein condensates. The expression obtained recently by V.V. Kocharovsky, Vl. V. Kocharovsky, and M.O. Scully for the characteristic function of the condensate occupation number [14] has led to great clarity, allowing us to understand the reason for the unexpected relation between the condensate fluctuations in the ideal and in the interacting Bose gas. In Sect. 4 we give a simplified derivation of this result, based on a disentangling theorem for angular momentum operators explained in Appendix 1, with the hope that this simplification will render this important contribution to the statistical physics of Bose–Einstein condensates more easily accessible.

2 Canonical Statistics of Ideal Bose–Einstein Condensates

2.1 The Emergence of Canonical Quasiparticles

Let us consider a gas of N ideal Bose particles stored in a trap with discrete single-particle energies ε_ν. The index $\nu \geq 0$ is used to denumerate the individual single-particle states, with $\nu = 0$ referring to the ground state. Moreover, let n_ν denote the occupation number of the ν-th level. Then the energy of a particular microstate of the gas, given by a particular configuration $\{n_\nu\}$ of occupation numbers with $\sum_{\nu=0}^{\infty} n_\nu = N$, is written as

$$\sum_{\nu=0}^{\infty} n_\nu \varepsilon_\nu = \sum_{\nu=1}^{\infty} n_\nu (\varepsilon_\nu - \varepsilon_0) + N\varepsilon_0$$
$$\equiv E + N\varepsilon_0 , \qquad (1)$$

where we have separated the true excitation energy E from the trivial N-particle ground-state energy $N\varepsilon_0$.

Next, let $\Omega(E, M)$ denote the number of microstates with a given excitation energy E distributed in an arbitrary manner among *up to* M particles, with $M = 1, 2, \ldots, N$. That is, $\Omega(E, M)$ also includes those configurations where E is actually shared among less than M particles, so that less than M particles are excited. In contrast, the differences

$$\Phi(E, M) \equiv \Omega(E, M) - \Omega(E, M-1) \qquad (2)$$

count those microstates where E is distributed over *exactly* M particles, so that M particles are excited, while the other $N - M$ particles remain in the ground state. Setting $\Omega(E, -1) = 0$, we then have

$$\sum_{M=0}^{N} \Phi(E, M) = \Omega(E, N) . \qquad (3)$$

Under conditions conforming to the *microcanonical* ensemble, that is, when the N-particle gas is thermally isolated from its surrounding and equipped with a fixed excitation energy E, the probability for finding M out of N particles in an excited state (and, accordingly, $N - M$ particles in the ground state) is given by

$$p_{\mathrm{mc}}(E, M) \equiv \frac{\Phi(E, M)}{\Omega(E, N)} , \qquad M = 1, \ldots, N . \qquad (4)$$

In contrast, under *canonical* conditions the gas is assumed to be in thermal contact with an environment of temperature T, so that its energy fluctuates. In this case, the probability for finding M excited particles becomes

$$p_{\mathrm{can}}(\beta, M) \equiv \frac{\sum_E \mathrm{e}^{-\beta E}\, \Phi(E, M)}{\sum_E \mathrm{e}^{-\beta E}\, \Omega(E, N)} , \qquad M = 1, \ldots, N , \qquad (5)$$

with inverse temperature $\beta = 1/(k_{\mathrm{B}}T)$, where k_{B} is Boltzmann's constant. While the microcanonical ensemble appears to come closer to present experiments with ultracold dilute Bose gases [1], from the theoretical viewpoint the canonical ensemble is considerably easier to deal with. Hence, we will restrict ourselves to canonical statistics in the following, and characterize the canonical distribution (5) in terms of its cumulants; in particular, we will compute the fluctuation of the number of Bose–Einstein condensed particles within the canonical ensemble.

The canonical M-particle partition functions $Z_M(\beta)$ now act as generating functions for the multiplicities $\Omega(E, M)$:

$$Z_M(\beta) = \sum_E \Omega(E, M) \exp(-M\beta\varepsilon_0 - \beta E) . \qquad (6)$$

In view of the distributions (4) and (5), however, one needs a generating function for the differences $\Phi(E,M)$. Such a function is constructed in the following manner [8]: First, we multiply each partition function $Z_M(\beta)$ by $(ze^{\beta\varepsilon_0})^M$, where z is an auxiliary variable, and then sum over M, thus obtaining the "grand" partition function [19–21]

$$\begin{aligned}\Xi(\beta,z) &\equiv \sum_{M=0}^{\infty}(ze^{\beta\varepsilon_0})^M Z_M(\beta) \\ &= \prod_{\nu=0}^{\infty}\frac{1}{1-z\exp[-\beta(\varepsilon_\nu-\varepsilon_0)]} \, . \end{aligned} \tag{7}$$

In view of (6), this function has the alternative representation

$$\Xi(\beta,z) = \sum_{M=0}^{\infty} z^M \sum_E \Omega(E,M)\exp(-\beta E) \, . \tag{8}$$

Therefore, by multiplying $\Xi(\beta,z)$ by $(1-z)$ and appropriately shifting the index in one of the resulting sums over M, we immediately arrive at the desired generating function for the differences $\Phi(E,M)$; this function is dubbed $\Xi_{\rm ex}(\beta,z)$:

$$\begin{aligned}(1-z)\,\Xi(\beta,z) &= \sum_{M=0}^{\infty}(z^M - z^{M+1})\sum_E \Omega(E,M)\exp(-\beta E) \\ &= \sum_{M=0}^{\infty} z^M \sum_E [\Omega(E,M)-\Omega(E,M-1)]\exp(-\beta E) \\ &= \sum_{M=0}^{\infty} z^M \sum_E \Phi(E,M)\exp(-\beta E) \\ &\equiv \Xi_{\rm ex}(\beta,z) \, . \end{aligned} \tag{9}$$

On the other hand, the product representation (7) tells us that multiplying $\Xi(\beta,z)$ by $(1-z)$ simply means to discard the ground-state factor $\nu=0$:

$$\Xi_{\rm ex}(\beta,z) = \prod_{\nu=1}^{\infty}\frac{1}{1-z\exp[-\beta(\varepsilon_\nu-\varepsilon_0)]} \, . \tag{10}$$

Thus, the grand partition function of a fictitious Bose gas which emerges from the actual gas by "amputating" the ground state serves as generating function for the differences $\Phi(E,M)$; these quantities, in their turn, determine the excited-particles distribution (5).

Taking up (9), we now have the identity

$$\left(z\frac{\partial}{\partial z}\right)^k \Xi_{\rm ex}(\beta,z)\Bigg|_{z=1} = \sum_E \exp(-\beta E)\sum_{M=0}^{\infty} M^k \Phi(E,M) \, , \tag{11}$$

which brings us back to the canonical distribution (5): If the sum over M did not extend to infinity, but if it rather terminated at $M = N$, i.e., at the actual number of particles constituting the gas, then the r.h.s. of this equation (11) would equal the k-th moment of the distribution (5), except for normalization. While exact equality is spoilt by the unrestricted summation, we still have at least approximate equality for temperatures below the onset of Bose–Einstein condensation: When there is a Bose–Einstein condensate, the probability for finding almost all particles in an excited state, given by (5) with M close to N, necessarily is negligible! Therefore, in the condensate regime we have the approximation

$$\sum_{M=0}^{\infty} M^k \Phi(E,M) \approx \sum_{M=0}^{N} M^k \Phi(E,M) , \tag{12}$$

and conclude that we can actually obtain the moments of the canonical distribution (5) by differentiating the "amputated" grand partition function (10) according to (11), provided we restrict ourselves to temperatures below the onset of Bose–Einstein condensation.

In physical terms, the approximation (12) means to replace the actual Bose–Einstein condensate, consisting of a finite number of particles, by an infinite supply of ground-state particles. Since the added particles do not appreciably change the statistical properties of the excited-states subsystem, the moments of the distribution (5) are not appreciably affected by this replacement, provided one remains in the condensate regime. This trick of augmenting the finite condensate to an infinite reservoir of ground-state particles had already been employed by Fierz in 1956 [22] for computing the fluctuations of a homogeneous ideal Bose gas; it also underlies the so-called "Maxwell's Demon ensemble" [7]. The consequences of this device become quite clear when considering (11) for $k = 0$: In that case, the approximation (12) takes the form

$$\sum_{M=0}^{\infty} \Phi(E,M) \approx \sum_{M=0}^{N} \Phi(E,M) = \Omega(E,N) , \tag{13}$$

see (3), so that (11) becomes

$$\prod_{\nu=1}^{\infty} \frac{1}{1 - \exp[-\beta(\varepsilon_\nu - \varepsilon_0)]} = \sum_E \exp(-\beta E)\, \Omega(E,N) . \tag{14}$$

Therefore, in the condensate regime the canonical partition function of the N-particle Bose gas equals a simple product of infinitely many geometrical series. Each of these series represents the canonical partition function of a harmonic oscillator; the fact that these individual partition functions are merely multiplied to obtain the partition function of the total system means that these fictitious oscillators are uncoupled, and obey Boltzmannian statistics. In other words, by means of the (exact) "ground-state amputation" procedure, which allowed us to switch from $\Omega(E,M)$ to $\Phi(E,M)$ in (9), and of the "infinite condensate" approximation (12), we have mapped the canonical statistics of the original gas of

Bose particles to an equivalent system of independent quasiparticles. Employing the language of second quantization, with creation operators $b_\nu^\dagger$ and annihilation operators b_ν referring to these "canonical quasiparticles", the Hamiltonian of the system then reads

$$H = \sum_{\nu=1}^{\infty} (\varepsilon_\nu - \varepsilon_0)\, b_\nu^\dagger b_\nu \; . \tag{15}$$

Again, the absence of the ground state $\nu = 0$ here reflects the previous shift from the grand partition function $\Xi(\beta, z)$ to the amputated function $\Xi_{\rm ex}(\beta, z)$.

Now we come back to the characterization of the canonical excited-particles distribution (5): Since the amputated function $\Xi_{\rm ex}(\beta, z)$ generates its moments in the condensate regime, it follows that the logarithm $\ln \Xi_{\rm ex}(\beta, z)$ generates its cumulants $\kappa_k(\beta)$:

$$\kappa_k(\beta) = \left(z \frac{\partial}{\partial z} \right)^k \ln \Xi_{\rm ex}(\beta, z) \Bigg|_{z=1} \; . \tag{16}$$

What are cumulants good for? — In general, the first cumulant κ_1 of a stochastic variable $\widehat{X}$ equals its expectation value, $\kappa_1 = \langle \widehat{X} \rangle$; the second cumulant gives its mean-square fluctuation, $\kappa_2 = \langle \widehat{X}^2 \rangle - \langle \widehat{X} \rangle^2$. Moreover, one has [23]

$$\begin{aligned} \kappa_3 &= \langle (\widehat{X} - \langle \widehat{X} \rangle)^3 \rangle \; , \\ \kappa_4 &= \langle (\widehat{X} - \langle \widehat{X} \rangle)^4 \rangle - 3\kappa_2^2 \; ; \end{aligned} \tag{17}$$

similar equations hold for the higher cumulants. If the underlying probability distribution is Gaussian, all cumulants higher than the second vanish [24]. Therefore, for any given distribution, such as the canonical excited-particles distribution (5), the numerical values of the cumulants κ_k with $k \geq 3$ quantify the deviation of that distribution from a Gaussian.

The main reason for focusing on the cumulants κ_k, rather than on the more familiar moments $\langle \widehat{X}^k \rangle$, lies in the following statement [24]: If $\widehat{X}$ is a sum of *independent* stochastic variables $\widehat{Y}_\nu$,

$$\widehat{X} = \sum_\nu \widehat{Y}_\nu \; , \tag{18}$$

then the k-th cumulant of the sum equals the sum of the individual k-th cumulants,

$$\kappa_k[\widehat{X}] = \sum_\nu \kappa_k[\widehat{Y}_\nu] \; . \tag{19}$$

In our case, if we take the occupation numbers of the individual excited levels $\nu \geq 1$ as stochastic variables $\widehat{n}_\nu$ and consider the total number $\widehat{N}_{\rm ex}$ of excited particles,

$$\widehat{N}_{\rm ex} = \sum_{\nu=1}^{\infty} \widehat{n}_\nu \; , \tag{20}$$

the independence of the "canonical quasiparticles" inferred from the partition function (14) implies the stochastic independence of the individual occupation numbers, so that the cumulants of $\widehat{N}_{\rm ex}$ are given by the sum of the cumulants of the $\widehat{n}_\nu$. Indeed, inserting the product representation (10) into the prescription (16), one finds

$$\kappa_1(\beta) = \sum_{\nu=1}^{\infty} \frac{1}{\exp[\beta(\varepsilon_\nu - \varepsilon_0)] - 1} \; . \tag{21}$$

Since the individual summands can be identified with the expectation values of the individual occupation numbers,

$$\frac{1}{\exp[\beta(\varepsilon_\nu - \varepsilon_0)] - 1} = \langle \widehat{n}_\nu \rangle \quad \text{in the condensate regime} \; , \tag{22}$$

this equation (21) reduces to the obvious identity

$$\langle \widehat{N}_{\rm ex} \rangle = \sum_{\nu=1}^{\infty} \langle \widehat{n}_\nu \rangle \; . \tag{23}$$

For $k = 2$, (16) then states that the mean-square fluctuation of $\widehat{N}_{\rm ex}$ equals the sum of the Bose fluctuations of the individual occupation numbers,

$$\kappa_2(\beta) = \sum_{\nu=1}^{\infty} \langle \widehat{n}_\nu \rangle \left(\langle \widehat{n}_\nu \rangle + 1 \right) \; . \tag{24}$$

Since the number of condensate particles is $\widehat{n}_0 = N - \widehat{N}_{\rm ex}$, this expression (24) equals the mean-square fluctuation of $\widehat{n}_0$, that is, the fluctuation of the number of particles comprising the condensate. Higher cumulants are obtained from (16) in the same manner; for instance, the third and the fourth cumulant take the form

$$\begin{aligned} \kappa_3(\beta) &= \sum_{\nu=1}^{\infty} \left(\langle \widehat{n}_\nu \rangle + 3\langle \widehat{n}_\nu \rangle^2 + 2\langle \widehat{n}_\nu \rangle^3 \right) \; , \\ \kappa_4(\beta) &= \sum_{\nu=1}^{\infty} \left(\langle \widehat{n}_\nu \rangle + 7\langle \widehat{n}_\nu \rangle^2 + 12\langle \widehat{n}_\nu \rangle^3 + 6\langle \widehat{n}_\nu \rangle^4 \right) \; . \end{aligned} \tag{25}$$

Finally, we emphasize once again that these results are restricted to the condensate regime: The approximation (12) holds only when there is a condensate, and it is this approximation which leads to the mutual independence of the canonical quasiparticles.

2.2 Evaluating the Sums

It is, of course, necessary to convert a sum over the individual levels $\nu \geq 1$, such as appearing in (21), (24), and (25), into a more explicit expression. This goal can be achieved by means of a Mellin transformation: Starting from the product representation (10), and utilizing the expansion

$$\ln(1-x) = -\sum_{n=1}^{\infty} \frac{x^n}{n} \quad \text{for } -1 \leq x < 1 \, , \tag{26}$$

we write

$$\begin{aligned} \ln \Xi_{\rm ex}(\beta, z) &= -\sum_{\nu=1}^{\infty} \ln(1 - z \exp[-\beta(\varepsilon_\nu - \varepsilon_0)]) \\ &= \sum_{\nu=1}^{\infty}\sum_{n=1}^{\infty} \frac{z^n \exp[-\beta(\varepsilon_\nu - \varepsilon_0)n]}{n} \, . \end{aligned} \tag{27}$$

Next, we observe [25] that e^{-a} is the Mellin transform of $\Gamma(t)$, namely

$$\mathrm{e}^{-a} = \frac{1}{2\pi \mathrm{i}} \int_{\tau - \mathrm{i}\infty}^{\tau + \mathrm{i}\infty} \mathrm{d}t \; a^{-t} \, \Gamma(t) \, , \tag{28}$$

valid for $\mathrm{Re}\, a > 0$ and $\tau > 0$. Hence, the cumulant-generating function takes the form

$$\begin{aligned} \ln \Xi_{\rm ex}(\beta, z) &= \sum_{\nu=1}^{\infty}\sum_{n=1}^{\infty} \frac{1}{2\pi \mathrm{i}} \int_{\tau - \mathrm{i}\infty}^{\tau + \mathrm{i}\infty} \mathrm{d}t \; \Gamma(t) \, \frac{z^n}{n} \, \frac{1}{(\beta[\varepsilon_\nu - \varepsilon_0]n)^t} \\ &= \frac{1}{2\pi \mathrm{i}} \int_{\tau - \mathrm{i}\infty}^{\tau + \mathrm{i}\infty} \mathrm{d}t \; \Gamma(t) \sum_{\nu=1}^{\infty} \frac{1}{(\beta[\varepsilon_\nu - \varepsilon_0])^t} \sum_{n=1}^{\infty} \frac{z^n}{n^{t+1}} \, . \end{aligned} \tag{29}$$

The interchange of summation and integration performed here is permissible when the resulting series under the integral are absolutely convergent. This requires that, after analytic continuation of these series, the real number τ be such that the path of integration parallel to the imaginary axis lies to the right of all their poles [25]. We then introduce the Bose function [21]

$$g_\alpha(z) = \sum_{n=1}^{\infty} \frac{z^n}{n^\alpha} \, , \tag{30}$$

and define the generalized Zeta function [26]

$$Z(\beta, t) \equiv \sum_{\nu=1}^{\infty} \frac{1}{(\beta[\varepsilon_\nu - \varepsilon_0])^t} \, . \tag{31}$$

Hence, we obtain the compact integral representation

$$\ln \Xi_{\rm ex}(\beta, z) = \frac{1}{2\pi \mathrm{i}} \int_{\tau - \mathrm{i}\infty}^{\tau + \mathrm{i}\infty} \mathrm{d}t \; \Gamma(t) \, Z(\beta, t) \, g_{t+1}(z) \, . \tag{32}$$

In order to calculate the canonical cumulants from this representation according to (16), we utilize the relations

$$z \frac{\mathrm{d}}{\mathrm{d}z} g_\alpha(z) = g_{\alpha-1}(z) \tag{33}$$

and

$$g_\alpha(1) = \zeta(\alpha) \; , \tag{34}$$

where $\zeta(\alpha) = \sum_{n=1}^{\infty} n^{-\alpha}$ is the Riemann Zeta function. In this way, we arrive at the desired cumulant formula [8,17]:

$$\kappa_k(\beta) = \frac{1}{2\pi \mathrm{i}} \int_{\tau-\mathrm{i}\infty}^{\tau+\mathrm{i}\infty} \mathrm{d}t \; \Gamma(t) \, Z(\beta,t) \, \zeta(t+1-k) \; . \tag{35}$$

This formula lends itself to systematic asymptotic expansion. The contributions to the cumulants, in the order of descending importance, are obtained by collecting the residues of the integrand in (35) from right to left. For a more exhaustive discussion of this procedure we refer to [17] and consider only an important and illustrative example here, namely an isotropic, three-dimensional harmonic oscillator trap with oscillator frequency ω. In this case, the single-particle energies are given by

$$\varepsilon_\nu = \hbar\omega(\nu + 3/2) \quad , \quad \nu = 0, 1, 2, \ldots \; ; \tag{36}$$

their degree of degeneracy is

$$g_\nu = \frac{1}{2}\nu^2 + \frac{3}{2}\nu + 1 \; . \tag{37}$$

Thus, the generalized Zeta function (31) can be written as a sum of Riemann Zeta functions [8,17],

$$\begin{aligned} Z(\beta,t) &= \sum_{\nu=1}^{\infty} \frac{g_\nu}{(\beta\hbar\omega\nu)^t} \\ &= (\beta\hbar\omega)^{-t} \left[\frac{1}{2}\zeta(t-2) + \frac{3}{2}\zeta(t-1) + \zeta(t) \right] \; . \end{aligned} \tag{38}$$

The only information required for evaluating the cumulant formula (35) is the pole structure of the integrand: $\zeta(z)$ possesses only one simple pole, located at $z = 1$ with residue $+1$, so that [27]

$$\zeta(z) \approx \frac{1}{z-1} + \gamma \qquad \text{for } z \approx 1 \; , \tag{39}$$

where $\gamma \approx 0.57722$ is Euler's constant. Because temperature enters into the cumulants (35) *only* via the prefactor $(\beta\hbar\omega)^{-t}$ of the spectral Zeta function (38), one sees that for temperatures T large compared to the level spacing temperature $\hbar\omega/k_{\mathrm{B}}$ (that is, for $\beta\hbar\omega \ll 1$, so that the gas occupies more than just the lowest few trap states) the temperature dependence of the k-th cumulant $\kappa_k(\beta)$

is governed by the factor $(\beta\hbar\omega)^{-p}$, where p is the position of the rightmost pole appearing in the integrand (35). Now $\Gamma(t)$ has poles at $t = 0, -1, -2, \ldots$, $Z(\beta, t)$ has poles at $t = 3, 2, 1$, and $\zeta(t + 1 - k)$ has a pole at $t = k$. Thus, for $k = 1$ and $k = 2$ the rightmost pole of the spectral Zeta function (38) dominates, so that, to leading order, both $\kappa_1(\beta) \propto (\beta\hbar\omega)^{-3}$ and $\kappa_2(\beta) \propto (\beta\hbar\omega)^{-3}$. For $k = 3$, the leading pole of $Z(\beta, t)$ at $t = 3$ coincides with the pole of $\zeta(t + 1 - k)$. Then $\kappa_3(\beta) \propto (\beta\hbar\omega)^{-3}$, with logarithmic corrections due to the double pole. For $k \geq 4$ the dominant pole is provided by $\zeta(t + 1 - k)$, giving $\kappa_k(\beta) \propto (\beta\hbar\omega)^{-k}$. Considering also the next-to-leading poles, we obtain the following asymptotic expressions, valid for $k_{\mathrm{B}}T/(\hbar\omega) \gg 1$ in the condensate regime:

- The first cumulant $\kappa_1(\beta)$ of the canonical distribution (5) equals the expectation value $\langle \widehat{N}_{\mathrm{ex}} \rangle$ of the number of excited particles. In the condensate regime, the expectation value $\langle \widehat{n}_0 \rangle$ of the number of condensed particles is then given by $N - \langle \widehat{N}_{\mathrm{ex}} \rangle = N - \kappa_1(\beta)$:

$$\begin{aligned} \langle \widehat{n}_0 \rangle \sim N &- \left(\frac{k_{\mathrm{B}}T}{\hbar\omega}\right)^3 \zeta(3) \\ &- \left(\frac{k_{\mathrm{B}}T}{\hbar\omega}\right)^2 \frac{3}{2}\zeta(2) \\ &- \frac{k_{\mathrm{B}}T}{\hbar\omega}\left[\ln\left(\frac{k_{\mathrm{B}}T}{\hbar\omega}\right) + \gamma - \frac{19}{24}\right] , \end{aligned} \tag{40}$$

  where we have taken into account the three poles at $t = 3, 2, 1$; the logarithm in the last term stems from the double pole at $t = 1$. Of course, the leading-order term coincides with the familiar result for "large" oscillator traps [28,29]; the next terms quantify the finite-size corrections.
- The second cumulant $\kappa_2(\beta)$ yields the mean-square fluctuation of the number of excited particles. Since the total particle number N is constant, this equals the mean-square fluctuation of the number of condensate particles, $\kappa_2(\beta) = \langle (\widehat{N}_{\mathrm{ex}} - \langle \widehat{N}_{\mathrm{ex}} \rangle)^2 \rangle = \langle (\widehat{n}_0 - \langle \widehat{n}_0 \rangle)^2 \rangle$. To the same accuracy as in (40), we find

$$\begin{aligned} \langle (\widehat{n}_0 - \langle \widehat{n}_0 \rangle)^2 \rangle \sim & \left(\frac{k_{\mathrm{B}}T}{\hbar\omega}\right)^3 \zeta(2) \\ &+ \left(\frac{k_{\mathrm{B}}T}{\hbar\omega}\right)^2 \left[\frac{3}{2}\ln\left(\frac{k_{\mathrm{B}}T}{\hbar\omega}\right) + \frac{3}{2}\gamma + \frac{5}{4} + \zeta(2)\right] \\ &+ \frac{k_{\mathrm{B}}T}{\hbar\omega}\left(-\frac{1}{2}\right) ; \end{aligned} \tag{41}$$

  the double pole now appearing at $t = 2$ in the integrand (35). The leading term here agrees with the result first stated by Politzer [5], but for moderate N the finite-size corrections are quite significant.
- The third cumulant, which determines the skewness of the distribution (5), equals the third central moment: $\kappa_3(\beta) = \langle (\widehat{N}_{\mathrm{ex}} - \langle \widehat{N}_{\mathrm{ex}} \rangle)^3 \rangle = -\langle (\widehat{n}_0 - \langle \widehat{n}_0 \rangle)^3 \rangle$.

Therefore,

$$\langle(\widehat{n}_0 - \langle\widehat{n}_0\rangle)^3\rangle \sim -\left(\frac{k_{\rm B}T}{\hbar\omega}\right)^3 \left[\ln\left(\frac{k_{\rm B}T}{\hbar\omega}\right) + \gamma + \frac{3}{2} + 3\zeta(2) + 2\zeta(3)\right] + \left(\frac{k_{\rm B}T}{\hbar\omega}\right)^2 \frac{3}{4} + \frac{k_{\rm B}T}{\hbar\omega}\frac{1}{12} \; . \tag{42}$$

- The calculation of all higher cumulants is simplified by the fact that no double poles appear, and, consequently, no logarithmic corrections like those in (40), (41), and (42). For example, the fourth cumulant, which is related to the fourth central moment through $\kappa_4(\beta) = \langle(\widehat{n}_0 - \langle\widehat{n}_0\rangle)^4\rangle - 3\kappa_2(\beta)^2$, takes the form

$$\kappa_4(\beta) \sim \left(\frac{k_{\rm B}T}{\hbar\omega}\right)^4 [3\zeta(2) + 9\zeta(3) + 6\zeta(4)] + \left(\frac{k_{\rm B}T}{\hbar\omega}\right)^3 \left(-\frac{1}{2}\right) + \left(\frac{k_{\rm B}T}{\hbar\omega}\right)^2 \left(-\frac{1}{8}\right) \; ; \tag{43}$$

  this fourth cumulant determines the flatness (excess) of the distribution (5) for the isotropic harmonic oscillator trap.

The corresponding evaluation of the lowest four cumulants for a homogeneous ideal Bose gas in a box with either periodic boundary conditions or hard walls can be found in [17]; it should be noted that in that case the cumulants $\kappa_k(\beta)$ with $k \geq 2$ remain sensitive to the boundary conditions even in the thermodynamic limit [30]. Nonetheless, the essence remains the same as in the harmonic-oscillator example: The pole structure of the generalized Zeta function (31) encodes, by means of the cumulant formula (35), the statistical properties of the condensate.

2.3 Rewriting the Sums

The analysis in the preceding subsection still is restricted to the ideal Bose gas. In order to find the "organizing principle" behind the coefficients appearing in the ideal-gas cumulants (25), and to cast these cumulants into a form suitable for comparison with the corresponding cumulants for a weakly interacting Bose gas, we now reformulate the results of Sect. 2.1 in the language of characteristic functions, following closely the original work by Kocharovsky, Kocharovsky, and Scully [14].

The characteristic function $G_{\widehat{X}}(y)$ pertaining to a stochastic variable $\widehat{X}$ is defined as the expectation value of the exponential $\exp({\rm i}y\widehat{X})$,

$$G_{\widehat{X}}(y) = \langle e^{iy\widehat{X}} \rangle = \sum_{M=0}^{\infty} \frac{(iy)^M}{M!} \langle \widehat{X}^M \rangle . \tag{44}$$

Therefore, $G_{\widehat{X}}(y)$ serves as the moment-generating function for $\widehat{X}$:

$$\langle \widehat{X}^k \rangle = \frac{d^k}{d(iy)^k} G_{\widehat{X}}(y) \bigg|_{y=0} , \tag{45}$$

implying that, in analogy to the previous equation (16), the logarithm of $G_{\widehat{X}}(y)$ generates the cumulants of $\widehat{X}$:

$$\kappa_k[\widehat{X}] = \frac{d^k}{d(iy)^k} \ln G_{\widehat{X}}(y) \bigg|_{y=0} . \tag{46}$$

If we now regard the occupation number of an individual excited single-particle state $\nu \geq 1$ as stochastic variable $\widehat{n}_\nu$, and recall that, within the canonical ensemble, the expectation value is computed by multiplying with the appropriate canonical density matrix $\widehat{\varrho}_\nu$ and forming the trace, we are led to the characteristic function

$$G_\nu(y) = \langle e^{iy\widehat{n}_\nu} \rangle = \mathrm{Tr}\left(e^{iy\widehat{n}_\nu} \widehat{\varrho}_\nu \right) . \tag{47}$$

What, then, is $\widehat{\varrho}_\nu$? — At this point we make use of the key result of Sect. 2.1: The individual excited levels effectively give rise, within the canonical ensemble, to independent quasiparticles, as expressed by the Hamiltonian (15). Therefore, each density matrix $\widehat{\varrho}_\nu$ equals the density matrix of the corresponding harmonic oscillator:

$$\widehat{\varrho}_\nu = \frac{\exp(-\beta \widehat{n}_\nu(\varepsilon_\nu - \varepsilon_0))}{\mathrm{Tr}\left[\exp(-\beta \widehat{n}_\nu(\varepsilon_\nu - \varepsilon_0))\right]} = \left(1 - e^{-\beta(\varepsilon_\nu - \varepsilon_0)}\right) \exp(-\beta \widehat{n}_\nu(\varepsilon_\nu - \varepsilon_0)) , \tag{48}$$

so that the characteristic function (47) becomes

$$G_\nu(y) = \left(1 - e^{-\beta(\varepsilon_\nu - \varepsilon_0)}\right) \mathrm{Tr}\left[e^{iy\widehat{n}_\nu} \exp(-\beta \widehat{n}_\nu(\varepsilon_\nu - \varepsilon_0)) \right] = \frac{1 - \exp(-\beta(\varepsilon_\nu - \varepsilon_0))}{1 - e^{iy} \exp(-\beta(\varepsilon_\nu - \varepsilon_0))} . \tag{49}$$

Since the individual quasiparticles are independent, the characteristic function $X(y)$ of the total excited states subsystem is now given by

$$X(y) = \prod_{\nu=1}^{\infty} G_\nu(y) . \tag{50}$$

Obviously, apart from normalization this function $X(y)$ is the same as the function $\Xi_{\mathrm{ex}}(\beta, z)$ constructed in Sect. 2.1, with the fugacity-type variable z in (10) replaced by $\mathrm{e}^{\mathrm{i}y}$ here. The cumulant-generating function then takes the form

$$\ln X(y) = -\sum_{\nu=1}^{\infty} \ln\left(\frac{1 - \mathrm{e}^{\mathrm{i}y}\exp(-\beta(\varepsilon_\nu - \varepsilon_0))}{1 - \exp(-\beta(\varepsilon_\nu - \varepsilon_0))}\right) . \tag{51}$$

Once again expanding the logarithms on the r.h.s. according to (26), we find the analogue of (27),

$$\begin{aligned} \ln X(y) &= \sum_{\nu=1}^{\infty}\sum_{n=1}^{\infty} \frac{1}{n}\left(\frac{\exp(\mathrm{i}y) - 1}{\exp(\beta(\varepsilon_\nu - \varepsilon_0)) - 1}\right)^n \\ &= \sum_{n=1}^{\infty} \frac{(\exp(\mathrm{i}y) - 1)^n}{n} \sum_{\nu=1}^{\infty} \langle \widehat{n}_\nu \rangle^n . \end{aligned} \tag{52}$$

Instead of transforming the sums into an integral, as done in the previous subsection, we have simply used the expression (22) for the expectation values $\langle \widehat{n}_\nu \rangle$ of the occupation numbers in the condensate regime. Next, we exploit the expansion

$$\left(\mathrm{e}^{\mathrm{i}y} - 1\right)^n = n! \sum_{k=n}^{\infty} \sigma_k^{(n)} \frac{(\mathrm{i}y)^k}{k!} , \tag{53}$$

where the $\sigma_k^{(n)}$ denote the so-called Stirling numbers of the second kind [23]; these numbers count the number of ways to partition a set of k elements into n non-empty subsets. Hence,

$$\begin{aligned} \ln X(y) &= \sum_{n=1}^{\infty}\sum_{k=n}^{\infty} (n-1)!\,\sigma_k^{(n)} \frac{(\mathrm{i}y)^k}{k!} \sum_{\nu=1}^{\infty} \langle \widehat{n}_\nu \rangle^n \\ &= \sum_{k=1}^{\infty}\left(\sum_{n=1}^{k} (n-1)!\,\sigma_k^{(n)} \sum_{\nu=1}^{\infty} \langle \widehat{n}_\nu \rangle^n\right) \frac{(\mathrm{i}y)^k}{k!} \\ &= \sum_{k=1}^{\infty} \kappa_k(\beta) \frac{(\mathrm{i}y)^k}{k!} , \end{aligned} \tag{54}$$

so that, by virtue of (46), we can directly read off a formal expression [14] for the cumulants pertaining to the number of particles occupying excited states, represented by the stochastic variable $\widehat{N}_{\mathrm{ex}} = \sum_{\nu=1}^{\infty} \widehat{n}_\nu$:

$$\kappa_k(\beta) = \sum_{n=1}^{k} (n-1)!\,\sigma_k^{(n)} \sum_{\nu=1}^{\infty} \langle \widehat{n}_\nu \rangle^n ; \tag{55}$$

once again, the validity of this result is restricted to the condensate regime. Let us check: For $k = 3$, one has [23] $\sigma_3^{(1)} = 1$, $\sigma_3^{(2)} = 3$, and $\sigma_3^{(3)} = 1$, so that (55) reduces to

$$\kappa_3(\beta) = \sum_{\nu=1}^{\infty} \left(\langle \widehat{n}_\nu \rangle + 3\langle \widehat{n}_\nu \rangle^2 + 2\langle \widehat{n}_\nu \rangle^3\right) , \tag{56}$$

which is precisely what we had found found in (25) by direct differentiation of $\ln \Xi_{\rm ex}(\beta, z)$. A similar check with $k = 4$, and $\sigma_4^{(1)} = 1$, $\sigma_4^{(2)} = 7$, $\sigma_4^{(3)} = 6$, and $\sigma_4^{(4)} = 1$ reproduces the second of the equations (25). Thus, we have reached our goal and linked the k-th cumulant $\kappa_k(\beta)$ in the condensate regime in a systematic manner to powers of $\langle \widehat{n}_\nu \rangle$; the results obtained here [14] are mathematically equivalent to those found in Sect. 2.1. Nonetheless, it has to be kept in mind that a key ingredient of the analysis, the independence of the canonical quasiparticles, has simply been assumed here, whereas in Sect. 2.1 it had been derived from first principles.

3 The Bogoliubov Transformation

3.1 The Emergence of Bogoliubov Quasiparticles

We now turn from the ideal to the weakly interacting Bose gas. Instead of treating an inhomogeneous gas in a trap, we restrict ourselves to a homogeneous Bose gas of N particles with mass m in a cubic volume $V = L^3$. Since we have the thermodynamic limit in mind, $N \to \infty$ and $V \to \infty$ such that the density N/V remains constant, we simply impose periodic boundary conditions, so that the single-particle states are plane waves with momentum $\boldsymbol{p}$, and the ground-state energy is $\varepsilon_0 = 0$. With respect to current experiments [1], this model certainly is oversimplified, but it will lend itself to a fairly detailed analytical study of the condensate statistics in Sect. 4.

The gas is considered as "weakly interacting" if the range of the interaction potential $U(\boldsymbol{r})$ is small in comparison with the mean distance $(V/N)^{1/3}$ between the particles. For low temperatures, that is, in the limit of slow collisions, the mutual scattering amplitude tends to a constant limit $-a_{\rm sc}$, where $a_{\rm sc}$ is known as the s-wave scattering length. For repulsive interaction, as assumed here, the scattering length is positive, $a_{\rm sc} > 0$. Then the criterion for "weak interaction" becomes

$$\left(\frac{N}{V} \, a_{\rm sc}^3 \right)^{1/3} \ll 1 \, . \tag{57}$$

To get a feeling for typical orders of magnitude, consider liquid ^{4}He with its density of $N/V = 2.2 \cdot 10^{28}$ m^{-3}, and an s-wave scattering length of $a_{\rm sc} = 0.21 \cdot 10^{-9}$ m, resulting in $(Na_{\rm sc}^3/V)^{1/3} \approx 0.59$. Hence, liquid ^{4}He is no weakly interacting substance. For ^{87}Rb, on the other hand, the scattering length is about 26 times larger, namely $a_{\rm sc} = 5.5 \cdot 10^{-9}$ m, but then a dilute vapour of ^{87}Rb atoms with a density of $N/V = 10^{20}$ m^{-3} gives $(Na_{\rm sc}^3/V)^{1/3} \approx 0.026$. That is, with dilute ultracold atomic gases one can actually satisfy the condition (57).

The Hamiltonian of such a gas then takes the form [18]

$$H = \sum_{\boldsymbol{p}} \frac{\boldsymbol{p}^2}{2m} a_{\boldsymbol{p}}^\dagger a_{\boldsymbol{p}} + \frac{1}{2} \sum \langle \boldsymbol{p}_1' \boldsymbol{p}_2' | U | \boldsymbol{p}_1 \boldsymbol{p}_2 \rangle \, a_{\boldsymbol{p}_1'}^\dagger a_{\boldsymbol{p}_2'}^\dagger a_{\boldsymbol{p}_2} a_{\boldsymbol{p}_1} \, , \tag{58}$$

where $a_{\boldsymbol{p}}^\dagger$ and $a_{\boldsymbol{p}}$ are the creation and annihilation operators for Bose gas particles with momentum $\boldsymbol{p}$. The summation in the second term extends over all momenta which appear as suffixes, subject to conservation of momentum in the collisions, $\boldsymbol{p}_1' + \boldsymbol{p}_2' = \boldsymbol{p}_1 + \boldsymbol{p}_2$. The matrix elements of the interaction potential read

$$\begin{aligned} \langle \boldsymbol{p}_1'\boldsymbol{p}_2'|U|\boldsymbol{p}_1\boldsymbol{p}_2\rangle &= \frac{1}{V^2}\int \mathrm{d}^3\boldsymbol{r}_1 \int \mathrm{d}^3\boldsymbol{r}_2\, \mathrm{e}^{-\mathrm{i}(\boldsymbol{p}_2'-\boldsymbol{p}_2)\cdot\boldsymbol{r}_2/\hbar}\, U(\boldsymbol{r}_1-\boldsymbol{r}_2)\, \mathrm{e}^{\mathrm{i}(\boldsymbol{p}_1-\boldsymbol{p}_1')\cdot\boldsymbol{r}_1/\hbar} \\ &= \frac{1}{V}\int \mathrm{d}^3\boldsymbol{r}\, \mathrm{e}^{\mathrm{i}\boldsymbol{q}\cdot\boldsymbol{r}/\hbar} U(\boldsymbol{r}) \;, \end{aligned} \tag{59}$$

where $\boldsymbol{q} = \boldsymbol{p}_1 - \boldsymbol{p}_1' = -(\boldsymbol{p}_2 - \boldsymbol{p}_2')$ is the momentum transferred in the collision. Since one is primarily interested in low temperatures, for which the relevant particle momenta are small, one usually approximates [18] all matrix elements by their values for zero momenta,

$$\langle \boldsymbol{p}_1'\boldsymbol{p}_2'|U|\boldsymbol{p}_1\boldsymbol{p}_2\rangle \approx \frac{U_0}{V} \;, \tag{60}$$

so that

$$U_0 = \int \mathrm{d}^3\boldsymbol{r}\, U(\boldsymbol{r}) \tag{61}$$

carries the dimension (energy × volume). We are then left with

$$H = \sum_{\boldsymbol{p}} \frac{\boldsymbol{p}^2}{2m} a_{\boldsymbol{p}}^\dagger a_{\boldsymbol{p}} + \frac{U_0}{2V}\sum a_{\boldsymbol{p}_1'}^\dagger a_{\boldsymbol{p}_2'}^\dagger a_{\boldsymbol{p}_2} a_{\boldsymbol{p}_1} \;. \tag{62}$$

The approximate calculation of the spectrum furnished by this Hamiltonian (62) rests on the following observation [18]: In the ground state of an ideal Bose gas all particles are in the condensate, that is, in the state $\boldsymbol{p} = \boldsymbol{0}$, so that the occupation number of the ground state is $n_0 = N \gg 1$, whereas all other occupation numbers are zero, $n_{\boldsymbol{p}} = 0$ for $\boldsymbol{p} \neq \boldsymbol{0}$. In the present case of a weakly interacting gas, the condensate is partially depleted even when the system is in its ground state, so that even for the ground state we have $n_{\boldsymbol{p}} \neq 0$ for $\boldsymbol{p} \neq \boldsymbol{0}$. However, "weak" interaction implies that these occupation numbers remain small in comparison with $n_0 \approx N$. The commutator

$$a_0 a_0^\dagger - a_0^\dagger a_0 = 1 \ll N \tag{63}$$

then is still small compared with the magnitudes of the individual operators $a_0^\dagger$ and a_0. Therefore, one can neglect the operator character of these quantities, and approximate both $a_0^\dagger$ and a_0 by the square root of the condensate occupation number n_0,

$$a_0 \approx \sqrt{n_0} \quad , \quad a_0^\dagger \approx \sqrt{n_0} \;. \tag{64}$$

We now decompose

$$N = \sum_{\boldsymbol{p}} a_{\boldsymbol{p}}^\dagger a_{\boldsymbol{p}} = n_0 + \sum_{\boldsymbol{p}\neq\boldsymbol{0}} a_{\boldsymbol{p}}^\dagger a_{\boldsymbol{p}} \;, \tag{65}$$

and treat $a_{\boldsymbol{p}}^\dagger$ and $a_{\boldsymbol{p}}$ for $\boldsymbol{p} \neq \boldsymbol{0}$ as small quantities when expanding the fourfold sum in the Hamiltonian (62).

To zeroth order, one then has

$$\sum a^\dagger_{\boldsymbol{p}'_1} a^\dagger_{\boldsymbol{p}'_2} a_{\boldsymbol{p}_2} a_{\boldsymbol{p}_1} = n_0^2 \,, \tag{66}$$

giving, with $n_0^2 \approx N^2$, the approximate ground-state energy

$$E_0 \approx \frac{U_0}{2V} N^2 \,. \tag{67}$$

Since conservation of momentum is impossible when only one of the four indices $\boldsymbol{p}_1$, $\boldsymbol{p}_2$, $\boldsymbol{p}'_1$, and $\boldsymbol{p}'_2$ differs from zero, there are no first-order contributions. To second order, one finds

$$\begin{aligned}\sum a^\dagger_{\boldsymbol{p}'_1} a^\dagger_{\boldsymbol{p}'_2} a_{\boldsymbol{p}_2} a_{\boldsymbol{p}_1} = n_0^2 + \sum_{\boldsymbol{p}\neq\boldsymbol{0}} \Big(& a^\dagger_{\boldsymbol{p}} a^\dagger_{-\boldsymbol{p}} a_0 a_0 + a^\dagger_0 a^\dagger_0 a_{\boldsymbol{p}} a_{-\boldsymbol{p}} + a^\dagger_0 a^\dagger_{\boldsymbol{p}} a_0 a_{\boldsymbol{p}} \\ & + a^\dagger_0 a^\dagger_{\boldsymbol{p}} a_{\boldsymbol{p}} a_0 + a^\dagger_{\boldsymbol{p}} a^\dagger_0 a_0 a_{\boldsymbol{p}} + a^\dagger_{\boldsymbol{p}} a^\dagger_0 a_{\boldsymbol{p}} a_0 \Big) \,. \end{aligned} \tag{68}$$

Equation (65) now requires the approximation

$$n_0^2 \approx N^2 - 2N \sum_{\boldsymbol{p}\neq\boldsymbol{0}} a^\dagger_{\boldsymbol{p}} a_{\boldsymbol{p}} \tag{69}$$

for the first term on the r.h.s. of (68), while we may set $n_0 \approx N$ under the sum, because of the additional small quantities appearing there. Hence, we arrive at the "quadratic" Hamiltonian

$$H = \frac{U_0}{2V} N^2 + \sum_{\boldsymbol{p}} \frac{\boldsymbol{p}^2}{2m} a^\dagger_{\boldsymbol{p}} a_{\boldsymbol{p}} + \frac{U_0}{2V} N \sum_{\boldsymbol{p}\neq\boldsymbol{0}} \Big(a^\dagger_{\boldsymbol{p}} a^\dagger_{-\boldsymbol{p}} + a_{\boldsymbol{p}} a_{-\boldsymbol{p}} + 2 a^\dagger_{\boldsymbol{p}} a_{\boldsymbol{p}} \Big) \,. \tag{70}$$

Note that the coefficient "2" accompanying $a^\dagger_{\boldsymbol{p}} a_{\boldsymbol{p}}$ in the last sum results from "4 − 2", with "4" originating from (68), and "−2" stemming from (69).

This Hamiltonian (70) still suffers from the drawback that for typical interatomic interaction potentials $U(\boldsymbol{r})$, such as the Lennard–Jones 6–12 potential, the zero-momenta matrix element U_0 diverges, because of the rapid increase of $U(\boldsymbol{r})$ at short distances. In order to circumvent this difficulty, one usually adopts the following strategy [18]: One first replaces the actual potential $U(\boldsymbol{r})$ by some other potential $\widetilde{U}(\boldsymbol{r})$ for which all formal manipulations are well defined, and which possesses the same s-wave scattering length a_{sc} as the original $U(\boldsymbol{r})$. If then the result of the calculations, performed with the auxiliary $\widetilde{U}(\boldsymbol{r})$, is expressed in terms of a_{sc}, it will be the same as the one provided by the true interaction potential.

Therefore, one now expresses U_0 through a_{sc}: While it is sufficient to employ the first Born approximation

$$\frac{4\pi a_{\mathrm{sc}} \hbar^2}{m} \approx U_0 \tag{71}$$

for the prefactor of the "nondiagonal" sum in (70), the leading term requires the second Born approximation

$$\frac{4\pi a_{\rm sc}\hbar^2}{m} \approx U_0 + \frac{1}{V}\sum_{\boldsymbol{p}\neq\boldsymbol{0}} \frac{U_0^2}{0+0-\frac{\boldsymbol{p}^2}{2m}-\frac{\boldsymbol{p}^2}{2m}} , \tag{72}$$

where "$0+0$" in the denominator is meant to indicate that two particles in the condensate collide, so that $\boldsymbol{p}_1 = \boldsymbol{0}$, $\boldsymbol{p}_2 = \boldsymbol{0}$, and $\boldsymbol{p}_1' = -\boldsymbol{p}_2' = \boldsymbol{p}$. Hence,

$$U_0 \approx \frac{4\pi a_{\rm sc}\hbar^2}{m}\left[1 + \frac{4\pi a_{\rm sc}\hbar^2}{V}\sum_{\boldsymbol{p}\neq\boldsymbol{0}}\frac{1}{\boldsymbol{p}^2}\right] , \tag{73}$$

and the Hamiltonian (70) adopts the form

$$\begin{aligned} H = {} & \frac{2\pi a_{\rm sc}\hbar^2}{m}\frac{N^2}{V}\left[1 + \frac{4\pi a_{\rm sc}\hbar^2}{V}\sum_{\boldsymbol{p}\neq\boldsymbol{0}}\frac{1}{\boldsymbol{p}^2}\right] + \sum_{\boldsymbol{p}}\frac{\boldsymbol{p}^2}{2m}a_{\boldsymbol{p}}^\dagger a_{\boldsymbol{p}} \\ & + \frac{2\pi a_{\rm sc}\hbar^2}{m}\frac{N}{V}\sum_{\boldsymbol{p}\neq\boldsymbol{0}}\left(a_{\boldsymbol{p}}^\dagger a_{-\boldsymbol{p}}^\dagger + a_{\boldsymbol{p}}a_{-\boldsymbol{p}} + 2a_{\boldsymbol{p}}^\dagger a_{\boldsymbol{p}}\right) . \end{aligned} \tag{74}$$

This Hamiltonian (74) will now be diagonalized by means of a Bogoliubov transformation [31]. In our case, this transformation starts, for each pair $(\boldsymbol{p}, -\boldsymbol{p})$ of momenta $\boldsymbol{p} \neq \boldsymbol{0}$, from a "quadratic" Hamiltonian of the form

$$H_{\boldsymbol{p}} = A_{\boldsymbol{p}} + B_{\boldsymbol{p}}a_{\boldsymbol{p}}^\dagger a_{-\boldsymbol{p}}^\dagger + B_{\boldsymbol{p}}a_{\boldsymbol{p}}a_{-\boldsymbol{p}} + C_{\boldsymbol{p}}a_{\boldsymbol{p}}^\dagger a_{\boldsymbol{p}} + (\boldsymbol{p}\to -\boldsymbol{p}) , \tag{75}$$

with

$$\begin{aligned} A_{\boldsymbol{p}} &= \frac{2\pi a_{\rm sc}\hbar^2}{m}\frac{N^2}{V}\cdot\frac{4\pi a_{\rm sc}\hbar^2}{V}\frac{1}{\boldsymbol{p}^2} \\ B_{\boldsymbol{p}} &= \frac{2\pi a_{\rm sc}\hbar^2}{m}\frac{N}{V} \\ C_{\boldsymbol{p}} &= \frac{4\pi a_{\rm sc}\hbar^2}{m}\frac{N}{V} + \frac{\boldsymbol{p}^2}{2m} , \end{aligned} \tag{76}$$

so that $A_{\boldsymbol{p}} = A_{-\boldsymbol{p}}$, etc. The goal is to bring such a Hamiltonian into the diagonal form

$$H_{\boldsymbol{p}} = \alpha(\boldsymbol{p}) + \varepsilon(\boldsymbol{p})\, b_{\boldsymbol{p}}^\dagger b_{\boldsymbol{p}} + (\boldsymbol{p}\to -\boldsymbol{p}) , \tag{77}$$

with new operators $b_{\boldsymbol{p}}^\dagger$, $b_{\boldsymbol{p}}$ which are required to obey Bose commutation rules,

$$[b_{\boldsymbol{p}}, b_{\boldsymbol{p}'}^\dagger] = \delta_{\boldsymbol{p},\boldsymbol{p}'} . \tag{78}$$

Then $b_{\boldsymbol{p}}^\dagger$ and $b_{\boldsymbol{p}}$ can be interpreted as creation and annihilation operators for noninteracting quasiparticles with energy $\varepsilon(\boldsymbol{p})$, which again obey Bose statistics.

The desired transformation from the original form (75) to the quasiparticle form (77) is achieved by a linear combination of the operators referring to the original particles, such that the annihilation operator $a_{\boldsymbol{p}}$ for a particle with momentum $\boldsymbol{p}$ is combined with the creation operator $a^\dagger_{-\boldsymbol{p}}$ for a particle with reverse momentum $-\boldsymbol{p}$:

$$\begin{aligned} b_{\boldsymbol{p}} &= u_{\boldsymbol{p}} a_{\boldsymbol{p}} + v_{\boldsymbol{p}} a^\dagger_{-\boldsymbol{p}} \\ b^\dagger_{\boldsymbol{p}} &= u_{\boldsymbol{p}} a^\dagger_{\boldsymbol{p}} + v_{\boldsymbol{p}} a_{-\boldsymbol{p}} , \end{aligned} \tag{79}$$

with real coefficients $u_{\boldsymbol{p}} = u_{-\boldsymbol{p}}$, $v_{\boldsymbol{p}} = v_{-\boldsymbol{p}}$ which still have to be determined. One then computes

$$[b_{\boldsymbol{p}}, b^\dagger_{\boldsymbol{p}}] = u^2_{\boldsymbol{p}} - v^2_{\boldsymbol{p}} , \tag{80}$$

so that the requirement (78) gives

$$u^2_{\boldsymbol{p}} - v^2_{\boldsymbol{p}} = 1 , \tag{81}$$

implying that there is only one independent transformation parameter. Thus, instead of the parameters $u_{\boldsymbol{p}}$ and $v_{\boldsymbol{p}}$ one can introduce a parameter $L_{\boldsymbol{p}}$ through

$$u_{\boldsymbol{p}} = \frac{1}{\sqrt{1 - L^2_{\boldsymbol{p}}}} , \quad v_{\boldsymbol{p}} = \frac{L_{\boldsymbol{p}}}{\sqrt{1 - L^2_{\boldsymbol{p}}}} . \tag{82}$$

Then

$$L_{\boldsymbol{p}} = \frac{v_{\boldsymbol{p}}}{u_{\boldsymbol{p}}} , \tag{83}$$

and (81) is satisfied automatically.

Inserting the ansatz (79) for the quasiparticle operators into the Hamiltonian (77), and reshuffling terms with indices $\boldsymbol{p}$ and $-\boldsymbol{p}$ as appropriate, one obtains

$$\begin{aligned} H_{\boldsymbol{p}} = \alpha(\boldsymbol{p}) &+ \varepsilon(\boldsymbol{p}) v^2_{\boldsymbol{p}} \\ &+ \varepsilon(\boldsymbol{p})\, u_{\boldsymbol{p}} v_{\boldsymbol{p}}\, a^\dagger_{\boldsymbol{p}} a^\dagger_{-\boldsymbol{p}} + \varepsilon(\boldsymbol{p})\, u_{\boldsymbol{p}} v_{\boldsymbol{p}}\, a_{\boldsymbol{p}} a_{-\boldsymbol{p}} + \varepsilon(\boldsymbol{p})(u^2_{\boldsymbol{p}} + v^2_{\boldsymbol{p}})\, a^\dagger_{\boldsymbol{p}} a_{\boldsymbol{p}} \\ &+ (\boldsymbol{p} \to -\boldsymbol{p}) . \end{aligned} \tag{84}$$

Comparison with the original form (75) then reveals

$$\begin{aligned} A_{\boldsymbol{p}} &= \alpha(\boldsymbol{p}) + \varepsilon(\boldsymbol{p}) v^2_{\boldsymbol{p}} \\ B_{\boldsymbol{p}} &= \varepsilon(\boldsymbol{p}) u_{\boldsymbol{p}} v_{\boldsymbol{p}} \\ C_{\boldsymbol{p}} &= \varepsilon(\boldsymbol{p})(u^2_{\boldsymbol{p}} + v^2_{\boldsymbol{p}}) , \end{aligned} \tag{85}$$

and inversion of these equations with the help of the relation (81) finally gives

$$\begin{aligned} \varepsilon(\boldsymbol{p})^2 &= C^2_{\boldsymbol{p}} - 4B^2_{\boldsymbol{p}} \\ \varepsilon(\boldsymbol{p}) - 2\alpha(\boldsymbol{p}) &= C_{\boldsymbol{p}} - 2A_{\boldsymbol{p}} . \end{aligned} \tag{86}$$

Applying this scheme, for each pair $(\boldsymbol{p}, -\boldsymbol{p})$ of nonzero momenta, to the approximate Hamiltonian (74) pertaining to the low-energy dynamics of the weakly interacting Bose gas, insertion of the parameters (76) into the first of the equations (86) immediately yields the famous Bogoliubov quasiparticle spectrum,

$$\varepsilon(\boldsymbol{p}) = \left[\left(\frac{\boldsymbol{p}^2}{2m} \right)^2 + \frac{4\pi a_{\rm sc}\hbar^2}{m} \frac{N}{V} \frac{\boldsymbol{p}^2}{m} \right]^{1/2} . \tag{87}$$

Thus, $\varepsilon(\boldsymbol{p})$ increases linearly with $p = |\boldsymbol{p}|$ for low momenta, $\varepsilon(\boldsymbol{p}) \approx cp$, where

$$c = \left(\frac{4\pi a_{\rm sc}\hbar^2}{m^2} \frac{N}{V} \right)^{1/2} \tag{88}$$

denotes the velocity of sound in the weakly interacting gas, so that the Bogoliubov quasiparticles are identified as phonons.

The second of the equations (86) then gives the energy shift

$$\begin{aligned} \alpha(\boldsymbol{p}) &= \frac{1}{2} \left(\varepsilon(\boldsymbol{p}) - C_{\boldsymbol{p}} + 2A_{\boldsymbol{p}} \right) \\ &= \frac{1}{2} \left[\varepsilon(\boldsymbol{p}) - \frac{\boldsymbol{p}^2}{2m} - \frac{4\pi a_{\rm sc}\hbar^2}{m} \frac{N}{V} + \left(\frac{4\pi a_{\rm sc}\hbar^2}{m} \frac{N}{V} \right)^2 \frac{m}{\boldsymbol{p}^2} \right] , \end{aligned} \tag{89}$$

while the transformation parameter $L_{\boldsymbol{p}} = v_{\boldsymbol{p}}/u_{\boldsymbol{p}}$ is expressed, by means of the first and the second of the equations (85), as

$$\begin{aligned} L_{\mathbf{p}} &= \frac{\varepsilon(\boldsymbol{p}) v_{\boldsymbol{p}}^2}{\varepsilon(\boldsymbol{p}) u_{\boldsymbol{p}} v_{\boldsymbol{p}}} \\ &= \frac{C_{\boldsymbol{p}} - \varepsilon(\boldsymbol{p})}{2B_{\boldsymbol{p}}} \\ &= 1 + \left(\frac{4\pi a_{\rm sc}\hbar^2}{m} \frac{N}{V} \right)^{-1} \left(\frac{\boldsymbol{p}^2}{2m} - \varepsilon(\boldsymbol{p}) \right) . \end{aligned} \tag{90}$$

Hence, the Bogoliubov transformation (79) brings the Hamiltonian (74) into the quasiparticle form

$$\begin{aligned} H &= \frac{2\pi a_{\rm sc}\hbar^2}{m} \frac{N^2}{V} + \sum_{\boldsymbol{p}\neq\boldsymbol{0}} H_{\boldsymbol{p}} \\ &= E_0 + \sum_{\boldsymbol{p}\neq\boldsymbol{0}} \varepsilon(\boldsymbol{p})\, b_{\boldsymbol{p}}^\dagger b_{\boldsymbol{p}} , \end{aligned} \tag{91}$$

where the ground-state energy E_0 now includes a correction to the previous crude estimate (67):

$$\begin{aligned} E_0 &= \frac{2\pi a_{\rm sc}\hbar^2}{m} \frac{N^2}{V} + \sum_{\boldsymbol{p}\neq\boldsymbol{0}} \alpha(\boldsymbol{p}) \\ &= \frac{2\pi a_{\rm sc}\hbar^2}{m} \frac{N^2}{V} + \frac{1}{2} \sum_{\boldsymbol{p}\neq\boldsymbol{0}} \left[\varepsilon(\boldsymbol{p}) - \frac{\boldsymbol{p}^2}{2m} - \frac{4\pi a_{\rm sc}\hbar^2}{m} \frac{N}{V} + \left(\frac{4\pi a_{\rm sc}\hbar^2}{m} \frac{N}{V} \right)^2 \frac{m}{\boldsymbol{p}^2} \right] . \end{aligned} \tag{92}$$

In order to recover the ideal gas from the weakly interacting Bose gas within the Bogoliubov theory, one has to stipulate that the interaction energy per particle be small compared to the nonzero unperturbed single-particle energies, so that

$$\frac{4\pi a_{\rm sc}\hbar^2}{m}\frac{N}{V} \ll \frac{\boldsymbol{p}^2}{2m} \tag{93}$$

for *all* momenta $\boldsymbol{p} \neq \boldsymbol{0}$. Since, in accordance with the periodic boundary conditions imposed on the volume $V = L^3$, one has $\boldsymbol{p} = \hbar(2\pi/L)\boldsymbol{n}$, where the vector $\boldsymbol{n}$ has integer components, the requirement (93) implies

$$\frac{4\pi a_{\rm sc}\hbar^2}{m}\frac{N}{L^3} \ll \frac{\hbar^2(2\pi)^2}{2mL^2} , \tag{94}$$

or

$$\frac{Na_{\rm sc}}{L} \ll 1 . \tag{95}$$

In this limiting case the quasiparticle energies (87) can be written as

$$\varepsilon(\boldsymbol{p}) = \frac{\boldsymbol{p}^2}{2m} + \frac{4\pi a_{\rm sc}\hbar^2}{m}\frac{N}{V} + \mathcal{O}(a_{\rm sc}^2) , \tag{96}$$

and thus differ from the unperturbed single-particle energies, to first order in $a_{\rm sc}$, by a constant ($\boldsymbol{p}$-independent) shift. As a consequence, (90) now yields

$$L_{\boldsymbol{p}} = \mathcal{O}(a_{\rm sc}) , \tag{97}$$

so that $L_{\boldsymbol{p}} \to 0$ when $a_{\rm sc} \to 0$. (Note that because of (95) the finite volume V has to be kept fixed when taking the limit $a_{\rm sc} \to 0$.) It follows that $u_{\boldsymbol{p}} \to 1$ and $v_{\boldsymbol{p}} \to 0$, enforcing, by virtue of (79), $b_{\boldsymbol{p}} \to a_{\boldsymbol{p}}$ and $b_{\boldsymbol{p}}^\dagger \to a_{\boldsymbol{p}}^\dagger$. Thus, the Bogoliubov quasiparticles properly reduce to the bare Bose particles when the scattering length vanishes.

However, even for dilute atomic vapours the ideal-gas condition (95) usually is not met: Considering, as in the example at the beginning of this section, a vapour of ^{87}Rb with a density of 10^{20} m^{-3}, and taking a linear extension of $L = 10$ µm, say, one finds $Na_{\rm sc}/L \approx 55$. Hence, under realistic conditions one explores the opposite limit

$$\frac{Na_{\rm sc}}{L} \gg 1 \tag{98}$$

even with weakly interacting Bose gases. In this case (98), we see from (90) that

$$L_{\boldsymbol{p}} \to 1 \tag{99}$$

for the low-lying momenta $\boldsymbol{p}$, for which the interaction energy per particle considerably exceeds the single-particle energies.

3.2 Ground-State Energy and Depletion

It is quite illustrative to evaluate the sum over the energy shifts $\alpha(\boldsymbol{p})$ appearing in (92) in the thermodynamic limit, and thus to calculate the ground-state energy explicitly within the present set of approximations. To this end, we utilize the density $\varrho(p)$ of states in momentum space,

$$\begin{aligned}\varrho(p)\,\mathrm{d}p &= \frac{V\,4\pi p^2\,\mathrm{d}p}{(2\pi\hbar)^3} \\ &= \frac{Vp^2\,\mathrm{d}p}{2\pi^2\hbar^3}\,.\end{aligned} \tag{100}$$

Since the natural energy scale of the weakly interacting Bose gas is set by the quantity $4\pi a_{\mathrm{sc}}\hbar^2 N/(mV)$, we define a dimensionless momentum x through the relation

$$\frac{\boldsymbol{p}^2}{2m} = \frac{4\pi a_{\mathrm{sc}}\hbar^2}{m}\frac{N}{V}x^2\,, \tag{101}$$

so that

$$\varepsilon(\boldsymbol{p}) = \frac{4\pi a_{\mathrm{sc}}\hbar^2}{m}\frac{N}{V}x\sqrt{x^2+2}\,. \tag{102}$$

According to (92), one then has

$$E_0 = \frac{2\pi a_{\mathrm{sc}}\hbar^2}{m}\frac{N^2}{V}\left[1+\sqrt{\frac{128}{\pi}}\left(\frac{N}{V}a_{\mathrm{sc}}^3\right)^{1/2} I\right]\,, \tag{103}$$

where the number I results from converting, with the help of the density (100), the sum over $\boldsymbol{p}\neq\boldsymbol{0}$ into an integral over x:

$$I = \int_0^\infty \mathrm{d}x\,x^2\left(x\sqrt{x^2+2}-x^2-1+\frac{1}{2x^2}\right)\,. \tag{104}$$

Convergence of this integral implies a subtle cancellation of the various diverging contributions. Indeed, one finds

$$\begin{aligned} I &= \lim_{a\to\infty}\left(\int_0^a \mathrm{d}x\,x^3\sqrt{x^2+2}-\frac{1}{5}a^5-\frac{1}{3}a^3+\frac{1}{2}a\right) \\ &= \frac{2}{15}2^{5/2} \\ &\quad + \lim_{a\to\infty}\left[-\frac{2}{15}a^5\left(1+\frac{2}{a^2}\right)^{5/2}+\frac{1}{3}a^5\left(1+\frac{2}{a^2}\right)^{3/2}-\frac{1}{5}a^5-\frac{1}{3}a^3+\frac{1}{2}a\right] \\ &= \frac{\sqrt{128}}{15}\,, \end{aligned} \tag{105}$$

leading to the "classic" result for E_0 first derived by Lee and Yang in 1957 [32]:

$$E_0 = \frac{2\pi a_{\mathrm{sc}}\hbar^2}{m}\frac{N^2}{V}\left[1+\frac{128}{15\sqrt{\pi}}\left(\frac{N}{V}a_{\mathrm{sc}}^3\right)^{1/2}\right]\,. \tag{106}$$

In particular, one now appreciates the significance of the second Born approximation (73), which had determined the coefficient $A_{\boldsymbol{p}}$ in (76): While that approximation does not influence the quasiparticle spectrum (since $\varepsilon(\boldsymbol{p})$ is linked to $B_{\boldsymbol{p}}$ and $C_{\boldsymbol{p}}$ only), it is responsible for the $\boldsymbol{p}^{-2}$-proportional contribution to the shift $\alpha(\boldsymbol{p})$. This contribution, in its turn, is necessary to ensure the convergence of the integral I (without it, the integral would be linearly divergent, and negative), and, hence, to provide a finite ground-state energy in the thermodynamic limit.

Let us also consider the depletion of the condensate at temperature $T = 0$, that is, the extent to which single-particle states other than the single-particle ground state are occupied when the weakly interacting gas is in its many-particle ground state $|\Psi_0\rangle$. To this end, we first invert the transformation formulas (79) with the help of (81), obtaining

$$\begin{aligned} a_{\boldsymbol{p}} &= u_{\boldsymbol{p}} b_{\boldsymbol{p}} - v_{\boldsymbol{p}} b^\dagger_{-\boldsymbol{p}} \\ a^\dagger_{\boldsymbol{p}} &= u_{\boldsymbol{p}} b^\dagger_{\boldsymbol{p}} - v_{\boldsymbol{p}} b_{-\boldsymbol{p}} \end{aligned} \tag{107}$$

or

$$\begin{aligned} a_{\boldsymbol{p}} &= \frac{b_{\boldsymbol{p}} - L_{\boldsymbol{p}} b^\dagger_{-\boldsymbol{p}}}{\sqrt{1 - L_{\boldsymbol{p}}^2}} \\ a^\dagger_{\boldsymbol{p}} &= \frac{b^\dagger_{\boldsymbol{p}} - L_{\boldsymbol{p}} b_{-\boldsymbol{p}}}{\sqrt{1 - L_{\boldsymbol{p}}^2}} \,, \end{aligned} \tag{108}$$

employing the transformation parameter $L_{\boldsymbol{p}} = v_{\boldsymbol{p}}/u_{\boldsymbol{p}}$; the inverse transformation thus differs from the "forward" scheme (79) only by the sign accompanying the negative-momentum admixture. Since the many-particle ground state $|\Psi_0\rangle$ acts as the vacuum for the quasiparticle operators, so that $b_{\boldsymbol{p}}|\Psi_0\rangle = 0$ for $\boldsymbol{p} \neq \boldsymbol{0}$, we then have

$$a_{\boldsymbol{p}}|\Psi_0\rangle = \frac{-L_{\boldsymbol{p}}}{\sqrt{1 - L_{\boldsymbol{p}}}} \, b^\dagger_{-\boldsymbol{p}}|\Psi_0\rangle \,, \tag{109}$$

implying

$$\begin{aligned} \langle\Psi_0|a^\dagger_{\boldsymbol{p}} a_{\boldsymbol{p}}|\Psi_0\rangle &= \frac{L_{\boldsymbol{p}}^2}{1 - L_{\boldsymbol{p}}^2} \, \langle\Psi_0|b_{-\boldsymbol{p}} b^\dagger_{-\boldsymbol{p}}|\Psi_0\rangle \\ &= \frac{L_{\boldsymbol{p}}^2}{1 - L_{\boldsymbol{p}}^2} \,. \end{aligned} \tag{110}$$

This is the expectation value for the occupation number $\widehat{n}_{\boldsymbol{p}}$ of the single-particle state with momentum $\boldsymbol{p}$ when the system is in the ground state. Invoking the dimensionless magnitude x of the momentum defined by (101), the result (90) for the Bogoliubov parameter $L_{\boldsymbol{p}}$ becomes

$$L_{\boldsymbol{p}} = 1 + x^2 - x\sqrt{x^2 + 2} \,, \tag{111}$$

giving

$$\langle \Psi_0 | a_{\boldsymbol{p}}^\dagger a_{\boldsymbol{p}} | \Psi_0 \rangle = \frac{1}{2} \frac{1+x^2}{x\sqrt{x^2+2}} - \frac{1}{2} \, . \tag{112}$$

Again employing the density of states (100) to convert the sum over $\boldsymbol{p}$ into an integral over x, the total depletion of the condensate is expressed as

$$\sum_{\boldsymbol{p} \neq \boldsymbol{0}} \langle \Psi_0 | a_{\boldsymbol{p}}^\dagger a_{\boldsymbol{p}} | \Psi_0 \rangle = N \left(\frac{32}{\pi} \frac{N}{V} a_{\rm sc}^3 \right)^{1/2} \int_0^\infty {\rm d}x \left(\frac{x + x^3}{\sqrt{x^2+2}} - x^2 \right) \, . \tag{113}$$

The integral appearing here is

$$\lim_{a \to \infty} \int_0^a {\rm d}x \left(\frac{x + x^3}{\sqrt{x^2+2}} - x^2 \right) = \frac{\sqrt{2}}{3} \, , \tag{114}$$

so that the depletion takes the final form

$$\sum_{\boldsymbol{p} \neq \boldsymbol{0}} \langle \Psi_0 | a_{\boldsymbol{p}}^\dagger a_{\boldsymbol{p}} | \Psi_0 \rangle = N \frac{8}{3\sqrt{\pi}} \left(\frac{N}{V} a_{\rm sc}^3 \right)^{1/2} \; ; \tag{115}$$

accordingly, the expectation value for the number of particles occupying the single-particle ground state at zero temperature reads [33]

$$\langle \Psi_0 | a_0^\dagger a_0 | \Psi_0 \rangle = N \left[1 - \frac{8}{3\sqrt{\pi}} \left(\frac{N}{V} a_{\rm sc}^3 \right)^{1/2} \right] \, . \tag{116}$$

Thus, the depletion of the condensate is only of the order $\mathcal{O}\big((N a_{\rm sc}^3/V)^{1/2}\big)$, which, according to the proposition (57), is small compared to unity. Therefore, the result (115) constitutes an *a posteriori*-justification of the key approximation (64), and expresses the consistency of the foregoing analysis.

4 Canonical Statistics of Weakly Interacting Bose–Einstein Condensates

4.1 The Pair Characteristic Function

We now investigate the canonical statistics of the number of particles in a weakly interacting, homogeneous Bose–Einstein condensate within the Bogoliubov approximation. Since particles with momentum $\boldsymbol{p}$ are then coupled, by virtue of (79), to particles with momentum $-\boldsymbol{p}$, we consider, in the spirit of Sect. 2.3, the characteristic function of the sum of the occupation numbers $\widehat{n}_{\boldsymbol{p}}$ and $\widehat{n}_{-\boldsymbol{p}}$ [14]. By immediate generalization of (47), this pair characteristic function is given by

$$\begin{aligned} G_{\boldsymbol{p}}(y) &= \langle {\rm e}^{{\rm i}y(\widehat{n}_{\boldsymbol{p}} + \widehat{n}_{-\boldsymbol{p}})} \rangle \\ &= {\rm Tr} \left({\rm e}^{{\rm i}y(\widehat{n}_{\boldsymbol{p}} + \widehat{n}_{-\boldsymbol{p}})} \, \widehat{\varrho}_{\boldsymbol{p}} \, \widehat{\varrho}_{-\boldsymbol{p}} \right) \, , \end{aligned} \tag{117}$$

generating the moments

$$\langle\, (\widehat{n}_{\boldsymbol{p}} + \widehat{n}_{-\boldsymbol{p}})^k \,\rangle = \left. \frac{\mathrm{d}^k}{\mathrm{d}(\mathrm{i}y)^k} G_{\boldsymbol{p}}(y) \right|_{y=0} . \tag{118}$$

Whereas the stochastic variables $\widehat{n}_{\boldsymbol{p}}$ and $\widehat{n}_{-\boldsymbol{p}}$ refer to the bare Bose particles, the canonical density matrix $\widehat{\varrho}_{\boldsymbol{p}}$ now quantifies the excitation of Bogoliubov quasiparticles at inverse temperature β,

$$\widehat{\varrho}_{\boldsymbol{p}} = \left(1 - \mathrm{e}^{-\beta\varepsilon(\boldsymbol{p})}\right) \exp(-\beta\varepsilon(\boldsymbol{p})\, b^\dagger_{\boldsymbol{p}} b_{\boldsymbol{p}}) \; ; \tag{119}$$

note that $\varepsilon(\boldsymbol{p}) = \varepsilon(-\boldsymbol{p})$. Since this density matrix is diagonal in a basis of quasiparticle Fock states, we also express the occupation numbers $a^\dagger_{\boldsymbol{p}} a_{\boldsymbol{p}}$ and $a^\dagger_{-\boldsymbol{p}} a_{-\boldsymbol{p}}$ for the actual particles in terms of quasiparticle operators. To this end, we utilize the inverse transformation formula (107) to write

$$\begin{aligned} &\widehat{n}_{\boldsymbol{p}} + \widehat{n}_{-\boldsymbol{p}} = a^\dagger_{\boldsymbol{p}} a_{\boldsymbol{p}} + a^\dagger_{-\boldsymbol{p}} a_{-\boldsymbol{p}} \\ &\qquad = (u^2_{\boldsymbol{p}} + v^2_{\boldsymbol{p}})(b^\dagger_{\boldsymbol{p}} b_{\boldsymbol{p}} + b^\dagger_{-\boldsymbol{p}} b_{-\boldsymbol{p}}) - 2u_{\boldsymbol{p}} v_{\boldsymbol{p}} (b^\dagger_{\boldsymbol{p}} b^\dagger_{-\boldsymbol{p}} + b_{\boldsymbol{p}} b_{-\boldsymbol{p}}) + 2v^2_{\boldsymbol{p}} \, , \end{aligned} \tag{120}$$

so that the characteristic function (117) acquires the somewhat messy form

$$\begin{aligned} G_{\boldsymbol{p}}(y) &= \left(1 - \mathrm{e}^{-\beta\varepsilon(\boldsymbol{p})}\right)^2 \\ &\mathrm{Tr}\left[\exp\Big(\mathrm{i}y\,(u^2_{\boldsymbol{p}} + v^2_{\boldsymbol{p}})(b^\dagger_{\boldsymbol{p}} b_{\boldsymbol{p}} + b^\dagger_{-\boldsymbol{p}} b_{-\boldsymbol{p}}) - 2\mathrm{i}y\, u_{\boldsymbol{p}} v_{\boldsymbol{p}}\, (b^\dagger_{\boldsymbol{p}} b^\dagger_{-\boldsymbol{p}} + b_{\boldsymbol{p}} b_{-\boldsymbol{p}}) + 2\mathrm{i}y\, v^2_{\boldsymbol{p}}\Big) \right. \\ &\quad \left. \cdot \exp\Big(-\beta\varepsilon(\boldsymbol{p})\, (b^\dagger_{\boldsymbol{p}} b_{\boldsymbol{p}} + b^\dagger_{-\boldsymbol{p}} b_{-\boldsymbol{p}})\Big)\right] . \end{aligned} \tag{121}$$

It is quite remarkable that the trace appearing here can be calculated explicitly, yielding a surprisingly simple expression for the characteristic function [14]. The mathematics underlying this calculation become more clear if we introduce the pair operators

$$\begin{aligned} B_{\boldsymbol{p}} &\equiv b_{\boldsymbol{p}} b_{-\boldsymbol{p}} \\ B^\dagger_{\boldsymbol{p}} &\equiv b^\dagger_{\boldsymbol{p}} b^\dagger_{-\boldsymbol{p}} \, , \end{aligned} \tag{122}$$

the commutator of which is

$$\begin{aligned} [B_{\boldsymbol{p}}, B^\dagger_{\boldsymbol{p}}] &= b^\dagger_{\boldsymbol{p}} b_{\boldsymbol{p}} + b^\dagger_{-\boldsymbol{p}} b_{-\boldsymbol{p}} + 1 \\ &\equiv C_{\boldsymbol{p}} \, . \end{aligned} \tag{123}$$

(Needless to say, these operators $B_{\boldsymbol{p}}$, $C_{\boldsymbol{p}}$ have nothing to do with the coefficients appearing in the model Hamiltonian (75).) In terms of these operators, the characteristic function reads

$$\begin{aligned} G_{\boldsymbol{p}}(y) &= \left(1 - \mathrm{e}^{-\beta\varepsilon(\boldsymbol{p})}\right)^2 \mathrm{e}^{\beta\varepsilon(\boldsymbol{p}) - \mathrm{i}y} \\ &\qquad \mathrm{Tr}\left[\exp\big(\mathrm{i}y(u^2_{\boldsymbol{p}} + v^2_{\boldsymbol{p}})\, C_{\boldsymbol{p}} - 2\mathrm{i}y u_{\boldsymbol{p}} v_{\boldsymbol{p}} (B_{\boldsymbol{p}} + B^\dagger_{\boldsymbol{p}})\big) \exp(-\beta\varepsilon(\boldsymbol{p}) C_{\boldsymbol{p}})\right] . \end{aligned} \tag{124}$$

It is now crucial to observe that the set of the three operators $B_{\boldsymbol{p}}$, $B_{\boldsymbol{p}}^\dagger$, and $C_{\boldsymbol{p}}$ obeys commutation relations which are characteristic for angular momentum operators: Together with (123), we also have

$$\begin{aligned} [B_{\boldsymbol{p}}, C_{\boldsymbol{p}}] &= 2B_{\boldsymbol{p}} \\ [B_{\boldsymbol{p}}^\dagger, C_{\boldsymbol{p}}] &= -2B_{\boldsymbol{p}}^\dagger \; ; \end{aligned} \tag{125}$$

another realization of the same algebra is obtained if we replace $C_{\boldsymbol{p}}$ by the 2×2 Pauli spin matrix σ_z, then $B_{\boldsymbol{p}}$ by the negative lowering operator $-\sigma_-$ defined through (168) in Appendix 1, and $B_{\boldsymbol{p}}^\dagger$ by the raising operator σ_+ given by (167). As explained in detail in Appendix 1, the commutation relations (123) and (125) allow us to "disentangle" the exponential under the trace in (124), and to write the characteristic function in the normal form

$$\begin{aligned} G_{\boldsymbol{p}}(y) = \left(1 - \mathrm{e}^{-\beta\varepsilon(\boldsymbol{p})}\right)^2 \mathrm{e}^{\beta\varepsilon(\boldsymbol{p}) - \mathrm{i}y} \\ \mathrm{Tr}\left[\exp(x_+ B_{\boldsymbol{p}}^\dagger) \exp(-x_- B_{\boldsymbol{p}}) \exp(\ln x_z C_{\boldsymbol{p}}) \exp(-\beta\varepsilon(\boldsymbol{p}) C_{\boldsymbol{p}})\right] \; , \end{aligned} \tag{126}$$

with coefficients x_+, x_-, and x_z which are related to $u_{\boldsymbol{p}}$, $v_{\boldsymbol{p}}$, and y through

$$\begin{aligned} \frac{1}{x_z} &= \cosh(\mathrm{i}y) - (u_{\boldsymbol{p}}^2 + v_{\boldsymbol{p}}^2) \sinh(\mathrm{i}y) \\ &= u_{\boldsymbol{p}}^2 \mathrm{e}^{-\mathrm{i}y} - v_{\boldsymbol{p}}^2 \mathrm{e}^{\mathrm{i}y} \; , \end{aligned} \tag{127}$$

together with

$$\begin{aligned} \frac{x_+}{x_z} &= -2u_{\boldsymbol{p}} v_{\boldsymbol{p}} \sinh(\mathrm{i}y) \\ &= -u_{\boldsymbol{p}} v_{\boldsymbol{p}} (\mathrm{e}^{\mathrm{i}y} - \mathrm{e}^{-\mathrm{i}y}) \end{aligned} \tag{128}$$

and

$$\begin{aligned} x_- x_z &= +2u_{\boldsymbol{p}} v_{\boldsymbol{p}} \sinh(\mathrm{i}y) \\ &= +u_{\boldsymbol{p}} v_{\boldsymbol{p}} (\mathrm{e}^{\mathrm{i}y} - \mathrm{e}^{-\mathrm{i}y}) \; . \end{aligned} \tag{129}$$

The calculation of the trace in (126) can then be managed with the help of Jacobi polynomials, as outlined in Appendix 2. Making use of (195) derived there, one obtains the appealingly simple result

$$G_{\boldsymbol{p}}(y) = \left(1 - \mathrm{e}^{-\beta\varepsilon(\boldsymbol{p})}\right)^2 \frac{\frac{1}{x_z} \, \mathrm{e}^{2\beta\varepsilon(\boldsymbol{p})} \, \mathrm{e}^{-\mathrm{i}y}}{\left(\frac{1}{x_z} \, \mathrm{e}^{\beta\varepsilon(\boldsymbol{p})} - 1\right)^2 + x_+ x_-} \; . \tag{130}$$

All that remains to be done now is to return from the auxiliary disentangling parameters x_+, x_-, and x_z to the Bogoliubov parameters $u_{\boldsymbol{p}}$ and $v_{\boldsymbol{p}}$, and to the actual argument y of the characteristic function. This is achieved by inserting (127) and the product of (128) and (129) into (130) and leads, after some tedious but elementary rearrangements, to the final form

$$G_{\boldsymbol{p}}(y) = \frac{(E(L_{\boldsymbol{p}}) - 1)\,(E(-L_{\boldsymbol{p}}) - 1)}{(E(L_{\boldsymbol{p}}) - \mathrm{e}^{\mathrm{i}y})\,(E(-L_{\boldsymbol{p}}) - \mathrm{e}^{\mathrm{i}y})} \; , \tag{131}$$

where

$$E(L_{\boldsymbol{p}}) = \frac{e^{\beta\varepsilon(\boldsymbol{p})} - L_{\boldsymbol{p}}}{1 - L_{\boldsymbol{p}} e^{\beta\varepsilon(\boldsymbol{p})}} , \tag{132}$$

and $L_{\boldsymbol{p}} = v_{\boldsymbol{p}}/u_{\boldsymbol{p}}$ has been introduced in (83).

4.2 Cumulants of the Condensate Occupation Number

For comparison with the weakly interacting Bose gas within the Bogoliubov approximation, let us recapitulate the key results of Sect. 2.3 for an ideal gas enclosed in a cube $V = L^3$ with periodic boundary conditions. In that case, the characteristic function (47) for an individual single-particle state with momentum $\boldsymbol{p} \neq \boldsymbol{0}$ reads

$$G_{\boldsymbol{p}}(y) = \frac{e^{\beta\varepsilon(\boldsymbol{p})} - 1}{e^{\beta\varepsilon(\boldsymbol{p})} - e^{iy}} , \tag{133}$$

the cumulant-generating function (51) is given by

$$\ln X(y) = \sum_{\boldsymbol{p}\neq\boldsymbol{0}} \ln\left(\frac{e^{\beta\varepsilon(\boldsymbol{p})} - 1}{e^{\beta\varepsilon(\boldsymbol{p})} - e^{iy}} \right) , \tag{134}$$

and the canonical cumulants of the total number of excited particles become

$$\kappa_k(\beta) = \sum_{n=1}^{k} (n-1)!\, \sigma_k^{(n)} \sum_{\boldsymbol{p}\neq\boldsymbol{0}} \frac{1}{(e^{\beta\varepsilon(\boldsymbol{p})} - 1)^n} . \tag{135}$$

In contrast, in the case of a weakly interacting Bose gas we encounter pairs $(\boldsymbol{p}, -\boldsymbol{p})$ of momentum states, with the pair characteristic function (131), so that the cumulant-generating function now takes the form [14]

$$\ln X(y) = \frac{1}{2} \sum_{\boldsymbol{p}\neq\boldsymbol{0}} \left[\ln\left(\frac{E(L_{\boldsymbol{p}}) - 1}{E(L_{\boldsymbol{p}}) - e^{iy}} \right) + \ln\left(\frac{E(-L_{\boldsymbol{p}}) - 1}{E(-L_{\boldsymbol{p}}) - e^{iy}} \right) \right] . \tag{136}$$

Thus, the weakly interacting Bose gas appears as a mixture of two ideal gases, such that the usual exponential $e^{\beta\varepsilon(\boldsymbol{p})}$ is replaced by $E(L_{\boldsymbol{p}})$ and $E(-L_{\boldsymbol{p}})$, as defined by (132). Therefore, one can immediately write down a formula for the canonical cumulants of the total number of excited particles in a weakly interacting Bose gas within the Bogoliubov approximation, valid for low temperatures such that interactions between the Bogoliubov quasiparticles can still be neglected [14]:

$$\kappa_k(\beta) = \frac{1}{2} \sum_{n=1}^{k} (n-1)!\, \sigma_k^{(n)} \sum_{\boldsymbol{p}\neq 0} \left(\frac{1}{(E(L_{\boldsymbol{p}}) - 1)^n} + \frac{1}{(E(-L_{\boldsymbol{p}}) - 1)^n} \right) . \tag{137}$$

Since the number $\widehat{N}_{\rm ex}$ of excited particles is the mirror image of the number $\widehat{n}_0$ of condensate particles, $\widehat{N}_{\rm ex} = N - \widehat{n}_0$, this formula (137) also governs the condensate statistics.

Specializing (137) to the first cumulant $\kappa_1(\beta) = \langle \widehat{N}_{\rm ex} \rangle$, and substituting for $E(L_{\boldsymbol p})$ and $E(-L_{\boldsymbol p})$ according to (132), one finds

$$\kappa_1(\beta) = \sum_{\boldsymbol{p}\neq\boldsymbol{0}} \left(\frac{1+L_{\boldsymbol p}^2}{1-L_{\boldsymbol p}^2} \frac{1}{\mathrm{e}^{\beta\varepsilon(\boldsymbol{p})}-1} + \frac{L_{\boldsymbol p}^2}{1-L_{\boldsymbol p}^2} \right) . \tag{138}$$

From the temperature-independent part here one recovers the condensate depletion (115), that is, the effect of quantum fluctuations at $T = 0$, whereas the temperature-dependent terms sum up to

$$\begin{aligned} &\sum_{\boldsymbol{p}\neq\boldsymbol{0}} \frac{1+L_{\boldsymbol p}^2}{1-L_{\boldsymbol p}^2} \frac{1}{\mathrm{e}^{\beta\varepsilon(\boldsymbol{p})}-1} \\ &= 8N \left(\frac{2}{\pi} \frac{N}{V} a_{\rm sc}^3 \right)^{1/2} \int_0^\infty \mathrm{d}x \, x^2 \left(\frac{1+x^2}{x\sqrt{x^2+2}} \right) \frac{1}{\exp(bx\sqrt{x^2+2})-1} \\ &= 8N \left(\frac{2}{\pi} \frac{N}{V} a_{\rm sc}^3 \right)^{1/2} \int_0^\infty \frac{\mathrm{d}y}{2b} \frac{x(y)}{\mathrm{e}^y - 1} . \end{aligned} \tag{139}$$

Here we have once again taken the thermodynamic limit, so that the sum over the momenta $\boldsymbol{p} \neq \boldsymbol{0}$ is replaced by an integral weighted with the density of states (100), and have made use of the dimensionless momentum variable x introduced previously in (101). In addition, in the second line of (139) we have employed the scaled (dimensionles) inverse temperature

$$b = \beta \, \frac{4\pi a_{\rm sc}\hbar^2}{m} \frac{N}{V} , \tag{140}$$

and made the substitution

$$y = bx\sqrt{x^2+2} \tag{141}$$

in the third, so that

$$x(y) = \sqrt{\sqrt{\left(\frac{y}{b}\right)^2 + 1} - 1} . \tag{142}$$

For sufficiently low temperatures, when the integral is dominated by contributions with $y/b \ll 1$, corresponding to low-energy modes with

$$\varepsilon(\boldsymbol{p}) \ll \frac{4\pi a_{\rm sc}\hbar^2}{m} \frac{N}{V} , \tag{143}$$

one may use the approximation

$$x(y) \approx \frac{y}{\sqrt{2b}} , \tag{144}$$

leading to the standard integral

$$\int_0^\infty \frac{\mathrm{d}y}{2b} \frac{\frac{y}{\sqrt{2b}}}{\mathrm{e}^y - 1} = \frac{1}{2^{3/2}b^2} \frac{\pi^2}{6} . \tag{145}$$

Therefore, at low T the temperature-dependent part of the number of excited particles in a Bogoliubov gas takes the form

$$\sum_{\boldsymbol{p}\neq\boldsymbol{0}} \frac{1+L_{\boldsymbol{p}}^2}{1-L_{\boldsymbol{p}}^2} \frac{1}{\mathrm{e}^{\beta\varepsilon(\boldsymbol{p})}-1} \approx \frac{mV(k_{\mathrm{B}}T)^2}{12c\hbar^3} , \tag{146}$$

where c is the velocity of sound, as encountered in (88). Interestingly, this thermal depletion (146) of the condensate grows proportionally to T^2 [34], whereas the number of excited particles in a homogeneous ideal Bose gas increases proportionally to $T^{3/2}$. Adding the depletion at $T = 0$, as given by (115), one finally obtains [11]

$$\langle \widehat{N}_{\mathrm{ex}} \rangle \approx N \left[\frac{m(k_{\mathrm{B}}T)^2}{12\frac{N}{V}c\hbar^3} + \frac{8}{3\sqrt{\pi}} \left(\frac{N}{V}\, a_{\mathrm{sc}}^3 \right)^{1/2} \right] . \tag{147}$$

As already remarked, the validity of this result is restricted to low temperatures, where the approximation (144) is viable and the Bogoliubov quasiparticles do not interact.

When evaluating the second cumulant $\kappa_2(\beta)$, which equals the canonical mean-square fluctuation of the number of condensate particles, (137) takes the suggestive form

$$\begin{aligned} \kappa_2(\beta) &= \kappa_1(\beta) + \sum_{\boldsymbol{p}\neq\boldsymbol{0}} \left(\frac{1}{(E(L_{\boldsymbol{p}})-1)^2} + \frac{1}{(E(-L_{\boldsymbol{p}})-1)^2} \right) \\ &= \sum_{\boldsymbol{p}\neq\boldsymbol{0}} \left[\frac{(L_{\boldsymbol{p}}^2+1)^2+4L_{\boldsymbol{p}}^2}{(1-L_{\boldsymbol{p}}^2)^2} \left(\frac{1}{\mathrm{e}^{\beta\varepsilon(\boldsymbol{p})}-1} + 1 \right) \frac{1}{\mathrm{e}^{\beta\varepsilon(\boldsymbol{p})}-1} + \frac{2L_{\boldsymbol{p}}^2}{(1-L_{\boldsymbol{p}}^2)^2} \right] . \end{aligned} \tag{148}$$

Once again, in the thermodynamic limit the mean-square fluctuation at $T = 0$ can be computed by transforming the sum into an integral, giving

$$\sum_{\boldsymbol{p}\neq\boldsymbol{0}} \frac{2L_{\boldsymbol{p}}^2}{(1-L_{\boldsymbol{p}}^2)^2} = 2N \left(\pi \frac{N}{V}\, a_{\mathrm{sc}}^3 \right)^{1/2} . \tag{149}$$

The temperature-dependent part of the fluctuation, however, cannot be treated in this manner, since the emerging integral would be infrared-divergent. This means that the sum over the momenta $\boldsymbol{p} \neq \boldsymbol{0}$ is dominated by the low-energy contributions, and has to be evaluated in discretized form [11]. Indeed, from the expression (111) for the Bogoliubov parameter $L_{\boldsymbol{p}}$ one deduces

$$1 - L_{\boldsymbol{p}}^2 \approx 2\sqrt{2}x \tag{150}$$

for low scaled momenta $x = x(\boldsymbol{p})$, as given by (101). Moreover, for low energies the Bose factors can be approximated as

$$\begin{aligned} \frac{1}{\mathrm{e}^{\beta\varepsilon(\boldsymbol{p})}-1} &= \frac{1}{\exp(bx\sqrt{x^2+2})-1} \\ &\approx \frac{1}{\sqrt{2}\,b\,x} , \end{aligned} \tag{151}$$

with scaled inverse temperature b as introduced in (140). Therefore, the dominant contribution to the thermal condensate fluctuation behaves as

$$\begin{aligned} & \sum_{\boldsymbol{p}\neq\boldsymbol{0}} \frac{(L_{\boldsymbol{p}}^2+1)^2+4L_{\boldsymbol{p}}^2}{(1-L_{\boldsymbol{p}}^2)^2}\left(\frac{1}{\mathrm{e}^{\beta\varepsilon(\boldsymbol{p})}-1}+1\right)\frac{1}{\mathrm{e}^{\beta\varepsilon(\boldsymbol{p})}-1} \\ \sim & \sum_{\boldsymbol{p}\neq\boldsymbol{0}} \frac{8}{8x^2}\,\frac{1}{2b^2x^2} \\ = & \frac{(k_{\mathrm{B}}T)^2}{2}\sum_{\boldsymbol{p}\neq\boldsymbol{0}}\frac{1}{\left(\frac{\boldsymbol{p}^2}{2m}\right)^2} \; . \end{aligned} \tag{152}$$

Hence, respecting the periodic boundary conditions and setting $\boldsymbol{p} = \hbar(2\pi/L)\boldsymbol{n}$, one finds

$$\kappa_2(\beta) \sim \frac{(2mk_{\mathrm{B}}T)^2}{2}\left(\frac{L}{2\pi\hbar}\right)^4 S(2) \; , \tag{153}$$

where $S(2)$ denotes the remaining lattice sum [17],

$$\begin{aligned} S(2) = & \sum_{n_1,n_2,n_3=-\infty}^{+\infty}{}' \frac{1}{(n_1^2+n_2^2+n_3^2)^2} \\ \approx & \; 16.532 \; . \end{aligned} \tag{154}$$

This result (153) for the condensate fluctuations is remarkable in several respects [11]. Firstly, it is independent of the interaction parameter a_{sc}, and thus of the velocity of sound. Secondly, the mean-square fluctuations are proportional to $L^4 = V^{4/3}$, and thus scale anomalously with the volume. Thirdly, they amount to exactly half of the condensate fluctuations in the ideal Bose gas: With

$$\frac{1}{\exp[\beta\boldsymbol{p}^2/(2m)]-1} \approx \frac{1}{\beta\frac{\boldsymbol{p}^2}{2m}} \tag{155}$$

for low $\boldsymbol{p}$, the ideal-gas formula (24) gives [22]

$$\kappa_2(\beta) \sim (k_{\mathrm{B}}T)^2 \sum_{\boldsymbol{p}\neq\boldsymbol{0}} \frac{1}{\left(\frac{\boldsymbol{p}^2}{2m}\right)^2} \; , \tag{156}$$

two times the expression obtained in (152) for the Bogoliubov case. One might have expected that the inclusion of two-body interactions would remove the anomalous condensate fluctuations of the homogeneous ideal gas, and lead to normal, V-proportional mean-square fluctuations in the Bogoliubov approximation, but this expectation turns out to be wrong: As a consequence of the interaction, the condensate fluctuations are reduced by a factor of two, but remain anomalous.

This unexpected relationship between the ideal and the weakly interacting Bose gas, which has been extended even to strongly interacting superfluids [12], is not fortuitous [14]. Namely, for $k \geq 2$, when the sum (137) is dominated by low-energy contributions, we may exploit (99) and consider the limit $L_{\boldsymbol{p}} \to 1$ for the relevant momenta $\boldsymbol{p}$, implying

$$\begin{aligned} E(L_{\boldsymbol{p}}) &\approx \frac{\mathrm{e}^{\beta\varepsilon(\boldsymbol{p})} - 1}{1 - \mathrm{e}^{\beta\varepsilon(\boldsymbol{p})}} = -1 \\ E(-L_{\boldsymbol{p}}) &\approx \frac{\mathrm{e}^{\beta\varepsilon(\boldsymbol{p})} + 1}{1 + \mathrm{e}^{\beta\varepsilon(\boldsymbol{p})}} = +1 \, . \end{aligned} \tag{157}$$

Therefore, the second half of the terms in the sum over $\boldsymbol{p} \neq \boldsymbol{0}$ in (137) has a small denominator and thus determines the result, so that

$$\kappa_k(\beta) \approx \frac{1}{2} \sum_{n=1}^{k} (n-1)!\, \sigma_k^{(n)} \sum_{\boldsymbol{p}\neq\boldsymbol{0}} \frac{1}{(E(-L_{\boldsymbol{p}}) - 1)^n} \qquad \text{for } k \geq 2 \, . \tag{158}$$

Next, assuming a temperature such that the dominant low-energy modes satisfy $\beta\varepsilon(\boldsymbol{p}) \ll 1$, we may set

$$\begin{aligned} E(-L_{\boldsymbol{p}}) &\approx \frac{1 + \beta\varepsilon(\boldsymbol{p}) + L_{\boldsymbol{p}}}{1 + L_{\boldsymbol{p}}[1 + \beta\varepsilon(\boldsymbol{p})]} \\ &\approx 1 + \frac{1 - L_{\boldsymbol{p}}}{2} \beta\varepsilon(\boldsymbol{p}) \end{aligned} \tag{159}$$

for these modes. Observing $\varepsilon(\boldsymbol{p}) \gg \boldsymbol{p}^2/(2m)$, the previous formula (90) then reduces to

$$1 - L_{\boldsymbol{p}} \approx \left(\frac{4\pi a_{\mathrm{sc}} \hbar^2}{m} \frac{N}{V} \right)^{-1} \varepsilon(\boldsymbol{p}) \, , \tag{160}$$

giving

$$\begin{aligned} E(-L_{\boldsymbol{p}}) &\approx 1 + \beta \left(\frac{4\pi a_{\mathrm{sc}} \hbar^2}{m} \frac{N}{V} \right)^{-1} \frac{\varepsilon(\boldsymbol{p})^2}{2} \\ &\approx 1 + \beta \frac{\boldsymbol{p}^2}{2m} \\ &\approx \exp[\beta \boldsymbol{p}^2/(2m)] \, ; \end{aligned} \tag{161}$$

again for the dominant modes. Inserting this approximate identity into (158), one concludes that all low-energy dominated cumulants $\kappa_k(\beta)$ (i.e., cumulants with $k \geq 2$) of the weakly interacting Bose gas differ by a factor of 1/2 from their ideal-gas counterpart [14]. Basically, this can be understood by observing that, as a consequence of (99) and (150), one has $u_{\boldsymbol{p}} = v_{\boldsymbol{p}} = (2\sqrt{2}x)^{-1/2}$ in (79), so that the quasiparticle operators $b_{\boldsymbol{p}}$ for the dominant modes are equal-weight superpositions of the bare particle operators $a_{\boldsymbol{p}}$ and $a_{-\boldsymbol{p}}^\dagger$. Hence, the Bogoliubov coupling between the members of the $(\boldsymbol{p}, -\boldsymbol{p})$-pairs effectively reduces, in comparison with the ideal Bose gas, the number of degrees of freedom by a factor of two [12].

Finally, we point out another interesting detail. From the definition (117) of the pair characteristic function $G_{\boldsymbol{p}}(y)$ it follows that the probability $P_{\boldsymbol{p}}(n)$ to find $n = n_{\boldsymbol{p}} + n_{-\boldsymbol{p}}$ particles in the coupled $(\boldsymbol{p}, -\boldsymbol{p})$-modes is given by the Fourier transform

$$\begin{aligned} P_{\boldsymbol{p}}(n) &= \frac{1}{2\pi} \int_{-\pi}^{\pi} \mathrm{d}y\, \mathrm{e}^{-\mathrm{i}yn} G_{\boldsymbol{p}}(y) \\ &= \frac{1}{2\pi\mathrm{i}} \oint \frac{\mathrm{d}z}{z^{n+1}} \frac{(E(L_{\boldsymbol{p}}) - 1)\,(E(-L_{\boldsymbol{p}}) - 1)}{(E(L_{\boldsymbol{p}}) - z)\,(E(-L_{\boldsymbol{p}}) - z)} , \end{aligned} \tag{162}$$

where we have set $z = \mathrm{e}^{\mathrm{i}y}$, and the contour of integration is led counter-clockwise on the unit circle in the complex z-plane. This gives

$$P_{\boldsymbol{p}}(n) = \frac{(E(L_{\boldsymbol{p}}) - 1)\,(E(-L_{\boldsymbol{p}}) - 1)}{E(-L_{\boldsymbol{p}}) - E(L_{\boldsymbol{p}})} \left(\frac{1}{E(L_{\boldsymbol{p}})^{n+1}} - \frac{1}{E(-L_{\boldsymbol{p}})^{n+1}} \right) , \tag{163}$$

so that there is a pronounced "even-odd effect" [14]: As a consequence of (157), the probability $P_{\boldsymbol{p}}(n)$ with even n is substantially higher than $P_{\boldsymbol{p}}(n)$ with odd n for low momenta $\boldsymbol{p}$, so that the excited particles have a tendency to appear or disappear in pairs [14].

5 Concluding Remarks

In this overview we have laid the focus on the fluctuation of the number of particles contained in a Bose–Einstein condensate within the canonical ensemble. Besides possessing their own, intrinsic interest, these fluctuations also play a role in practical considerations concerning the coherence of the condensate; for instance, they determine the "line width of an atom laser" [35,36]. Although some clarity has been achieved by now, several questions related to condensate fluctuations remain: For example, in the case of an ideal Bose gas "in a box" the fluctuations do depend on the particular boundary condition even in the thermodynamic limit [30]. It is tempting to assume that this unusual behaviour is a pathology of the ideal gas which is cured by the inclusion of interactions, but a detailed proof of this surmise still needs to be given. More importantly, the canonical ensemble certainly is a convenient theoretical tool, but in typical laboratory experiments the condensates are thermally isolated, calling for a microcanonical treatment. In the practically important case of a harmonic oscillator trap, the microcanonical condensate fluctuations pertaining to an ideal gas are substantially smaller than their canonical counterparts: While the canonical fluctuation formula (41) gives, to leading order,

$$\langle (\widehat{n}_0 - \langle \widehat{n}_0 \rangle_{\mathrm{can}})^2 \rangle_{\mathrm{can}} \sim 1.64493 \left(\frac{k_{\mathrm{B}}T}{\hbar\omega} \right)^3 , \tag{164}$$

a microcanonical analysis yields [7,8]

$$\langle (\widehat{n}_0 - \langle \widehat{n}_0 \rangle_{\mathrm{mc}})^2 \rangle_{\mathrm{mc}} \sim 0.64366 \left(\frac{k_{\mathrm{B}}T}{\hbar\omega} \right)^3 , \tag{165}$$

so that the two ensembles predict notably different fluctuations even for large numbers of trapped particles. Such differences between ideal Bose gases in the canonical and in the microcanonical ensemble have recently also been explored in a number-theoretical context [37]. It is now an important question whether or not these differences will persist in weakly interacting gases. Since the canonical analysis summarized in this contribution, in particular the introduction of the canonical quasiparticles, relies crucially on the presence of a heat bath, a full-fledged microcanonical study seems to require significant further developments. We hope that the material compiled and explained here will help to tackle these and other "simple" questions in the near future.

Appendix 1: Disentangling Theorem for Angular Momentum Operators

In this appendix we explain in detail how to derive the normal form (126) of the pair characteristic function (117) from the entangled form (124); this is essentially an application of the disentangling theorem for exponential angular momentum operators as put forward in [38].

We start by considering the 2×2 matrix representation of the algebra of the rotation group provided by the Pauli spin matrices

$$\sigma_x = \begin{pmatrix} 0 & 1 \\ 1 & 0 \end{pmatrix} \quad , \quad \sigma_y = \begin{pmatrix} 0 & -\mathrm{i} \\ \mathrm{i} & 0 \end{pmatrix} \quad , \quad \sigma_z = \begin{pmatrix} 1 & 0 \\ 0 & -1 \end{pmatrix} , \tag{166}$$

and introduce the raising operator

$$\sigma_+ = \frac{1}{2}\left(\sigma_x + \mathrm{i}\sigma_y\right) = \begin{pmatrix} 0 & 1 \\ 0 & 0 \end{pmatrix} \tag{167}$$

and the lowering operator

$$\sigma_- = \frac{1}{2}\left(\sigma_x - \mathrm{i}\sigma_y\right) = \begin{pmatrix} 0 & 0 \\ 1 & 0 \end{pmatrix} . \tag{168}$$

These matrices obey the familiar commutation relations

$$\begin{aligned} [\sigma_z, \sigma_\pm] &= \pm 2\sigma_\pm \\ [\sigma_+, \sigma_-] &= \sigma_z \; . \end{aligned} \tag{169}$$

The goal now is to bring an entangled operator

$$\exp(w_+\sigma_+ + w_-\sigma_- + w_z\sigma_z) \tag{170}$$

into the disentangled form [38]

$$\exp(x_+\sigma_+)\, \exp(x_-\sigma_-)\, \exp(\ln x_z \sigma_z) \; . \tag{171}$$

Calculating the matrix elements of the entangled operator (170), one obtains

$$\exp\left(w_+\sigma_+ + w_-\sigma_- + w_z\sigma_z\right) = \begin{pmatrix} \cosh w + \frac{w_z}{w}\sinh w & \frac{w_+}{w}\sinh w \\ \frac{w_-}{w}\sinh w & \cosh w - \frac{w_z}{w}\sinh w \end{pmatrix} \tag{172}$$

with

$$w = \left(w_+w_- + w_z^2\right)^{1/2} , \tag{173}$$

whereas the disentangled operator (171) leads to

$$\begin{aligned} \mathrm{e}^{x_+\sigma_+}\mathrm{e}^{x_-\sigma_-}\mathrm{e}^{\ln x_z\sigma_z} &= (1 + x_+\sigma_+)\,(1 + x_-\sigma_-)\,(\cosh\ln x_z + \sigma_z\sinh\ln x_z) \\ &= \begin{pmatrix} x_z + x_+x_-x_z & \frac{x_+}{x_z} \\ x_-x_z & \frac{1}{x_z} \end{pmatrix} . \end{aligned} \tag{174}$$

Equating the matrix elements in (172) and (174), we first find

$$x_z = \left(\cosh w - \frac{w_z}{w}\sinh w\right)^{-1} \tag{175}$$

from the lower right corners, then

$$x_+ = \frac{\frac{w_+}{w}\sinh w}{\cosh w - \frac{w_z}{w}\sinh w} \tag{176}$$

and

$$x_- = \frac{w_-}{w}\sinh w\left(\cosh w - \frac{w_z}{w}\sinh w\right) \tag{177}$$

from the off-diagonal elements. Since the spin matrices are traceless, the determinant of both matrices (172) and (174) is unity, so that equating also the matrix elements in the upper left corners yields no further information. Indeed, from (175) – (177) one obtains after a short calculation

$$x_z(1 + x_+x_-) = \cosh w + \frac{w_z}{w}\sinh w , \tag{178}$$

as expected. Hence, we have proven the identity

$$\exp(w_+\sigma_+ + w_-\sigma_- + w_z\sigma_z) = \exp(x_+\sigma_+)\,\exp(x_-\sigma_-)\,\exp(\ln x_z\sigma_z) , \tag{179}$$

where the transformation from the given set (w_+, w_-, w_z) to the new coefficients (x_+, x_-, x_z) is achieved through (175), (176), and (177), together with (173) for w.

The validity of this disentangling theorem (179) is not restricted to the 2×2 representation of the rotation group: Any relation between exponential operators (that is, between group operations) which is valid for one particular faithful representation of the rotation group remains valid for all others [38]. Even more, since (179) is a consequence of the commutation relations (169) only, it follows that the spin matrices σ_-, σ_+, and σ_z can be replaced by other operators which obey the same relations [39,40]. This observation brings us back to the pair operators $B_{\boldsymbol{p}}$, $B^\dagger_{\boldsymbol{p}}$, and $C_{\boldsymbol{p}}$ introduced in (122) and (123) in Sect. 4.1: With the replacements

$$\begin{aligned}\sigma_- &\longrightarrow -B_{\boldsymbol{p}}\\ \sigma_+ &\longrightarrow B^\dagger_{\boldsymbol{p}}\\ \sigma_z &\longrightarrow C_{\boldsymbol{p}} = [B_{\boldsymbol{p}}, B^\dagger_{\boldsymbol{p}}]\end{aligned} \tag{180}$$

one sees immediately that the commutation relations (123) and (125) for $B_{\boldsymbol{p}}$, $B^\dagger_{\boldsymbol{p}}$, and $C_{\boldsymbol{p}}$ are the same as the relations (169) for the spin matrices. Therefore, with these substitutions (180) we can apply the disentangling theorem (179) directly to the entangled form (124) of the pair characteristic function $G_{\boldsymbol{p}}(y)$, identifying the parameters

$$\begin{aligned}w_+ &= -2\mathrm{i}y\, u_{\boldsymbol{p}}v_{\boldsymbol{p}}\\ w_- &= +2\mathrm{i}y\, u_{\boldsymbol{p}}v_{\boldsymbol{p}}\\ w_z &= \mathrm{i}y\,(u^2_{\boldsymbol{p}} + v^2_{\boldsymbol{p}})\ .\end{aligned} \tag{181}$$

From (173) and (81) one then finds

$$w = \mathrm{i}y\ , \tag{182}$$

and the expressions (127), (128), and (129) for the coefficients of the disentangled characteristic function (126) follow directly from (175), (176), and (177).

Appendix 2: Computing the Trace

In this appendix we outline how to compute the trace occurring in the characteristic function (126), following closely the original strategy by Kocharovsky, Kocharovsky, and Scully [14]. Since, according to (123), the operator $C_{\boldsymbol{p}}$ is the number operator for the quasiparticles with momentum $\boldsymbol{p}$ and $-\boldsymbol{p}$, plus one, we employ a Fock basis $\{\,|N_{\boldsymbol{p}}, N_{-\boldsymbol{p}}\rangle\,\}$ of quasiparticle states, so that $N_{\boldsymbol{p}} = 0,1,2,3,\ldots$ denotes the eigenvalues of $b^\dagger_{\boldsymbol{p}}b_{\boldsymbol{p}}$, and $N_{-\boldsymbol{p}} = 0,1,2,3,\ldots$ those of $b^\dagger_{-\boldsymbol{p}}b_{-\boldsymbol{p}}$. Expanding the exponentials $\exp\big(x_+B^\dagger_{\boldsymbol{p}}\big)$ and $\exp(-x_-B_{\boldsymbol{p}})$, we then have

$$\begin{aligned}&\mathrm{Tr}\left[\exp\big(x_+B^\dagger_{\boldsymbol{p}}\big)\,\exp(-x_-B_{\boldsymbol{p}})\,\exp((\ln x_z - \beta\varepsilon(\boldsymbol{p}))C_{\boldsymbol{p}})\right]\\ &= \sum_{N_{\boldsymbol{p}}=0}^{\infty}\sum_{N_{-\boldsymbol{p}}=0}^{\infty}\left(x_z\mathrm{e}^{-\beta\varepsilon(\boldsymbol{p})}\right)^{(N_{\boldsymbol{p}}+N_{-\boldsymbol{p}}+1)}\\ &\quad\cdot\sum_{m_1=0}^{\infty}\sum_{m_2=0}^{\infty}\frac{(x_+)^{m_1}}{m_1!}\frac{(-x_-)^{m_2}}{m_2!}\langle N_{\boldsymbol{p}}N_{-\boldsymbol{p}}|(b^\dagger_{\boldsymbol{p}}b^\dagger_{-\boldsymbol{p}})^{m_1}(b_{\boldsymbol{p}}b_{-\boldsymbol{p}})^{m_2}|N_{\boldsymbol{p}}N_{-\boldsymbol{p}}\rangle\ .\end{aligned} \tag{183}$$

Obviously, the double sum over m_1 and m_2 yields nonvanishing contributions only for $m_1 = m_2 = m$, as long as $0 \le m \le M = \min\{N_{\boldsymbol{p}}, N_{-\boldsymbol{p}}\}$. It is easily shown by induction that

$$\langle N_{\boldsymbol{p}}|\,(b^\dagger_{\boldsymbol{p}})^m(b_{\boldsymbol{p}})^m\,|N_{\boldsymbol{p}}\rangle = \frac{N_{\boldsymbol{p}}!}{(N_{\boldsymbol{p}}-m)!}\ , \tag{184}$$

so that

$$\begin{aligned}
&\mathrm{Tr}\left[\exp\left(x_+ B_{\boldsymbol{p}}^\dagger\right)\,\exp(-x_- B_{\boldsymbol{p}})\,\exp((\ln x_z - \beta\varepsilon(\boldsymbol{p}))C_{\boldsymbol{p}})\right] \\
&= x_z \mathrm{e}^{-\beta\varepsilon(\boldsymbol{p})} \sum_{N_{\boldsymbol{p}}=0}^{\infty} \sum_{N_{-\boldsymbol{p}}=0}^{\infty} \left(x_z \mathrm{e}^{-\beta\varepsilon(\boldsymbol{p})}\right)^{(N_{\boldsymbol{p}}+N_{-\boldsymbol{p}})} \sum_{m=0}^{M} \binom{N_{\boldsymbol{p}}}{m}\binom{N_{-\boldsymbol{p}}}{m}(-x_+x_-)^m \,. \qquad (185)
\end{aligned}$$

The sum over a product of binomial coefficients emerging here is closely related to the Jacobi polynomials $P_M^{(\alpha,\beta)}(x)$: According to their definition [23], one has

$$P_M^{(\alpha,\beta)}(x) = \frac{(x-1)^M}{2^M} \sum_{m=0}^{M} \binom{M+\alpha}{m}\binom{M+\beta}{M-m}\left(\frac{x+1}{x-1}\right)^m \qquad (186)$$

with $\alpha > -1$ and $\beta > -1$; hence

$$P_M^{(\alpha,0)}\left(\frac{z+1}{z-1}\right) = \left(\frac{1}{z-1}\right)^M \sum_{m=0}^{M} \binom{M+\alpha}{m}\binom{M}{m} z^m \,. \qquad (187)$$

Utilizing this expression in (185), with $M = \min\{N_{\boldsymbol{p}}, N_{-\boldsymbol{p}}\}$ and $\alpha = |N_{\boldsymbol{p}} - N_{-\boldsymbol{p}}| \equiv |\Delta N|$, then writing $N_{\boldsymbol{p}} + N_{-\boldsymbol{p}} = 2M + |\Delta N|$ and rearranging the double sum over $N_{\boldsymbol{p}}$ and $N_{-\boldsymbol{p}}$ accordingly, we arrive at

$$\begin{aligned}
&\mathrm{Tr}\left[\exp\left(x_+ B_{\boldsymbol{p}}^\dagger\right)\,\exp(-x_- B_{\boldsymbol{p}})\,\exp((\ln x_z - \beta\varepsilon(\boldsymbol{p}))C_{\boldsymbol{p}})\right] \\
&= x_z \mathrm{e}^{-\beta\varepsilon(\boldsymbol{p})} \sum_{\Delta N=-\infty}^{\infty} \sum_{M=0}^{\infty} \left(x_z \mathrm{e}^{-\beta\varepsilon(\boldsymbol{p})}\right)^{(2M+|\Delta N|)} \\
&\quad \cdot (-x_+x_- - 1)^M \, P_M^{(|\Delta N|,0)}\left(\frac{-x_+x_- + 1}{-x_+x_- - 1}\right) \,. \qquad (188)
\end{aligned}$$

When evaluating the sum over M, the generating function for the Jacobi polynomials comes into play: In general, one has [23]

$$\sum_{M=0}^{\infty} 2^{-\alpha-\beta} P_M^{(\alpha,\beta)}(x)\, z^M = \frac{1}{R(1-z+R)^\alpha(1+z+R)^\beta} \qquad (189)$$

with

$$R = \sqrt{1 - 2xz + z^2}\,; \qquad (190)$$

giving, when applied to (188),

$$\begin{aligned}
&\sum_{M=0}^{\infty} \left(x_z^2\, \mathrm{e}^{-2\beta\varepsilon(\boldsymbol{p})}\,(-x_+x_- - 1)\right)^M P_M^{(|\Delta N|,0)}\left(\frac{-x_+x_- + 1}{-x_+x_- - 1}\right) \\
&= \frac{1}{R}\left(\frac{2}{1-z+R}\right)^{|\Delta N|} , \qquad (191)
\end{aligned}$$

with

$$z = x_z^2\, \mathrm{e}^{-2\beta\varepsilon(\boldsymbol{p})}\,(-x_+x_- - 1) \qquad (192)$$

and

$$R = \sqrt{(1-z)^2 - (2\,x_z\,\mathrm{e}^{-\beta\varepsilon(\boldsymbol{p})})^2}\;. \tag{193}$$

The remaining sum over ΔN in (188) then is a simple geometric series, leading, after some elementary manipulations, to

$$\begin{aligned}
&\frac{1}{R}\sum_{\Delta N=-\infty}^{+\infty}\left(\frac{2x_z\mathrm{e}^{-\beta\varepsilon(\boldsymbol{p})}}{1-z+R}\right)^{|\Delta N|} \\
&= \frac{1}{R}\,\frac{1-z+2x_z\mathrm{e}^{-\beta\varepsilon(\boldsymbol{p})}+R}{1-z-2x_z\mathrm{e}^{-\beta\varepsilon(\boldsymbol{p})}+R} \\
&= \frac{1}{(1-x_z\mathrm{e}^{-\beta\varepsilon(\boldsymbol{p})})^2 + x_z^2 x_+ x_- \mathrm{e}^{-2\beta\varepsilon(\boldsymbol{p})}}\;.
\end{aligned} \tag{194}$$

Thus, we finally obtain the desired trace in the form

$$\begin{aligned}
&\mathrm{Tr}\left[\exp\left(x_+ B_{\boldsymbol{p}}^\dagger\right)\exp(-x_- B_{\boldsymbol{p}})\exp((\ln x_z - \beta\varepsilon(\boldsymbol{p}))C_{\boldsymbol{p}})\right] \\
&= \frac{x_z\mathrm{e}^{-\beta\varepsilon(\boldsymbol{p})}}{(1-x_z\mathrm{e}^{-\beta\varepsilon(\boldsymbol{p})})^2 + x_z^2 x_+ x_- \mathrm{e}^{-2\beta\varepsilon(\boldsymbol{p})}}\;.
\end{aligned} \tag{195}$$

References

1. For a review, see W. Ketterle, D.S. Durfee, D.M. Stamper-Kurn: 'Making, Probing and Understanding Bose–Einstein Condensates'. In: *Proceedings of the International School of Physics "Enrico Fermi", Course CXL*, ed. by M. Inguscio, S. Stringari, and C.E. Wiemann (IOS Press, Amsterdam, 1999), pp. 67–176
2. M. Wilkens, F. Illuminati, M. Krämer: J. Phys. B **33**, L779 (2000), and references therein
3. R.M. Ziff, G.E. Uhlenbeck, M. Kac: Phys. Rep. **32**, 169 (1977)
4. S. Grossmann, M. Holthaus: Phys. Rev. E **54**, 3495 (1996)
5. H.D. Politzer: Phys. Rev. A **54**, 5048 (1996)
6. C. Weiss, M. Wilkens: Opt. Express **1**, 272 (1997)
7. P. Navez, D. Bitouk, M. Gajda, Z. Idziaszek, K. Rzążewski: Phys. Rev. Lett. **79**, 1789 (1997)
8. M. Holthaus, E. Kalinowski, K. Kirsten: Ann. Phys. (N.Y.) **270**, 198 (1998)
9. M. O. Scully: Phys. Rev. Lett. **82**, 3927 (1999)
10. V.V. Kocharovsky, M.O. Scully, S.-Y. Zhu, M.S. Zubairy: Phys. Rev. A **61**, 023609 (2000)
11. S. Giorgini, L.P. Pitaevskii, S. Stringari: Phys. Rev. Lett. **80**, 5040 (1998)
12. F. Meier, W. Zwerger: Phys. Rev. A **60**, 5133 (1999)
13. V.V. Kocharovsky, Vl.V. Kocharovsky, M.O. Scully: Phys. Rev. Lett. **84**, 2306 (2000)
14. V.V. Kocharovsky, Vl.V. Kocharovsky, M.O. Scully: Phys. Rev. A **61**, 053606 (2000)
15. Z. Idziaszek, M. Gajda, P. Navez, M. Wilkens, K. Rzążewski: Phys. Rev. Lett. **82**, 4376 (1999)
16. F. Illuminati, P. Navez, M. Wilkens: J. Phys. B **32**, L461 (1999)

17. M. Holthaus, K.T. Kapale, V.V. Kocharovsky, M.O. Scully: Physica A **300**, 433 (2001)
18. L.D. Landau, E.M. Lifshitz: *Course of Theoretical Physics, Vol. 9: Statistical Physics, Part 2* (Butterworth-Heinemann, Oxford, 1998)
19. L.D. Landau, E.M. Lifshitz: *Course of Theoretical Physics, Vol. 5: Statistical Physics, Part 1* (Butterworth-Heinemann, Oxford, 2001)
20. K. Huang: *Statistical Mechanics* (John Wiley, New York, 1963)
21. R.K. Pathria: *Statistical Mechanics* (Pergamon Press, Oxford, 1985)
22. M. Fierz: Helv. Phys. Acta **29**, 47 (1956)
23. *Handbook of Mathematical Functions*, ed. by M. Abramowitz and I.A. Stegun (Dover, New York, 1972)
24. C.W. Gardiner: *Handbook of Stochastic Methods* (Springer Verlag, Berlin, 1985)
25. K. Kirsten, D.J. Toms: Phys. Rev. E **59**, 158 (1999)
26. A. Voros: Commun. Math. Phys. **110**, 439 (1987)
27. E.T. Whittaker, G.N. Watson: *A Course of Modern Analysis* (Cambridge University Press, Cambridge, 1969)
28. S.R. de Groot, G.J. Hooyman, C.A. ten Seldam: Proc. Roy. Soc. London A **203**, 266 (1950)
29. V. Bagnato, D.E. Pritchard, D. Kleppner: Phys. Rev. A **35**, 4354 (1987)
30. M. Holthaus, K.T. Kapale, M.O. Scully: Phys. Rev. E **65**, 036129 (2002)
31. N. Bogoliubov: J. Phys. USSR **11**, 23 (1947)
32. T.D. Lee, C.N. Yang: Phys. Rev. **105**, 1119 (1957)
33. T.D. Lee, K. Huang, C.N. Yang: Phys. Rev. **106**, 1135 (1957)
34. R.A. Ferell, N. Meynhard, H. Schmidt, F. Schwabl, P. Szépfalusy: Ann. Phys. (N.Y.) **47**, 565 (1968)
35. R. Graham: Phys. Rev. Lett. **81**, 5256 (1998)
36. R. Graham: Phys. Rev. A **62**, 023609 (2000)
37. C. Weiss, M. Holthaus: Europhys. Lett., to appear (2002)
38. F.T. Arecchi, E. Courtens, R. Gilmore, H. Thomas: Phys. Rev. A **6**, 2211 (1972)
39. J. Schwinger: In *Quantum Theory of Angular Momentum*, ed. by L.C. Biedenharn and H. van Dam (Academic, New York, 1965)
40. B.L. Schumaker, C.M. Caves: Phys. Rev. A **31**, 3093 (1985)

New Regimes in Cold Gases via Laser-Induced Long-Range Interactions

Gershon Kurizki[1], Stefano Giovanazzi[1,2], Duncan O'Dell[3], and Alexandre I. Artemiev[1,4,5]

[1] Department of Chemical Physics, Weizmann Institute of Science, 76100 Rehovot, Israel
[2] Fakultat Physik & Physikalisches Institut, Universitat Stuttgart, Pfaffenwaldring 57, D-70550, Stuttgart, Germany
[3] Sussex Centre for Optical and Atomic Physics, C.P.E.S., University of Sussex, Brighton BN1 9QH, UK
[4] Department of Physics, Texas A&M University, College Station, Texas, 77843-4242, USA
[5] General Physics Institute, 38 Vavilov St., Moscow, 119991, Russia

Abstract. The modification of the properties of a Bose-Einstein or a Fermi-Dirac atomic gas due to laser induced dipole dipole interactions between the atoms are considered. Nearly-isotropic illumination of the sample by spectrally-fluctuating laser beams averages out the static r^{-3} dipole-dipole interaction, leaving the retarded r^{-1} "self-gravitating" attraction in the near zone. The analogies of ultracold many-atom systems, self-bound by such laser-induced "gravity", with compact stars ("Bose stars" or "White Dwarfs") are emphasized. Even a single plane-wave laser induces dipole-dipole interactions capable of causing a cigar-shaped Bose condensate to exhibit self binding and density modulations.

1 Introduction

In the atomic Bose-Einstein condensates (BECs), Maxwell-Boltzmann or Fermi-Dirac gases (MBGs or FDGs) investigated until recently [1]–[5], the atoms interact only at very *short distances*. Most of the properties of these *dilute* gases can be understood by taking into account only two-body collisions which are characterized by the s-wave scattering length [3,4] (in FDGs a mixture of different spin states is needed to provide s-wave scattering [5]). A number of groups [3] have investigated the fascinating possibility of changing the magnitude and sign of the s-wave scattering length, particularly near Feshbach resonances, using external fields. The resulting atom-atom interactions retain their essentially hard-sphere, s-wave, nature.

By contrast, here we focus upon the basic modifications of atom-atom scattering due to *long-range* laser-induced dipole-dipole interactions. These modifications are predicted to give rise to hitherto unexplored regimes in cold atomic gases: new forms of dynamics, thermodynamics and equilibrium configurations.

Our general approach is based on two coupled equations: one for the matter distribution and the other for the electromagnetic (EM) field, which must be solved self-consistently [6]. For cold atoms and off-resonant EM fields the matter-field interaction simplifies, because the internal atomic states respond instantaneously to the field and therefore can be decoupled from the atomic translational states (the spatial and momentum distributions), which evolve much more slowly. This decoupling results in a Lippmann-Schwinger type of an integral equation for the local electric field, in terms of the sum of the incident field and that scattered by all the atoms. In Sect. 2 we demonstrate that, for large detuning, this integral equation can be reduced to its Born-approximation limit, leading to the laser-induced non-local Hartree-Fock potential, which describes the sum over all atom pairs of their induced dipole-dipole interaction. The laser-induced potential derived in Sect. 2 is associated with electrostriction, which is the tendency of matter to become compressed in the presence of an electric field [7]. In an optically (dipole-)trapped multi-atom system, electrostriction is provided by the *gradient* of the incident electric field (one-body dipole forces). However, it can arise from *retarded* dipole-dipole interactions induced by off-resonant EM fields, even if these external fields are *homogeneous*.

In his work on EM-induced dipole-dipole interactions, Thirunamachandran [8] noted that when *an average over all orientations* of an interacting atom pair is taken, the static r^{-3} part of the fully retarded dipole-dipole interaction cancels out in the *near zone*, leaving an isotropic, attractive, r^{-1} dependence, caused by retardation. In our papers [9,10] we suggested a scheme, detailed in Sect. 3, for realising the necessary orientational averaging for cold-atom collisions, using perpendicular, mutually detuned (or spectrally fluctuating) laser beams. The long-range, retarded, r^{-1} attractive potential resulting from this averaging is different in character from either its static r^{-3} counterpart or the short-range s-wave scattering, in that it allows atomic BECs or FDGs to be self-bound *without external traps*, as detailed below.

In Sect. 4 we incorporate the light-induced long-range potentials of Sects. 2 and 3 into the dynamical and thermodynamical treatment of MBGs and FDGs. Using a mean-field approximation, we examine the interplay between the laser-induced electrostrictive potentials and the kinetic energy or a hard-core inter-atomic potential (in a BEC), or between laser-induced potentials and fermionic statistical repulsion (in a FDG), identifying qualitatively new regimes. The r^{-1} attractive, near-zone, laser-induced potential provides the fascinating possibility to study gravitational effects, normally only important on the stellar scale, in the laboratory. Particularly interesting is the possibility of experimentally emulating compact Bose stars [11] (Sects. 5, 6) or "White Dwarfs" [12], [13] (Sect. 7): gravitationally-bound degenerate "clouds" of finite volume, in which the zero-point quantum kinetic pressure (in a BEC) or the quantum-statistical pressure (in a FDG) balances the gravitational attraction and thus *stabilizes* the system against collapse. The laser-induced "gravitational" constant can be some 21 orders of magnitude (depending on the laser intensity and wavelength) greater in strength than its genuine counterpart, leading to a *submicron-sized* self-bound atomic "cloud".

The treatment in Sect. 5 reveals the threshold laser intensity (at a given detuning from resonance) and number of atoms for a "gravitationally" self-bound BECs. This treatment pertains also to atomic "clouds" that extend beyond the near-zone (i.e., can exceed the laser wavelength), so that the fully retarded dependence of the interatomic potential is taken into account. Once again, self-bound systems are predicted [10], with properties closely related to those found for the simpler r^{-1} potential.

The collective excitations of a laser-induced "gravitationally" bound BEC and its Jeans instability, known from compact stars, are analyzed in Sect. 6, following [14]. In Sect. 7 we investigate the formation of a degenerate "core" in a non-degenerate, externally-trapped, spin-polarized fermionic gas, as the strength of the laser-induced "gravity" increases. This core is a genuine analog of a "White-Dwarf" star, [12,13].

In Sect. 8 we show that even in the simplest case of a plane-wave, off-resonant laser beam, a BEC can be self-bound by electrostriction, again due to the *retarded* behavior of the induced dipole-dipole interaction [15]. By contrast, the purely static (r^{-3}) dipole-dipole interaction does not allow for such self-binding [16]. The self-bound condensate is predicted [15] to be elongated and compressed (pencil-shaped) along the axis of a circularly polarized laser beam, in principle even for arbitrarily small intensity. As the laser intensity increases, the electrostriction can give rise to a remarkable one-dimensional density modulation of the condensate in its ground state. Such a condensate bears similarity to a supersolid, i.e., a long-range crystalline-like density modulation imposed upon a superfluid by interparticle forces [17].

Finally, in Sect. 9 we discuss various loss mechanisms, such as Rayleigh scattering, by which atoms are ejected from the system or heated up by the action of the lasers. The conclusions are presented in Sect. 10.

2 Light-Induced Potentials in a Classical Dielectric

A defining feature of the atomic BECs or cold MBGs which have been realised thus far is that they are very dilute when compared to the s-wave scattering length a, obeying $a \ll \rho^{-1/3}$, ρ being the atomic number density. This diluteness leads to the familiar description of the interatomic interactions as contact (delta-function) pseudo-potentials. However, an external electromagnetic field introduces a new density scale into the problem, which, within certain limits, we are free to choose. Here we consider the high-density regime $\rho \gg \lambda_{\mathrm{L}}^{-3}$, where λ_{L} is the vacuum wavelength of the incident light.

Many of the early studies [6] of light incident upon an atomic BEC considered the incident light to be nearly resonant to an atomic optical transition. By contrast, our treatment assumes the extreme off-resonant case.

If the incident EM field, of wavenumber $q = 2\pi/\lambda_{\mathrm{L}}$, is very far from resonance, then the dynamical atomic polarizability, $\alpha(q)$, approaches its static limit, α_0, which can be many orders of magnitude smaller than the on-resonance value. The high density $\rho \gg \lambda_{\mathrm{L}}^{-3}$ and low polarizability α are counterbalanced in the dimensionless quantity $\alpha\rho/\epsilon_0$, which largely governs the response of the system to the laser. For sodium, the static polarizability is $\alpha_0 = 2.679 \times 10^{-39}$ Coulomb m^2/Volt (in CGS units: $\alpha_0(\mathrm{cm}^3) = 24.08 \times 10^{-24}\mathrm{cm}^3$), and so even at the higher densities achieved experimentally (see above), one has $\alpha_0\rho/\epsilon_0 \approx 10^{-7}$. Thus, we shall consider the limit

$$\frac{\alpha(q)\rho(\mathbf{r})}{\epsilon_0} \ll 1, \tag{1}$$

in which the response of the atoms to the external radiation is very weak. Such a system behaves as a *nearly transparent* classical dielectric medium with dielectric constant $\epsilon(q) \approx 1$. The polarization density, $P(\mathbf{r}, t)$, and the macroscopic electric field in the medium, $\boldsymbol{\mathcal{E}}(\mathbf{r}, t)$, are related via $P(\mathbf{r}, t) = (\epsilon(q) - 1)\,\epsilon_0\,\boldsymbol{\mathcal{E}}(\mathbf{r}, t)$.

Using classical arguments we shall now obtain an expression for the field-induced energy of the condensate. We begin from the classical Lorentz-Lorenz (LL) formula [7], [18], which can be expressed as a relation between the dielectric constant, $\epsilon(q)$, and the dynamic polarizability, $\alpha(q)$, of the single atom

$$\epsilon(q) - 1 = \frac{\alpha(q)\rho}{\epsilon_0}\left(1 - \frac{\alpha(q)\rho}{3\epsilon_0}\right)^{-1}, \tag{2}$$

where ρ is the atomic density. The dielectric constant $\epsilon(q)$ describes elastic scattering, i.e., the light frequency is not shifted. Inelastic light scattering turns out to be negligible (Sect. 9). The LL formula implies that the actual field which drives an atom in a medium is not the macroscopic field (which is an average over some small volume) but the *local* field, $\boldsymbol{\mathcal{E}}_{\mathrm{loc}}(\mathbf{r}, t)$, so that

$$\mathbf{P}(\mathbf{r}, t) = \alpha(q)\rho(\mathbf{r})\boldsymbol{\mathcal{E}}_{\mathrm{loc}}(\mathbf{r}, t) \equiv \alpha(q)\rho(\mathbf{r})\left(\boldsymbol{\mathcal{E}}_{\mathrm{L}}(\mathbf{r}, t) + \frac{\mathbf{P}(\mathbf{r}, t)}{3\epsilon_0}\right), \tag{3}$$

where $\boldsymbol{\mathcal{E}}_{\mathrm{L}}(\mathbf{r}, t)$ is the external (incident) laser field. The local field accounts for the fact that the effective single-atom polarizability already self-consistently contains the influence of the atom's own field [18,19].

Note that in (3) the time dependence of the atom density is ignored. This is because *for cold atoms and a heavily detuned field* the internal degrees of freedom of the atoms evolve much faster than their external ones [20], so that *the field adjusts to the atomic distribution instantaneously.* Observe also that we take $\alpha(q)$ to be a real scalar, that is we assume an *isotropic, real* polarizability. One expects this to be the case for large detuning from all transitions.

The local electric field inside a medium, $\boldsymbol{\mathcal{E}}_{\mathrm{loc}}(\mathbf{r}, t)$, induced by an external electric field $\boldsymbol{\mathcal{E}}_{\mathrm{L}}(\mathbf{r}, t)$, satisfies the following Lippmann-Schwinger integral equation [18,19]

$$\boldsymbol{\mathcal{E}}_{\mathrm{loc}}(\mathbf{r}, t) = \boldsymbol{\mathcal{E}}_{\mathrm{L}}(\mathbf{r}, t) + \frac{\alpha(q)}{\epsilon_0}\int d^3r'\; \mathbf{G}^0(\mathbf{r} - \mathbf{r}')\;\rho(\mathbf{r}')\boldsymbol{\mathcal{E}}_{\mathrm{loc}}(\mathbf{r}', t)\,, \tag{4}$$

which is sometimes referred to as the 'coupled dipole equation' (CDE) [21]. Here $\mathbf{G}^0$ is the propagator for the *local* electric field, namely, the electric field is modified so that there is no interaction between coinciding (overlapping) atoms, i.e.

$\mathbf{G}^0(0) = 0$, Ref. [22,23]. For the positive-frequency part (varying as $\exp(-\mathrm{i}\omega t)$) of $\mathbf{E}(\mathbf{r}, t)$, $\mathbf{G}^0$ takes the form

$$G^0_{ij}(\mathbf{r}) = -\frac{\exp{(iqr)}}{4\pi r^3} \times \left[(\delta_{ij} - 3\hat{r}_i\hat{r}_j)(1 - \mathrm{i}qr) - (\delta_{ij} - \hat{r}_i\hat{r}_j)(qr)^2 \right] , \tag{5}$$

where $\hat{r}_j = r_j/r$. Retardation effects are contained in the *spatial dependence* of the propagator. The propagator $\mathbf{G}$ for the full electric field is related to $\mathbf{G}^0$ by $G_{ij}(\mathbf{r}) = G^0_{ij}(\mathbf{r}) - \frac{1}{3}\delta_{ij}\delta(\mathbf{r})$. The propagator $\mathbf{G}^0$, given by (5), is now understood as the propagator for the local electric field, that includes the Lorentz-Lorenz correction: since $\mathbf{G}^0(0) = 0$, the local electric field $\boldsymbol{\mathcal{E}}_{\mathrm{loc}}(\mathbf{r}, t)$ is independent of the polarization $\mathbf{P}(\mathbf{r}, t)$ at the same point [22,23].

Solving the CDE (4) is in general a difficult problem. Morice et al [22] developed a hierarchy of equations from the CDE relating successively higher correlation functions between the atomic dipoles. For the large detuning considered here our task is simpler, since condition (1) suggests an iterative solution to the CDE (cf. [24] where a similar perturbative expansion is used for the polarization of a quantum gas). We shall work within *the first Born approximation* (in $\alpha\rho/\epsilon_0 \sim (\epsilon - 1)$ and thus iterate (4) only once

$$\boldsymbol{\mathcal{E}}_{\mathrm{loc}}(\mathbf{r}, t) \approx \boldsymbol{\mathcal{E}}_{\mathrm{L}}(\mathbf{r}, t) + \frac{\alpha(q)}{\epsilon_0} \int d^3r' \, \mathbf{G}^0(\mathbf{r} - \mathbf{r}') \, \rho(\mathbf{r}')\boldsymbol{\mathcal{E}}_{\mathrm{L}}(\mathbf{r}', t) \, . \tag{6}$$

The Born approximation is applicable in a high-density system because of the weak-response condition (1).

The expression for the energy E_{ind} of a classical ideal dielectric (linear) medium induced by an external field $\boldsymbol{\mathcal{E}}_{\mathrm{L}} = \hat{\mathbf{e}}\,\mathcal{E}_{\mathrm{L}} \exp(\mathrm{i}\mathbf{q}\cdot\mathbf{r} - \mathrm{i}\omega t)$ oscillating at frequency $\omega = cq$ is given by [7,18,19]

$$E_{\mathrm{ind}} = -\frac{1}{2} \int d^3r \, \langle \mathrm{Re}[\mathbf{P}(\mathbf{r}, t)] \, \mathrm{Re}[\boldsymbol{\mathcal{E}}_{\mathrm{L}}(\mathbf{r}, t)] \, \rangle \, , \tag{7}$$

where $\langle \ldots \rangle$ denotes the time average. In order to evaluate the energy (7) we require the polarization, which in turn depends on the local electric field (4). Using the Born approximation (6) for the local field in (3), $\mathbf{P}(\mathbf{r}, t)$, the energy functional (7) can be expanded to second order in $\alpha\rho/\epsilon_0$ of (2), as

$$E_{\mathrm{ind}} = -\frac{1}{4} \, \alpha \, \boldsymbol{\mathcal{E}}_{\mathrm{L}}^2 \int d^3r \, \rho(\mathbf{r}) + \frac{1}{2} \int d^3r \int d^3r' \rho(\mathbf{r}) U(\mathbf{r} - \mathbf{r}') \rho(\mathbf{r}') \, . \tag{8}$$

The light-induced perturbative energy shift in (8) that depends on the inter-atom separation vector, $\mathbf{r}$, is associated with the spatially-retarded induced dipole-dipole potential, $U(\mathbf{r})$. Written in terms of cartesian components i, j, this potential has the following form in S.I. units

$$U(\mathbf{r}) = \left(\frac{I}{4\pi c \varepsilon_0^2} \right) \alpha^2(q) \hat{e}_i^* \hat{e}_j \mathcal{V}_{ij}(q, \mathbf{r}) \cos(\mathbf{q}\cdot\mathbf{r}). \tag{9}$$

Here $\alpha(q)$ is the isotropic, dynamic polarizability of the atoms at frequency $c\,q$, $\hat{e}_i$, $\hat{e}_j$ are the components of the laser polarization vector $\hat{\mathbf{e}}$, and $\mathcal{V}_{ij}$ is the retarded dipole-dipole interaction tensor. It is simply related to the real part of $\mathbf{G}^0$:

$$\mathcal{V}_{ij}(\mathbf{r}) = -4\pi \mathrm{Re}\left[G^0_{ij}(\mathbf{r})\right], \tag{10}$$

and has the form

$$\mathcal{V}_{ij}(q,\mathbf{r}) = \frac{1}{r^3}\Big[\big(\delta_{ij} - 3\hat{r}_i\hat{r}_j\big)\big(\cos qr + qr\sin qr\big) \\ -\big(\delta_{ij} - \hat{r}_i\hat{r}_j\big)q^2r^2\cos qr\Big]. \tag{11}$$

Equation (11) gives the well-known static (non-retarded) r^{-3} variation of the interaction energy at near-zone separations ($qr \ll 1$). The near-zone limit of $U(\mathbf{r})$ (9) – (11) is *strongly anisotropic.*

3 Electromagnetically Induced "Gravity"

3.1 Near-Zone Orientational Average

Thirunamachandran [8] showed that when an *average over all orientations* of the interatomic axis with respect to the incident radiation is taken in (9) and (11), the static dipolar part of the interatomic coupling (i.e., the instantaneous, non-retarded part $r^{-3}(\delta_{ij} - 3\hat{\mathbf{r}}_i\hat{\mathbf{r}}_j)$) vanishes. The remaining 'transverse' part (in the multipolar representation, see [8]) is, in the near zone ($qr \ll 1$), an *attractive* r^{-1} potential. It is weaker by a factor of $(qr)^2$ than the r^{-3} term. This remarkable result reflects the contrasting symmetries of the two parts: the r^{-3} and r^{-1} parts have odd and even angular symmetry, respectively. However, until recently, no scheme has been suggested wherein an average over all orientations is guaranteed for dipole-dipole interactions in cold gases, for which the atomic collisions are slow. In what follows we revisit the spatial configuration of external fields suggested in [9,10], which *enforces the 'averaging out'* of the r^{-3} interactions.

A simple setup which ensures the suppression of the r^{-3} interaction while retaining the weaker r^{-1} attraction in the near zone, uses three orthogonal circularly polarized laser beams pointing along $\hat{\mathbf{x}}, \hat{\mathbf{y}}, \hat{\mathbf{z}}$ ('a triad' – see Fig. 1) [9]. Let us momentarily ignore interference between the three beams, and only consider the sum of their intensities. Using the identity

$$\hat{\mathbf{e}}_i^{*(\pm)}(\mathbf{q})\hat{\mathbf{e}}_j^{(\pm)}(\mathbf{q}) = \frac{1}{2}\left((\delta_{ij} - \hat{\mathbf{q}}_i\hat{\mathbf{q}}_j) \pm i\epsilon_{ijk}\hat{\mathbf{q}}_k\right) \tag{12}$$

with $+(-)$ corresponding to left(right) circular polarizations, and retaining the lowest power of qr in (9) and (11), in the near zone ($qr \ll 1$), we find that the triad configuration then gives rise to an r^{-1} attractive pair potential

$$U^{\mathrm{NZ}}_{\mathrm{triad}}(\mathbf{r}) = -\frac{3Iq^2\alpha^2}{(16\pi c\varepsilon_0^2)}\times \\ \frac{1}{r}\left[\frac{7}{3} + (\sin\theta\cos\phi)^4 + (\sin\theta\sin\phi)^4 + (\cos\theta)^4\right]. \tag{13}$$

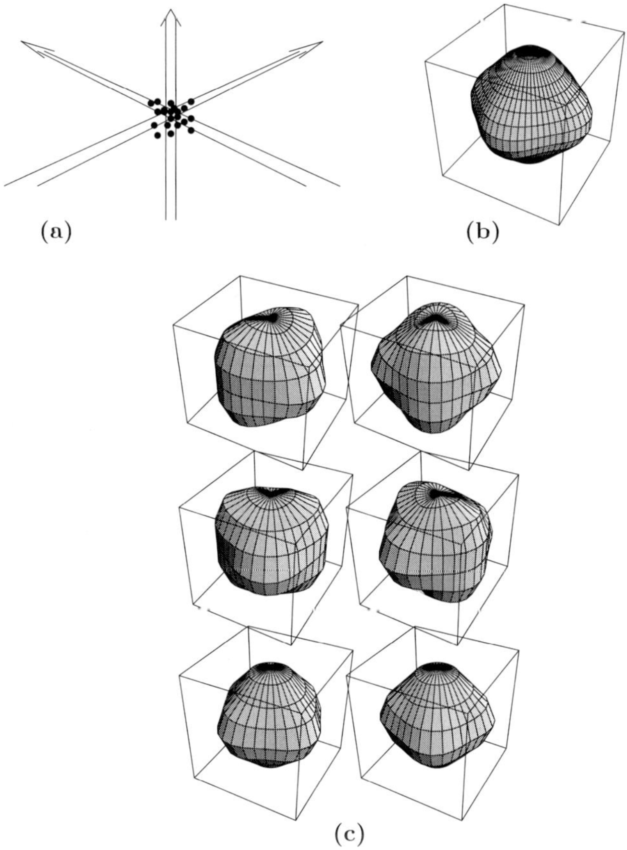

Fig. 1. (a) Schematic depiction of a triad of lasers incident upon an ensemble of atoms. This triad generates the attractive r^{-1} potential (13), whose magnitude has the angular dependence shown in **(b)**.
(c) The angular dependence of the r^{-1} potential generated by each of the six rotated triads of lasers. Together the six triads give an *isotropic* r^{-1} potential in the near zone.

Note that this interaction is *attractive for any orientation* (θ, ϕ) of $\mathbf{r}$ relative to the beams, as long as the polarizability $\alpha(q)$ is real.

If one wishes, the angular anisotropy in (13) can be cancelled to give a purely radial r^{-1} potential, by combining a number of such triads with different orientations. It is convenient to define the orientation of each triad by the Euler angles (α, β, γ) [25], namely, a rotation of α about the $\hat{\mathbf{z}}$ axis, followed by a rotation of β about the new $\hat{\mathbf{y}}$ axis and finally a rotation of γ about the final $\hat{\mathbf{z}}$ axis. One configuration that cancels the anisotropy completely uses 6 triads (18 laser beams) rotated through the following Euler angles: $(0, \pi/4, \pi/8)$, $(0, \pi/4, -\pi/8)$, $(0, \pi/4, 3\pi/8)$, $(0, \pi/4, -3\pi/8)$, $(0, 0, \pi/8)$, $(0, 0, -\pi/8)$. The last two triads should be of half the intensity I of the others. In fact, five of the directions are duplicated, so that only 13 lasers are necessary, four with twice

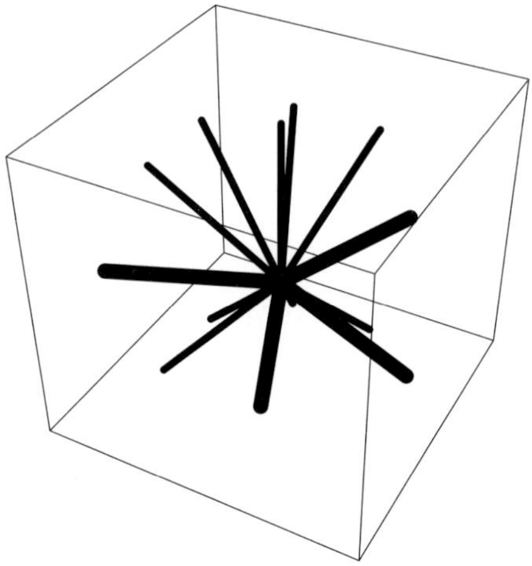

Fig. 2. The configuration of 13 lasers necessary to produce a pure isotropic near-zone r^{-1} potential. An approximately isotropic r^{-1} (always attractive) potential can be obtained with 3 lasers, see [10].

the intensity of the others, see Fig. 2. Then the interatomic potential becomes

$$U^{\mathrm{NZ}}_{\mathrm{isotropic}}(\mathbf{r}) = -\frac{11}{4\pi}\frac{Iq^2\alpha^2}{c\varepsilon_0^2}\frac{1}{r} = -\frac{u}{r} \,. \qquad (14)$$

The main difficulty in realizing the near-zone r^{-1} potential is that the r^{-3} interaction survives due to the *interference* between different pairs of beams, whose contribution is proportional to the product of their respective field amplitudes. This difficulty can be overcome if one introduces frequency shifts between the laser beams. Spreading the frequencies ω_n of the E_n laser fields ($n = 1, 2, 3$ for one triad or $n = 1, 2 \ldots 18$ for six triads) in intervals about the central frequency makes the r^{-3} interference terms in the interaction energy $\propto E_n E^*_{n'}$, $(n \neq n')$, oscillate at the difference frequencies $|\omega_n - \omega_{n'}|$. If these difference frequencies are much higher than the other relevant frequencies (e.g., collective oscillation frequencies of the gas) then the contribution of the r^{-3} interference terms to the mean field potential averages to zero in time. Typically, these conditions hold for $|\omega_n - \omega_{n'}| \geq 10\,\mathrm{MHz}$. Alternatively, the perpendicular laser beams can be chosen to be *spectrally-noisy* (phase-fluctuating), whereupon the suppression of interference also requires a Lorentzian noise spectrum much broader than 10 MHz. In either case, the oscillatory time-dependence of the r^{-3} interference terms gives rise to losses, i.e., ejection of atoms from the mean-field r^{-1} potential (see Sect. 9.3).

Angular misalignment errors, δ, between the orthogonal beams should satisfy $\delta \ll q\,L$ (where L is the mean radius of the "cloud") and intensity fluctuations should satisfy $\Delta I/I \ll q\,L$, in order to ensure the r^{-3} cancellation for the non-interfering terms $\propto \sum_n |E_n|^2$.

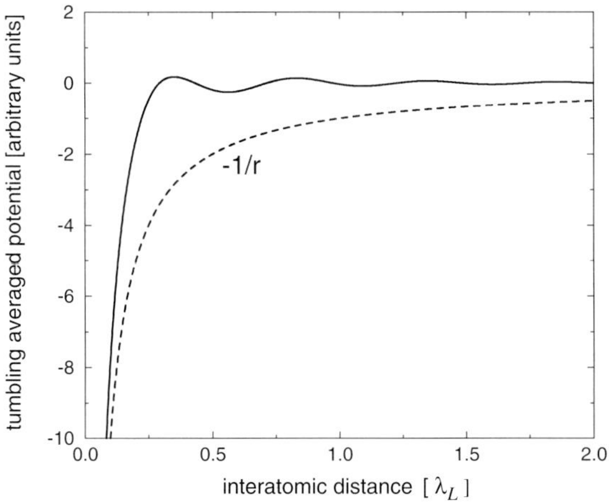

Fig. 3. The "tumbling" averaged interatomic potential (solid line) compared its asymptotic near zone $-1/r$ behavior (long dashed line). The radius is expressed in terms of λ_{L}.

3.2 Far-Zone Orientational Averaging

In the previous sub-section we have discussed the limit in which all of the interatomic distances are much smaller than the wavelength λ_{L} of the laser beams (near-zone limit). Now we consider the case in which such a condition is relaxed and the dipole-dipole induced interaction is no longer an attractive $1/r$ potential. For simplicity we treat only the ideal case of a purely radial (isotropic) interaction. The natural spatial scale for the following discussion is λ_{L}, so we define $\tilde{r} = r/\lambda_{\mathrm{L}}$. If *an orientational "tumbling" average* (pair-orientational average) is performed upon the expression (9), (11) at arbitrary $\tilde{r}$, one gets the isotropic potential (see Fig. 3)

$$\begin{aligned} U_{\mathrm{isotropic}}(\tilde{r}) &= -\frac{15\pi u}{11\lambda_{\mathrm{L}}}\left(\frac{\sin(4\pi\tilde{r})}{(2\pi\tilde{r})^2} + 2\frac{\cos(4\pi\tilde{r})}{(2\pi\tilde{r})^3}\right. \\ &\quad \left. - 5\frac{\sin(4\pi\tilde{r})}{(2\pi\tilde{r})^4} - 6\frac{\cos(4\pi\tilde{r})}{(2\pi\tilde{r})^5} + 3\frac{\sin(4\pi\tilde{r})}{(2\pi\tilde{r})^6}\right) \end{aligned} \tag{15}$$

with the following asymptotic values for the near zone (NZ) and far zone (FZ)

$$\tilde{r} \ll 1 \qquad U_{\mathrm{isotropic}}(\tilde{r}) \longrightarrow U^{\mathrm{NZ}}_{\mathrm{isotropic}} = -\frac{u}{\lambda_{\mathrm{L}}}\frac{1}{\tilde{r}} \tag{16}$$

$$\tilde{r} \gg 1 \qquad U_{\mathrm{isotropic}}(\tilde{r}) \longrightarrow U^{\mathrm{FZ}}_{\mathrm{isotropic}} = -\frac{15\pi u}{11\lambda_{\mathrm{L}}}\frac{\sin(4\pi\tilde{r})}{(2\pi\tilde{r})^2}, \tag{17}$$

where u is given by the expression

$$u = \frac{11\pi}{15}\frac{I\alpha^2}{c\epsilon_0^2\lambda_{\mathrm{L}}^2}. \tag{18}$$

4 State Equations for Fermions

4.1 Preliminaries

Consider a system of N atoms, each of mass m, possibly in a trapping potential $V_{\rm trap}(\mathbf{r})$ interacting via light-induced dipole-dipole forces described by the potential $U(\mathbf{r}_i, \mathbf{r}_j)$ and/or via short-range forces (s-wave scattering). Let $f(\mathbf{r}, \mathbf{p}, t)$ denote the distribution function of the system, $f(\mathbf{r}, \mathbf{p}, t)d^3\mathbf{r}d^3\mathbf{p}$ gives the number of particles whose position and momentum are in the cell $(\mathbf{r}, \mathbf{p}; \mathbf{r}+d^3\mathbf{r}, \mathbf{p}+d^3\mathbf{p})$ at time t (the quantum treatment requires cells to exceed $(2\pi\hbar)^3$ in phase space). The integral of f over momenta determines the number density ρ

$$\rho(\mathbf{r}) = \int f(\mathbf{r}, \mathbf{p})\, d^3\mathbf{p}. \tag{19}$$

The total number of particles N is given by

$$N = \int \rho(\mathbf{r})\, d^3\mathbf{r}\ , \tag{20}$$

and the mean energy can be expressed as

$$\begin{aligned} E &\equiv \int f(\mathbf{r}, \mathbf{p})\, E(\mathbf{r}, \mathbf{p})\, d^3\mathbf{r}\, d^3\mathbf{p} \\ &= \int f(\mathbf{r}, \mathbf{p})\, \frac{p^2}{2m}\, d^3\mathbf{r}\, d^3\mathbf{p} + \int \rho(\mathbf{r}) V_{\rm tot}(\mathbf{r})\, d^3\mathbf{r} \end{aligned} \tag{21}$$

with

$$V_{\rm tot}(\mathbf{r}) = V_{\rm trap}(r) + g\rho(\mathbf{r}) + \int U(\mathbf{r}, \mathbf{r}')\rho(\mathbf{r}')\, d^3\mathbf{r}'\ , \tag{22}$$

where g is determined by the strength of the s-wave scattering in BECs (see below), whereas in FDGs $g = 0$.

4.2 Non-degenerate Gas

In the non-degenerate Maxwell-Boltzmann gas (MBG), the equilibrium configuration of the system maximizes the Boltzmann entropy [26]–[28]

$$S = -k_B \int f \ln f\ d^3\mathbf{r}d^3\mathbf{p}\ , \tag{23}$$

while conserving the number of particles and energy (microcanonical ensemble). Using the method of Lagrange multipliers, we find that the critical points of the MBG entropy are given by the function $f(\mathbf{r}, \mathbf{p})$, which, upon integrating over the momenta yields the following number density distribution

$$\rho(\mathbf{r}) = \rho_0\, \mathrm{e}^{-\Phi}\ .$$

Here ρ_0 is the density in the center of the cloud and we have introduced the function

$$\Phi = \beta(V_{\rm tot}(\mathbf{r}) - V_0), \tag{25}$$

where $\beta = 1/k_B T$ measures the inverse temperature and $V_0 = V_{\rm tot}(0)$ is the potential at the cloud center. The integral equations (22), (24), and (25) for the density $\rho(r)$ and the potential $V_{\rm tot}(\mathbf{r})$ can be solved iteratively.

4.3 Hydrostatic Equilibria for Quantum Fermi Gases

For a degenerate Fermi-Dirac gas (FDG), the equilibrium configuration of the system is determined by maximizing the entropy [26]–[28]

$$S_{\rm Fermi} = -k_B \int (f \,\ln f \,+\, (1-f)\,\ln(1-f))\, d^3\mathbf{r} d^3\mathbf{p} \,, \tag{26}$$

while conserving the number of particles and energy (i.e., assuming a microcanonical ensemble). The method of Lagrange multipliers now yields the critical points of entropy, which are given by the following distribution functions

$$f_{\rm Fermi} = \frac{1}{{\rm e}^{\beta(E(\mathbf{r},\mathbf{p})\, -\, \mu_0)} + 1}\,,$$

where $E(\mathbf{r},\mathbf{p})$ is given by (21) and μ_0 is the "global" chemical potential of the cloud, which is determined from the normalization condition (20). Following [28] we introduce the local chemical potential, which is the difference between μ_0 and the local potential energy:

$$\mu(\mathbf{r}) = \mu_0 - V_{\rm tot}(\mathbf{r})\,. \tag{27}$$

The main parameter of interest here is the local fugacity Z [27]

$$Z(\mathbf{r}) = {\rm e}^{\beta\mu(\mathbf{r})}, \tag{28}$$

By integrating the distribution function $f(\mathbf{r},\mathbf{p})$ over the momentum, we obtain the number density distribution

$$\rho_{\rm Fermi}(\mathbf{r}) = \sum_{\mathbf{p}} f_{\rm Fermi}(\mathbf{r},\mathbf{p}) \longrightarrow \int \frac{f_{\rm Fermi}(\mathbf{r},\mathbf{p})}{(2\pi\hbar)^3}\, d^3\mathbf{p} \;=\; \rho_0\, \frac{{\rm f}_{3/2}(Z(\mathbf{r}))}{{\rm f}_{3/2}(Z(0))}\,, \tag{29}$$

where $Z(\mathbf{r}) = e^{\beta\mu(\mathbf{r})}$ is the local fugacity [27], ρ_0 is as in (24) and $f_{3/2}(Z)$ is the index-3/2 polylogarithmic function $f_n(Z)$ [29,30], which is represented as:

$$\begin{aligned} f_n(Z) &\equiv \frac{1}{\Gamma(n)} \int_0^\infty \frac{x^{n-1}}{1 + Z^{-1} e^x} dx \\ &= \sum_{\alpha=1}^\infty (-1)^{\alpha+1} \frac{Z^\alpha}{\alpha^n} , \end{aligned} \tag{30}$$

$$\begin{aligned} f_{3/2}(Z) &= \frac{4}{\sqrt{\pi}} \int_0^\infty x^2 \frac{Z}{Z + e^{x^2}} dx \\ &= \sum_{\alpha=1}^\infty (-1)^{\alpha+1} \frac{Z^\alpha}{\alpha^{3/2}} . \end{aligned} \tag{31}$$

The integral equation (20), (24) for the density $\rho(r)$ and the potential $V_{\rm tot}$ (22) can be solved iteratively.

The equation of state relates the density $\rho_{\rm Fermi}(\mathbf{r})$ and pressure $P_{\rm Fermi}(\mathbf{r})$. It is given implicitly for a FDG as

$$\rho_{\rm Fermi}(\mathbf{r}) = \frac{1}{\lambda_T^3} f_{3/2}(Z) \tag{32}$$

$$P_{\rm Fermi}(\mathbf{r}) = \frac{1}{\lambda_T^3} \frac{1}{\beta} f_{5/2}(Z) , \tag{33}$$

where the thermal de Broglie wavelength

$$\lambda_T = \left(\frac{2\pi\hbar^2\beta}{m} \right)^{1/2} . \tag{34}$$

The polylogarithmic functions [27,29,30] $f_{3/2}(Z)$, $f_{5/2}(Z)$ are related to each other via $f_{3/2}(Z) = Z \frac{d}{dZ} f_{5/2}(Z)$, leading to the following relation between the particle number density and the pressure for a FDG [12]:

$$\rho_{\rm Fermi}(Z) = \beta Z \frac{d}{dZ} P_{\rm Fermi}(Z). \tag{35}$$

The hydrostatic equilibrium condition

$$\nabla P_{\rm Fermi}(\mathbf{r}) = -\rho_{\rm Fermi}(\mathbf{r}) \nabla V_{\rm tot}(\mathbf{r}) \tag{36}$$

combined with the equation of state (32), (33), (35), determines the density distribution of the trapped FDG.

5 Self-bound Bose Condensates with Laser-Induced "Gravitation"

We use the mean-field approximation (MFA), as embodied in the following generalized Gross-Pitaevskii equation (GGPE) [9,10], to calculate the ground-state order parameter $\Psi(\mathbf{r})$ of a zero-temperature BEC subject to a laser-induced interatomic interaction

$$i\hbar\frac{\partial}{\partial t}\,\Psi(\mathbf{r},t) = \left[-\frac{\hbar^2}{2m}\,\nabla^2 + V_{\rm tot}(\mathbf{r})\right]\Psi(\mathbf{r},t)\,. \tag{37}$$

Here $V_{\rm tot}(\mathbf{r})$ is the self-consistent total potential (22), with the isotropic trap potential

$$V_{trap}(\mathbf{r}) = m\omega_0^2 r^2/2 \tag{38}$$

which can be set to zero in certain cases – see below, $\rho(\mathbf{r}) = |\Psi(\mathbf{r})|^2$ is the density, $g = 4\pi a\hbar^2/m$, a being the s-scattering length, and $U(\mathbf{R})$ is the laser-induced electrostricive (dipole-dipole) potential of (9), (13), or (15), depending on the laser beam geometry.

In cold, dilute, atomic BECs with short-range s-wave scattering, the validity of the MFA (i.e., the Gross-Pitaevskii equation [2]) is well established, provided $\rho a^3 \ll 1$. However, the MFA is also valid for the long-range repulsive Coulomb-like potential, $+u/r$ [31], provided many atoms lie within an interaction sphere with a Bohr-type radius, $a_* = \hbar^2/mu$, so that $\rho a_*^3 \gg 1$. This condition means that the potential must be weak. Remarkably, self-gravitating BECs simultaneously satisfy *both* of these MFA validity conditions, as can be readily verified using the ensuing expressions.

There are two limiting regimes for self-gravitating BECs [9,10]: the purely gravitational G regime, where the kinetic energy is balanced by the gravitational-like potential and the s-wave scattering is negligible, and the Thomas-Fermi gravitational (TF-G) regime, where the kinetic energy is negligible and repulsive s-wave scattering balances the gravitational-like potential.

The condensate radius can be studied using the variational gaussian wave function

$$\Psi_w(\mathbf{r}) = \sqrt{N}\exp(-R^2/2w^2\lambda_{\rm L}^2)/(\pi w^2\lambda_{\rm L}^2)^{3/4}\;, \tag{39}$$

where w is a dimensionless variational parameter giving the width of the condensate. The variational solution in the limit of negligible kinetic energy (Thomas-Fermi limit) yields a self-bound condensate, i.e., finite w (see Fig. 4), if the laser intensity exceeds the following threshold value (in SI units):

$$I_0 = \frac{48\pi}{7}\frac{\hbar^2\pi c\epsilon_0^2}{m\alpha^2}a \approx 21.5423\,\frac{\hbar^2 c\epsilon_0^2}{m\alpha^2}\,a\,. \tag{40}$$

Here I_0 is the total intensity supplied by all the laser beams: for a triad each laser should have 1/3 of the above value and for the six-triad configuration [9]; 12 of the lasers should have 1/15, and the remaining 6 should have 1/30, of the above value. The threshold I_0 signifies the equality of the gravitational-like potential and the s-wave scattering potential.

With an intensity 1.5 times the threshold intensity of (40) (arrow in Fig. 5), the expectation value of the rms condensate radius $R_{\rm rms} = \sqrt{\langle R^2\rangle}$ is a fraction of the laser wavelength $\lambda_{\rm L}$ ($R_{\rm rms} \approx 0.43 \times \lambda_{\rm L}$). The condensate is less and less confined as the intensity approaches the threshold (40) from above

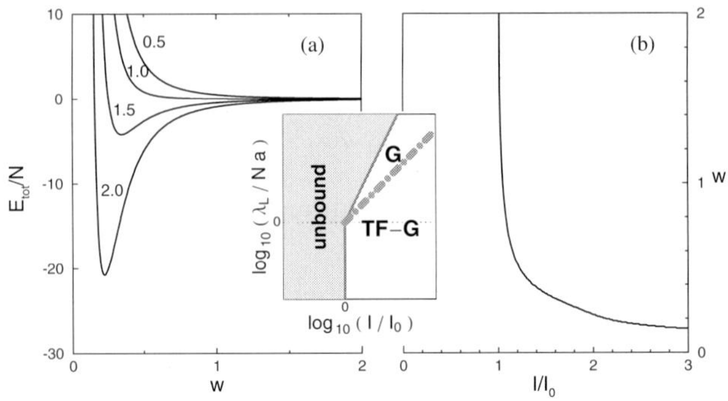

Fig. 4. (a) Variational mean field energies per particle in the case of negligible kinetic energy (TF-G regime) $\lambda_{\rm L}/Na \ll 1$ (where $\lambda_{\rm L}/Na$ represents the ratio of the kinetic energy to the interaction energy) plotted versus the trial size w for different values of I/I_0. (b) Equilibrium value of w versus I/I_0 in the limit of negligible kinetic energy (Thomas-Fermi limit). Only for $I/I_0 > 1$ are *self-bound* variational solutions (having a minimum at finite w) observed. Inset – Schematic phase portrait of the transition from unbound to self-bound regime for negligible external trapping is plotted versus $\log_{10}(\lambda_{\rm L}/Na)$ and $\log_{10}(I/I_0)$.

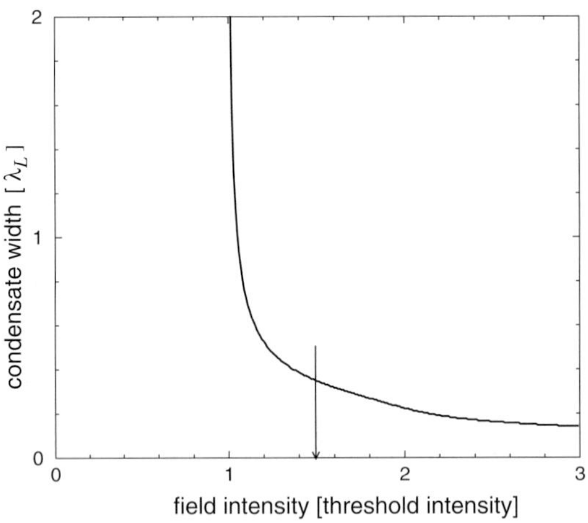

Fig. 5. Condensate radius w (in units of λ_L) calculated by the variational method as a function of the normalized intensity I/I_0. Solutions are *self-bound* (finite $w\lambda_{\rm L}$) above the threshold intensity.

(see Figs. 4, 5). Increasing the intensity I reduces the condensate radius, which becomes, in the asymptotic limit, proportional to $1/\sqrt{I}$. Thus *the dependence* $R_{\rm rms} \sim (I_0/I)^{1/2}\lambda_{\rm L}$ *is a distinct experimental signature of self-binding.*

Above the threshold intensity, the effect of an external harmonic trap becomes negligible when $\rho\, l_0 \lambda_{\rm L} a \gg 1$, where $l_0^2 = \hbar/m\omega_0$. As the laser intensity is increased beyond this value the trap becomes increasingly "irrelevant" – *it is not necessary* to turn it off to access the TF-G regime, where r^{-1} and s-wave scattering dominate.

The threshold I_0 (see (40)) is evaluated neglecting the kinetic energy. The role of kinetic energy can be discussed in terms of λ_L/Na (approximately the ratio between the kinetic energy $N\hbar^2/m\lambda_L^2$ and the scattering energy $N^2\hbar^2 a/m\lambda_L^3$), as shown schematically in the phase portrait in Fig. 4 (drawn for negligible external trapping) and can modify the threshold for self-binding. The G regime, representing the purely "gravitational" counterpart of the TF-G regime, where only "self-gravitation" and kinetic energy play a role [9,10] (as in a Bose star [11]) is accessed when

$$\frac{\lambda_L}{Na} \leq \frac{I}{I_0} \leq \left(\frac{\lambda_L}{Na}\right)^2 , \tag{41}$$

that implies $1 \leq \lambda_L/Na$.

In recent experiments on superradiance [32] the laser was red detuned by 1.7 GHz from the $3S_{1/2}$, F=1 $\rightarrow$ $3P_{3/2}$, F = 0,1,2, transition frequency of sodium. With this detuning, moderate from resonance, the polarizability in CGS units is $\alpha = 3.534 \times 10^{-18} cm^3$, which is $\approx 1.5 \times 10^5$ times the static value of the polarizability. The threshold intensity (40) is then reduced by a factor $\approx 2.3 \times 10^{10}$ compared to the static polarizability case, becoming $I_0 \approx 262$ mW/cm^2 for sodium, which is close to the values used in [32]. With this value of threshold intensity the gradient forces can be negligible if the focal spots of the lasers are much wider than λ_L.

A key experimental restriction on self-binding is that the atoms should be in the near zone to feel the $1/r$ potential: a condensate smaller than the laser wavelength severely limits the number of atoms involved. At moderately detuned wavelengths ($\approx 0.589\mu$m for sodium) and an intensity $I \approx 1.5 I_0$, the atom cloud (Fig. 6) contains $\approx$ 40 to 100 atoms as the peak density ranges from 10^{15} to 10^{16} atoms/cm^3. Although this number is small, it is *sufficient to demonstrate the self-binding effect.*

For given values of I, α, a, and m, we are either in the G regime or the TF-G regime, depending on whether the number of atoms N is smaller or larger than the number [9] $N_{\text{border}} \approx \sqrt{3\pi\hbar^2/(2mua)}$ which corresponds to the line separating the two regions in the inset of Fig. 4. It so happens that 40, the lower estimate of the number of self-bound sodium atoms obtainable in the moderate-detuning regime, is very close to N_{border} (Fig. 6). This is an interesting region, because both the kinetic energy and the s-wave scattering are significant and together with the r^{-1} attraction determine the condensate properties.

6 Collective Excitations of a "Gravitationally" Self-bound Bose Gas

6.1 Dynamical Equations

It is known that the long-wavelength collective excitations of a *spatially-uniform* system of both fermions (e.g., the electron gas) and bosons [31], interacting via

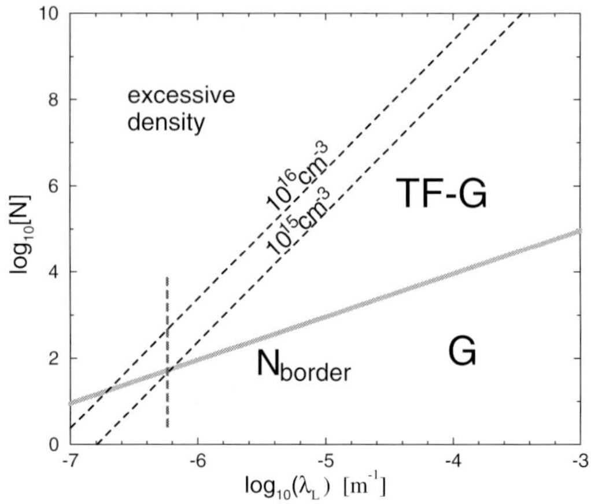

Fig. 6. Range of numbers N of Na condensate atoms as a function of λ_L that are compatible with a TF-G or G solution. The density is 10^{15}–10^{16} atoms/cm^3 and the intensity is 1.5 times the threshold intensity (40). The region above 10^{16} cm^{-3} corresponds to excessive density. The vertical long-dashed line corresponds to the moderate-detuning choice discussed for Na.

a *repulsive* potential of the form u/r, are characterized by a spectral gap, fixed by the "plasma" frequency

$$\omega_p^2 = 4\pi u\rho/m \ , \tag{42}$$

where ρ is the particle density. In this Section we show, following [14], that also in the case of a *self-bound* atomic BEC or FDG, interacting via an *attractive* $-u/r$ potential, the collective oscillations are characterized by an effective "plasma" frequency ω_p of the form (42), with a spatially-averaged density ρ and the positive coupling constant u [see (14)]. Its correspondence with the fundamental oscillation frequency of "White Dwarfs" and Neutron Stars [12] is pointed out and the analog of the Jeans gravitational instability is explicitly analyzed.

By expressing the complex BEC order parameter through the condensate density ρ and phase ϕ as $\Psi = \sqrt{\rho}\exp(i\phi)$, and introducing the superfluid (mean) velocity $\mathbf{v}$, we can write the GGPE (37) as a coupled set of collisionless hydrodynamic equations:

$$\frac{\partial\rho}{\partial t} + \nabla(\rho\mathbf{v}) = 0, \tag{43}$$

$$m\frac{\partial\mathbf{v}}{\partial t} + \nabla\left(\frac{mv^2}{2} - \frac{\hbar^2\nabla^2\sqrt{\rho}}{2m\sqrt{\rho}} + \rho + V_{\text{tot}}\right) = 0 \ . \tag{44}$$

Here V_{tot} given by (22) includes the *isotropic* light-induced mean-field potential term $\int d^3\mathbf{r}\ U_{\text{isotropic}}(\mathbf{r}-\mathbf{r}')\rho(\mathbf{r}')$, where $U_{\text{isotropic}}(\mathbf{r}-\mathbf{r}')$ is given by (13). We shall focus here on its "self-gravitational" near-zone limit, (14).

Equation (44) establishes the irrotational property of the superfluid velocity field defined by $\mathbf{v} = \hbar\nabla\phi/m$. Similar equations have been derived in the case of a harmonically trapped BEC interacting via short-range forces [33].

6.2 Jeans-Like Instability

In the absence of external potentials, $V_{\text{trap}} = 0$ in V_{tot}, (43) and (44) admit a stationary solution with constant density ρ. Such a solution is, however, dynamically unstable, as can be seen directly from the Bogoliubov dispersion relation [26,34] corresponding to (43) and (44):

$$\omega_k^2 = \frac{\hbar^2 k^4}{4m^2} + c_s^2 k^2 - \omega_p^2 \,. \tag{45}$$

In (45) $c_s = \sqrt{g\rho/m}$ is the Bogoliubov speed of sound and the plasma frequency ω_p is given by (42). The occurrence of imaginary frequencies ω_k for small values of k reveals the existence of an instability, which resembles the Jeans gravitational instability [11]–[13]. Assuming

$$\hbar\omega_p \ll mc_s^2, \tag{46}$$

the minimal wavelength (Jeans wavelength) ensuring a real value for the frequency (45) is given by

$$\lambda_J = 2\pi c_s/\omega_p = 2\pi(\hbar^2 a/mu)^{1/2} \,. \tag{47}$$

In a "self-gravitating" BEC, both ω_p and c_s scale as $\rho^{1/2}$, so that the critical wavelength λ_J is *independent of the density*, as explicitly revealed by (47).

Besides these unstable solutions, (43) and (44) admit a *stable self-bound solution* for any finite number N of atoms. In this case long-wavelength fluctuations, which can trigger the Jeans instability, are excluded by the finite size of the system. Remarkably, the ground-state density of a self-bound Bose cloud, in the limit in which the kinetic energy can be neglected (Thomas-Fermi limit for Bose gases), has the analytic profile [9,14]

$$\rho_0(\mathbf{r}) = \frac{N}{4R_0^2}\frac{\sin(\pi r/R_0)}{r}\theta(R_0 - r) = 0, \tag{48}$$

where the *radius* $R_0 = \lambda_J/2$ is *fixed by the Jeans wavelength* and θ is the Heaviside function.

In the Thomas-Fermi "gravitational" (TF-G) regime, the equations of motion (43) and (44) can be simplified , since the quantum pressure term, proportional to $\hbar^2$, can be safely ignored. After linearization they yield the equation of motion for the density fluctuations $\delta\rho$ [14]

$$\frac{\partial^2}{\partial t^2}\delta\rho = \nabla\left[c_s^2\nabla\left(\delta\rho + \delta V_{\text{tot}}/g\right)\right] \,, \tag{49}$$

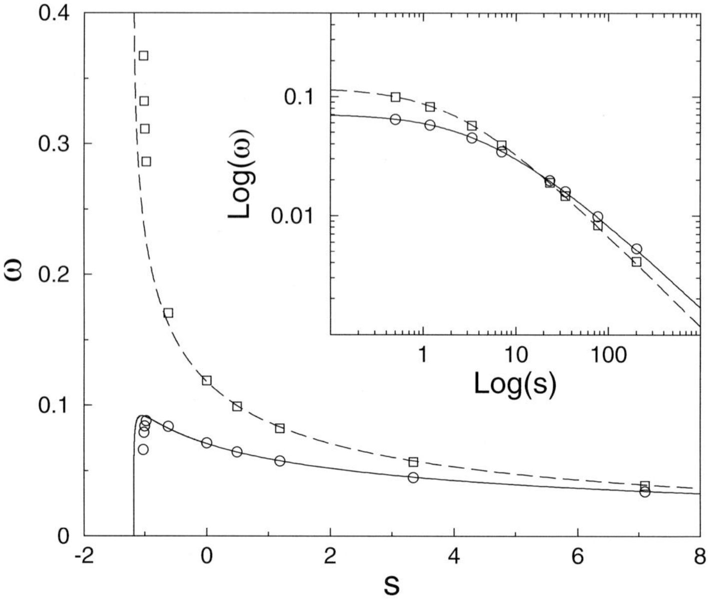

Fig. 7. Monopole and quadrupole frequencies of a self-bound condensate (in $mu^2N^2/\hbar^3$ units) versus the scattering parameter $s = muN^2a/\hbar^2$, for both positive and negative values of a. Circles: monopole frequencies; squares: quadrupole frequencies, calculated numerically. Inset: the corresponding Log-Log plot for $a > 0$.

where the "local" sound velocity $c_s(r) = \sqrt{g\rho(r)/m}$ is calculated at the equilibrium density (48). The time dependence of $\delta\rho$ is chosen in the form of $\exp(-i\omega t)$, and (49) is transformed, with the help of the Poisson equation for the laser-induced "gravitation" potential (see (53)), into

$$\tilde{\omega}^2\tilde{\nabla}^2\xi + \tilde{\nabla}\left[\frac{\sin(2\pi\tilde{r})}{2\pi\tilde{r}}\tilde{\nabla}\left(\frac{\tilde{\nabla}^2\xi}{4\pi^2} + \xi\right)\right] = 0\,, \tag{50}$$

where we have introduced the "gravitational" fluctuation variable $\xi = \delta V_{\mathrm{tot}}/g$ and used the dimensionless coordinate $\tilde{\mathbf{r}} = \mathbf{r}/\lambda_J$ and frequency $\tilde{\omega} = \omega/\omega_p$.

To solve this equation, we set $\xi = \xi_l(\tilde{r})Y_{lm}(\vartheta,\varphi)$, where Y_{lm} are the usual spherical harmonics. The boundary conditions can be derived from the tailoring of the inner and outer solutions on the condensate surface, i.e. at $\tilde{r} = \pi$ [35]. The conservation of the total number of atoms implies that the radial derivative of the potential perturbation vanishes at $\tilde{r} \le \pi$ for the monopole mode. For the modes with $l = 1, 2, 3, \ldots$ the function ξ_l instead decreases as $r^{-(l+1)}$. Hence, the boundary conditions are $(d\xi_0/d\tilde{r})|_{\tilde{r}=\pi} = 0$ for $l = 0$ and $(d\xi_l/d\tilde{r})|_{\tilde{r}=\pi} = -(l+1)\xi_l/\pi$ for $l > 0$ [35].

The numerical solutions of (50) for the monopole and quadrupole frequencies yield $\omega_M = 0.6168\,\omega_p$, $\omega_Q = 0.392\,\omega_p$ with ω_p as in (42). These frequencies turn out to be *very close* (within 1%) to the sum-rule estimates derived in [14] (Fig. 7).

7 "White Dwarfs"

7.1 Hydrostatic Equilibrium of a Self-Gravitating Fermi Gas in a Parabolic Trap at $T \neq 0$

We follow Chandrasekhar [12] and Sect. 4, relating the pressure $P(r)$ and number density $\rho(r)$ profiles via the hydrostatic equilibrium equation (36) and the equation of state (32), (33), (35). We assume that the fermionic spin-polarized gas is confined in a spherically symmetric parabolic trap of strength b, $V_{\text{trap}} = br^2/2$, acting on the atoms. In addition, it is illuminated by the laser beam configuration of Sect. 3.1. The spin-polarized interatomic potential is then free of the s-wave scattering effect ($g = 0$ in (22)) and has the form of $U^{\text{NZ}}_{\text{isotropic}}$ (14), yielding gravity-like r^{-1} attraction, with the interaction constant changed from G to $u/(2\pi)$.

The total potential (22) is then

$$V_{\text{tot}}(\mathbf{r}) = br^2/2 + V(r), \tag{51}$$

with the mean-field light-induced potential

$$V(\mathbf{r}) = \int U^{\text{NZ}}_{\text{isotropic}}(\mathbf{r} - \mathbf{r}_1)\, \rho(\mathbf{r}_1)\, d^3\mathbf{r}_1 = \int -\frac{u}{\mathbf{r} - \mathbf{r}_1}\, \rho(\mathbf{r}_1)\, d^3\mathbf{r}_1\,, \tag{52}$$

which obeys the Poisson equation

$$\Delta V(\mathbf{r}) = \frac{u}{2\pi} 4\pi\, \rho(\mathbf{r}) = 2u\, \rho(\mathbf{r})\,. \tag{53}$$

It is valid if the atomic cloud is within the near zone of the laser wavelength (see Sect. 3 above).

Within the local density approximation, the gas pressure then changes as

$$\frac{dP}{dr} = -\frac{u}{2\pi}\frac{N(r)\rho(r)}{r^2} - b\,\rho(r)\, r, \tag{54}$$

where $N(r) = \int_0^r \rho(r) 4\pi r^2 dr$ is the number of particles within the sphere of radius r. It follows that

$$\frac{1}{r^2}\frac{\mathrm{d}}{\mathrm{d}r}\left(\frac{r^2}{\rho(r)}\frac{dP}{dr}\right) = -\,2\,u\,\rho(r) - 3\,b. \tag{55}$$

The boundary conditions at the center of the cloud are

$$\begin{aligned} &\rho(r=0) \equiv \rho_0\,; \quad \rho'(r=0) = 0 \\ &P|_{r=0} = P(\rho_0)\,; \quad P'|_{r=0} = 0. \end{aligned} \tag{56}$$

We introduce the dimensionless radius $\xi = r/R_0$ with the characteristic radial scale R_0 of the "self-gravitating" cloud

$$R_0 = \frac{\hbar}{m^{1/2}\, u^{1/2}\, \rho_0^{1/6}} \left(\frac{3\pi^2}{4}\right)^{1/3} \tag{57}$$

and the scaled strength δ of the trap

$$\delta = \frac{2b}{3\rho_0 u} \tag{58}$$

The degeneracy of the atomic gas is characterized [27] by the parameter $\mathcal{D}$, which is the amount of particles in a cubic thermal (de Broglie) wavelength:

$$\mathcal{D} = \rho_0 \lambda_T^3 \equiv \frac{\rho_0 \beta^{3/2} \hbar^3}{m^{3/2}} (2\pi)^{3/2} . \tag{59}$$

Both the gas density ρ and pressure P are related to the fugacity $Z(r) = \exp(\beta\mu(r))$, where μ is the local chemical potential. Then the equation for the fugacity Z follows

$$\frac{1}{\xi^2} \frac{\mathrm{d}}{\mathrm{d}\xi} \left(\frac{\xi^2}{Z} \frac{dZ}{d\xi} \right) = \frac{(9\pi/16)^{1/3}}{\mathcal{D}^{1/3}} \left(f_{3/2}(Z) - \delta \cdot \mathcal{D} \right) \tag{60}$$

with the boundary conditions

$$f_{3/2}(Z(0)) = \mathcal{D}, \quad Z'(0) = 0. \tag{61}$$

The cloud profile is determined (see (55)–(61)) by the dimensionless trap strength δ and the degeneracy parameter $\mathcal{D}$.

7.2 The Condition for a Laser-Induced "White Dwarf"

We consider the change in size and density profile of a spin-polarized Fermi gas as the temperature T and the ratio of the trap potential $V_{\text{trap}} = br^2/2$ to the laser-induced "self-gravitating" potential $U = -u/r$ are varied. In Fig. 8 we plot the density profile $\rho(\xi)$ for a strong and a weak trap ($\delta = 4$ and $\delta = 0.04$ for Fig. 8a and 8b, respectively). In both cases the density profile is calculated for the degeneracy parameter $\mathcal{D}$ in the center of the cloud ranging from $\mathcal{D} = 40$ (cold and compressed cloud) down to $\mathcal{D} = 1$ (hot and/or dilute cloud).

At high temperatures, $kT \gg \varepsilon_F$, the size of the self-gravitating cloud follows from $kT \equiv 1/\beta \sim \frac{1}{2} b r_{\text{cloud}}^2$ as

$$\frac{r_{\text{cloud}}}{R_0} \sim \frac{2}{\sqrt{\delta}\, \mathcal{D}^{1/3}} . \tag{62}$$

For a *weak trap* ($\delta \ll 1$) *and high temperature*, $kT \gg \epsilon_F$, the number of atoms in the core of the cloud is small as compared to the total number of atoms in the trap. Then the characteristic size of the cloud is $\langle R \rangle \sim R_0$ (see (57)), whereas the tail of the distribution is confined by the trap to $r_{\text{cloud}} \sim 2R_0/\sqrt{\delta}$. This limit is *favorable for adiabatic compression*: a small number of atoms is compressed to degeneracy while the temperature of the system is almost unchanged [36].

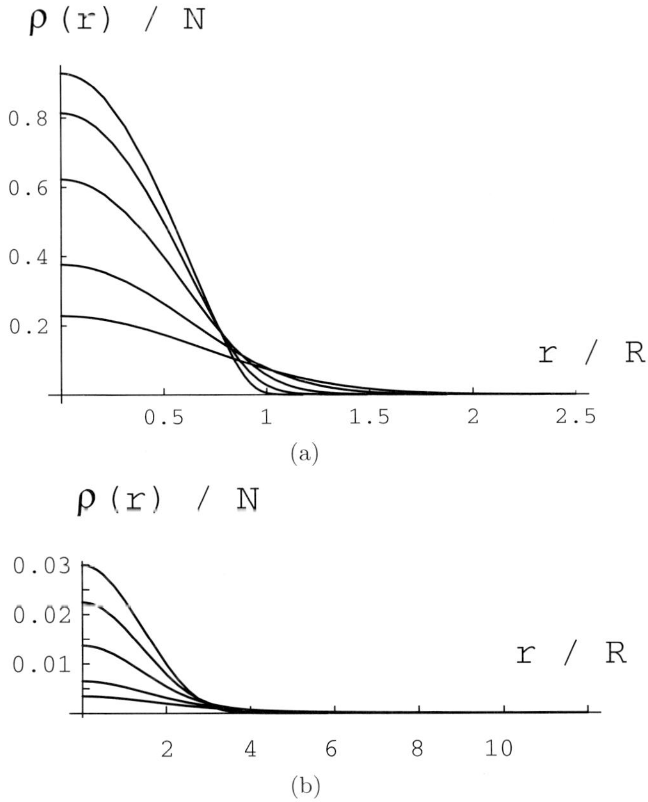

Fig. 8. Scaled density profile $\rho(r)/N$ for a degenerate Fermi gas in a trap at finite temperature. The trap strength $\delta = 4$. Normalization is chosen so that the total number of particles in the trap N is constant. The two plots correspond to (a) strong trap, $\delta = 4$; (b) weak trap, $\delta = 0.04$.
Different density profiles (from top to bottom) correspond to degeneracy $\mathcal{D} = 40$ (high degeneracy), $\mathcal{D} = 12$, $\mathcal{D} = 5$, $\mathcal{D} = 2$, $\mathcal{D} = 1$ (low degeneracy).

The characteristic size of the cloud *at zero temperature* is estimated from the condition $\frac{1}{2}br_{\text{cloud}}^2 \sim \varepsilon_F \equiv \hbar^2(6\pi^2\rho)^{2/3}/2m$, as

$$\frac{r_{\text{cloud}}}{R_0} \sim \min\left(\frac{2}{\sqrt{\delta}}\ ;\ 1\right) . \tag{63}$$

As shown below, for *a weak trap at zero temperature* the entire cloud is confined within a size $r_{\text{cloud}} \sim R_0$, without a trap-dependent tail, thus constituting the analog of a "White Dwarf".

The local chemical potential of the spin-polarized FDG at $T = 0$ equals the local Fermi energy ε_F and is related to the density profile $\rho(r)$ by:

$$\mu(T=0) \equiv \varepsilon_F \equiv \frac{p_F^2}{2m} = \frac{\hbar^2}{2m}\left(6\pi^2\rho(r)\right)^{2/3} . \tag{64}$$

To find the density profile $\rho(r)$ we use the relation [28]

$$\mu(r) + V(r) + br^2/2 = const\ . \tag{65}$$

With the Poisson equation for $\Delta V(\mathbf{r})$ we get:

$$\Delta\mu = -\Delta V - 3b = -2u\,\rho - 3b\ , \tag{66}$$

or

$$\frac{1}{r^2}\frac{\mathrm{d}}{\mathrm{d}r}\left(r^2\frac{d\mu}{dr}\right) = -\mu^{3/2}\kappa - 3b\,, \tag{67}$$

where [28]

$$\kappa = \frac{u}{3\pi^2}\left(\frac{2m}{h^2}\right)^{3/2}\ . \tag{68}$$

The dimensionless chemical potential F is introduced by setting $\mu(r) = \mu_0\,F(\xi)$, where μ_0 is the chemical potential at the center of the cloud. In terms of F and ξ, the equilibrium condition leads to the equation, which is similar to the Lane-Emden equation [12] of index 3/2:

$$\frac{1}{\xi^2}\frac{\mathrm{d}}{\mathrm{d}\xi}\left(\xi^2\frac{dF(\xi)}{d\xi}\right) = -F(\xi)^{3/2} - \delta\ . \tag{69}$$

The regular solution of this equation is found numerically [29] for the boundary conditions

$$F(\xi=0) = 1, \qquad F'(\xi=0) = 0. \tag{70}$$

The $T = 0$ solution $F(\xi)$ is totally confined, without any tail, within the sphere of radius R_{max}, which is of the order of R_0. Once a self-confined solution is achieved, the trap can be switched off. In the absence of the trap, the characteristic radius R_0 is determined [12], [28] in terms of the particle number density ρ_0 at the cloud center or the total number of particles in the cloud $N \equiv N(R_{max})$:

$$\begin{aligned} R_0 &= \frac{\hbar}{m^{1/2}\,u^{1/2}\,\rho_0^{1/6}}\left(\frac{3\pi^2}{4}\right)^{1/3}, \\ R_0 &= \frac{\hbar^2}{muN^{1/3}}\left(\frac{3}{2}\right)^{2/3}\pi^{5/3}\left(\xi_1^2F'(\xi_1)\right)^{1/3}, \end{aligned} \tag{71}$$

where $\xi_1 \approx 3.65$ is the first positive zero of $F(\xi)$ and $\xi_1^2F'(\xi_1) \approx 2.71$ [12]. The radius R_{max} of this self-gravitating "White Dwarf" follows:

$$R_{max} = R_0\,\xi_1 \approx 3.65R. \tag{72}$$

The density profile of the cloud is then

$$\rho(r) = \rho_0\,\left(F(r/R_0)\right)^{3/2}\ , \tag{73}$$

where the central density ρ_0 equals

$$\rho_0 = \frac{N^2u^3m^3}{\hbar^6}\,\frac{F(\xi_1)^{3/2}}{9\pi^6(\xi_1^2F'(\xi_1))^2}\ . \tag{74}$$

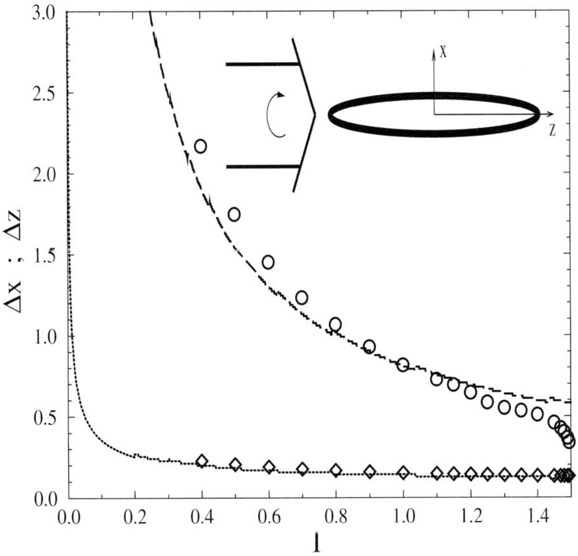

Fig. 9. Inset – the laser beam and condensate geometry. Mean radii (normalized to λ_L) of the self-bound condensate versus the intensity I (normalized to $8\pi\hbar^2 c\epsilon_0^2 a/m\alpha^2$) in the TF-limit. Long-dashed (dotted) lines represent the expectation value of Δz (Δx) obtained from the Gaussian ansatz. Circles (diamonds) represent the expectation value of Δz (Δx) calculated from the quasi-1D ansatz, (76).

8 Density Modulations of a Cigar-Shaped BEC via Laser-Induced Interactions

Following [15], we consider a far-off-resonant circularly polarized laser beam propagating along the positive $\hat{z}$ direction with wavevector q (Fig. 9 – inset). The dipole-dipole induced interatomic potential [8] then becomes, using (9 – 11)

$$U(\boldsymbol{r}) = \frac{3}{4}\hbar\Gamma_{ray}\Big[\frac{2z^2 - x^2 - y^2}{q^3 r^5}\big(\cos qr + qr\sin qr\big) - \frac{2z^2 + x^2 + y^2}{qr^3}\cos qr\Big]\cos(qz)\,, \tag{75}$$

where x, y, z are the components of the interatomic separation $\mathbf{r}$ and $\hbar\Gamma_{ray} = \alpha^2 q^3 I/6\pi c\varepsilon_0^2$ is the single-atom (spontaneous) Rayleigh scattering rate, which is proportional to the laser intensity I and to α^2, i.e. to the saturation parameter that scales as the inverse square of the detuning from the nearest atomic resonance [9,10]. We assume in the following that the condensate contains many atoms per cubic wavelength to ensure the validity of the mean-field approach. The condensate is taken to be at zero temperature and the rate of heating due to incoherent Rayleigh scattering is found to be negligible [15]. The Gross-Pitaevskii equation [2] for the order parameter $\Psi(\mathbf{r}, t)$ has now the form (37) with the potential (22) having $U(\mathbf{r}, \mathbf{r}') = V(\mathbf{r} - \mathbf{r}')$ of (22) in the integrand.

The static near-zone $qr \ll 1$ limit of the dipole-dipole potential (75) is positive (repulsive) along the z-axis and negative in the radial direction, thereby precluding stable binding. By contrast, the far-zone (retarded) behavior of this potential $\propto -\cos(qz)^2/qz$ is *negative* for any z, provided that the atoms are aligned along the z-axis. Nonaligned atoms feel less effective attraction, or even repulsion, at far-zone separations. Hence, an extended cloud of atoms subject to interaction (75) is expected to have a cigar-shaped ground state. We therefore adopt the following cylindrically symmetric (about the axial z-direction) variational ansatz for the macroscopic order parameter

$$\Psi(r_\perp, z) = \psi(z)\,\exp\left(-r_\perp^2/2w_r^2\right)/\pi^{1/2}w_r \tag{76}$$

where w_r is the variational radial width. It is reasonable to approximate the radial profile of Ψ by a Gaussian when its width w_r is less than half the laser wavelength $\lambda_{\mathrm{L}} = 2\pi/q$; in which case the oscillations of (75) are not manifest along x or y. We then obtain $\psi(z)$, as well as w_r, by numerically minimizing the mean-field energy expressed by (21) with the first (kinetic energy) term replaced by

$$E_{\mathrm{kin}} = \int d\mathbf{r}\frac{\hbar^2}{2m}\left|\nabla\Psi\right|^2 . \tag{77}$$

The numerical solutions for the ground state, based on (76), reveal the existence of a *self-bound* condensate, without external confinement, $V_{\mathrm{trap}} = 0$. Such solutions are obtained in the Thomas-Fermi (TF) limit of negligible kinetic energy compared to the energy of the s-wave scattering E_{sc}: $E_{\mathrm{kin}} \ll E_{\mathrm{sc}}$, which characterizes condensates with many atoms N, as specified below. Then, upon minimizing the total mean-field energy, we obtain a negative value of its expectation value (bound state) and, concurrently, finite mean quadratic radii $(\Delta x)^2 \equiv \langle x\rangle^2 = w_r^2/2$ and $(\Delta z)^2 \equiv \langle z\rangle^2$ (plotted in Fig. 9), We have verified that the *static* form of the dipole-dipole potential $U \propto 1/r^3$ *does not allow* for variational bound-state solutions.

The control parameter that allows one to play with the ratio of the induced dipole-dipole forces to the s-wave scattering is the dimensionless intensity

$$I_D = \frac{I\alpha^2 m}{8\pi\epsilon_0^2 c\hbar^2 a} . \tag{78}$$

The laser intensities corresponding to self-binding in Fig. 9 are below the threshold of the instability caused by the static $1/r^3$ part of the dipole-dipole potential (75), [16], $I \le 12\pi\hbar^2 c\epsilon_0^2 a/m\alpha^2$, Hence, the required intensity can be, in principle, arbitrarily small for any positive scattering length a.

Using a Gaussian ansatz for $\psi(z)$ in (76), we obtain analytical approximations for the radii of the condensate, normalized to the laser wavelength λ, in the limit $I_D \ll 1$, $\Delta x = 0.1125\,I_D^{-1/2}$ and $\Delta z = 0.7847\,I_D^{-1}$. For laser intensity $I_D > 0.1$ the condensate is strongly confined in the radial direction with $\Delta x \sim 0.2$, and less confined in the longitudinal direction, with typical size larger or of the order of the wavelength, $\Delta z > 0.6$.

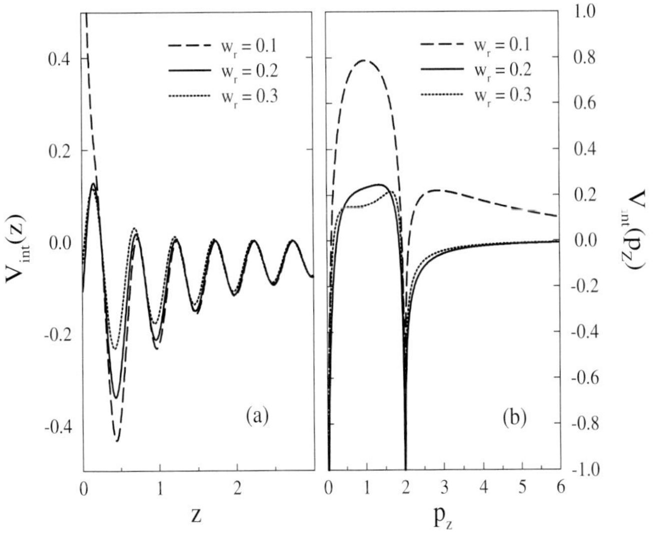

Fig. 10. (a) the induced 1D dipole-dipole interaction in units of $\hbar\Gamma_{ray}$, as a function of z (normalized to λ_L), (b) its Fourier transform. The dashed, solid and dotted lines correspond to increasing values of the variational parameter $w_r = 0.1$, 0.2 and 0.3, respectively.

The oscillatory long-range behavior of the potential (75) is manifested along the z-axis, due to the large extension of the condensate in this direction. To gain more insight into this behavior, we introduce the 1D-reduced form of the electromagnetically induced mean-field energy

$$E^{1D} = \frac{1}{2}\int dz\, dz'\, U^{1D}(z-z')\, |\psi_{1D}(z)|^2 |\psi_{1D}(z')|^2, \tag{79}$$

expressed through the 1D-reduced dipole-dipole potential

$$U^{1D}(z) = \frac{1}{2\pi w_r^2}\int d^2 r_\perp \exp\left(-\frac{r_\perp^2}{2w_r^2}\right) U(r_\perp, z)\,. \tag{80}$$

This 1D-reduced interaction contains an *attractive* singular part $V^{1D}(z)\big|_{\text{singular}} = -(\hbar\Gamma_{ray}/q^3 w_r^2)\,\delta(z)$, which arises from the interplay of $-\frac{x^2+y^2}{q^3 r^5}$ and $\frac{2z^2}{q^3 r^5}$ in the static limit of (75). For $I_D < 1.5$ this attractive singular part is balanced by a similar repulsive term $U^{1D}_{scat}(z) = (2a\hbar^2/mw_r^2)\delta(z)$ which arises from the (positive) s-wave scattering length, thereby stabilizing the condensate. For $I_D > 1.5$ the system is therefore unstable because of the static part of the $1/r^3$. Figure 10a shows the nonsingular part of U^{1D} for three different values of the variational parameter w_r. The near zone $1/z^3$ on-axis repulsion of the potential (75) is evident in the dashed curve (for small $w_r \leq 0.1$), as opposed to the near-zone attraction (solid

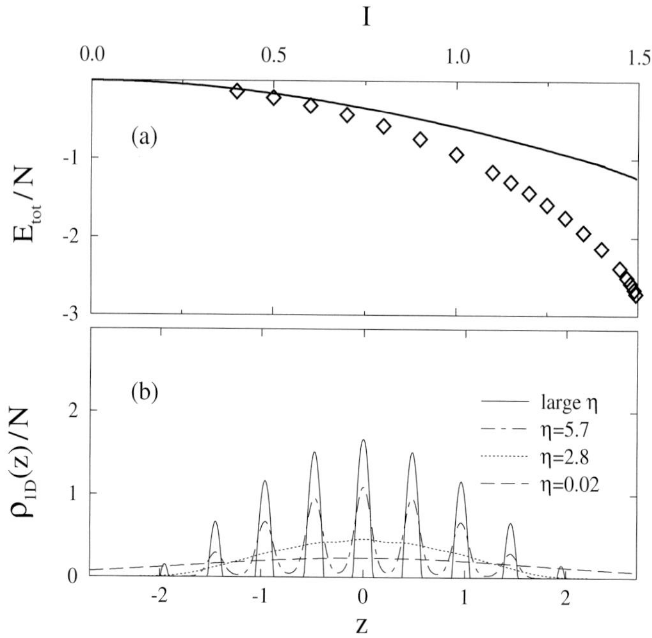

Fig. 11. (a) TF mean-field energies per particle (normalized to ηE_R) versus the intensity I (normalized to $8\pi\hbar^2 c\epsilon_0^2 a/m\alpha^2$) as obtained from the Gaussian ansatz (solid-line) and numerically (diamonds). (b) Kinetic energy effects: longitudinal equilibrium densities for different values of η and for $I = 1$. A radial external confinement is used to keep the radial size of the condensate constant and equal to that obtained for η very large without confinement.

line) sufficiently far off axis (for $0.2 \leq w_r \leq 0.5$). For large values of z the $1D$-reduced potential in Fig. 10a oscillates as $U^{1D}(z) \approx -(3\hbar\Gamma_{\rm ray}/2q)\cos(qz)^2/z$, but remains *attractive on average*, indicating a macroscopic electrostrictive force.

A conspicuous feature in the TF limit of negligible $E_{\rm kin}$ is the abrupt modulation of the density along z, namely the formation of *isolated condensate droplets* of size less than $\lambda_{\rm L}/4$ at approximately $\lambda_{\rm L}/2$ separations. The corresponding variational values for the ground-state energies, $E_{\rm tot}$, are substantially below the ones obtained by a Gaussian ansatz for $\psi(z)$ in (76), as displayed in Fig. 11a. This indicates that such longitudinal density modulation further stabilizes the condensate, which may seem counterintuitive. This density modulation is due to the interference of the backscattered and the incident fields, which creates a series of attractive traps at $\lambda_{\rm L}/2$ separations (Fig. 10a).

The changes in the density profile as the kinetic energy becomes non-negligible can be discussed in terms of the dimensionless parameter

$$\eta = Na/\lambda_{\rm L} \ , \tag{81}$$

which is approximately the ratio between the TF-ground state energy per particle for $I_D = 1$ and the recoil energy $E_R = \hbar^2 q^2/2m$ (see Fig. 11a). When the product

$\eta\, I_D$ decreases and becomes comparable to one, the radii of the condensate both increase with respect to those obtained in the TF-limit. This can be understood comparing the TF-energy, $E_{\rm tot} \sim N\,\eta\, E_R\, I_D^2$, with the kinetic energy, $E_{\rm kin} \sim N\, E_R\, I_D$, estimated using the TF-Gaussian solutions.

We can compensate the kinetic energy pressure in the radial direction, when necessary, by adding external radial confinement, e.g., by an appropriate choice of the focus of the laser beam. Fig. 11b shows the results of fixing the radial width w_r and decreasing the parameter η. As soon as the kinetic energy associated with the density modulation exceeds the energy reduction caused by density modulation, we find the oscillatory pattern to be washed out. Thus, for η smaller than 2.8 at $I_D = 1$, the condensate is still self-confined along the longitudinal direction, with a similar value of the longitudinal size Δz, but the density oscillations almost disappear. Only for much smaller values of η, for instance ~ 0.02 at $I_D = 1$, does the kinetic energy also affect Δz (long dashed line in Fig. 11b). Remarkably, when $\eta I_D^2 \sim 1$, the single condensate droplets overlap, self-organizing into a *spatially-coherent long-range density modulation* – a Bose "supersolid" [17] is formed. In this novel regime the atoms are coherently distributed among the wells, thus allowing the establishment of a phase between the overlapping condensates. An example is given by the dot-dashed curve in Fig. 11b.

Using the Gaussian ansatz in its time-dependent form [15], we then obtain analytically, in the small-intensity limit ($I_D \ll \eta I_D^2 \ll 1$), the following values of the radial and longitudinal oscillation frequencies of a self-bound condensate:

$$\Omega_r^2 = 2.88\left(\frac{E_R}{\hbar}\right)^2 I_D^3\eta\,; \qquad \Omega_z^2 = 0.0148\left(\frac{E_R}{\hbar}\right)^2 I_D^4\eta\,. \tag{82}$$

It is noteworthy that, although the condensate radii are much larger than the laser wavelength (*far-zone*), the frequency Ω_r^2 of the compressional mode along the radial direction is close to the "plasma" frequency (42)

$$\omega_p^2 = 4\pi u\rho(0)/m = 9.86\,(E_R/\hbar)^2\,(I_D^3\eta)\,, \tag{83}$$

introduced in the near zone "self-gravity" regimes [14]. The longitudinal frequency Ω_z is expected to be substantially modified in the presence of the density modulations.

For the density modulations discussed above, most of the photons are *nearly elastically* and *coherently* scattered. Their total cross section is given by $N^2 f_{cm}$ times the single-atom cross section, where the fraction f_{cm} is defined as an appropriate average over all the possible directions of scattering (in the Born approximation)

$$f_{cm} = \frac{3}{2N^2}\int_{-1}^{1} d(\cos(\theta))\,(1+\cos(\theta)^2)\,|\rho(\mathbf{k})|^2\,. \tag{84}$$

Here $\rho(\mathbf{k})$ is the Fourier component of atomic density corresponding to $\mathbf{k} = 2q(\sin(\theta)\hat{x}, 0, \cos(\theta)\hat{z})$, the momentum transferred by a single photon in the x-z plane and θ is the angle between the incident beam and the scattered direction. Only the (complementary) fraction $(1 - f_{cm})$ of the rate of energy change, that is related to the incoherent part of the scattering cross-section, is absorbed by the interatomic degree of freedom in the center-of-mass frame, thereby providing a partial suppression of heating (see Sect. 9).

The center of mass is therefore subject to a constant *radiation force* that is enhanced by a factor $N f_{cm}$ due to the density modulation. Correspondingly, the electromagnetic field is strongly back-scattered (diffracted) in a coherent fashion. The radiation force can either shift the equilibrium position of the condensate, if it is located in a longitudinal trap, or else accelerate it uniformly. In the latter case, the scattered light will be Doppler shifted. The effects described here are due to the *same* matter-field interactions as those responsible for stimulated Rayleigh scattering [19] and collective atomic recoil (CARL) [37]. In fact, f_{cm} will lead to similar effects. However, the change of atomic energy, unaccounted for by CARL, is the essence of our effects.

A central prediction of this section is that induced dipole-dipole forces result in a *stationary* density modulation for the ground state of the condensate. Density modulations in a condensate can also occur in the presence of phonon excitations, as observed in the experiment by Inouye et al [32] that has demonstrated the superradiant Rayleigh scattering [37] of a *pulse* of light by an atomic BEC. However, travelling phonons represent excitations of the system and so are quite distinct from the ground-state density modulation proposed here. The two situations can, for instance, be distinguished by diffracting a nearly perpendicular probe laser with z-component of the wavevector, $k_z \approx 2q$. Sound waves at frequency $\Omega = v_s 2q$, where v_s is the sound velocity, lead to a density modulation of the form $\cos(2qz - \Omega t)$ so that the first diffraction orders (i.e. Brillouin peaks) of the probe beam (wavenumber k), at angles $\theta \approx \pm 2q/k$, will be frequency shifted to $ck \pm \Omega$ [19]. By contrast, diffraction from the stationary density modulation discussed above will be *elastic*, with no frequency shift.

An example of the experimental conditions required for the predicted effects involves the following parameters for a cloud of $N \sim 10^3$ sodium atoms: a circularly polarized laser beam, red-detuned by 1.7 GHz from the $3S_{1/2}$ (F=1) $\rightarrow 3P_{3/2}$ (F = 0,1,2) transition, gives the threshold for $1/r^3$ instability at $\approx 525\ \mathrm{mW/cm^2}$ [9]. For such intensity, the acceleration of the center of mass is approximately $\approx 500 N f_{cm}\ [\mathrm{m/s^2}]$, where f_{cm} is a non-negligible fraction (for instance $f_{cm} \sim 0.1$ for $I \sim 1$ and $f_{cm} \sim 1$ for $I \ll 1$). It is possible to balance this force by combining longitudinal harmonic confinement with a magnetic field gradient, so that the condensate will not accelerate.

9 Losses and Limitations

9.1 Spontaneous Rayleigh Losses: General Considerations

Incoherent Rayleigh scattering from the far-off-resonant laser beams leads to heating and depletes the quantum degenerate BEC or FDG. The rate of incoherent scattering, which is $N\Gamma_{\mathrm{ray}}$ for N noninteracting atoms, is suppressed by the interatomic interaction [38]. However, the corresponding rate of change of the total energy is still given by the simple universal expression [39]

$$\frac{\mathrm{d}}{\mathrm{d}t}E_{\mathrm{tot}} = 2E_R N \Gamma_{\mathrm{ray}}, \tag{85}$$

which is *independent* of the interaction properties and corresponds to ρ times the single-atom rate of energy change $2E_R\Gamma_{\mathrm{ray}}$. From the rate of change of the ground-state energy we can estimate the effective evaporation or heating time of the self-bound atomic cloud to be

$$\tau_{\mathrm{heat}} = E_{\mathrm{tot}}/(\frac{\mathrm{d}}{\mathrm{d}t}E_{\mathrm{tot}}) \ . \tag{86}$$

We proceed to evaluate the probability of inelastic scattering by single-photon scattering. The matrix element for scattering (elastic as well as inelastic) is proportional to [20] $\langle\psi_f|\rho_{(\mathbf{q}-\mathbf{q}')}|\psi_i\rangle$, where ψ_f, ψ_i are the final and initial wave functions of the atomic translational motion, $\mathbf{q}$ and $\mathbf{q}'$ are the initial and final photon wavenumbers. We expand the density operator $\rho_{\mathbf{q}}$ as follows:

$$\rho_{\mathbf{q}} = \sum_{n,m}\langle n|\exp(i\mathbf{q}\mathbf{r})|m\rangle a_n^\dagger a_m \ , \tag{87}$$

where $a_n^\dagger$ and a_n are the creation and annihilation operators for the quasiparticle excitation in mode $|n\rangle$. The probability that a photon $\mathbf{q}$ is scattered to a photon $\mathbf{q}'$ for a system initially prepared in excitation state $|i\rangle$, is proportional to [39]

$$\sum_f \langle i|\rho^\dagger_{\mathbf{q}-\mathbf{q}'}|f\rangle\langle f|\rho_{\mathbf{q}-\mathbf{q}'}|i\rangle = \langle i|\rho^\dagger_{\mathbf{q}-\mathbf{q}'}\rho_{\mathbf{q}-\mathbf{q}'}|i\rangle \ , \tag{88}$$

where $|i\rangle$, $|f\rangle \equiv |n_i\rangle$, $|n_f\rangle$.

9.2 Rayleigh Scattering from a Small-Size BEC

For a small system confined to a near zone of the laser, $qR \ll 1$, the probability (88) is proportional to N^2, whereas for an extended system ($qR \gg 1$) it is proportional to the static structure factor $N\,S(q)$ [39,40]. Considering as initial state $|0\rangle$ the macroscopically occupied non-interacting ground state $|\psi_0\rangle$, and allowing for to singly-excited states $|n\rangle = a_n^\dagger a_0|0\rangle$, we have from (87)

$$\rho_{\mathbf{q}}|0\rangle = N\langle\psi_0|\exp(i\mathbf{q}\mathbf{r})|\psi_0\rangle|0\rangle + \sqrt{N}\sum_{n\neq 0}\langle\psi_n|\exp(i\mathbf{q}\mathbf{r})|\psi_0\rangle|n\rangle \tag{89}$$

For a near-zone confined system the above expression can be approximated to

$$\rho_{\mathbf{q}}|0\rangle \approx N|0\rangle + \sqrt{N}i\mathbf{q}\sum_{n\neq 0}\langle\psi_n|\mathbf{r}|\psi_0\rangle|n\rangle \tag{90}$$

The first and second terms on the right represent the collective elastic ("superradiant") scattering and the inelastic ("spontaneous") scattering, respectively.

Note that the elastic part is enhanced with respect to its inelastic counterpart by the usual Bose factor ($\sqrt{N}$). The probability ratio between an inelastic excitation and elastic (collective) scattering with the same momentum transfer $\mathbf{q}$, is given by

$$\frac{P^{\mathrm{exc}}}{P^{\mathrm{elas}}} \approx \frac{1}{2N} (qR_{\mathrm{rms}})^2 , \tag{91}$$

where R_{rms} is the mean radius of the condensate. For each scattering event, P^{exc} (the probability to lose one atom from the condensate) is thus suppressed by the Lamb-Dicke factor $(qR_{\mathrm{rms}})^2$ [40,41].

The lifetime of the system, when determined from spontaneous Rayleigh scattering alone, is therefore estimated to be

$$\tau_{\mathrm{ray}} \geq \left(\Gamma_{\mathrm{ray}} \, (q \, R_{\mathrm{rms}})^2 \right)^{-1} . \tag{92}$$

Since $\Gamma_{\mathrm{ray}} = Iq^3\alpha^2/(3\hbar\epsilon_0^2 c)$ [10], it can be expressed in terms of the laser-induced "gravitational" energy of a *single* pair of atoms separated by the laser wavelength

$$\Gamma_{\mathrm{ray}} = \left(\frac{20\pi}{11} \right) \frac{u}{\hbar \lambda_{\mathrm{L}}} \tag{93}$$

where u is defined in (18). Using this relation, we can compare the upper bound on the condensate lifetime set by Rayleigh scattering with the time scale of the dynamics, the requirement being that the system exists long enough to equilibrate.

In "gravitationally" self-bound regions, a characteristic time scale for the dynamics is provided by the "plasma" frequency ω_{p} (see (42)). We can express ω_{p} in terms of the recoil energy $E_{\mathrm{R}} = \hbar^2 q^2/2m$ (q being the mean laser wavelength) and the Rayleigh scattering rate Γ_{ray} using (93)

$$\omega_{\mathrm{p}} \approx 0.25 \, \frac{\hbar \Gamma_{\mathrm{ray}}^2}{E_{\mathrm{R}}} \, N^2 F^{-3/2} \tag{94}$$

where the factor

$$F = \frac{1}{2} + \sqrt{\frac{1}{4} + \frac{N^2}{N_{\mathrm{border}}^2}} \tag{95}$$

is asymptotically equal to 1 in the G region and N/N_{border} in the TF-G region. It follows from (94) that the characteristic oscillation frequency ω_{p} can be much bigger than Γ_{ray}, by a factor proportional to N^2 or $N^{1/2}$ in the G or TF-G region, respectively. Thus the Rayleigh lifetime can be considerably longer than the characteristic time scale of the dynamics.

Even for the small number of 40 sodium atoms in the "gravitationally" self-bound moderate-detuning regime ($I = 1.5 \times I_0$, $\delta = 1.7\,\mathrm{GHz}$), for which the recoil energy is $E_{\mathrm{R}}/\hbar = 1.57 \times 10^5\ \mathrm{s}^{-1}$ and $\Gamma_{\mathrm{ray}} = 1.58 \times 10^4\ \mathrm{s}^{-1}$, we find $\omega_{\mathrm{p}} \approx 20 \times \Gamma_{\mathrm{ray}}$. This implies that at least several oscillation periods of the "gravitationally" self-bound condensate can occur within the Rayleigh lifetime.

In the cigar-shaped condensate (Sect. 8) we estimate

$$\tau_{\text{heat}} = \eta I_D^2 \Gamma_{\text{ray}}^{-1}, \tag{96}$$

where we used (81) and the variational ansatz of Sect. 8 for the TF limit $\eta I_D^2 \gg 1$. The product $\tau_{heat}\omega_p$, where ω_p is given by (83), can be expressed in terms of the number of atoms per cubic wavelength as

$$\tau_{heat}\,\omega_p = 0.11\,\left(a/\lambda_{\text{L}}\right)^{1/2}\left(\lambda_{\text{L}}^3\,\rho(0)\right)^{3/2} I^{-1/2}. \tag{97}$$

This expression can satisfy $\tau_{heat}\omega_p \gg 1$, which proves that heating due to spontaneous Rayleigh scattering events does not provide a fundamental limitation on the observability of the discussed effects for sufficient atomic density.

9.3 Losses due to the r^{-3} Oscillating *Interfering* Terms

In the following we give a simple estimation of the losses that may be generated by the interference of intersecting laser beams in the triad of Fig. 1. Such interference gives rise to the oscillating terms in the laser-induced potential, such as

$$A_{xy}(\boldsymbol{r},t) = -3\frac{u}{q^2}\frac{\hat{x}\hat{y}}{r^3}\cos(\omega_{ij}t) \tag{98}$$

where u is the laser-induced "gravitational" constant, $\hat{x}$, $\hat{y}$ are the x-component and y-component of the unit vector $\hat{r} = \boldsymbol{r}/r$ and ω_{ij} is the difference in frequency between a pair of interfering lasers i, j. The resulting loss rate Γ_{interf} [9] can be expressed in terms of the recoil energy E_{R} and the Rayleigh scattering rate Γ_{ray} as in Sect. 9.2:

$$\Gamma_{\text{interf}} \approx 0.05\left(\frac{\hbar\Gamma_{\text{ray}}N}{E_{\text{R}}}\right)^4\sqrt{\frac{\hbar\omega_{ij}}{E_{\text{R}}}}\,\Gamma_{\text{ray}}\,F^{-3}\,, \tag{99}$$

where F is given by (95). In the numerical example given in Sect. 9.2 above, Γ_{interf} turns out to be few times bigger than Γ_{ray}, when ω_{ij} is chosen to be of the order of ω_{p}.

9.4 Moderate-Detuning Saturation and Repulsion

The potential (15) is the result of a resonant fourth-order, two-atom, QED process [8], valid when the laser is *far detuned* from any atomic transitions. This means that the initial absorption of a laser photon and the subsequent intermediate steps are virtual processes (which are most significant in the near zone), followed by photon emission back into the original laser mode. This is consistent with the polarizability α being real, as assumed in Sect. 2. A different process can take place when the laser is on-resonance. Genuine absorption of a laser photon by a single atom (measured by the saturation), followed by spontaneous emission of this real photon is a process that radiates energy. If another atom absorbs this radiation then in the far-zone it feels a repulsive Coulomb-like force

$F_{\text{repuls}} = K/r^2$ [41], which has been recently measured in rubidium molasses [42]. For moderate detuning, can this force counteract our attractive gravitation-like force $F_{grav} = -u/r^2$?

For detuning δ much larger than both the Rabi frequency Ω and the linewidth γ of the resonance, the saturation parameter $s = Id^2/(\epsilon_0 c\hbar^2\delta^2)$, where d is the dipole matrix element, becomes *independent of the detuning* when calculated at the threshold intensity (40)

$$s(I = I_0) = \frac{48\pi}{7} \frac{a\epsilon_0\hbar^2}{md^2} . \qquad (100)$$

It is then found that [41,42] $K \approx \sigma_0^2 I_s \Omega^4/(16c\delta^2)$, where σ_0 is the resonant absorption cross section and I_s is the corresponding saturation intensity. On comparing this expression with u (see (18)), we find that, in terms of the saturation parameter s,

$$K \approx su . \qquad (101)$$

For the sodium transition and 1.7 GHz detuning referred above, (100) yields very small value $s \approx 0.0003$. This implies that under the moderate-detuning conditions discussed above, the repulsive force has a *negligible* effect on self-binding.

10 Conclusions

The laser-induced attractive r^{-1} interaction can give rise to stable self-bound BECs and FDGs with distinct equilibrium, dynamic, and thermodynamic properties (Sects. 4–6). Their fascinating analogy with gravitating quantum systems, such as Bose stars (Sects. 5 and 6) or White Dwarfs (Sect. 7), whose gravitational interaction can be enormously enhanced by the laser field, merits further research.

The stability, low loss-rates incurred by the r^{-3} oscillating terms, and lack of sensitivity to alignment errors or amplitude noise of the laser beams makes the experimental realization of such self-bound objects rather likely. At least the TF-G regime of a self-bound BEC is experimentally accessible (Sect. 5), although such an experiment would be challenging. Moderate detuning is preferable to the longer-wavelength case, due to the huge enhancement in the polarizability, but it allows the self-binding of few (less that 100) atoms. If the scattering length were reduced via a Feshbach resonance, then this would further facilitate the self-trapping of many more atoms using longer-wavelength (near-infrared) lasers. A cold molecular gas [43] can be advantageous for creation a self-confined degenerate cloud because of the polarization resonances in molecules in the infrared domain.

We have also demonstrated a new regime of self-confined and self-organized ground-state density modulations in a cigar-shaped BEC illuminated by a single circular polarized laser beam, which induces quasi-1D retarded long-range interactions (Sect. 8). We have shown that non-linear scattering of light may then arise, even in the small-saturation limit, and have estimated the enhancement of the total "elastic" cross section. This regime is inherently possible even for a plane-wave laser, although it is facilitated by the radial focusing of the beam.

The losses (atom ejection from the self-bound system) due to Rayleigh scattering and multi-beam interference of lasers in the near zone have been analyzed (Sect. 9). They have been estimated not to preclude the observability of the distinct new regimes discussed in this article.

Acknowledgements

D.H.J.O. is grateful to C. Eberlein and J. Steinhauer for illuminating discussions.

This work have been supported by the UK Engineering and Physical Sciences Research Council (EPSRC) (D.H.J.O.), the EC (QUACS RTN) and the US-Israel Binational Science Foundation (BSF) (G.K. and A.I.A.), as well as the Russian Foundation for Basic Research grant 02-02-17135 (A.I.A.).

References

1. M.H. Anderson et al: Science **269**, 198 (1995); C.C. Bradley, C. A. Sackett et al: Phys. Rev. Lett. **75**, 1687 (1995); K.B. Davis, M.-O. Mewes et al: Phys. Rev. Lett. **75**, 3969 (1995)
2. L.P. Pitaevskii: Sov. Phys. JETP **13**, 451 (1961); E.P. Gross: Nuovo Cimento **20**, 454 (1961); J. Math. Phys. **4**, 195 (1963); F. Dalfovo, S. Giorgini, L. P. Pitaevskii and S. Stringari: Rev. Mod. Phys. **71**, 463 (1999); A. Parkins and D. Walls, Phys. Rep. **303**, 1 (1998)
3. E. Tiesinga, B.J. Verhaar, and H.T.C. Stoof: Phys. Rev. A **46**, R1167 (1992); P.O. Fedichev et al: Phys. Rev. Lett. **77**, 2913 (1996); Yu. Kagan, E.L. Surkov and G.V. Shlyapnikov: Phys. Rev. Lett. **79**, 2604 (1997); J.L. Bohn and P.S. Julienne: Phys. Rev. A **56**, 1486 (1997); S. Inouye et al: Nature **392**, 151 (1998); Ph. Courteille et al: Phys. Rev. Lett. **81**, 69 (1998); M. Marinescu and L. You: Phys. Rev. Lett. **81**, 4596 (1998)
4. K. Burnett: Nature **392**, 125 (1998); J. Anglin and W. Ketterle: Nature, **416**, 211 (2002)
5. B. DeMarco, D.S. Jin: Science **285**, 1703 (1999); Z. Hadzibabic, C.A. Stan et al: *Two-species mixture of quantum degenerate Bose and Fermi gases*. Preprint cond-mat/0112425 at http://www.arxiv.org (2001); S.R. Granade, M.E. Gehm, K.M. O'Hara, J.E. Thomas: *All-Optical Production of a Degenerate Fermi Gas*. Preprint cond-mat/0111344 at http://www.arxiv.org (2001)
6. G. Lenz, P. Meystre, E.M. Wright: Phys. Rev. Lett. **71**, 3271 (1993); W. Zhang, D.F. Walls, B.C. Sanders: Phys. Rev. Lett. **72**, 60 (1994)
7. L.D. Landau, E.M. Lifshitz: *Electrodynamics of continuous media* (Pergamon Press, New York, 1960)
8. T. Thirunamachandran: Mol. Phys. **40**, 393 (1980); D.P. Craig and T. Thirunamachandran: *Molecular Quantum Electrodynamics* (Academic Press, London, 1984), Sec. 7.12
9. D. O'Dell, S. Giovanazzi, G. Kurizki, V.M. Akulin: Phys. Rev. Lett. **84**, 5687 (2000); see commentary by J. Anglin: Nature **406** 29 (2000)
10. S. Giovanazzi, D. O'Dell, G. Kurizki: Phys. Rev. A **63**, 031603 (2001)

11. 11 R. Ruffini, S. Bonazzola: Phys. Rev. **187**, 1767 (1969); G. Ingrosso, D. Grasso, R. Ruffini: Astron. Astrophys. **248**, 481 (1991); P. Jetzer: Phys. Rep. **220**, 163 (1992)
12. S. Chandrasekhar: *An introduction to the study of stellar structure* (Dover Publications, New York 1958), Chapters III, IV; S.L. Shapiro, S.A. Teukolsky: *Black Holes, White Dwarfs, and Neutron Stars* (Wiley, New York 1983)
13. P.-H. Chavanis: *Phase Transitions in self-gravitating systems. Self gravitating fermions and hard sphere models* Phys. Rev. E, **65**, 056123 (2002), also Preprint cond-mat/0109294 at http://www.arxiv.org (2002)
14. S. Giovanazzi, G. Kurizki, I.E. Mazets, S. Stringari: Europhys. Lett. **56**, 1 (2001)
15. S. Giovanazzi, D. O'Dell, G. Kurizki: Phys. Rev. Lett. **88**, 130402 (2002)
16. K. Góral, K. Rzążewski, T. Pfau: Phys. Rev. A **61**, 051601(R) (2000); L. Santos, G.V. Shlyapnikov, P. Zoller, M. Lewenstein: Phys. Rev. Lett. **85**, 1791 (2000)
17. Y. Pomeau and S. Rica: Phys. Rev. Lett. **72**, 2426 (1994)
18. J. Schwinger, L.L. DeRaad, Jr., K.A. Milton, W. Tsai: *Classical Electrodynamics* (Perseus Books, Reading, MA, 1998)
19. M. Born, E. Wolf: *Principles of Optics: Electromagnetic Theory of Propagation, Interference and Diffraction of Light* (Pergamon, London, 1959)
20. C. Cohen Tannoudji, J. Dupont-Roc, G. Grynberg: *Atom-Photon Interactions* (Wiley, New York, 1992)
21. V.A. Markel: J. Opt. Soc. Am. B **12**, 1783 (1995)
22. Y. Castin, K. Mølmer: Phys. Rev. A. **51**, R3426 (1995); O. Morice, Y. Castin, J. Dalibard: Phys. Rev. A **51**, 3896 (1995)
23. J. Javanainen, J. Ruostekoski: Phys. Rev. A **52**, 3033 (1995); J. Ruostekoski, J. Javanainen: Phys. Rev. A **55**, 513 (1997); J. Ruostekoski, J. Javanainen: Phys. Rev. A **56**, 2056 (1997)
24. L. Jansen, P. Mazur: Physica **21**, 193 (1955); **21**, 208 (1955); P. Mazur, M. Mandel: Physica **22**, 289 (1956); **22**, 299 (1956)
25. J. Mathews, R.L. Walker: *Mathematical Methods of Physics*, (Addison-Wesley, New York, 2nd Edition, 1970)
26. E.M. Lifshitz, L.P. Pitaevskii: *Statistical Physics Part 2* (Butterworth-Heinemann, Oxford, 1998)
27. L.E. Reichl: *A modern course in statistical physics* (John Wiley & Sons 1988), Chapter 7
28. L.D. Landau, E.M. Lifshitz: *Statistical physics. Course in theoretical physics*, Vol. V
29. D. Reiss: *Stellar Structure and the Lane-Emden Function: Introduction*, at http://www.scientificarts.com/laneemden/laneemden.html
30. H. Haber: *Mathematical Notes on the Fermi-Dirac and Bose-Einstein Integrals*, at http://scipp.ucsc.edu/~haber/ph112/fermigas2.pdf ;
31. L.L. Foldy: Phys. Rev. **124**, 649 (1961); A.S. Alexandrov, W.H. Beere: Phys. Rev. B **51**, 5887 (1995)
32. S. Inouye et al: Science **285**, 571 (1999); D.M. Stamper-Kurn et al: Phys. Rev. Lett. **83**, 2876 (1999)
33. A. Brunello, F. Dalfovo, L. Pitaevskii, S. Stringari: Phys. Rev. Lett. **85**, 4422 (2000)
34. N.N. Bogoliubov: J. Phys. (U.S.S.R.) **11**, 23 (1947)
35. I.E. Mazets: *Collective excitation frequencies of a Bose - Einstein condensate with electromagnetically induced 1/r attraction.* Preprint cond-mat/0007209 at http://www.arxiv.org (2000)
36. L. Viverit, S. Giorgini, L.P. Pitaevskii, S. Stringari: Phys. Rev. A, **63** 033603 (2001); J. Oliva: Phys. Rev. B **39**, 4204-4209 (1989)

37. M.G. Moore, P. Meystre: Phys. Rev. Lett. **83**, 5202 (1999); N. Piovella, R. Bonifacio, B.W.J. McNeil, G.R.M. Robb: Opt. Commun. **187**, 165 (2001)
38. V.M. PerezGarcia et al: Phys. Rev. Lett. **77**, 5320 (1996)
39. D. Pines, P. Nozieres: *The Theory of Quantum Liquids* (Benjamin, 1966)
40. R.H. Dicke: Phys. Rev. **89**, 472 (1953)
41. T. Walker, D. Sesko, C. Wiemann: Phys. Rev. Lett. **64**, 408 (1990)
42. L. Pruvoust, I. Serre, H.T. Duong, and J. Jortner: Phys. Rev. A **61**, 053408 (2000)
43. M. Mackie: *Feshbach-Stimulated Photoproduction of a Stable Molecular Condensate*. Preprint physics/0202041 at http://www.arxiv.org (2002)

Part IV

Nonlinear Dynamics

Dynamics and Self-consistent Chaos in a Mean Field Hamiltonian Model

Diego del-Castillo-Negrete

Oak Ridge National Laboratory, Oak Ridge, TN 37831-8071, USA

Abstract. We study a mean field Hamiltonian model that describes the collective dynamics of marginally stable fluids and plasmas in the finite N and $N \to \infty$ kinetic limit (where N is the number of particles). The linear stability of equilibria in the kinetic model is studied as well as the initial value problem including Landau damping. Numerical simulations show the existence of coherent, rotating dipole states. We approximate the dipole as two macroparticles and show that the $N = 2$ limit has a family of rotating integrable solutions that provide an accurate description of the dynamics. We discuss the role of self-consistent Hamiltonian chaos in the formation of coherent structures, and discuss a mechanism of "violent" mixing caused by a self-consistent elliptic-hyperbolic bifurcation in phase space.

1 Introduction

The study of many-body interacting systems has been a problem of interest in science for many years. Examples of particular interest include plasmas, hydrodynamics, gravitational systems, and coupled dynamical systems. The dynamics of these systems is self-consistent in the sense that the evolution of a given member of the system is determined by the collective effect of all the other members of the system. Self-consistent dynamics is ubiquitous in nature. For example, in plasma physics the dynamics of an ensemble of charged particles is determined by the electromagnetic fields generated by the particles themselves. In gravitational systems, the dynamics of an ensemble of masses is determined by the gravitational potential created by all the particles in the system. In hydrodynamics, an ensemble of point vortices evolves under the advection of the velocity field generated by the vortices themselves. Although the equations governing the above mentioned systems have been known for quite a long time (e.g., the Lorentz force and Maxwell's equations in the case of a plasma) their self-consistent dynamics is not well-understood. For example, a lot is known about the chaotic dynamics of a single charged particle in a given, time-dependent electrostatic wave. But, much less is known about the problem of self-consistent chaos in an ensemble of charged particles. One of the main goals of this chapter is to study the problem of self-consistent dynamics and chaos in particular in the context of a simple Hamiltonian mean field model known as the single wave model (SWM).

The SWM has its origins in the study of the beam plasma instability [27–29]. More recently, the model has been derived under more general conditions to describe the weakly nonlinear dynamics of marginally stable Vlasov-Poisson systems [10,11]. Other derivations and applications of the SWM include [9], the work in [1] where the mean field–particle Lagrangian of the model was derived from the full N-body classical mechanics Lagrangian for Coulomb interactions, the study of the model to treat Langmuir turbulence [19], and Landau damping[17,20]. In a more general setting, single wave models have been used to study self-consistent Lagrangian transport [12,13], the relationship between self-consistent chaos and phase space coherent structures [14], finite-amplitude non-axisymmetric perturbation of vortices [4], critical layer dynamics in shear flows [5], and free electron lasers [18].

The organization of this chapter is as follows. In the next section we define the SWM in the finite N and in the $N \to \infty$ limit and discuss the application of the model to fluids and plasmas. Also, in this section we discuss the relationship between the SWM and the Hamiltonian mean field model proposed in [2,24,6,16]. Section 3 discusses the integrability of the SWM in the finite N case. In particular, using the symmetries and the conservation laws of the model, it is shown that the model is integrable in the $N = 1$ case and in the $N = 2$ symmetric case. These integrable solutions are important because they provide the basic understanding of the coherent structures found in the large N and kinetic limits of the model. Section 4 presents a symplectic map version of the SWM. Section 5 studies the linear theory of the SWM in the kinetic limit. In particular, we derive the dispersion relation and present a criterion that gives necessary and sufficient conditions for the stability of a general equilibrium with a single extremum. In addition, we discuss the initial value problem and Landau damping. Section 6 shows the construction of exact nonlinear solutions of the SWM, which are the analogue of the BGK (Bernstein-Greene-Kruskal) modes in plasma physics [26]. Section 7 is devoted to the study of the role of self-consistent chaos in the formation of coherent structures . Section 8 discusses a self-consistent dynamical bifurcation in the SWM that leads to the destruction of coherent structures and violent mixing of the phase space. The conclusions are presented in Sect. 8, and a derivation of the SWM from the Vlasov-Poisson system is presented in Appendix A.

2 The Single Wave Model

The single wave model (SWM) is a Hamiltonian system consisting of an ensemble of N particles in one-dimension with phase space coordinates (x_j, u_j), $j = 1, 2, \ldots N$, such that

$$\frac{dx_j}{dt} = \frac{\partial H}{\partial u_j}, \qquad \frac{du_j}{dt} = -\frac{\partial H}{\partial x_j}, \tag{1}$$

where

$$H = \sum_{k=1}^{N} \left[\frac{1}{2} u_k^2 - a(t)\, e^{ix_k} - a^*(t)\, e^{-ix_k} \right], \tag{2}$$

$$\frac{da}{dt} - iUa = \frac{i}{N} \sum_{k=1}^{N} \Gamma_k\, e^{-ix_k}, \tag{3}$$

and σ, U and Γ_j are parameters.

Equation (2) is the Hamiltonian of an ensemble of particles in the potential of a time-dependent wave with a single harmonic with wave-number $\kappa = 1$. The dynamics of Hamiltonians of this type is well understood in two limits: when $a(t)$ =constant, and when $a(t)$ is a *given* function of time. The first case is fully integrable. In particular, the N particles problem decouples and reduces to the pendulum problem. In the second case, when $a(t)$ is a given function, the system also decouples, but, due to the time dependence, the system is in general not integrable. For example, the case $a = \cos(\omega t)$ leads to the extensively studied periodically perturbed pendulum, which has both integrable and chaotic solutions [25]. The nontrivial aspect of the SWM lays in the fact that the function $a(t)$, rather than being given a priori, is determined by the dynamics itself. That is, according to (3), the dynamics of the particles, $x_j(t)$, determines the dynamics of the field, $a(t)$. In this respect, the SWM is a *self-consistent* model.

The SWM is a *mean field* model; it treats the interacting elements of the system as independent entities moving in an average field determined self-consistently from the dynamics of all the elements. Defining

$$a = \sqrt{J}\, e^{-i\theta} \,, \tag{4}$$

and writing the SWM as

$$\frac{d^2 x_j}{dt^2} = -\frac{\partial V}{\partial x_j} \,, \qquad V = -2\sqrt{J}\, \cos(x_j - \theta) \,, \tag{5}$$

it is clearly seen that the equations governing the dynamics of the N interacting particles reduce to the equation governing the motion of a single particle under the effective potential V. That is, the only influence a particle has in the rest is through its contribution to the effective potential which is determined from (3).

In the kinetic limit ($N \to \infty$), the system is described by a phase distribution function f, evolving according to the Vlasov equation

$$\partial_t f + \partial_u H \; \partial_x f - \partial_x H \, \partial_u f \; = 0 \,, \tag{6}$$

with

$$H = \frac{u^2}{2} - a(t)\, e^{i\,x} - a^*(t)\, e^{-i\,x} \,, \tag{7}$$

$$\sigma \frac{da}{dt} - iUa = \frac{i}{2\pi} \int e^{-ix}\, dx \int du\, f \,. \tag{8}$$

The extra parameter σ (which in the finite N case was taken to be $\sigma = 1$) has been introduced for convenience when comparing with the HMF (Hamiltonian mean field model) which corresponds to $\sigma = 0$ [2,16].

Most of the studies of the SWM so far have limited attention to the case $\Gamma_k > 0$. In fact, early derivations of the model in the context of the beam-plasma instability considered compact electron equilibrium distributions for which only positive Γ_k are acceptable [27,28,1]. However, in the description of generic instabilities of marginal stable plasmas it is possible to have positive and negative Γ_k's [10,13]. In this case the starting point is a marginally stable background plasma equilibrium density. Perturbations that locally increase the equilibrium density have $\Gamma_k > 0$ and are called "clumps," while perturbations that locally deplete the equilibrium density have $\Gamma_k < 0$ and are called "holes." In the fluid dynamics context, the interpretation of the Γ_k's is straightforward: Γ_k is the circulation of the point vortex k that can be negative or positive. Consideration of Γ_k's with different signs opens a new chapter in the study of mean field coupled Hamiltonian systems from the physical and the dynamical systems point of view.

The SWM is a universal model describing the dynamics of marginally stable Vlasov-Poisson systems. By universal we mean that the model describes the dynamics of a large class of instabilities, and it does not depend on the details of the long-range interactions involved. In particular the SWM describes the weakly nonlinear dynamics of Vlasov-Poisson system of the form

$$\partial_T F + u\,\partial_X F + \partial_u (F_0 + F)\,\partial_X \Phi = 0\,, \tag{9}$$

$$G(k)\,\tilde{\Phi}(k,T) = -\int_{-\infty}^{\infty} \tilde{F}(k,u,T)du\;, \tag{10}$$

where the tilde denotes Fourier transform. Appendix A presents a derivation of the SWM from (9)–(10), further details can be found in [10,11].

In the plasma physics case, according to Poisson equation,

$$G(k) = k^2\,, \tag{11}$$

and (9) is the kinetic equation for the electron distribution function in a uniform neutralizing ion background with $F(X,u,T)$ denoting the departure from the equilibrium $F_0(u)$ [26]. Here time and space have been normalized using the electron plasma frequency, $\omega_e^2 = 4\pi n_0 e^2/m$, and the Debye length, $\lambda_D = V/\omega_e$, respectively, where e is the electron charge, m is the electron mass, n_0 is the equilibrium plasma density, and V is a chosen velocity scale. The distribution function, and potential are nondimensionalized with V/n_0 and $e/(mV^2)$ respectively, and thus $\int_{-\infty}^{\infty} F_0\,du = 1$.

However, the validity of the SWM does not depend on the specific functional form of $G(k)$. Another important case is

$$G(k) = 2k\coth(k)\,, \tag{12}$$

which corresponds to the vorticity defect model that describes the dynamics of localized vorticity perturbations in a strong shear flow [3]. In this context, the Vlasov equation plays the role of the vorticity advection equation, with $F + F_0$ representing the vorticity, and the (x,u) phase space coordinates playing the role of the (x,y) spatial coordinates of a two-dimensional incompressible fluid. The specific form of the function G in this case arises from a matched asymptotic expansion that divides the fluid in two regions; an outer region dominated by a constant vorticity Couette flow and an inner region over which the vorticity varies rapidly. The vorticity defect model provides an interesting link between the SWM model and fluid dynamics, passing through an analogy with the Vlasov-Poisson equation. However, as discussed in [8,21,4,5] the SWM, and generalizations of it, can be derived directly from the fluid dynamics equations including the two-dimensional Euler equation and the β-plane equation describing quasigeostrophic flows.

The HMF model [2,16]

$$\frac{d^2 x_j}{dt^2} = -\frac{\epsilon}{N} \sum_{k=1}^{N} \sin(x_j - x_k) \,, \tag{13}$$

is a special case of the single wave model when the first term, da/dt, on the left hand side of (3) vanishes, and

$$U = -\frac{2}{\epsilon} \,, \qquad \Gamma_k = 1 \,, \tag{14}$$

where in this case, the mean field a corresponds to the magnetization

$$M = \frac{1}{N} \sum_{k=1}^{N} e^{i x_k} = \frac{2}{\epsilon}\, a \,. \tag{15}$$

In the kinetic description the HMF model is recovered from (6)–(8) when $v = 0$. In this special case, the Hamiltonian in (6) can be rewritten as

$$H = \frac{u^2}{2} + V(x,t) \,, \qquad \partial_x^2 V = \frac{\epsilon}{2\pi} \int dx' \int du \cos(x - x')\, f(x', u, t) \,, \tag{16}$$

which is the kinetic model in [2], except for a trivial $1/(2\pi)$ normalization factor.

3 Symmetries, Conservation Laws, and Integrability

Defining

$$p_k = \Gamma_k\, u_k \,, \qquad a = \sqrt{\frac{J}{N}}\, e^{-i\theta} \,, \tag{17}$$

the SWM model can be written as the $N+1$ Hamiltonian system

$$\frac{dx_k}{dt} = \frac{\partial \mathcal{H}}{\partial p_k} \,, \qquad \frac{dp_k}{dt} = -\frac{\partial \mathcal{H}}{\partial x_k} \,, \tag{18}$$

$$\frac{d\theta}{dt} = \frac{\partial \mathcal{H}}{\partial J} \,, \qquad \frac{dJ}{dt} = -\frac{\partial \mathcal{H}}{\partial \theta} \,, \tag{19}$$

where the Hamiltonian

$$\mathcal{H} = \sum_{j=1}^{N} \left[\frac{1}{2\Gamma_j} p_j^2 - 2\Gamma_j \sqrt{\frac{J}{N}} \cos(x_j - \theta) \right] - U J \,, \tag{20}$$

is a function of the canonical conjugate coordinates of the particles (x_j, p_j), and the canonically conjugate coordinates of the mean field (θ, J). The invariance of

Hamiltonian in (20) under time translations, $t \to t+\tau$, implies the conservation of the energy $\mathcal{E}$, which in (x_j, u_j, a) coordinates can be written as

$$\mathcal{E} = \frac{1}{N}\sum_{j=1}^{N} \Gamma_j \left(\frac{u_j^2}{2} - a\, e^{i x_j} - a^*\, e^{-i x_j} \right) + U\, |a|^2 \,. \tag{21}$$

The Hamiltonian is also invariant under space translations $(x_k,\theta) \mapsto (x_k+\alpha,\theta+\alpha)$, and this leads to the conservation of the total momentum $\mathcal{P}$

$$\mathcal{P} = \frac{1}{N}\sum_{j=1}^{N} \Gamma_j\, u_j + |a|^2 \,. \tag{22}$$

The first term on the right-hand side of this equation is the momentum of the particles, and the second term is the momentum of the mean field.

Under the time-dependent canonical transformation

$$x_k \to x_k + Ut\,, \qquad p_k \to p_k + \Gamma_k\, U\,, \tag{23}$$

$$\theta \to \theta + Ut\,, \qquad J \to J\,, \tag{24}$$

the Hamiltonian transforms as $\mathcal{H} \to \mathcal{H} + U\,J$. Therefore, without loss of generality we can assume that $U = 0$. In noncanonical variables, this transformation corresponds to a Galilean transformation $(x_k, u_k) \to (x_k + Ut, u_k + U)$, $a \to a\exp(-iUt)$.

When $U = 0$, the Hamiltonian (18)–(19) are invariant with respect to the transformation

$$(x_k, \theta, \Gamma_k) \longmapsto (-x_k, -\theta, -\Gamma_k)\,. \tag{25}$$

In noncanonical coordinates this transformation corresponds to a reflection of the particles with respect to the origin in the (x_j, u_j) space, accompanied by a change in sign in Γ_k, and the complex conjugation of the mean field

$$(x_k, u_k, \Gamma_k, a) \longmapsto (-x_k, -u_k, -\Gamma_k, a^*)\,. \tag{26}$$

Initial conditions invariant under the transformation in (26) are called *symmetric states*. In general, an N-particles symmetric state consists of $N/2$ clumps (i.e., particles with $\Gamma_k > 0$) symmetrically opposed to $N/2$ holes (i.e., particles with $\Gamma_k < 0$) that is

$$(x_j, u_j) = -(x_{j+N/2}, u_{j+N/2})\,, \qquad \Gamma_j = -\Gamma_{j+N/2}\,, \qquad a_i = 0 \tag{27}$$

where $j = 1, \ldots N/2$, and a_i is the imaginary part of the mean field that vanishes because $a = a^*$. Because the equations are invariant under this transformation, symmetric states remain symmetric for all time, and thus a symmetric state with N particles can be effectively regarded as a system with $N/2$ particles.

In general, a state with N particles is a $N+1$ degrees-of-freedom Hamiltonian system, and according to the Liouville-Arnold theorem, integrability requires $N+1$ constants of motion. The only known constants of motion are the momentum and the energy and thus the $N+1=2$ system consisting of one particle in the mean field is integrable [29]. Because in general there are no other known constants of motion, this seems to be the only fully integrable case of the single wave model. The following subsection discusses this solution. However, when limiting attention to symmetric states the $N=2$ system is also integrable [14]. In this case, the system has $N/2+1=2$ degrees of freedom and, as described in Sect. 3.2, use of the momentum conservation law allows the integration of the solution. The main role of SWM is to describe the mean field interaction of systems composed of a large, or even infinite, number of particles. In this regard the integrability of the model in the $N=1$ and $N=2$ cases might seem like an academic issue. However, this is not the case. As we will discuss in Sect. 7, these low degree of freedom integrable solutions provide the key elements for the description of coherent states with $N \gg 1$ and $N \to \infty$.

3.1 Integrability of the $N=1$ Model

When there is only one particle, the single wave Hamiltonian has two degrees of freedom, one for the particle and one for the wave. Using the momentum conservation law, the problem can be reduced to one degree of freedom [29]. To do this, we define $q = x - \theta$, substitute the expression for the momentum $\mathcal{P}$ in (20) with $N=1$ and get

$$\mathcal{H} = \frac{1}{2\Gamma} p^2 - 2\Gamma \sqrt{\mathcal{P} - p} \cos q \,. \tag{28}$$

Since this Hamiltonian is time-independent, and it has only one degree of freedom, it is completely integrable. In particular, the particle orbits in the phase space follow the constant energy contours, $\mathcal{H} = $ constant.

The fixed points of the Hamiltonian (28), defined by the condition

$$\dot{q} = \frac{\partial H}{\partial p} = 0 \,, \qquad \dot{p} = -\frac{\partial H}{\partial q} = 0 \,, \tag{29}$$

are determined by the solutions of

$$\sin q = 0 \,, \qquad p\sqrt{\mathcal{P} - p} + \Gamma^2 \cos q = 0 \,. \tag{30}$$

The first equation has two solutions $q=0$ and $q=\pi$. In the first case there is only one solution of the second equation in (30) for p. In the second case, when $q=\pi$, there are no solutions for p, except when

$$\frac{3}{4^{1/3}} < \frac{\mathcal{P}}{\Gamma^{4/3}} \,, \tag{31}$$

in which case there are two solutions for p. Thus, as shown in Fig. 1, there is a fixed point at $(0, p_0)$ with $p_0 < 0$; and, if condition (31) is satisfied, there are two other fixed points at (π, p_1), and (π, p_2).

The stability of the fixed points is determined by linearizing the equations of motion. Doing this, we conclude that (π, p_1) is stable and (π, p_2) unstable. The

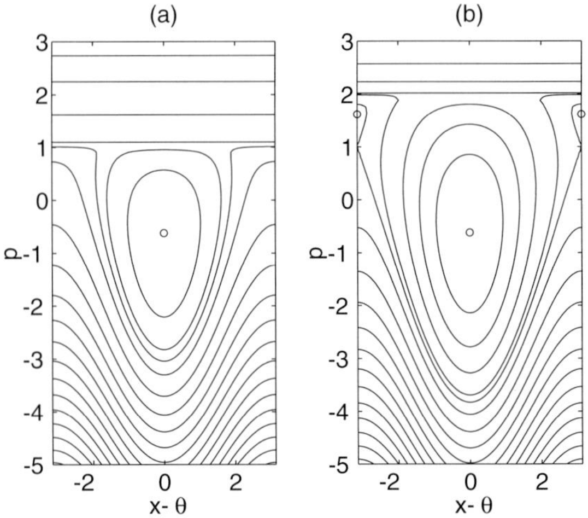

Fig. 1. Phase space of reduced, $N = 1$, single wave model Hamiltonian in (28). Panel (a) shows the case $\Gamma = 1$, $\mathcal{P} = 1$, and panel (b) shows the case $\Gamma = 1$, $\mathcal{P} = 2$. In both cases, the stable fixed point at $x - \theta = 0$ corresponds to the particle at the bottom of the single-wave potential. Consistent with (31), in case (b) there is a stable and an unstable fixed point at $x - \theta = \pi$.

fixed point at $(0, p_1)$ is stable, and it corresponds to a particle oscillating at the bottom of the potential with oscillation frequency

$$\omega^2 = 2\sqrt{\mathcal{P} - p_0}\left[1 + \frac{\Gamma^2}{2(\mathcal{P} - p_0)^{3/2}}\right] . \tag{32}$$

3.2 Integrability of $N = 2$ Symmetric States

To construct symmetric integrable solutions, consider an $N = 2$ state consisting of one clump and one hole with $\Gamma_1 = -\Gamma_2 = \Gamma$, $U = 0$, and symmetric initial conditions $x_1(0) = -x_2(0)$, $u_1(0) = -u_2(0)$, and $a(0) = a^*(0)$. Substituting $x_1(t) = -x_2(t) = x(t)$, $u_1(t) = -u_2(t) = u(t)$ and $a(t) = a^*(t)$ into (1)–(3), we get

$$\frac{dx}{dt} = u , \qquad \frac{du}{dt} = -2\, a \sin x , \qquad \frac{da}{dt} = \Gamma \sin x . \tag{33}$$

This system of three ordinary differential equations can be reduced to a two-dimensional integrable system using the conservation of momentum

$$\mathcal{P} = \Gamma u + a^2 . \tag{34}$$

In this case, the conservation of energy is of no use because for symmetric states the energy vanishes identically, $\mathcal{E} = 0$.

For the two-dimensional reduction, we take x and a as basic variables. Solving for u in (34) and substituting the result in (33), we conclude that the dynamics

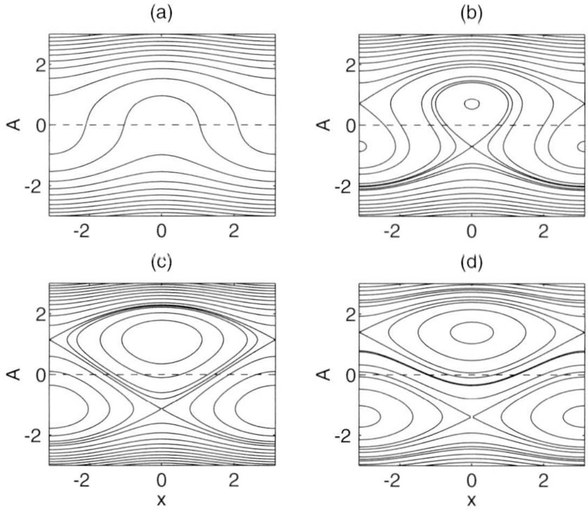

Fig. 2. Phase space of reduced, $N = 2$, single wave model Hamiltonian in (36). Panel (a) $\alpha = -0.2$, (b) $\alpha = 0.5$, (c) $\alpha = (3/2)^{2/3}$, and (d) $\alpha = 1.95$. In (a) there are no fixed points, case (c) corresponds to the separatrix reconnection threshold, case (b) shows the homoclinic topology, and case (d) the heteroclinic topology.

of the (x, a) variables is governed by the Hamiltonian system

$$\frac{dx}{d\tau} = \frac{\partial H}{\partial A}\,, \qquad \frac{dA}{d\tau} = -\frac{\partial H}{\partial x}\,, \tag{35}$$

with Hamiltonian

$$H = \alpha\, A - \frac{A^3}{3} + \cos x\,, \tag{36}$$

where we have introduced the rescaled variables

$$A = a\Gamma^{-2/3}\,, \qquad \tau = \Gamma^{1/3}\, t\,, \qquad \alpha = \mathcal{P}\Gamma^{-4/3}\,. \tag{37}$$

Once A and x are found, u is determined from

$$u = \frac{\mathcal{P}}{\Gamma}\left(1 - \frac{A^2}{\alpha}\right)\,. \tag{38}$$

Since the Hamiltonian (36) is time independent, the system is completely integrable and the orbits follow the contours H =constant.

As shown in Fig. 2, when $\alpha < 0$, the system (35) has no fixed points, and when $\alpha > 0$ the system has four fixed points: $(x_0, A_0) = (0, \sqrt{\alpha})$, $(0, -\sqrt{\alpha})$, $(\pi, \sqrt{\alpha})$, and $(\pi, -\sqrt{\alpha})$, with eigenvalues $\lambda = \pm i\omega_0$, $\lambda = \pm\omega_0$, $\lambda = \pm\omega_0$, and $\lambda = \pm i\omega_0$, respectively, where

$$\omega_0 = \sqrt{2}\alpha^{1/4}. \tag{39}$$

That is the elliptic and hyperbolic fixed points come in pairs since the equations of motion (35) are invariant under the transformation $(x, a) \longmapsto (x + \pi, -a)$.

The value $\alpha = 0$ is the bifurcation point at which an elliptic-hyperbolic pair is created at $x = 0$, and another elliptic-hyperbolic pair is created at $x = \pm\pi$. A perturbative solution of the equations of motion of the Hamiltonian (36) gives a a nonlinear correction to the rotation frequency

$$\omega = \left[1 - \left(\frac{5+9\lambda}{12}\right)\varepsilon^2\right]\omega_0\,, \qquad \lambda = \left(\frac{\alpha}{\alpha_*}\right)^{3/2}\,, \tag{40}$$

and

$$A(\tau) = \frac{\omega_0^2}{2}\left[1 - 2\,\varepsilon\,\cos\omega\,\tau - \varepsilon^2\left(1 - \frac{2}{3}\,\cos\omega\tau - \frac{1}{3}\,\cos 2\,\omega\,\tau\right)\right]\,, \tag{41}$$

$$x(\tau) = \omega_0^3\left[\varepsilon\,\sin\omega\,\tau - \frac{\varepsilon^2}{3}\,(\sin\omega\,\tau + \sin 2\,\omega\,\tau)\right]\,, \tag{42}$$

where

$$\varepsilon = \frac{\sqrt{\alpha} - A(0)}{\omega_0^2} \ll 1\,, \tag{43}$$

is a small parameter measuring the departure of the initial condition $(x(0), A(0))$ from the $(\sqrt{\alpha}, 0)$ elliptic fixed point. The interested reader should refer to [14] for the details of the calculation.

As observed in Fig. 2, for different values of α, the phase space topology of the H =constant contours change. These global bifurcations are caused by separatrix reconnection. In a Hamiltonian system a separatrix is a distinguished orbit that joins the unstable and stable manifolds of hyperbolic points. Separatrix reconnection corresponds to the different ways in which the hyperbolic points can connect. In panel (b) of Fig. 2 the separatrices join the unstable and stable manifolds of the same point, whereas in panel (c) the separatrices reconnect and join different hyperbolic points. The reconnection threshold can be obtained from the condition $H(P_0) = H(P_\pi)$, where $P_0 = (0, -\sqrt{\alpha})$, and $P_\pi = (\pi, \sqrt{\alpha})$. This condition gives $\alpha_* = (3/2)^{2/3}$. For $\alpha < \alpha_*$, the H =constant contours have the homoclinic topology of Fig. 2b, and for $\alpha > \alpha_*$ the heteroclinic topology of Fig. 2c. Using (37) the separatrix reconnection bifurcation can be parametrized in terms of the momentum of the system according to

$$\mathcal{P} < \left(\frac{3\,\Gamma^2}{2}\right)^{2/3} \quad \text{Homoclinic}\,, \qquad \mathcal{P} > \left(\frac{3\,\Gamma^2}{2}\right)^{2/3} \quad \text{Heteroclinic}\,, \tag{44}$$

with $\mathcal{P} = (3\,\Gamma^2/2)^{2/3}$ giving the reconnection threshold. Further discussion on separatrix reconnection can be found in [15].

4 Symplectic Map Formulation

Symplectic maps arise naturally in the study of Hamiltonian systems. A well-known example is the standard map which is an area preserving transformation of the (q, p) plane defined as

$$\begin{aligned} q^{n+1} &= q^n + p^{n+1} \\ p^{n+1} &= p^n - \kappa \sin q^n \ , \end{aligned} \tag{45}$$

where (q^n, p^n) are phase space coordinates at the discrete time n, and κ is a parameter. The main advantage of maps rests in their relative mathematically simplicity, compared with the time-continuous Hamiltonian systems. Whereas continuous dynamical systems can be time-consuming to integrate numerically, especially for very long times and many initial conditions, maps are very efficient. Another advantage is that maps are relatively simpler to study analytically than their continuous analogues. Because of this, maps are useful to carry out numerical studies of asymptotic properties of Hamiltonian systems.

In this section we discuss a map that is the discrete version of the single wave model. Given a set of differential equations, a map can be obtained by discretizing the time derivatives. However, in the case of Hamiltonian systems, special care must be taken to preserve the symplectic nature of the dynamics. For example, the standard map can be obtained from the time-discretization of the equations of motion of a pendulum, but whereas the map in (45) is an area preserving transformation, other discretization might not be. One way of to guarantee that a map will be symplectic is to define it using a generating function. Let $(\mathbf{q}^n, \mathbf{p}^n)$ denote the canonical conjugate coordinates of an M-dimensional Hamiltonian system at time n. Then, the transformation $(\mathbf{q}^n, \mathbf{p}^n) \rightarrow (\mathbf{q}^{n+1}, \mathbf{p}^{n+1})$ defined by

$$\mathbf{q}^{n+1} = \frac{\partial S}{\partial \mathbf{p}^{n+1}} \, , \qquad \mathbf{p}^n = \frac{\partial S}{\partial \mathbf{q}^n} \tag{46}$$

defines a symplectic map with generating function $S = S(\mathbf{q}^n, \mathbf{p}^{n+1})$. Equation (46) defines a canonical transformation, consistent with the fact the evolution of a Hamiltonian system is a canonical transformation.

Let $\mathbf{q}^n = (x_1^n, x_2^n, \dots x_N^n, \theta)$ and $\mathbf{p}^n = (p_1^n, p_2^n, \dots p_N^n, J)$ denote the canonical conjugate coordinates of the single wave model Hamiltonian at time n, and define

$$S = \sum_{j=1}^{N} \left[x_j^n \, p_j^{n+1} + \frac{\tau}{2\,\Gamma_j} \left(p_j^{n+1}\right)^2 - 2\,\tau \sqrt{J^{n+1}}\, \Gamma_j \, \cos\left(x_j^n - \theta\right) \right] + \theta^n \, J^{n+1} \, . \tag{47}$$

Substituting this generating function into (46) we get the symplectic map

$$X_k^{n+1} = X_k^n + \left(\frac{\tau}{\Gamma_k}\right) p_k^{n+1} \tag{48}$$

$$p_k^{n+1} = p_k^n - 2\tau\, \Gamma_k \sqrt{J^{n+1}}\, \sin(X_k^n - \Theta^n) \tag{49}$$

$$\Theta^{n+1} = \Theta^n - \frac{\tau}{\sqrt{J^{n+1}}} \sum_{j=1}^{N} \Gamma_j \cos(X_j^n - \Theta^n) \tag{50}$$

$$J^{n+1} = J^n + 2\tau \sqrt{J^{n+1}} \sum_{j=1}^{N} \Gamma_j \sin(X_j^n - \Theta^n) \ , \tag{51}$$

for $k = 1, 2, \ldots N$. As expected, this map is a symplectic, time-discretization of the single wave model equations of motion defined in (18)–(20), where without loss of generality we have assumed that $U = 0$.

The momentum of the system at time n is

$$\mathcal{P}^n = \sum_{j=1}^{N} p_j^n + J^n , \tag{52}$$

and the conservation of momentum in the SMW implies the conservation of the momentum in the map

$$\mathcal{P}^{n+1} = \mathcal{P}^n . \tag{53}$$

On the other hand, whereas the continuous model conserves the energy $\mathcal{E}$, in the map the energy is only conserved to first order in the map parameter τ. In general, energy conservation is problematic with maps because of the time discretization.

To simplify the notation we define:

$$x_k^n = X_k^n \quad , y_k^n = \left(\frac{\tau}{\Gamma_k}\right) p_k^n \quad , \theta^n = \Theta \quad , \kappa^n = 2\tau^2 \sqrt{J^n} , \tag{54}$$

$$\gamma_k = 2\tau^3 \,\Gamma_k , \qquad \Omega = U\tau , \qquad \eta^n = \sum_{j=1}^{N} \gamma_j \sin\left(x_j^n - \theta^n\right) , \tag{55}$$

and write the SWM map as

$$x_k^{n+1} = x_k^n + y_k^{n+1} , \tag{56}$$

$$y_k^{n+1} = y_k^n - \kappa^{n+1} \sin\left(x_k^n - \theta^n\right) , \tag{57}$$

$$\kappa^{n+1} = \sqrt{\left(\kappa^n\right)^2 + \left(\eta^n\right)^2} + \eta^n , \tag{58}$$

$$\theta^{n+1} = \theta^n - \Omega + \frac{1}{\kappa^{n+1}} \frac{\partial \eta^n}{\partial \theta^n} . \tag{59}$$

Note that in doing this we have gone from the implicit J^n map in (51) to the explicitly κ^n map in (59). Comparing (45) with (56)–(59) one observes that the SWM map has the structure of an ensemble of standard maps coupled through a mean field with amplitude κ^n and phase θ. Whereas in the standard map, the parameter κ is fixed, in the SWM map this parameter is self-consistently coupled to the dynamics of the particles [13].

5 Kinetic Linear Theory: Stability and Landau Damping

In this section we return to the kinetic description of the single wave model. An important problem in kinetic theory is the stability of solutions. The simplest problem is the linear stability of solutions of the form $f = f_0(u)$, $a = 0$. To study this problem we substitute

$$f = f_0(u) + \sum_{n=-\infty}^{\infty} \psi_n(u,t)\, e^{inx} , \qquad a = a_1(t) \tag{60}$$

into (6)–(8), neglect the nonlinear terms, and get the linear stability equations

$$\partial_t \psi_n + inu\, \psi_n + i\delta_{n,\pm 1}\, f_0'\, a_1 = 0 , \tag{61}$$

$$\sigma \frac{da_1}{dt} - iU a_1 = i \int_{-\infty}^{\infty} \psi_1(u,t) du . \tag{62}$$

Assuming a normal mode solution,

$$\psi_n(u,t) = \xi_n(u)\, e^{-ict} \qquad a_1(t) = \rho\, e^{-ict} , \tag{63}$$

we get from (61) the eigenfunctions

$$\xi_n(u) = -\frac{\rho f_0'}{u-c}\, \delta_{n,\pm 1} , \tag{64}$$

and the *dispersion relation*

$$D(c) \equiv \sigma c + U - \int_{-\infty}^{\infty} \frac{f_0'}{u-c} du = 0 , \tag{65}$$

where $c = c_r + ic_i$.

An equilibrium solution $f = f_0(u)$, $a = 0$ is linearly stable with respect to normal mode perturbations if and only if there are no solutions of (65) with $c_i > 0$. As an illustrative example, a normalized Lorentzian equilibrium with "temperature" $T = 1/\beta$

$$f_0 = \frac{\sqrt{\beta}}{\pi} \frac{1}{1+\beta u^2} . \tag{66}$$

In this case the integral can be calculated using the Cauchy residue theorem, and it is concluded that the equilibrium is unstable if and only if $\epsilon > 2/\beta$, in which case

$$c = \pm i \left(\sqrt{\frac{\epsilon}{2}} - \frac{1}{\sqrt{\beta}} \right) . \tag{67}$$

In the previous examples, the integral of the dispersion relation was easy to compute. However, in general this is not the case, and it is important to derive general stability criteria to determine the stability of equilibria without the need of explicitly solving the dispersion relation. In [11] the Nyquist method was used to derive stability criteria for general equilibria. Here we quote the result for the case of equilibria with a single extremum; the interested reader can find further details in [11]. An equilibrium f_0 with a single maximum at $u = u_1$ is unstable if and only if

$$u_1 + U + \int_{-\infty}^{\infty} \frac{f_0(u_1) - f_0(u)}{(u_1 - u)^2} du > 0 . \tag{68}$$

If the extremum is a minimum the inequality sign is reversed. In the HMF model case, (16), an equilibrium $f_0(u)$ with a single maximum at $u = 0$ is unstable if and only if

$$2 < \epsilon \int_{-\infty}^{\infty} \frac{f_0(0) - f_0(u)}{u^2} \, du \,. \tag{69}$$

As a particular example, consider a Gaussian distribution with temperature $T = 1/\beta$

$$f_0(u) = \sqrt{\frac{\beta}{2\pi}} \exp\left(-\beta \frac{u^2}{2}\right) . \tag{70}$$

In this case we have that the equilibrium is unstable if and only if

$$\epsilon > \frac{2}{\beta} \,. \tag{71}$$

That is, decreasing the temperature or increasing the coupling leads to instability.

5.1 High Frequency Oscillations

In a way similar to what is done in a Vlasov-Poisson plasma, one can look for pure real solutions of the dispersion relation for a general distribution $f_0(u)$ with temperature $T = \langle u^2 \rangle$ in the limit

$$T \ll c^2 \,. \tag{72}$$

This is the analogue of the Langmuir waves in plasma physics. In this case, assuming $f_0(u) = 0$ for $|u| > u_*$, the integral in the dispersion relation can be expanded as

$$\int_{-\infty}^{\infty} \frac{f_0'}{u-c} \, du \approx \int_{-u_*}^{u_*} \frac{f_0}{(u-c)^2} \, du \approx \frac{1}{c^2} \int_{-u_*}^{u_*} f_0 \left(1 + \frac{2u}{c} + \frac{3u^2}{c^2} \dots \right) du \,, \tag{73}$$

doing the integrals

$$\sigma c^3 + U c^2 - 1 - 3(T/c^2) = 0 \,. \tag{74}$$

In the SWM case, $\sigma = 1$, $U = 0$, we can solve this equation by a perturbation expansion in T/c^2 to get

$$c = (3T + 1)^{1/3} \,. \tag{75}$$

In the $\sigma = 0$, HMF model case, (16), a perturbative solution of the dispersion relation gives

$$c = 3T - \epsilon/2 \,, \tag{76}$$

which generalizes the result reported in [2] for the case of a "water-bag" equilibrium distribution. Equations (75) and (76) are the analogue of the Langmuir waves dispersion relation.

5.2 Initial Value Problem and Landau Damping

When $n \neq \pm 1$ the solution of (61) with initial condition $\psi_n(u, t=0) = \psi_n(u, 0)$, $a_1(t=0) = a(0)$, is $\psi_n(u,t) = \psi_n(u,0)\exp(-inut)$, and $f(x,u,t)$ simply phase mixes. The general solution for $n = \pm 1$ is slightly more complicated but it can be obtained in the same way as it is done for the Vlasov-Poisson equation by introducing the Laplace transforms

$$\hat{\psi}(u,p) = \int_0^\infty \psi(u,t)e^{-pt}dt \,, \qquad \hat{a}(p) = \int_0^\infty a(t)e^{-pt}dt \,, \tag{77}$$

where $p = p_r + ip_i$. The convergence of the integrals in (77) requires $p_r \geq p_0$, where p_0 is a real number.

Substituting (77) into (61) and solving for $\hat{a}$, we get

$$\hat{a}(p) = \frac{N(p)}{D(p)} \,, \tag{78}$$

where

$$N(p) = i\sigma a(0) + i\int_{-\infty}^{\infty} \frac{\psi(u,0)}{u-ip}du \,, \qquad D(p) = i\sigma p + U - \int_{-\infty}^{\infty} \frac{f_0'(u)}{u-ip}du \tag{79}$$

Knowing $N(p)$ and $D(p)$, the mean field a is obtained by inverting the Laplace transform

$$a(t) = \frac{1}{2\pi i}\int_{p_0-i\infty}^{p_0+i\infty} \frac{N(p)}{D(p)}e^{pt}dp \,, \tag{80}$$

where the integration is along the Bromwich contour, which is a straight line parallel to the imaginary axis in the complex p-plane and located to the right of the abscissa of convergence. The evaluation of the integral in (80) is simplified by deforming the contour to the left-half of the complex p plane and using the Cauchy residue theorem. However, doing this requires the analytic continuation of the functions $D(p)$ and $N(p)$, originally defined for $p \geq p_0$, into the entire complex plane. One way of performing this analytic continuation is following the Landau prescription according to which the integrals in D and N should be viewed as integral in the complex u-plane and evaluated along a contour that should always pass below the singularity $u = ip$ regardless of the sign of p_r. As in the case of the Vlasov-Poisson system, this prescription allows the determination of the asymptotic value of the integral in (80) and leads to the conclusion that for stable equilibria, $f_0(u)$, the mean field exhibits decay known as Landau damping [23,26]. However, as in the plasma and fluid dynamics cases, one should be aware of technical difficulties that can lead to more general time asymptotic behavior [30,3]. Here we limit attention to an analytical tractable example that illustrates Landau damping in the context of the HMF model.

The equilibrium considered is the normalized Lorentzian distribution in (66). In this case, the integral in (79) can be easily computed and analytically continued giving:

$$\mathcal{D}(p) = -\frac{2\,\beta}{\epsilon}\,\frac{(p-p_+)(p-p_-)}{(\sqrt{\beta}\,p+1)^2} \,, \qquad p_\pm = \pm\sqrt{\frac{\epsilon}{2}} - \frac{1}{\sqrt{\beta}} \,. \tag{81}$$

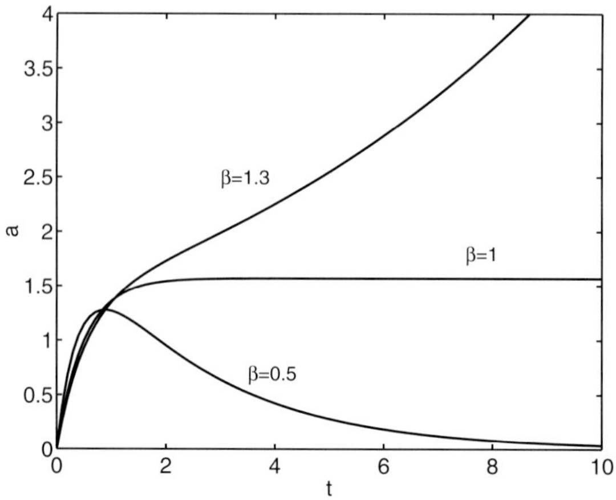

Fig. 3. Instability and Landau damping in the HMF model limit of the SWM model for the Lorentzian equilibrium in (66) and Lorentzian perturbation of the form in (82) with $\epsilon = 2$ and $\Delta = \beta$. The critical value of β in this case is $\beta = 1$. For $\beta > 1$ the equilibrium is linearly unstable. For $\beta < 1$ the equilibrium is stable, and the perturbation exhibits Landau damping.

For a perturbation of the form

$$\psi_1(u, 0) = \frac{1}{1 + \Delta u^2}, \tag{82}$$

a similar calculation leads to

$$\mathcal{N}(p) = -\frac{\pi}{1 + \sqrt{\Delta}\, p}. \tag{83}$$

Substituting (81) and (83) into (80) we get

$$a(t) = -\frac{i\,\epsilon}{4\,\beta} \int_{p_0 - i\infty}^{p_0 + i\infty} \frac{(\sqrt{\beta}\, p + 1)^2\, e^{pt}}{(1 + \sqrt{\Delta}\, p)(p - p_+)(p - p_-)}\, dp, \tag{84}$$

which can easily be evaluated using the Cauchy residue theorem giving, for $\beta = \Delta$,

$$a(t) = \frac{\epsilon\,\pi}{4\sqrt{\beta}} \left(e^{p_+ t} - e^{p_- t}\right). \tag{85}$$

Figure 3 shows $a(t)$ in (85) for $\epsilon = 2$ and $\beta = 0.5$, 1 and 1.3. According to (67) the Lorentzian equilibrium (66) is unstable if and only if $\epsilon > 2/\beta$. Thus, as expected, case $\beta = 1.3$ in Fig. 3 exhibits a transient followed by an exponential growth due to the linear instability. For $\beta = 0.5$ the equilibrium is stable, and following a transient growth, the mean field Landau damps. The case $\beta = 1$ is the marginal value and gives rise to a neutral oscillation of the mean field.

6 Kinetic Nonlinear Solutions

The kinetic limit of the SWM, (6)–(8), admits a large class of exact nonlinear solutions, which are the analogue of the BGK modes for the Vlasov-Poisson system [10,11]. To construct these solutions, we substitute

$$f = f(\xi, u), \qquad a = \rho\, e^{i\delta - ict}\,, \tag{86}$$

into (6)–(8), where $\xi = x - ct$ and get

$$\frac{\partial(H, f)}{\partial(\xi, u)} = 0 \tag{87}$$

$$\int_{-\infty}^{\infty} du \int_0^{2\pi} d\xi\, f(H)\, e^{-i\xi} = -2\pi(U + \sigma c)\,\rho\,; \tag{88}$$

where

$$H = \frac{1}{2}(u - c)^2 - 2\rho \cos\xi\ . \tag{89}$$

Any function of the form $f = f(H)$ is a solution of (87). However, $f = f(H)$ will be a self-consistent solution of (8), only if (88) is also satisfied. As an example, consider the normalized Gibbs distribution

$$f(H) = \frac{e^{-\beta H}}{\lambda}\,, \tag{90}$$

where $\lambda = \langle e^{-\beta H} \rangle$ is a normalization factor. In this case the self-consistent condition leads to

$$\rho(U + \sigma c) = -\frac{I_1(2\beta\, \rho)}{I_0(2\beta\, \rho)}\,, \tag{91}$$

where $I_1(z)$ is the modified Bessel function of order one [10,11]. In the $\sigma = 0$, HMF model case, (16), the solution (91) reduces to the one reported in [24].

$$M = \frac{I_1(\epsilon\beta M)}{I_0(\epsilon\beta M)}\,, \tag{92}$$

where M is the magnitude of the magnetization.

7 Self-consistent Chaos and Coherent Structures

The spontaneous formation of coherent structures is commonly observed in experiments and numerical simulations in hydrodynamics and plasmas. Coherent structures are also routinely observed in planetary flows (e.g., the Jupiter's red spot.) These structures typically coexist with a turbulent background, and a problem of interest is to explain their robustness with respect to the chaotic transport induced by the background. The goal of this section is to study this problem in the context of the SWM.

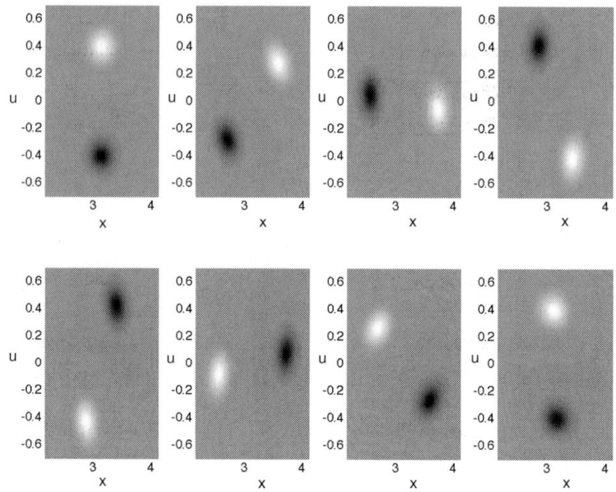

Fig. 4. Rotating dipole obtained from the numerical integration of the kinetic single wave model (6)–(7) with initial condition in (95)–(96). White corresponds to $f > 0$ (clump), and black to $f < 0$ (hole). The panels show the dipole at $t = 0, \Delta, 2\Delta, \dots 7\Delta$, where $\Delta = 1.237$ and $7\Delta = 8.66$ is the rotation period of the dipole [14].

The existence of phase space coherent structures in large degrees-of-freedom Hamiltonian systems is in principle puzzling because coherence requires some degree of integrability, which one might expect to lose as the number of degrees of freedom increases. However, as we will discuss in this section, there are situations in which mean field self-consistent systems exhibit an effective reduction of the number of degrees of freedom, which leads to coherent behavior.

As in the finite N case, the dependence on U in the kinetic description can be removed by a Galilean transformation, and when $U = 0$ the system is invariant under the transformation

$$(x, u, f, a) \longmapsto (-x, -u, -f, a^*) . \tag{93}$$

In the kinetic limit, a symmetric state is defined by the condition

$$f(x, u, t) = -f(-x, -u, t) , \qquad a(t) = a^*(t) , \tag{94}$$

Figure 4 shows an example of a rotating dipolar coherent structure. This state was obtained by integrating numerically the SWM in the kinetic limit, (6)–(8), with a symmetric initial condition consisting of a Gaussian distributed hole-clump pair:

$$f(x, u, t = 0) = f_1 - f_2 , \qquad a(t = 0) = a_0 , \tag{95}$$

where

$$f_j = \gamma \exp\left[-\left(\frac{x - x_j}{\sigma_x}\right)^2 - \left(\frac{u - u_j}{\sigma_u}\right)^2 \right] , \tag{96}$$

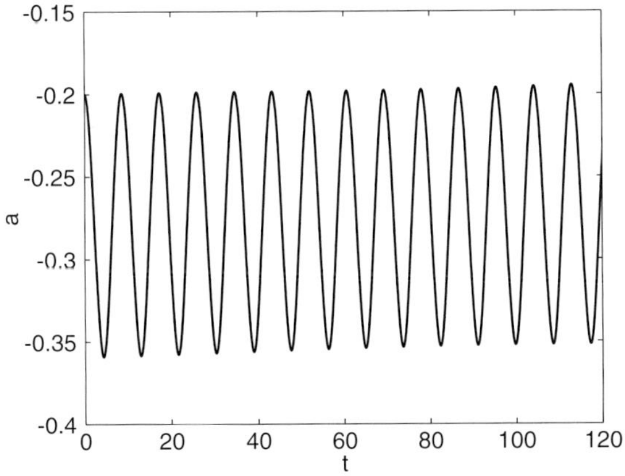

Fig. 5. Real part of mean field a as function of time corresponding to the coherent rotating dipole solution shown in Fig. 4. The imaginary part of a vanishes identically for all t in this case [14].

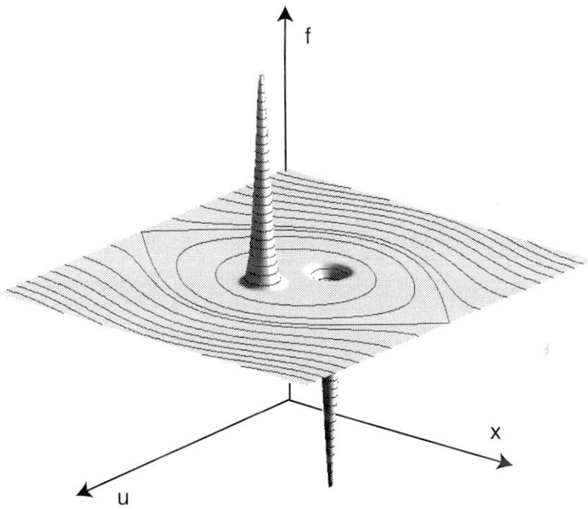

Fig. 6. Three dimensional perspective of a hole-clump dipole in the single wave model. The dipole is trapped in the mean-field potential and rotates around the stable fixed point $(x, u) = (\pi, 0)$ of the mean-field Hamiltonian.

$\gamma = 5$, $x_1 = x_2 = \pi$, $u_1 = -u_2 = 0.4$, $a_0 = -0.2$, $\sigma_x = 0.2$, $\sigma_u = 0.1$, and $U = 0$. As the figure illustrates, the structure rotates maintaining its coherence. Figure 5 shows the evolution of the real part of the mean field $a(t)$. Because this is a symmetric state, the imaginary part of a is identically zero. Figure 6 offers a three-dimensional perspective of the hole-clump pair together with a contour plot of the mean field Hamiltonian

$$H = \frac{u^2}{2} - a(t)e^{ix} - a^*(t)e^{-ix}\,, \tag{97}$$

at a fixed time. As the figure shows, the dipole is trapped in the mean-field potential and rotates around the stable fixed point $(x, u) = (\pi, 0)$. As expected by self-consistency, the rotation period of the dipole is close to the period of the mean field.

Kinetic simulations always introduce some sort of dissipation due to the coarse graining of filamentary structures with a scale beyond the resolution of the numerical method. Because of this, the relationship between very large N discrete particle simulations, which are in principle conservative, and kinetic results might not always be trivial. To explore this issue in the dipole problem, we integrated the finite N SWM model with an initial condition consisting of a discretization of f in (95)–(96) with $N = 1000$ (half holes and half clumps) and the appropriate values of Γ_k. In this case, there is not any significant difference between the finite N and the kinetic phase space evolution. However, in the finite N case, the rotation period of the mean field is slightly smaller [14].

7.1 Macroparticle Description

The coherence of the dipole suggests the description of the system as two "macroparticles," one representing the hole and another the clump. The macroparticle representation is a sort of renormalization process in which a group of particles with different values of Γ_k are replaced by one with an effective Γ. For the kinetic initial condition in (95)–(96) the effective Γ for the clump is simply

$$\Gamma = \frac{1}{\pi} \int\!\!\int f_1 \, dx du = \gamma \, \sigma_x \sigma_u \,, \tag{98}$$

and for the hole is $-\Gamma$. The macroparticle representation provides a link between systems with large (or infinite) N, and systems with small N, say $N = 1$, or 2. In particular, according to the results discussed in Sect. 3.2, the rotation period of a symmetric hole-clump system is $T = \sqrt{2}\pi/\mathcal{P}^{1/4}$ where $\mathcal{P}$ is the momentum of the $N = 2$ system. The momentum of the kinetic initial condition in (95)–(96) is

$$\mathcal{P} = \frac{1}{2\,\pi} \int\!\!\int f \, u \, dx du \, + |a|^2 = \gamma \, u_1 \sigma_x \sigma_u + |a_0|^2 \,, \tag{99}$$

which, according to the macroparticle description, gives the following expression for the rotation period of the kinetic dipole

$$T = \frac{\sqrt{2}\,\pi}{\mathcal{P}^{1/4}} = \frac{\sqrt{2}\,\pi}{\left(\gamma \, u_1 \sigma_x \sigma_u + |a_0|^2\right)^{1/4}} \,. \tag{100}$$

For the parameter values used in Fig. 4, (100) gives $T = 8.35$ close to the period $T = 8.66$ according to Fig. 5. Equation (40) provides a second order correction of the kinetic dipole rotation period in (100). To compute this correction note that for the initial condition in (95)–(96)

$$\lambda = \frac{2}{3\gamma^2 \sigma_x^2 \sigma_u^2} \left(\gamma u_1 \sigma_x \sigma_u + |a_0|\right)^{3/2} \,, \tag{101}$$

which for the parameter values used in Fig. 4, gives $\lambda = 1.5084$. In the evaluation of ε in (43) we have to take the negative sign of $\sqrt{a}$ because $a(0) = -0.2 < 0$. Doing this, we get $\varepsilon = 0.146$ and $\omega = 1.566$, which gives $T = 8.64$; a value remarkably close to the period $T = 8.66$ according to the numerical results in Fig. 5. Using (98), the value of $\alpha = \mathcal{P}\Gamma^{-4/3}$ can be computed, and the topology of the reduced nontwist Hamiltonian of the macroparticles determined. In particular, for the kinetic simulation in Fig. 4, $\alpha = 1.7235 > \alpha_*$, which corresponds to the heteroclinic topology.

A rotating dipole creates a time periodic self-consistent mean field. In dynamical systems, it is well known that time periodic Hamiltonian systems typically give rise to chaos[1] and an important problem is to understand the effect of chaos in the dipole. To this end, consider the evolution of "test" particles with phase space coordinates $(q, \dot{q})$ in the mean field of the rotating dipole

$$\frac{d^2 q}{d\tau^2} = 2A(\tau)\,\sin q\,, \tag{102}$$

where A is give by (41).

Panel (a) of Fig. 7 shows a snapshot of the rotating dipole at time τ_0, and panel (b) shows a Poincare section of the system in (102) with $A(\tau_0+T) = A(\tau_0)$ where T is the fundamental period of the self-consistent mean field. As it is usually the case in Hamiltonian systems, the perturbation gives rise to chaos in the separatrix region. However, in this region the distribution function f is negligible and so is the effect of chaos. On the other hand the perturbation creates islands of stability that "shield" the hole and the clump from the chaotic region. As the dipole rotates, the phase of the Poincare section changes and the islands of integrability follow the motion of the coherent structure.

7.2 Parametric Resonance

The frequency of the self-consistent mean field is close to the rotation frequency of the dipole, and there is the possibility of parametric resonance. To explore the effect of this resonance in the coherence of the dipole, consider a symmetric state with N particles, $N/2$ holes, and $N/2$ clumps, and define the center of mass coordinates

$$\bar{x} = \frac{2}{N}\sum_{j=1}^{N/2} \Gamma_j\, x_j\,, \qquad \bar{u} = \frac{2}{N}\sum_{j=1}^{N/2} \Gamma_j\, u_j\,. \tag{103}$$

Substituting (103) into (1), and assuming $x_j \ll 1$ we have

$$\frac{d\bar{x}}{dt} = \bar{u}\,, \qquad \frac{d\bar{u}}{dt} = 2\,a\sin\bar{x}\,, \qquad \frac{da}{dt} = \sin\bar{x}\,. \tag{104}$$

[1] We say typically because there are special cases in which the time dependence in the Hamiltonian can be eliminated with an appropriate time-dependent canonical transformation.

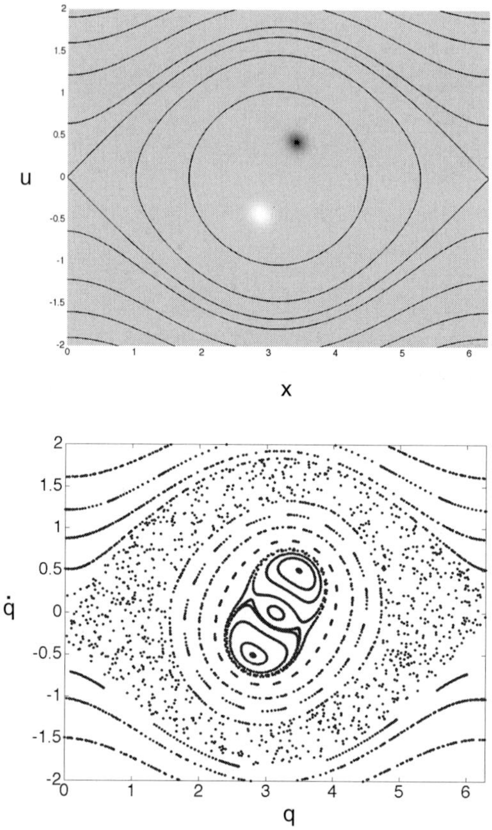

Fig. 7. The top panel show contour plots of the rotating dipole solution at a fixed time along with the contour lines of the Hamiltonian in (2). The panel on the bottom shows a Poincare section in the mean field of the rotating dipole solution.

Comparing with (33), it is concluded that, in this approximation, the evolution of the center of mass of the hole-clump dipole is described by the integrable $N = 2$ system. Let

$$q_j = x_j - \bar{x}\,, \qquad p_j = u_j - \bar{u} \tag{105}$$

be the relative coordinates with respect to the center of mass. Substituting (105) into (1) we have

$$\frac{dq_j}{dt} = p_j\,, \qquad \frac{dp_j}{dt} = -\left(2a\,\cos\bar{x}\right)\,q_j\;. \tag{106}$$

Approximating $\bar{x} \approx 0$, and using the mean field equation, we conclude that the relative position evolves according to the Mathieu equation

$$\frac{d^2 q_j}{d\hat{\tau}^2} + \left[1 + 2\left(\frac{5+9\lambda}{3}\right)\,\varepsilon^2 \; - 2\,\varepsilon\,\cos\hat{\tau}\right]\,q_j = 0\,. \tag{107}$$

The stability properties of this Mathieu equation are well-understood. In particular, the equilibrium solution $(q, dq/d\hat{\tau}) = (0, 0)$ is stable if $\lambda > 5/9$ and unstable

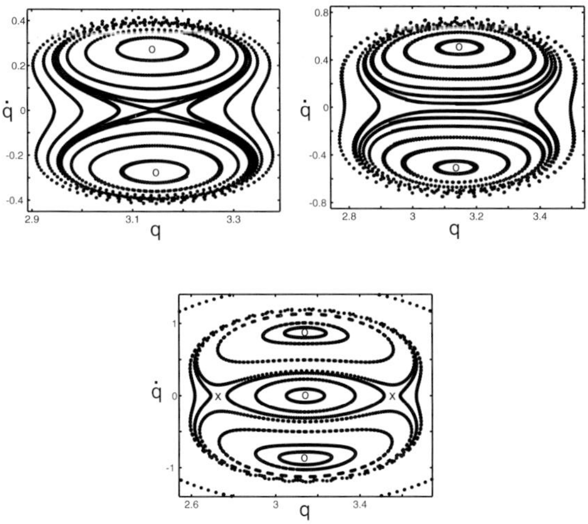

Fig. 8. Poincare sections in the $(q_j, \dot{q}_j)$ –plane of (107), illustrating the bifurcation of the stability properties of the origin due to parametric resonance. In (a) $\alpha = 2$, and in (c) $\alpha = 0.2$. Panel (b) corresponds to the bifurcation point $\alpha = \alpha_b = (5/6)^{2/3} = 0.8855$ according to (108). To ease the comparison with the Fig. 7, we have shifted $q_j \to q_j + \pi$

if $\lambda < 5/9$; see for example [22]. In terms of the parameter α, the dynamics of the relative coordinates is stable if and only if

$$\alpha > \alpha_b = (5/6)^{2/3} . \tag{108}$$

Figure 8 illustrates this transition.

8 Violent Mixing

The SWM Hamiltonian for a symmetric state

$$H = \sum_{j=1}^{N} \left[\frac{u_j}{2} - 2a(t) \cos x_j \right] , \tag{109}$$

has fixed points at $(x_j, u_j) = (0, 0)$ and $(\pi, 0)$. The stability of these fixed points depends on the sign of the mean field a. If $a > 0$, $(0, 0)$ is stable (elliptic) and $(\pi, 0)$ is unstable (hyperbolic). However, if $a < 0$, $(0, 0)$ is unstable (hyperbolic) and $(\pi, 0)$ is stable (elliptic). Thus, a change in the sign of a implies an elliptic-hyperbolic bifurcation. An example of this bifurcation is illustrated in Fig. 9 where we have plotted the evolution of an $N = 2$ hole-clump pair together with the instantaneous location of the separatrix of the Hamiltonian in phase space. Because of the mean field nature of the model, it is irrelevant which phase space plane one chooses because (x_k, u_k) evolves under the same Hamiltonian for all k. As the hole-clump rotates, the width of the separatrix decreases and

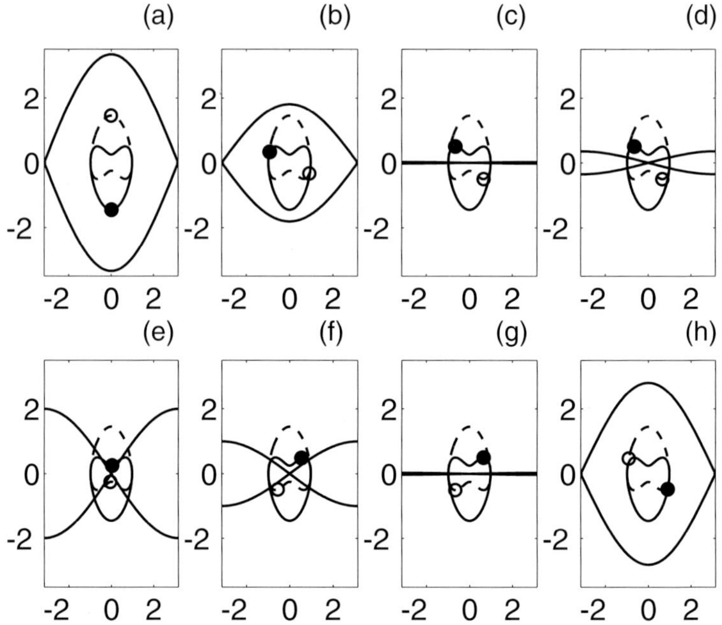

Fig. 9. Elliptic-hyperbolic bifurcation generated by a trapped $N = 2$ rotating hole-clump system.

eventually vanishes. At this point, panel (c), an elliptic-hyperbolic bifurcation occurs leading to the creation of an hyperbolic point at the origin as shown in panel (d). As the hole-clump rotates the separatrix grows, and once again eventually decreases and vanishes, leading to another bifurcation that restores the elliptic point at the origin. In this case, this sequence of bifurcations repeats indefinitely because the mean-field is periodic and changes sign twice during one rotation period of the dipole.

An elliptic-hyperbolic bifurcation occurs whenever the mean field vanishes. Whether or not this happens during the evolution of the system depends on the initial conditions and the parameter values of the system. The symmetric $N = 2$ system is particularly simple to understand. As discussed previously, this system is integrable, and the clump orbits follow the iso-contours of the nontwist Hamiltonian (36) in the (x, A) space shown in Fig. 2. In this case, the orbits that exhibit elliptic-hyperbolic bifurcations are those that cross the $A = 0$ dashed line. Consider an orbit with initial condition (x_0, u_0) that exhibits an elliptic-hyperbolic bifurcation when it is at (x_b, u_b). Because the Hamiltonian is a constant of motion, $H = \alpha A_0 - A_0^3/3 + \cos x_0 = \cos x_b$. Using the fact that $-1 < \cos x_b < 1$, we get the following condition for the bifurcation to occur [14]:

$$-1 < \alpha\, A_0 - \frac{A_0^3}{3} + \cos x_0 < 1\,. \tag{110}$$

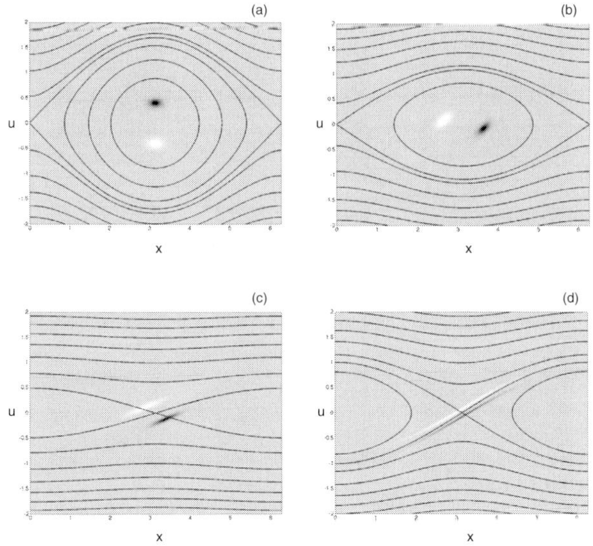

Fig. 10. Dipole destruction due to mean-field elliptic-hyperbolic bifurcation.

Assuming for the sake of simplicity that $x_0 = 0$ and using the definition of α in (37) we can write this condition as in terms of the initial conditions $a_0 = a(0)$, $u_0 = u(0)$, and the value of Γ

$$-6\,\Gamma^2 < 3\,\Gamma u_0\, a_0 + 2\,a_0^3 < 0\,. \tag{111}$$

In the case $N \gg 1$, and in the kinetic limit, the elliptic-hyperbolic bifurcation can lead to the destruction of phase space coherent structures. For example, panel (a) in Fig. 10 shows the initial condition in (95) consisting of two symmetric Gaussian-distributed holes and clumps with $\gamma = 12$, $x_1 = \pi$, $x_2 = \pi$, $u_1 = -u_2 = -0.4$, and $a_0 = -0.36$. This is the same initial condition as the one used in Fig. 4 except that the value of γ has been increased. Note also that the dipole in Fig. 10a is rotated by π, and that there is a corresponding phase shift in $a(t)$. At the beginning the dipole rotates around $(x, y) = (\pi, 0)$, but eventually an elliptic-hyperbolic bifurcation occurs and the dipole is destroyed and filaments along the unstable manifolds of the hyperbolic point. At this point, to follow the evolution of this system, it is more convenient to resort to finite–N particle simulations. In these simulations, it is observed that the dipole filamentation due to the elliptic–hyperbolic bifurcation is followed by the chaotic mixing of the holes and clumps in the stochastic layer of the mean field [14].

Hyperbolic-elliptic bifurcations can occur in more general settings and they lead to a very rapid mixing of far from equilibrium initial conditions. An example is shown in Fig. 11 where, after several successive elliptic-hyperbolic bifurcations, the system relaxes to a coherent state consisting of two untrapped dipoles traveling in opposite directions. In this case, the mean field, shown in Fig. 12, vanishes several times during a transient phase, and then it relaxes to a

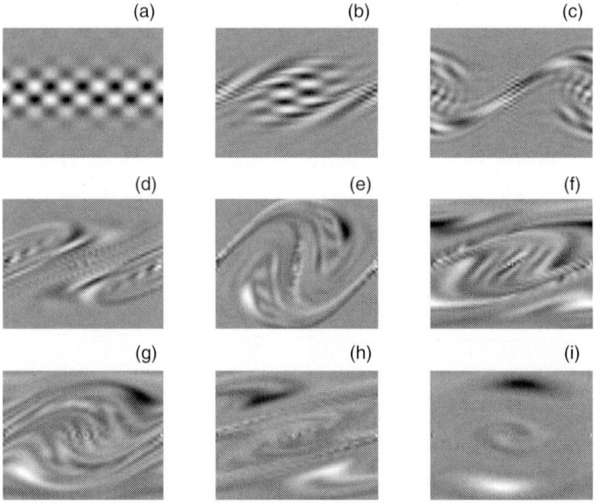

Fig. 11. Violent relaxation of far from equilibrium initial condition. The different panels, show the phase space at $t = 0$, 2.5, 5, 10, 17.5, 30, 32.5, 40, 100.

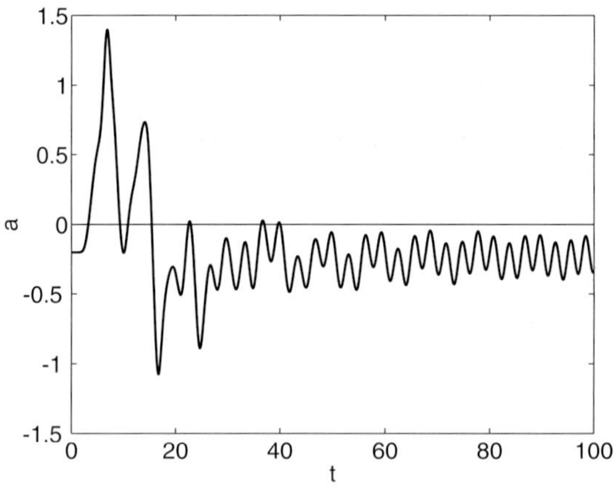

Fig. 12. Mean field $a(t)$ corresponding to the simulation in Fig. 11. The vanishing of a gives rise to elliptic-hyperbolic bifurcations in the phase space.

near periodic state. The relaxation of far from equilibrium states in systems with long-range interactions is a problem of interest in hydrodynamics, plasmas and galactic dynamics, and several models have been discussed in the literature as potential mechanisms. The elliptic-bifurcation discussed above is a novel mixing mechanism which in the SWM leads to "violent relaxation". We use here the term violent relaxation (a term originally introduced in galactic dynamics) to

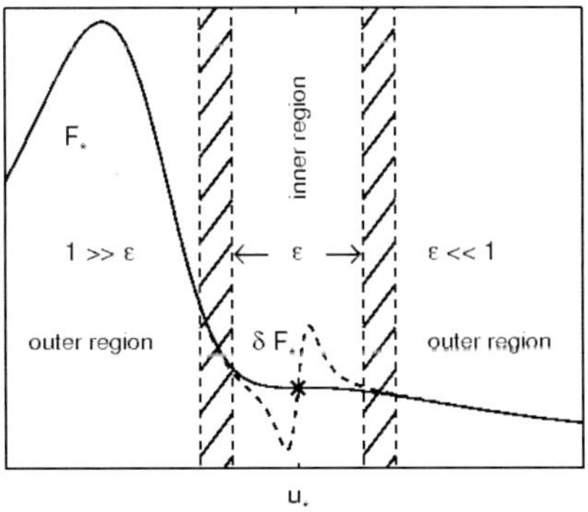

Fig. 13. Sketch of a localized perturbation (dashed line) on a marginally stable equilibrium (solid line). The matched asymptotic expansion discussed in the appendix reduces the Vlasov-Poisson system (9)–(10) to the single wave model (6)–(8) that describes the nonlinear evolution in the inner, $O(\varepsilon)$, region. In the outer, $1 \gg \varepsilon$, region the system responds linearly. The shaded area is the asymptotic matching region [11].

distinguish this kind of mixing from other milder phase space mixing mechanisms including phase mixing and chaotic mixing.

9 Conclusions

We have presented a study of the single wave model which is a mean field Hamiltonian model that describes the weakly nonlinear dynamics of fluids and plasmas. The relative mathematical simplicity of the model makes it a tractable system to study the dynamics of large degrees-of-freedom, long-range interacting systems. Understanding the coexistence of coherent structures and self-consistent Hamiltonian chaos is a problem of interest to fluid dynamics, plasma physics, galactic dynamics, and statistical mechanics. Two-dimensional, incompressible fluids and plasmas are known to develop large–scale coherent structures that live in a turbulent background. From a dynamical systems point of view, the problem is to explain the self-consistent formation of islands of integrability that shield the coherent structures from the chaotic mixing of the background. In this chapter we have shown that in the SWM there is a self-consistent parametric resonance between the rotation frequency of the dipole and the frequency of the mean field. This resonance creates islands of stability that shield the dipole from the chaotic transport of the background. A problem complementary to the formation of coherent structures is the problem of "violent" mixing and relaxation of far from equilibrium initial conditions. We have shown that in the SWM the mean field can exhibit hyperbolic-elliptic bifurcations in the phase space that lead to very rapid or "violent" mixing of phase space structures. By "violent" we mean that the time scale of the mixing is faster than a diffusive or chaotic advection mixing process. We also discussed the linear stability of equilibria and the problem of Landau damping in the kinetic limit of the single wave model.

Acknowledgments

I would like to give special thanks to my collaborator in the study of self-consistent chaos and coherent structures, Marie-Christine Firpo. Also, it is a pleasure to acknowledge useful conversations with P. J. Morrison, A. Vulpiani, G. Boffetta, S. Ruffo, and Y. Elskens. Finally, I want to thank the organizers of the Les Houches Winter School, Thierry Dauxois, Stefano Ruffo, Ennio Arimodo, and Martin Wilkens for inviting me to participate in a very enlightening workshop that, hopefully, will promote a long-range interaction among its participants. This work was sponsored by the Oak Ridge National Laboratory, managed by UT-Battelle, LLC, for the U.S. Dept. of Energy under contract DE-AC05-00OR22725.

Appendix A: Derivation of the SWM for Marginally Stable Vlasov-Poisson Systems

In this appendix we derive the single wave model in (6)–(8) as a weakly nonlinear description of Vlasov-Poisson systems of the form in (9)–(10). We skip technical details that the interested reader can find in 0[10,11].

The starting point is the assumption that the equilibrium is marginally stable with a stationary inflexion point at $u = c_0$, that is $F_0'(c_0) = F_0''(c_0) = 0$. The destabilization of F_0 is caused by perturbations that turn the locally flat region around the stationary inflection point into a region of adverse gradient as shown in Fig. 13.

We introduce the slow time variable

$$t \equiv (2\pi/L)\,\varepsilon\, T \ , \tag{112}$$

where ε, the small parameter of the perturbation expansion, is proportional to the growth rate of the instability, and L is the spatial length of the system. The quantization condition implies

$$k = \frac{2\pi}{L} = k_0(1 - \varepsilon \Lambda) \ , \tag{113}$$

where k is the wavelength of the instability, k_0 is the wavelength of the neutral inflexion point mode, and Λ is a constant. In addition, we assume the trapping scaling according to which the amplitude of the nonlinear saturated fields is of order ε^2,

$$F = \varepsilon^2\, f\,(x, u, t) \ , \qquad \Phi = \varepsilon^2\, \phi\,(x, t) \, , \tag{114}$$

and do a Galilean transformation to a reference frame moving at the speed of the inflection point mode

$$x \equiv (2\pi/L)(X - c_0 T) \ . \tag{115}$$

In terms of the variable x, the domain has period 2π, and ϕ, and f can be expanded as a Fourier series:

$$\phi(x,t) = \sum_{n=-\infty}^{\infty} \tilde{\phi}(n,t)\, e^{inx} \,, \qquad f(x,u,t) = \sum_{n=-\infty}^{\infty} \tilde{f}(n,u,t)\, e^{inx} \,. \tag{116}$$

Substituting (112)–(116) into (9) and (10) we get

$$\varepsilon\, \partial_t f + (u - c_0)\, \partial_x f + \left(F_0' + \varepsilon^2 \partial_u f\right) \partial_x \phi = 0 \,, \tag{117}$$

$$G\left[nk_0\left(1 - \varepsilon \Lambda\right)\right] \tilde{\phi}(n,t) = -\int_{-\infty}^{\infty} \tilde{f}(k,u,t) du \,. \tag{118}$$

These equations are exact, and the derivation of the SWM proceeds by a perturbative solution of them.

In the perturbation expansion, we write

$$\phi(x,t) = \phi_0 + \varepsilon\, \phi_1 \ldots \tag{119}$$

and divide the (x,u) space in two regions; an inner region where $u - c_0 = O(\varepsilon)$ and an outer region where $u - c_0 = O(1)$. In each region we solve (117) by expanding f in powers of ε and match the solution in the intermediate region. Once a global solution of f is found, ϕ is determined self-consistently from (118).

To treat the inner region, we introduce the stretched coordinate $v = (u - c_0)/\varepsilon$, substitute $f = f_0^i + \varepsilon f_1^i + \ldots$ into (117), and get at first order a Vlasov equation describing the distribution function in the inner region

$$\partial_t f_0^i + v \partial_x f_0^i + \partial_x \phi_0\, \partial_v f_0^i = 0 \,. \tag{120}$$

In the outer region, $u - c_0 = O(1)$, we substitute $\tilde{f}(n,u,t) = \tilde{f}_0^o + \varepsilon\, \tilde{f}_1^o + \ldots$ into (117) and get at $O(1)$ the linear eigenmode of the instability

$$\tilde{f}_0^o = -\frac{F_0'}{u - c_0}\, \tilde{\phi}_0 \tag{121}$$

and at $O(\varepsilon)$

$$\tilde{f}_1^o = -\frac{i}{n} \frac{F_0'}{(u - c_0)^2}\, \partial_t \tilde{\phi}_0 - \frac{F_0'}{u - c_0}\, \tilde{\phi}_1 \,. \tag{122}$$

Note that, because of the inflexion point condition $F_0'(c_0) = F_0''(c_0) = 0$, there are no singularities in (121)–(122). The last step in the derivation is to derive a self-consistent equation for the potential by solving (118) perturbatively. At $O(1)$ (118) reduces to the dispersion relation of the neutral model, and at $O(\varepsilon)$

$$G\left(nk_0\right) \tilde{\phi}_1 - nk_0 \Lambda\, G'\left(nk_0\right) \tilde{\phi}_0 = -\int_{-\infty}^{\infty} \tilde{f}_1^o\, du - \int_{-\infty}^{\infty} \tilde{f}_0^i\, dv \,.$$

Because $\int \tilde{f}^i du = \varepsilon \int \tilde{f} dv$, the zeroth order term in the inner field, f^i, contributes to the first order term of ϕ. Substituting (121) into (123), we conclude $n = \pm 1$, and

$$\phi_0 = \tilde{\phi}_0(1,t)\, e^{ix} + \tilde{\phi}_0(-1,t)\, e^{-ix} \,. \tag{124}$$

That is, the single-wave spatial structure of the potential arises naturally from the leading order balance in (118). To determine the time evolution of ϕ_0, we substitute (121) into (123) and get

$$\gamma \frac{d\tilde{\phi}_0}{dt} + i\lambda\,\tilde{\phi}_0 = \frac{i}{2\pi}\int_{-\infty}^{\infty}\int_0^{2\pi} e^{-ix} f_0^i\, dx dv \ . \qquad (125)$$

Equations (120) and (125) form a closed, self-consistent system of equations describing the nonlinearly saturated state in the vicinity of the stationary inflection point. Defining $\sigma \equiv \mathrm{sign}\,(\gamma)$, $\ell \equiv \mathrm{sign}\,(\Lambda)$ where "sign" denotes the sign, and $\mathrm{sign}\,(0) = 0$, and rescaling as $v \to |\lambda/\gamma|\, v$, $t \to |\gamma/\lambda|\, t$, $f \equiv |\gamma|/\lambda^2\, f_0^i$, $a \equiv (\gamma/\lambda)^2\ \tilde{\phi}_0$, (120) and (125) become the SWM in (6)–(8).

References

1. M. Antoni, Y. Elskens, and D. F. Escande, Phys. Plasmas **5**, 841 (1998)
2. M. Antoni and S. Ruffo, Phys. Rev. E **52**, 2361 (1995)
3. N. J. Balmforth, D. del-Castillo-Negrete, and W. R. Young, J. Fluid Mech., **333**, 197-230 (1997).
4. N. J. Balmforth, S. L. Smith and W. R. Young, J. Fluid Mech., **10**, 95-133 (2001).
5. N. J. Balmforth and C. Piccolo, J. Fluid Mech. **449**, 85-114 (2001).
6. J. Barré, T. Dauxois, and S. Ruffo, Physica A **295**, 254-60 (2001).
7. H. L. Berk, B. N. Breizman, et. al., Phys. Plasmas **6**, 3102 (1999).
8. S. M. Churilov, and G. Shukhman, Astrophys. Fluid Dynamics, **38**, 145 (1987)
9. J. D. Crawford and A. Jayaraman, Phys. Plasmas **6**, (1999).
10. D. del-Castillo-Negrete, Phys. Lett. A **241**, 99 (1998).
11. D. del-Castillo-Negrete, Phys. Plasmas **5**, 3886 (1998).
12. D. del-Castillo-Negrete, Physica A **280**, 10 (2000).
13. D. del-Castillo-Negrete, Chaos **10**, 75 (2000).
14. D. del-Castillo-Negrete and M.-C. Firpo, Chaos **12**, 496 (2002).
15. D. del-Castillo-Negrete, J. M. Greene and P. J. Morrison, Physica D **91**, 1 (1996).
16. T. Dauxois, V. Latora, A. Rapisarda, S. Ruffo and A. Torcini, *The Hamiltonian Mean Field Model: from Dynamics to Statistical Mechanics and back*, in "Dynamics and Thermodynamics of Systems with Long-Range Interactions", T. Dauxois, S. Ruffo, E. Arimondo, M. Wilkens Eds., Lecture Notes in Physics Vol. 602, Springer (2002), (in this volume)
17. D. F. Escande, S. Zekri and Y. Elskens, Phys. Plasmas **3**, 3534 (1996).
18. D. Farina, F. Casagrande, U. Colombo, R. Pozzoli, Phys. Rev. E, **49**, 1603 (1994).
19. M.-C. Firpo and Y. Elskens, J. Stat. Phys. **93**, 193 (1998)
20. M.-C. Firpo and Y. Elskens, Phys. Rev. Lett. **84**, 3318 (2000).
21. M. E. Goldstein, and L. S. Hultgren, J. Fluid Mech. **197**, 295 (1988).
22. E. J. Hinch, *Perturbation Methods*, Cambridge University Press, New York, (1991).
23. L. Landau, J. Phys. USSR (Moscow) **10**, 85-93 (1946).
24. V. Latora, A. Rapisarda, and S. Ruffo, Physica D, **131**, 38 (1999).
25. A. J. Lichtenberg, and M.A. Lieberman, *Regular and stochastic motion*, Springer-Verlag, second edition (1983).
26. D. R. Nicholson, *Introduction to Plasma Theory*, Wiley (1983).
27. T. M. O'Neil, J. H. Winfrey and J. H. Malmberg, Phys. Fluids **14**, 1204 (1971).
28. I. N. Onischenko *et al*, JETP Lett. **12**, 281 (1970).
29. J. L. Tennyson, J. D. Meiss and P. J. Morrison, Physica D **71** , 1 (1994).
30. H. Weitzner, Phys. Fluids **6**, 1123-1127 (1963).

Kinetic Theory for Plasmas and Wave-Particle Hamiltonian Dynamics

Yves Elskens

UMR 6633 CNRS–Université de Provence, Equipe turbulence plasma, case 321, campus Saint-Jérôme, F-13397 Marseille cedex 20 elskens@newsup.univ-mrs.fr

Abstract. The plasma limit is characterized by the fact that a particle interacts with many partners. The dynamics leads to the Vlasov equation in the mean-field limit. Collective behaviour in the $\mathcal{N}$-body system is naturally described by $M \ll \mathcal{N}$ wavelike degrees of freedom, which behave as harmonic oscillator interacting with a population of $N \ll \mathcal{N}$ particles. The self-consistent hamiltonian dynamics with M waves is also mean-field for the N particles, and a Vlasov equation applies for $N \to \infty$. The careful understanding of the limit $N \to \infty$ requires taking into account a continuous spectrum of singular excitations of van Kampen-Case type. For $M \to \infty$, the motion of a single particle obeys a Fokker-Planck equation in the strong resonance overlap limit, and the full particle-wave system obeys the quasilinear evolution equations in the weak turbulence regime.

1 The Plasma Limit

Plasmas and self-gravitating systems are dominated by long-range interactions. In a non-relativistic, non-quantum description, the prototype of these systems is described by the hamiltonian

$$H = K + V \tag{1}$$

$$K = \sum_{r=1}^{N} \frac{p_r^2}{2m} \tag{2}$$

$$V = \frac{1}{2} \sum_{i \neq r} q^2 \phi(x_i - x_r) \tag{3}$$

where all $N \gg 1$ particles have the same mass m, the same charge q and interact with the two-body potential ϕ. Then the kinetic energy is the sum of N terms while the potential energy is the sum of $N(N-1)/2$ terms, so that the scaling of H with N obviously depends on how individual terms in the sums depend on N. Let $\alpha \to \infty$ control the scaling and assume that

$$\frac{N}{N_0} = \frac{m_0}{m} = \frac{q_0}{q} = \alpha \tag{4}$$

with some fixed N_0, m_0, q_0. Moreover, introduce the velocity

$$v_r = p_r/m \tag{5}$$

so that, if $m \to 0$, the momentum associated with an O(1) velocity scales like $1/\alpha \to 0$. The dynamical equations generated by (1) read

$$\dot{x}_r = v_r \tag{6}$$

$$\dot{v}_r = -\frac{q}{m}\partial_x W(x_r) \tag{7}$$

where we introduce the total potential

$$W(x) = q\sum_r \phi(x - x_r). \tag{8}$$

With the joint scaling (4), the dynamical equations are formally independent from α. More precisely, if ϕ is sufficiently regular (e.g. twice continuously derivable, with bounded second derivatives), then the dynamics are invariant under the substitution of each particle (x_i, v_i) by several smaller particles, say α with the same data (x_r, v_r) for $(i-1)\alpha + 1 \leq r \leq i\alpha$.

In 3 space dimensions, for Coulomb interaction, $\phi(x) = (4\pi\epsilon_0|x|)^{-1}$, while for Newton gravitation $q = m$ and $\phi(x) = -G|x|^{-1}$. These potentials are singular at $|x| = 0$, and the sharp control of the above procedure is more subtle. However, they provide the main inspiration to major work on these long-range systems. In particular, the characteristic time scale for the collective effects in a plasma with spatial density n_0 is the reciprocal of the plasma frequency

$$\omega_{\mathrm{p}} = \sqrt{\frac{q^2 n_0}{m\epsilon_0}} \tag{9}$$

which is independent of α.

Given a characteristic thermal velocity $v_T = \sqrt{k_{\mathrm{B}}T/m}$, this frequency also defines a length scale

$$\lambda_{\mathrm{D}} = v_T/\omega_{\mathrm{p}} \tag{10}$$

known as the Debye length for the plasma. Phenomena on a characteristic spatial scale larger than λ_{D} are not sensitive to the granular[1] nature of the system but mainly to its collective aspects. The large parameter

$$N_{\mathrm{D}} = n_0\lambda_{\mathrm{D}}^3 \propto \alpha \tag{11}$$

is the plasma parameter, which counts the effective number of particles with which a single body interacts. As N_{D} is also the ratio between the thermal energy $k_{\mathrm{B}}T$ of a particle and its typical potential energy $q^2/(4\pi\epsilon_0\lambda_{\mathrm{D}})$ in the pair interaction, the plasma limit is a weak coupling limit.

The scaling (4) and regularity of the limit equilibrium are signatures of plasma-like behaviour, in contradistinction to gases or liquids, where the number of interaction partners remains small in the limit $N \to \infty$ (6 in two space dimensions, 12 in three space dimensions for hard spheres), or to systems defined on a

[1] In this contribution, granular must be understood as referring to the finiteness of the number of interacting bodies – not to the hard bodies undergoing inelastic collisions.

lattice (where the coordination number is kept constant in the thermodynamic limit). This plasma-like behaviour results in the validity of mean-field limits, where the sums as (8) are turned into integrals (à la Riemann, or better à la Lebesgue).

In the gravitational case, the attraction makes ω_p^2 negative, and one calls $|\omega_\mathrm{p}|$ and $v_T/|\omega_\mathrm{p}|$ respectively the Jeans frequency and length. We refer to Kiessling [18] for a discussion of subtleties related to their definition.

2 Equilibrium and Collective Behaviour

The plasma limit $N_\mathrm{D} \to \infty$ ensures that Gibbs equilibrium states are properly described by the mean-field distributions. This is most easily seen for systems with smooth interactions such as the hamiltonian XY model [8,22], with $\phi(y) = \eta \cos y$ (and $\eta = \pm 1$ makes the interaction either attractive or repulsive).

The microscopic state of the system is described by the empirical distribution

$$f^{(N)}(x,p) = \frac{1}{N} \sum_r \delta(x - x_r)\delta(p - p_r) \tag{12}$$

over the single particle phase space, i.e. Boltzmann μ-space. This distribution is not a statistical object but merely a convenient tool for expressing the properties of the N-body system in such a way that the limit $N \to \infty$ can be taken. In particular, the force on a particle reads $\dot{p} = qE(x)$, with the field

$$E(x) = - \int \partial_x \phi(x - x') f(x', p') \mathrm{d}x' \mathrm{d}p', \tag{13}$$

provided that ϕ is smooth enough (at least C^2) for this integral to be well-defined. The lack of smoothness of ϕ for Coulomb and Newton potentials is a major difficulty in the definition of $E(x)$, which is classically solved by regularizing ϕ at short range, e.g. replacing $|x|^{-1}$ by $(|x|^2 + a^2)^{-1/2}$ (a more radical solution would be calling on quantum theory to get rid of $\delta(x)$ altogether...).

The probabilistic interpretation of f applies to the sampling of the particle distribution. Then one uses f to draw at random one particle from the system – but this is not the same (in principle) as drawing one realisation of the many-body system from the Gibbs equilibrium distribution.

3 Vlasov Equation and Its Linearisation

In the mean-field limit, the evolution of the distribution f over μ-space is easily deduced from the microscopic, finite-N dynamics. It obeys the Vlasov equation [3], formally deduced from (6)-(7)-(8) as

$$\partial_t f + (p/m) \cdot \partial_x f + qE(x) \cdot \partial_p f = 0, \tag{14}$$

coupled with (13). Note that (14) is quadratic with respect to f, because E is linear in f by (13). Thus the Vlasov equation might develop singularities after a finite time (just as $\dot{u} - u^2 = 0$ does for any initial data $u_0 > 0$).

Establishing rigorously that the N-body dynamics converges to the system (13)-(14) requires that the Cauchy problem for the latter system admit a unique solution, not only for smooth initial data f but also for distributions like $f^{(N)}$. This can be ensured provided that the microscopic dynamics stems from a smooth interaction with a bounded potential. A key estimate in this argument is in the form

$$\|f_1(.,.,t) - f_2(.,.,t)\| \leq \|f_1(.,.,0) - f_2(.,.,0)\| \mathrm{e}^{c|t|} \tag{15}$$

where $f_i(.,.,t)$, $i = 1, 2$, are two solutions of (13)-(14) and c is a constant determined by initial data $f_i(.,.,0)$. In essence, c is an upper bound on the largest Lyapunov exponent of the dynamics (6)-(7). In (15) the norm $\|.\|$ is such that $\|\delta(x-a)\| = 1$ and $\|\delta(x-a) - \delta(x-b)\| = \min(1, |a-b|)$.

The mean-field nature of the coupling is crucial in turning sums like (8) into integrals. If a particle of interest keeps interacting with only a few other ones in the limit $N \to \infty$, the limiting kinetic theory may be the Boltzmann equation, which has dramatically different properties [3,9,23]. In particular, (14) conserves the Casimir invariants $C_n[f] = \int [f(x,p)]^n \mathrm{d}x\mathrm{d}p$ and the entropy $S[f] = \int f(x,p) \ln[f(x,p)/a] \mathrm{d}x\mathrm{d}p$ (with a normalizing constant a) – whereas the Boltzmann equation does not preserve these quantities.

The complete proof of the Vlasov equation is given by Spohn [23] for the case of smooth interactions such as for the XY model. The non-relativistic Coulomb case is more difficult but a similar result holds [19,23]. As of February 2002, no rigorous derivation has been published yet of the Vlasov-Maxwell system from microscopic dynamics of point particles interacting with electromagnetic fields. This is because the Lorentz-invariant interaction requires any model of an extended particle to be covariant, which is more demanding than the Galileo-invariant model of solid bodies. However, the recent construction of such models [2] brings a relativistic foundation for the Vlasov equation coupled with field equations within the reach of analysis.

The proof of the Vlasov equation amounts to a law of large numbers, as the resulting fields $E(x,t)$ are deterministic functions. Finite-N corrections generate microscopic fluctuations around these deterministic expressions. The most straightforward corrections stem from the propagation in time of initial "fluctuations": the latter being small evolve according to the linearized Vlasov equation, and the expression (8) yields initial data as a sum of $N \to \infty$ identically distributed random function, independent in the limit, so that initial data are gaussian. Again, the proof of this fluctuation theorem requires smooth interactions (which excludes Coulomb potential) [6,23]. However, for one-dimensional systems with periodic boundary conditions, the finite-N effects lead to essentially the same behaviour for the Coulomb interaction (with $\phi(x) = |x|$ for $-\pi \leq x \leq \pi$) and for its truncations to a finite number of Fourier components – including the hamiltonian XY model [1,20,21].

Note that the time scale over which these fluctuations do remain small is finite. Indeed, typically the solution to the linearized Vlasov equation must be expected to grow exponentially, soon entering the nonlinear regime of the perturbations. For a rough estimate, take f_1 as a finite-N approximation to a smooth f_2, so that $\|f_1(.,.,0) - f_2(.,.,0)\| \sim N^{-1/2}$, while c is typically estimated independently from N. Then the time τ_N over which (15) remains small scales like $(2c)^{-1} \ln N$ (which implies that limits $N \to \infty$ and $t \to \infty$ need not commute).

This regime must be analysed with new tools, taking into account the actual motion of each particle. This is a regime where the chaotic dynamics may play a major role, i.e. a turbulent regime.

4 Wave-Particle Self-consistent Dynamics

For simplicity we now specialize the models to a single space dimension, with 2π-periodic boundary conditions.

An equilibrium state of the N-body system may be hard to construct in general. However, in the limit $N \to \infty$ the Vlasov equation may admit stationary solutions with vanishing field. The simplest instance is found for the XY model, for which it suffices to group particles (labeled by $1 \leq r \leq N$) in b beams of $N_s \geq 2$ ($1 \leq s \leq b$, $\sum_s N_s = N$) with velocity $v_r = v_s$ and positions $x_r = \phi_s + v_s t + x_{r0} \bmod (2\pi)$, with $x_{r0} = 2\pi r / N_s$. For a Coulomb plasma, this "multibeam" type of particle distribution can make the long-wavelength components of the field vanish, but the short-wavelength spatial Fourier components are merely small but nonzero for finite N [10]. This small intensity regime is typical of near-thermal equilibrium [1,10,20,21].

Small perturbations around such an equilibrium can be expressed in terms of eigenmodes. In plasmas at equilibrium temperature T, these are Langmuir waves, and their dispersion relation established by Bohm and Gross [10] reads

$$\omega^2 = \omega_{\mathrm{p}}^2 + 3 v_T^2 k^2 + ... \tag{16}$$

for the spatial Fourier component of the Coulomb field with wavevector k. The total field in this small perturbation regime reads $E(x,t) = \sum_n A_n \mathrm{e}^{\mathrm{i}(k_n x - \omega_n t)}$.

A Langmuir wave propagates at phase velocity $v = \omega_{\mathrm{p}}/k + ...$ For a particle with velocity p, this wave oscillates rapidly and induces a small perturbation of its motion provided

$$|p - v| \gtrsim 2 |A|^{1/2} \tag{17}$$

(with $m = q = 1$). Then the averaging theory for fast-oscillating perturbations of the motion applies. This holds well for long-wavelength waves, as $\lim_{k \to 0} \omega_{\mathrm{p}}/|k| = \infty$. Then criterion (17) is fulfilled for all particles except possibly a few very rapid ones, which we may call the tail of the particles velocity distribution. Physically, the cross-over wavenumber around which Langmuir waves become resonant with many particles is the Debye wavenumber $k_{\mathrm{D}} = \lambda_{\mathrm{D}}^{-1}$. Thus physically the Debye length marks the scale above which the plasma excitations are collective and below which granular effects may be important.

It is then possible to reformulate the dynamics of the many-body system as a reduced dynamics, with two types of physical objects interacting, M Langmuir waves and N (possibly resonant)[2] particles. The waves are characterized by their intensity I_j and phase θ_j; they are spatially delocalized (hence they may interact with many particles) and have a natural phase velocity ω_{j0}/k_j (where ω_{j0} is the Bohm-Gross eigenfrequency), favouring interaction with particles at nearby velocity. The particles are characterized by their position x_r and momentum p_r; they are spatially localized and typically interact with many waves, favouring those with nearby phase velocity. The effective dynamics is generated by the self-consistent wave-particle hamiltonian

$$H_{\text{sc}} = \sum_{r=1}^{N} \frac{p_r^2}{2} + \sum_{j=1}^{M} \omega_{j0} I_j + \sum_{r=1}^{N} \sum_{j=1}^{M} \lambda_j \sqrt{2I_j} \cos(k_j x_r - \theta_j), \tag{18}$$

with coupling constants λ_j determined by the original $\mathcal{N}$-body system. The first term generates the free motion of particles, the second one generates the waves eigenoscillations, and the third one couples particles with the sinusoidal potential of the waves. Feedback from particles on the waves is granted by the hamiltonian nature of the model, which also makes the particles appear as a source for the wave evolution. This model is a paradigm of many-body, self-consistent dynamics, and similar models are encountered in various fluid and plasma physics applications [7].

Explicitly, the dynamics generated by (18) reads

$$\dot{x}_r = p_r \tag{19}$$

$$\dot{p}_r = \Re\Big(\sum_{j=1}^{M} \mathrm{i} k_j \lambda_j Z_j \mathrm{e}^{\mathrm{i} k_j x_r}\Big) \tag{20}$$

$$\dot{Z}_j = -\mathrm{i}\omega_{j0} Z_j + \mathrm{i}\lambda_j \sum_{r=1}^{N} \mathrm{e}^{-\mathrm{i} k_j x_r}, \tag{21}$$

where $Z_j = X_j + \mathrm{i}Y_j = \sqrt{2I_j}\mathrm{e}^{-\mathrm{i}\theta_j}$. It is convenient to define the complex wave envelope $z_j = Z_j \exp(\mathrm{i}\omega_{j0} t)$. The evolution equation for z_j is

$$\dot{z}_j = \mathrm{i}\lambda_j \sum_{l=1}^{N} \mathrm{e}^{\mathrm{i}(\omega_{j0} t - k_j x_l)}. \tag{22}$$

The dynamics obviously preserves both the total energy H_{sc} and total momentum $P = \sum_r p_r + \sum_j k_j I_j$.

Moreover, the most important contribution to $\dot{p}_r$ stems from waves with $\omega_{j0} \sim k_j p_r$, and the most important contribution to $\dot{z}_j$ stems from particles with $p_r \sim \omega_{j0}/k_j$. These resonant wave-particle interactions essentially exchange

[2] With respect to the initial system with, say $\mathcal{N}$ particles, the reduced dynamics has $M \ll \mathcal{N}$ and $N \ll \mathcal{N}$.

momentum between particles and waves at almost the same velocity, so that momentum is conserved locally on the average [10].

The limit $N \to \infty$ for this system is also a mean-field one. In this limit, the natural scaling for I_j is $I_j = N\iota_j$. For given, finite M, the interaction is smooth, and the kinetic equation for $f^{(N)}$ is the Vlasov equation, coupled now to wave envelope evolution equations (22) [15]. It is worth noting that the Gibbs equilibrium of the single wave case ($M = 1$), which is specially relevant also to the XY model, exhibits a second order phase transition [16] – though the Coulomb plasma does not : this fact stresses the importance of identifying the relevant physical variables of the dynamics to properly formulate the effective thermodynamics. Finite-N effects must be considered carefully, because the dynamics amplify them through instabilities [13].

The microscopic reason why the Gibbs ensemble description of the system accounts of the behaviour of a single realisation is the chaotic motion of a particle in a many-degrees-of-freedom dynamics, which provides an "effective ergodicity" [14].

5 Linear Perturbations and Van Kampen-Case Modes

For finite M and large enough finite N, the self consistent wave-particle dynamics admits also vanishing-field solutions of the multibeam type, say with b beams. Their small perturbations may again be analysed in terms of eigenmodes. Two types of excitations are identified :

- ballistic-like excitations, where $I_j = 0$ for all times : the number of such excitations is $2N - 4Mb$;
- wave-like excitations, where particles and one wave oscillate at one frequency ω solution of a dispersion relation, which is recovered in the limit $N \to \infty$ from the Vlasov equation : the number of such excitations is $2M(2b + 1)$.

However, the wave-like excitations do not all reduce to genuine smooth perturbations of a smooth reference distribution function in the kinetic limit. Actually most of them turn out to involve particles of beams with velocities in a narrow range. Such excitations are termed van Kampen-Case eigenmodes in the kinetic limit [10,17]. They play a fundamental role in ensuring that the smooth functions of time, analytic in complex plane, describing the finite-number-of-degrees-of-freedom dynamics, tend to the functions typical of initial-value problems of kinetic theory, which behave like $e^{\gamma|t|}$ and are not differentiable at $t = 0$ [10].

In particular, the van Kampen-Case-like eigenmodes generate Landau damping by phase mixing, thanks to the Fourier-transform identity for $\gamma < 0$

$$e^{\gamma|t|} = \frac{1}{\pi} \int \frac{|\gamma|}{u^2 + \gamma^2} du \tag{23}$$

where the use of a continuous spectrum $u \in]-\infty, \infty[$ is crucial[3]. Landau damping is a fundamental process in plasmas, where an excitation damps in a non-

[3] The continuous spectrum cannot exist for a finite number of degrees of freedom, but is natural for partial differential equations.

dissipative system : a purely hamiltonian description of the process cannot account for damping in terms of a single eigenmode, because the reversibility of hamiltonian dynamics requires that the existence of an eigenmode with eigenvalue $\sigma = \gamma + \mathrm{i}\omega$ ($\gamma \neq 0$) implies the existence of an eigenmode with opposite eigenvalue $-\sigma$; hence a reference solution of hamiltonian ordinary differential equations cannot be attractive (in the sense of almost all neighbouring solutions converging towards it). Damping of typical perturbations in hamiltonian dynamics requires a continuous spectrum, as provided by kinetic theory.

Moreover, van Kampen-Case-like eigenmodes are also essential to the description of linear perturbations of unstable reference states, with the identity for $\gamma > 0$

$$\mathrm{e}^{\gamma|t|} = \mathrm{e}^{\gamma t} + \mathrm{e}^{-\gamma t} - \frac{1}{\pi}\int \frac{|\gamma|}{u^2+\gamma^2}\mathrm{d}u \tag{24}$$

providing a representation of the non-differentiable solution to the kinetic equation in terms of two analytic functions of time (corresponding to genuine eigenmodes) and a continuum of oscillating functions of t. This decomposition applies to the Landau instability, which kinetic theory describes by formulæ similar to Landau damping – though the eigenmode counting and underlying dynamics are quite different [10].

6 Chaos and Transport

The above linear theory is encouraging, but the actual regimes of interest involve nonlinear dynamics. Indeed, only nonlinear processes can halt the growth of instabilities. As the envelope equation (22) is linear, with a source, the nonlinearity will stem essentially from the particle equations.

For the particles, the motion in the field of a single wave amounts to the pendulum problem, and the nonlinear regime corresponds to having both sides of (17) with the same order of magnitude. If there is a single mode in the dynamics (i.e. $M = 1$), the resulting regime will be one where particles bounce more or less coherently in the wave potential well [10,13].

The regime where a particle interacts with many waves has a different nature. This is the weak turbulence regime. The chaoticity of a particle motion may be characterized by the number of waves for which condition (17) is violated. Assume that waves are labeled according to increasing phase velocity and let $\Delta v_j = v_{j+1} - v_j$. Considering a wave j with phase velocity v_j and intensity $I_j = A_j^2/(2\lambda_j^2)$ and the next wave $j+1$, their resonance overlap parameter is defined as

$$s_{\mathrm{ov}\,j} = 2\frac{\sqrt{A_j} + \sqrt{A_{j+1}}}{|\Delta v_j|} \tag{25}$$

so that a particle typically interacts with $\mathrm{O}(s)$ waves. If the waves have amplitudes A_j of the same order of magnitude and independent phases, the force they apply on a particle is, in first approximation for $s_{\mathrm{ov}} \gg 1$, a white noise. Therefore one expects the particle velocity to diffuse, in such a way that

$$\langle [p(t) - p(0)]^2 \rangle = 2D_{\mathrm{QL}}t \tag{26}$$

with the so-called quasilinear diffusion coefficient $D_{\mathrm{QL}} = \pi A_j^2 k_j / (2|\Delta v_j|)$. The particle position distribution then spreads so that $\langle [q(t) - q(0) \quad p(0)t]^2 \rangle = \frac{2}{3} D_{\mathrm{QL}} t^3$.

This argument cannot hold for times shorter than the field correlation time $\tau_{\mathrm{c}} = 1/[\max(\omega_{j0}) - \min(\omega_{j0})]$, but it seems reasonable over the time range $\tau_{\mathrm{c}} \lesssim t \lesssim \tau_{\mathrm{spread}}$, where the upper bound corresponds to $\langle [q(t) - q(0) - p(0)t]^2 \rangle = \mathrm{O}(1)$. However, the force on the particle is not a stochastic process. Rather, it is given by (20), which is deterministic. Therefore, it may exhibit correlations in time, forbidding the Markov description assumed in the white noise estimate.

To account properly of the time evolution of the particle, one shows that, in the limit of large resonance overlap $s_{\mathrm{ov}} \to \infty$, statistical properties of the dynamics are universal [4,5]. Then, for times longer than τ_{spread}, the very chaotic nature of the dynamics forbids the correlations and ensures that the particle velocity follows the quasilinear behaviour to first approximation [10,11]. Among the key properties of the dynamics enabling this result stands the fact that the particle is only weakly influenced by any set of a few waves – and its trajectory does not enable it to identify accurately many waves. Hence the particle motion appears on the long timescales to be governed by the Fokker-Planck equation

$$\partial_t f = \partial_p J(p) \tag{27}$$

where $J(p) = D_{\mathrm{QL}} \partial_p f$. In contrast to the Vlasov equation, the Fokker-Planck equation describes a microscopically stochastic process. Because of the "large" accelerations implied by the strong overlap regime ($M \gg 1$, $A \gg \Delta v^2$), the position variable has been integrated out from f in the Fokker-Planck equation, while it is a relevant argument for the distribution function in the Vlasov equation.

Moreover, the full N-particle, M-wave self-consistent dynamics lends itself to the same estimates, for several reasons : the weak correlation of any wave with any particle, the even weaker correlation between any two waves or any two particles, the local conservation on the average of wave-particle momentum, and the fact that the diffusion is quasilinear in the non self-consistent case for both the initial non-chaotic regime, and for the chaotic regime in the limit of a continuous spectrum (strong resonance overlap).

This leads to the quasilinear set of coupled equations for the particle velocity distribution function $f(p)$ and the rescaled wave intensity spectrum $\psi(v)$ interpolating the function $\psi(v_j) = k_j I_j / (N \Delta v_j)$,

$$\partial_t f = \partial_p \partial_t \psi \tag{28}$$

$$\partial_t \psi = \pi N \lambda^2 \psi \partial_p f - F_{\mathrm{c}} f \tag{29}$$

where the friction coefficient $F_{\mathrm{c}}(v_j) = -\pi \lambda_j^2 / |\Delta v_j|$ accounts for wave spontaneous emission by particles when they accelerate, in agreement with local conservation of momentum on the average [10,12].

7 Final Remarks

The main property on which our arguments rest is the mean-field nature of the couplings. This is a natural scaling for plasmas, where a particle interacts with many partners, in contrast with gases or liquids. Two types of kinetic equations may be derived.

The Vlasov equation is established systematically from the $N \to \infty$ limit, provided that the interaction is smooth. This equation preserves essentially hamiltonian properties of the dynamics (witnessed by the Casimir invariants).

The Fokker-Planck equation and the quasilinear system of equations are obtained in a strongly chaotic regime, characteristic of turbulent motion. Though it may seem surprising to establish the basic equations of turbulence for a plasma rather than for a fluid with short-range interactions, the proof is easier in the plasma case, thanks to the existence of two types of physical actors, each of which interacting only with the other species, thereby preventing the formation of strong correlations [10].

Acknowledgements

It is a pleasure to thank E. Arimondo, Th. Dauxois, S. Ruffo and M. Wilkens for organizing this school and smoothly running it, and CNRS, université J. Fourier and the European science foundation for supporting it. I am grateful to lecturers for their teaching, and to participants for fruitful and pleasant discussions. The contents of these notes benefited from discussions with F. Doveil, D.F. Escande, M-C. Firpo and M. K-H. Kiessling.

References

1. M. Antoni, Y. Elskens and C. Sandoz: Weak turbulence and structure evolution in N-body hamiltonian systems with long range force, *Phys. Rev. E* **57** (1998) 5347-5357
2. W. Appel and M.K-H. Kiessling: Mass and spin renormalization in Lorentz electrodynamics, *Ann. Phys. (N.Y.)* **289** (2001) 24-83
3. R. Balescu: *Statistical dynamics – Matter out of equilibrium* (Imperial college press, London, 1997)
4. D. Bénisti and D.F. Escande: Origin of diffusion in hamiltonian dynamics *Phys. Plasmas* **4** (1997) 1576-1581
5. D. Bénisti and D.F. Escande: Finite range of large perturbations in hamiltonian dynamics *J. Stat. Phys.* **92** (1998) 909-972
6. W. Braun and K. Hepp: The Vlasov dynamics and its fluctuations in the $1/N$ limit of interacting classical particles, *Commun. Math. Phys.* **56** (1977) 101-113
7. D. del-Castillo-Negrete and M-C. Firpo: Coherent structures and self-consistent transport in a mean field hamiltonian model *Chaos* **12** (2002) 496-507
8. T. Dauxois, V. Latora, A. Rapisarda, S. Ruffo and A. Torcini: *The hamiltonian mean field model : from dynamics to statistical mechanics and back* in "Dynamics and Thermodynamics of Systems with Long Range Interactions", T. Dauxois, S. Ruffo, E. Arimondo, M. Wilkens Eds., Lecture Notes in Physics Vol. 602, Springer (2002), (in this volume)

9. Y. Elskens: Dynamical and kinetic aspects of collisions *Singularities in gravitational systems – Applications to chaotic transport in the solar system* D. Benest and C. Froeschlé eds *Lect. Notes Phys.* (Springer, Berlin, 2002)
10. Y. Elskens and D.F. Escande: *Microscopic dynamics of plasmas and chaos* (IoP Publishing, Bristol, 2002)
11. D.F. Escande and Y. Elskens: Quasilinear diffusion for the chaotic motion of a particle in a set of longitudinal waves *Acta Phys. Pol. B* **33** (2002) 1073-1084
12. D.F. Escande and Y. Elskens: Proof of quasilinear equations in the chaotic regime of the weak warm beam instability (preprint)
13. M-C. Firpo, F. Doveil, Y. Elskens, P. Bertrand, M. Poleni and D. Guyomarc'h: Long-time discrete particle effects versus kinetic theory in the self-consistent single-wave model *Phys. Rev. E* **64** (2001) 026407
14. M-C. Firpo and F. Doveil: Velocity width of the resonant domain in wave-particle interaction *Phys. Rev. E* **65** (2002) 016411
15. M-C. Firpo and Y. Elskens: Kinetic limit of N-body description of wave-particle self-consistent interaction *J. Stat. Phys.* **93** (1998) 193-209
16. M-C. Firpo and Y. Elskens: Phase transition in the collisionless damping regime for wave-particle interaction *Phys. Rev. Lett.* **84** (2000) 3318-3321
17. N.G. van Kampen and B.U. Felderhof: *Theoretical methods in plasma physics* (North-Holland, Amsterdam, 1967)
18. M. K-H. Kiessling: Mathematical vindication of the "Jeans swindle" `arXiv:astro-ph/9910247`
19. H. Neunzert: An introduction to the nonlinear Boltzmann-Vlasov equation *Kinetic theories and the Boltzmann equation* C. Cercignani ed. *Lect. Notes Math.* **1048** (Springer, Berlin, 1984) 60-110
20. J.L. Rouet and M.R. Feix: Relaxation for a one-dimensional plasma : test particles versus global distribution behaviour *Phys. Fluids B* **3** (1991) 1830-1834
21. J.L. Rouet and M.R. Feix: Computer experiments on dynamical cloud and space time fluctuations in one-dimensional meta-equilibrium plasmas *Phys. Plasmas* **3** (1996) 2538-2545
22. S. Ruffo: Hamiltonian dynamics and phase transition *Transport, chaos and plasma physics* S. Benkadda, F. Doveil and Y. Elskens eds (World Scientific, Singapore, 1994) 114-119
23. H. Spohn: *Large scale dynamics of interacting particles* (Springer, New York, 1991)

Emergence of Fractal Clusters in Sequential Adsorption Processes

Ofer Biham and Ofer Malcai

Racah Institute of Physics, The Hebrew University, Jerusalem 91904, Israel

Abstract. The emergence of fractal clusters in a model of sequential adsorption of particles on a surface is studied using theoretical and numerical techniques. A phase transition that separates between a nearly homogeneous phase and a phase which exhibits strong clustering is identified. The order parameter of the transition is examined and the fractal dimension of the clusters at the transition is calculated. While for finite systems the transition is smoothed out, in the infinite system limit it is a sharp second order transition.

1 Introduction

Clustering phenomena have been observed in recent years in a great variety of natural and social systems over a vast range of scales. These include galaxy clusters dominating the large scale distribution of mass in the universe [1] as well as dynamically formed clusters in dissipative granular gas [2,3]. On microscopic scales clusters of atoms (also called islands) as well as clusters of atomic steps are commonly observed on surfaces [4,5]. Examples from social systems include the human population distribution on the surface of Earth which is dominated by dynamically evolving clusters called cities [6–8], while the mobility in these cities and on the highways connecting them is dominated by clusters of cars, called traffic jams [9]. A common feature which appears in these systems is the tendency of the objects to form clusters of high density (often of fractal shape), surrounded by low density areas or voids. The phenomenon of cluster formation is therefore generic in spite of the fact that the pattern forming dynamical processes may vary substantially from system to system.

In surface adsorption systems, pattern formation is often dominated by particle diffusion and sticking [10], giving rise to the formation of diffusion limited aggregation (DLA) like fractal clusters [11]. In a different class of models, describing random sequential adsorption (RSA) processes [12] particles stick at the point in which they hit the surface and diffusion is suppressed. The pattern formation mechanism in these systems is dominated by the dependence of the sticking probability on the structure in the vicinity of the point in which the new particle hits the surface. In some of the RSA systems particles cannot overlap. These systems tend to reach a jamming limit, in which the sticking probability of a new particle vanishes [13–15], describing various packing and car parking problems of considerable practical importance. Models which allow overlap [16] or growth of more than one layer describe processes such as deposition of colloids, liquid crystals [17], polymers and fiber particles [18,19].

The Power-Law Sequential Adsorption (PLSA) model describes a variety of adsorption and clustering processes [20]. This model generates clustered structures, which resemble some of the experimentally observed morphologies. It exhibits a clustering phase transition that separates between a weakly correlated phase in which the adsorbed particles are distributed nearly homogeneously on the surface and a strongly correlated phase in which they form clusters.

In this paper we study the emergence of clusters in the PLSA model. We identify the order parameter associated with the clustering transition and examine the clusters geometry using fractal analysis. We show that finite size effects play an important role in smoothing the transition, while in the infinite system limit it develops into a sharp second order transition. The critical exponents that characterize the transition are calculated analytically, using some approximations. The fractal dimension at the transition is calculated both analytically and numerically. In case that overlaps between particles are allowed the order parameter is given by the overlapping areas between particles. In case that overlaps are suppressed, the order parameter, that coincides with the fractal dimension, is extracted from the distribution of the gap sizes between adjacent particles.

The paper is organized as follows. In Sect. 2 we introduce the model. The clustering transition is studied in Sect. 3 and the results are summarized in Sect. 4.

2 The Model

In the two dimensional (2D) PLSA model circular particles (or disks) of diameter d are randomly deposited on the surface one at a time. The deposition process starts from an initial state where there is one seed particle on the surface. The sticking probability $p(r)$ of a newly deposited particle depends on the distance r from its center to the center of the nearest particle which is already on the surface. This probability is given by

$$p(r) = \begin{cases} P_o : & r < d \\ (d/r)^\alpha : & r \geq d, \end{cases} \tag{1}$$

where the exponent $\alpha \geq 0$ is a parameter of the model. We distinguish between two cases: (a) particles are allowed to overlap, with a sticking probability $P_o = 1$; (b) particles are not allowed to overlap, and the sticking probability of a particle deposited on top of another particle, namely with $r < d$ is $P_o = 0$ (see (1)). In both cases the model exhibits higher sticking probabilities around regions that are already dense, possibly leading to clustering. The random deposition process is repeated until the desired number of particles, M, stick to the surface. Since the sticking probability is given by a power law function, which is of a long-range nature, it is expected that finite-size corrections will play an important role. Numerical simulations are done on a finite system of area $L \times L$, where $L/d \gg 1$. The coverage is given by $\eta = A/L^2$ where A is the total area covered by the particles. In case that overlaps are not allowed η is equal to $\eta_0 = \pi(d/2)^2 \cdot M/L^2$, while if overlaps are allowed $\eta \leq \eta_0$.

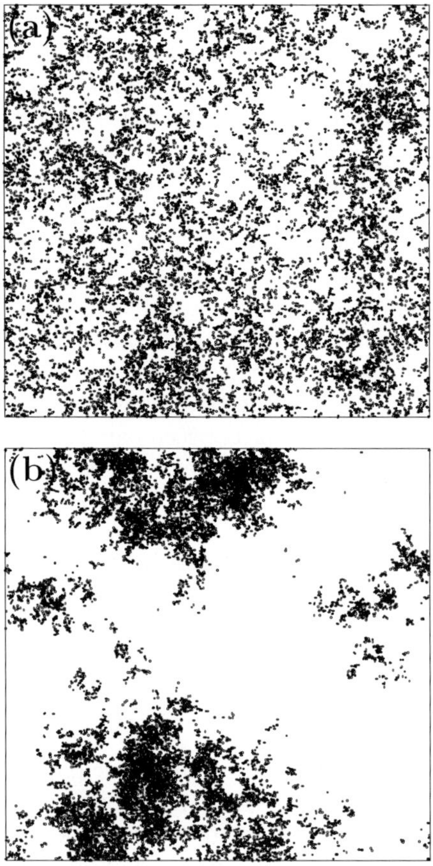

Fig. 1. Particles adsorbed on the surface of a unit square ($L = 1$), with periodic boundary conditions, according to the PLSA model for $M = 15,916$ particles of diameter $d = 0.004$ (namely, $\eta_0 = 0.2$): (a) for $\alpha = 1.6$ we observe a rather homogeneous distribution of particles that extends over the entire system; (b) for $\alpha = 2.4$ we observe a strongly clustered structure and vacant area elsewhere.

In the limit $\alpha \to 0$ the adsorption process is completely uncorrelated for $P_o = 1$, while for $P_o = 0$, some short range correlation is still maintained due to the suppressed overlaps. This limit was studied using fractal analysis, and the box-counting and Minkowski functions were calculated analytically [21]. For large values of α the adsorption process is strongly correlated since the sticking probability vanishes except for the immediate vicinity of existing particles. In this case only a single dense cluster is generated on the surface. The perimeter of this cluster grows slowly when new particles are deposited and stick near the cluster edge. The cluster also becomes more dense inside, particularly for $P_o = 1$ [18,19], while for $P_o = 0$ the core may reach the jamming limit [13–15]. In Fig. 1 we present two configurations of particles, randomly deposited and adsorbed according to (1) onto the unit square ($L = 1$) for $P_o = 1$.

For $\alpha = 1.6$ the particle distribution exhibits local density fluctuations but on larger scales it is rather homogeneous and extends over the entire system (Fig. 1a). For $\alpha = 2.4$ we observe a strongly clustered structure (Fig. 1b). This structure resembles the set of turning points of a Lévy flight random walker which exhibits a power law distribution of hopping distances [22]. In fact, a Lévy flight corresponds to the special case in which the sticking probability of the next deposited particle depends only on the position of the latest particle adsorbed on the surface according to (1), ignoring the previously adsorbed particles.

3 The Clustering Transition

3.1 Clustering in Two Dimensions

Consider a seed particle located at the origin of an infinite plane. Particles are randomly deposited one at a time according to the PLSA model (with $P_o = 1$) until one particle sticks to the surface. The probability that the distance r between the center of the first particle which sticks and the seed particle at the origin will be larger than some value r_f, where $r_f > d$, is given by:

$$P(r > r_f) = \frac{\displaystyle\int_{r_f}^{\infty} 2\pi r \left(\frac{d}{r}\right)^{\alpha} dr}{\displaystyle\int_0^d 2\pi r dr + \int_d^{\infty} 2\pi r \left(\frac{d}{r}\right)^{\alpha} dr}. \tag{2}$$

Performing the integration, one can verify that for $\alpha < 2$ the probability $P(r > r_f) = 1$ for any finite r_f. For $\alpha > 2$, on the other hand, this probability (for $r_f > d$) is given by

$$P(r > r_f) = \frac{2}{\alpha}\left(\frac{d}{r_f}\right)^{\alpha-2}. \tag{3}$$

Therefore, in the infinite system limit there is a transition at $\alpha_c = 2$. For $\alpha < \alpha_c$ the probability that the next particle will stick within a given distance from an existing seed particle or cluster is always zero. For $\alpha > \alpha_c$ the probability that the next particle will stick within a given finite distance r_f from the seed particle or the cluster edge can be made arbitrarily close to one, by an adjustment of r_f according to (3). In general, the value of α_c for which the clustering transition takes place is $\alpha_c = n$, where n is the space dimension.

In case that overlaps are allowed ($P_o = 1$), the order parameter of the clustering transition is

$$V = (\eta_0 - \eta)/\eta_0, \tag{4}$$

namely the fraction of the total area of the adsorbed particles which is lost due to overlapping particles. For $\alpha < \alpha_c$, in the infinite system limit, the probability that the next particle will stick at a distance $r < d$ from an existing particle vanishes. Therefore, the order parameter V which is proportional to the probability of overlap is zero. The order parameter V increases gradually above α_c.

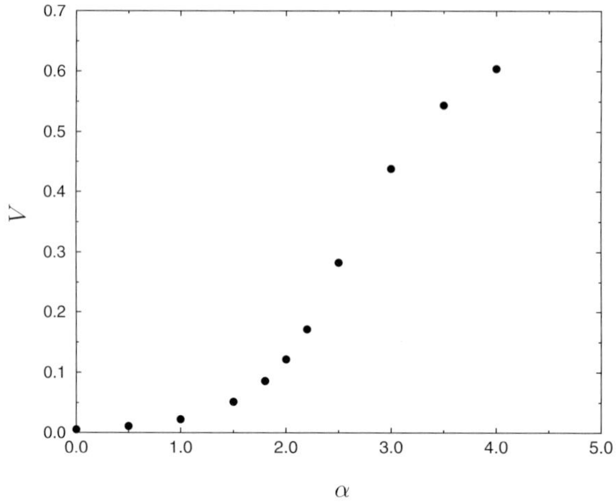

Fig. 2. The order parameter $V = (\eta_0 - \eta)/\eta_0$, for the clustering transition in the 2D PLSA model is shown as a function of α for $d = 0.002$ and $M = 3184$ (namely, $\eta_0 = 0.01$). It represents the fraction of the total area of the adsorbed particles which is lost due to overlaps.

In two dimensions, the clusters form complicated shapes with multiple overlaps. Analytical calculation of η is thus difficult. Furthermore, even exact numerical evaluation of η, based on the location of the centers r_i, $i = 1, \ldots, M$, becomes time consuming. In the case of a finite system size L^2, an often more efficient way to calculate η numerically is by sampling a large number of random test points on the plane. The coverage η is estimated as the fraction of the random test points that fall within the area covered by at least one particle. One should be careful, though, and assure that the resolution of the random number generator is much finer than the particle size d. The order parameter V vs. α for a finite two dimensional system of size L^2 is shown in Fig. 2. The transition is smoothed out due to the finite size of the system and a non-vanishing (but small) value of V is obtained below the transition. This is due to the fact that in a finite system all particles are deposited within a distance L from each other and there is always a finite probability for an overlap to occur.

3.2 Clustering in One Dimension

In the one dimensional (1D) case the geometry is simpler and allows us to analyze the clustering transition both analytically and numerically. In 1D the particles take the form of rods of length d that are deposited randomly on the $[-L/2, L/2]$ interval, of length L. The clustering transition takes place at $\alpha_c = 1$. An advantage of the 1D case is that the linear ordering of the rods immediately provides the identity of the nearest rod to the newly deposited rod. Thus the distance r and the sticking probability can be easily calculated. In 1D $\eta_0 = Md/L$, where M is the number of rods. In case where overlaps are allowed ($P_\mathrm{o} = 1$) we define $\eta = d_s/L \leq \eta_0$, where d_s is the total length covered by the rods. In the infinite system limit ($L \to \infty$), for $\alpha < \alpha_c$, and any finite number of rods, M, overlaps are negligible and $V = 0$. For $\alpha > \alpha_c$ clusters appear and become denser as α increases, giving rise to more overlaps.

Consider a configuration of M rods in an interval of size L. The centers of the rods are positioned at x_i, $i = 1, \dots, M$ and the indices are ordered according to their positions from left to right, namely $x_1 < \cdots < x_M$. The set is fully specified by the ordered list of $M - 1$ distances, $\ell_i = x_{i+1} - x_i$, $i = 1, \dots, M - 1$ between their centers in addition to the distances ℓ_0 and ℓ_M between the peripheral rod centers and the edges of the system. In the infinite system limit the two peripheral distances are semi-infinite, namely $\ell_0 = \ell_M = \infty$. We will also define the gaps between the edges of adjacent rods to be

$$g_i = \begin{cases} \ell_i - d: & \ell_i > d \\ 0: & \ell_i \leq d, \end{cases} \tag{5}$$

namely, the gap between two rods that overlap is $g_i = 0$. In case where overlaps are not allowed $g_i > 0$ for all $1 \leq i \leq M - 1$. Also, the rods are not allowed to intersect with the edges of the system, at $\pm L/2$, and therefore $g_0 > 0$ and $g_M > 0$.

To analyze the clustering transition we consider a one dimensional system in which overlaps are not allowed, in the infinite system limit. We will set the rod size to be $d = 1$ and position a seed rod at the origin. The distance between the center of the seed rod and the first one that will stick is taken from the distribution of (1) with $P_o = 0$. We proceed in the construction of the set by depositing more particles (rods) according to the procedure described above. When the set already includes M rods, the next rod may stick either in one of the $M - 1$ gaps or beyond the system edge, namely in one of the semi-infinite gaps. We will now calculate the probability that the rod will stick in each one of these gaps. For the gap g_i this probability will be

$$P(x_{\text{new}} \in g_i) = \frac{1}{Z} \int_1^\infty \left(\frac{1}{x}\right)^\alpha dx, \tag{6}$$

for $i = 0$ or $i = M$, and

$$P(x_{\text{new}} \in g_i) = \frac{2}{Z} \int_1^{\ell_i/2} \left(\frac{1}{x}\right)^\alpha dx, \tag{7}$$

for $i = 1, \dots, M - 1$, where

$$Z = 2 \int_1^\infty \left(\frac{1}{x}\right)^\alpha dx + 2 \sum_{i=1}^{M-1} \int_1^{\ell_i/2} \left(\frac{1}{x}\right)^\alpha dx \tag{8}$$

is the normalization factor. Note that there is a reflection symmetry in the sticking probability function with respect to the center of each finite gap, giving rise to the factors of 2. Solving the integrals we find that

$$P(x_{new} \in g_i) = \frac{m(i) \left[\left(\frac{\ell_i}{2}\right)^{\alpha_c - \alpha} - 1 \right]}{\sum_{j=0}^{M} m(j) \left[\left(\frac{\ell_j}{2}\right)^{\alpha_c - \alpha} - 1 \right]}. \tag{9}$$

where $m(i) = 2$ for $i = 1, \ldots, M-1$, $m(i) = 1$ for $i = 0,\ M$, and $\alpha_c = 1$. In the infinite system limit the edge gaps $g_0 \to \infty$ and $g_M \to \infty$. In this case, for $\alpha < \alpha_c$

$$\left(\frac{\ell_i}{2}\right)^{\alpha_c - \alpha} - 1 \to \infty. \tag{10}$$

for $i = 0$ and $i = M$. As a result, for $\alpha < \alpha_c$ the probability of sticking in one of the finite gaps vanishes and the next rod must be added in one of the two semi-infinite gaps. Moreover, the expectation value for the distance between this rod and the nearest one (located at x_1 or x_M) diverges. In this case it is not really possible to construct the set in the infinite system limit. For any finite number of rods, M, the resulting set will be of zero density and its fractal dimension will be $D = 0$. The results in the case of a finite system size, L are markedly different. In this case the tail of the sticking probability distribution, that was dominant in the infinite system case is trimmed by the boundaries. Due to the slow decay of $p(r)$, particles will stick all over the (finite) system and create a rather homogeneous distribution, except for some short range correlations.

To analyze the clustering transition we consider the infinite system limit with $\alpha = \alpha_c + \epsilon$ where $\epsilon \to 0$. In this limit the possibility of sticking in one of the semi-infinite gaps is still dominant. Thus, in this limit all the gaps are expected to be generated by a new rod sticking in one of the two semi-infinite gaps. The probability distribution of the distance between a new rod sticking in one of the semi-infinite gaps and its nearest neighbor is given by (1). Since no rods are expected to stick within existing gaps, the lengths of these gaps remain unchanged. As a result, the distribution of distances between the centers of the rods is given by $p(\ell) \sim \ell^{-\alpha}$. Since these distances tend to be large, namely $\ell \gg 1$, the distribution of the gaps $g_i = \ell_i - 1$, $i = 1, \ldots, M-1$ also follows, to a very good approximation, a similar power-law behavior of the form

$$P(g) \sim g^{-\alpha}. \tag{11}$$

Standard fractal analysis for the 1D case shows that in this case the fractal dimension of the set of rods is $D = \alpha - \alpha_c$ (where $\alpha_c = 1$). To show this, consider a set of rods with gaps distributed according to (11). To calculate the Minkowski dimension we will draw a one dimensional Minkowski sphere of radius R around each point in the set of rods. We will then calculate the (one dimensional) volume $V_M(R)$ of the union of all spheres and examine how it varies as a function of R.

In each gap g_i, $i = 1, \ldots, M-1$ there are two Minkowski spheres, one on each side. The volume of each of these spheres is R as long as $R < g_i/2$. For larger values of R the two spheres start to overlap. The sum of their volumes remains fixed at g_i even when R increases. Taking into account the distribution of the gap sizes (see (11)) we obtain

$$V_M(R) \sim \int_0^R gP(g)dg + \int_R^\infty RP(g)dg. \tag{12}$$

Performing the integrations we obtain

$$V_M(R) \sim R^{2-\alpha}. \tag{13}$$

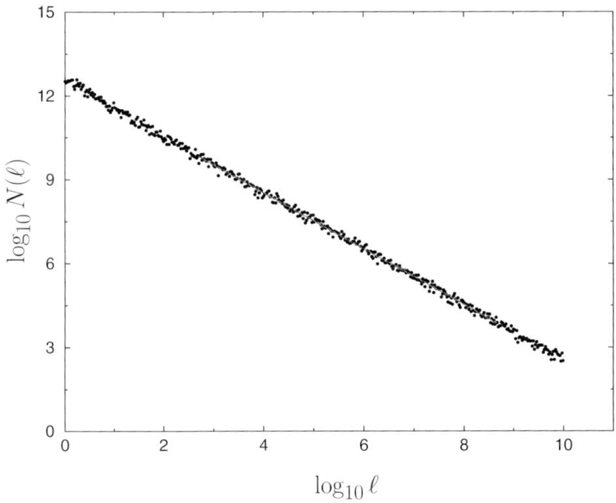

Fig. 3. The distribution of the distances ℓ between the centers of nearest neighbor rods in the 1D PLSA model in the infinite system limit for $\alpha = 1 + \epsilon$ and $\epsilon = 10^{-5}$. Here, $\ell_i = g_i + d$, where g_i is the gap size and the rod size is $d = 1$. $N(\ell)$ is the number of nearest neighbor pairs separated by a distance within a narrow bin around size ℓ. The data was obtained from an ensemble of configurations with $M = 100$ rods in each one of them. The straight line on the log-log scale exhibits a power-law distribution and the slope is -1.0 ± 0.01.

In the Minkowski analysis, a set is considered fractal with dimension D over a range of scales if

$$V_M(R) \sim R^{n-D} \tag{14}$$

where n is the space dimension (here, $n = 1$). The Minkowski dimension is known to be identical from the mathematical point of view to other dimensions such as the box counting dimension if the limit $R \to 0$ can be taken [23]. The resulting set, which exhibits a power-law distribution of gaps, is thus a fractal of dimension $D = \alpha - 1$. The fractal dimension thus satisfies

$$D = (\alpha - \alpha_c)^{\beta} \tag{15}$$

where $\beta = 1$. However, this relation is maintained only within a very narrow range of $0 < \epsilon \ll 1$. In Fig. 3 we show the distribution of gaps $P(\ell)$ vs. ℓ on a log-log scale for $\epsilon = 10^{-5}$. The slope of the line is found to be -1.0 ± 0.01, namely consistent with (11) for $\alpha = 1$. This indicates that the configuration just above α_c corresponds to a fractal of dimension $D = 0$. For larger values of ϵ the probability of sticking inside the finite gaps increases and causes deviations in the distribution $P(g)$.

We will now examine the critical behavior in the case that rods are allowed to overlap. In this case, for $L \to \infty$, the probability that the next rod will stick inside ℓ_i, $i = 0, \ldots, M$ is

$$P(x_{new} \in l_i) = \frac{m(i)\left[\alpha - \left(\frac{\ell_i}{2}\right)^{\alpha_c - \alpha}\right]}{\sum_{j=0}^{M} m(j)\left[\alpha - \left(\frac{\ell_j}{2}\right)^{\alpha_c - \alpha}\right]}. \tag{16}$$

The probability that the new rod inside l_i will overlap with one of the two neighboring rods is given by

$$P(\text{overlap}|\ell_i) = \frac{(\alpha - \alpha_c)}{\left[\alpha - \left(\frac{l_i}{2}\right)^{\alpha_c - \alpha}\right]}. \tag{17}$$

Therefore, the probability that the new rod will overlap with another rod is

$$P(\text{overlap}) = \frac{M(\alpha - \alpha_c)}{\sum_i \left[\alpha - \left(\frac{l_i}{2}\right)^{\alpha_c - \alpha}\right]}. \tag{18}$$

For rods that exhibit such overlaps the distribution of the overlap lengths is a homogeneous one in the range $[0, d]$. Therefore, the average overlap size is $d/2$. As a result, the expectation value of the overlap length is linearly proportional to $P(\text{overlap})$. The right hand side of (18) is linearly proportional to $\alpha - \alpha_c$. Thus, the order parameter V (see (4)) satisfies

$$V \sim (\alpha - \alpha_c)^{\gamma} \tag{19}$$

where $\gamma = 1$.

4 Summary

We have studied the clustering transition in the PLSA model in one and two dimensions. In this model the sticking probability of a new particle exhibits a power-law dependence on the distance to its nearest adsrobed particle with the exponent α. The transition is of second order and occurs at $\alpha_c = n$, where n is the space dimension. Below the transition the resulting set is nearly uncorrelated, while strong correlations appear above the transition and the particles exhibit strong tendency to cluster. We have studied the critical behavior near the transition. In case that overlaps are allowed we identify the order parameter as the fraction of the rod length that overlaps. In case that overlaps are suppressed the order parameter coincides with the fractal dimension of the set, characterized by the power-law distribution of the gaps between the rods.

We thank D. Avnir for many helpful and stimulating discussions.

References

1. P.J.E. Peebles, *Principles of Physical Cosmology* (Princeton University Press, Princeton, 1993).
2. I. Goldhirsch and G. Zanetti, Phys. Rev. Lett. **70**, 1619 (1993).
3. A. Kudrolli,M. Wolpert and J.P. Gollub, Phys. Rev. Lett. **78**, 1383 (1997).
4. D. Kandel and J.D. Weeks, Phys. Rev. B **49**, 5554 (1994).
5. Y.-N. Yang, E.S. Fu and E.D. Williams, Surf. Sci. **356**, 101 (1996).
6. M. Batty and P. Longley, *Fractal Cities: A Geometry of Form and Function* (Academic Press, London, 1994).
7. H.A. Makse, S. Havlin and H.E. Stanley, Nature **377**, 608 (1995).
8. H.A. Makse, J.S. Andrade, M. Batty, S. Havlin and H.E. Stanley, Phys. Rev. E **58**, 7054 (1998).
9. *Traffic and Granular Flow*, edited by D. Wolf, M. Schreckenberg and A. Bachem (World Scientific, Singapore, 1995).
10. Z. Zhang and M.G. Lagally, Science **276**, 377 (1997).
11. T.A. Witten, Jr. and L.M. Sander, Phys. Rev. Lett. **47**, 1400 (1981).
12. J.W. Evans, Rev. Mod. Phys. **65**, 1281 (1993).
13. P. Viot, G. Tarjus, S.M. Ricci and J. Talbot, J. Chem. Phys. **97**, 5212 (1992).
14. S.M. Ricci, J. Talbot, G. Tarjus and P. Viot, J. Chem. Phys. **97**, 5219 (1992).
15. N.V. Brilliantov, Yu.A. Andrienko, P.L. Krapivsky and J. Kurths, Phys. Rev. Lett. **76**, 4058 (1996).
16. H. Hermann, Phys. Stat. Sol. **163**, 329 (1991).
17. L. A. Chick and C. Viney, Molec. Cryst. Liq. Cryst. **226**, 25 (1993).
18. N. Provatas, T. Ala-Nissila and M.J. Alava, Phys. Rev. Lett. **75**, 3556 (1995).
19. N. Provatas, M. Haataja, E. Seppala, S. Majaniemi, J. Astrom, M. Alava and T. Ala-Nissila, J. Stat. Phys. **87**, 385 (1997).
20. O. Biham, O. Malcai, D.A. Lidar and D. Avnir, Phys. Rev. E **59**, R4713 (1999).
21. D.A. Hamburger, O. Biham and D. Avnir, Phys. Rev. E **53**, 3342 (1996).
22. *Lévy Flights and Related Topics in Physics*, edited by M.F. Shlesinger, G.M. Zaslavsky and U. Frisch (Springer, Berlin, 1995).
23. K. Falconer, *Fractal Geometry: Mathematical Foundations and Applications* (Wiley, Chichester, 1990).

The Hamiltonian Mean Field Model: From Dynamics to Statistical Mechanics and Back

Thierry Dauxois[1], Vito Latora[2], Andrea Rapisarda[2], Stefano Ruffo[1,3], and Alessandro Torcini[3,4]

[1] Laboratoire de Physique, UMR CNRS 5672, ENS Lyon, 46, allée d'Italie, F-69007 Lyon, France
[2] Dipartimento di Fisica, Universitá di Catania, and Istituto Nazionale di Fisica Nucleare, Sezione di Catania, Corso Italia 57, I-95129 Catania, Italy
[3] Dipartimento di Energetica "S. Stecco", Università di Firenze, via S. Marta, 3, INFM and INFN, I-50139 Firenze, Italy
[4] UMR-CNRS 6171, Université d'Aix-Marseille III, Av. Esc. Normandie-Niemen, F-13397 Marseille Cedex 20, France

Abstract. The thermodynamics and the dynamics of particle systems with infinite-range coupling display several unusual and new features with respect to systems with short-range interactions. The Hamiltonian Mean Field (HMF) model represents a paradigmatic example of this class of systems. The present study addresses both attractive and repulsive interactions, with a particular emphasis on the description of clustering phenomena from a thermodynamical as well as from a dynamical point of view. The observed clustering transition can be first or second order, in the usual thermodynamical sense. In the former case, ensemble inequivalence naturally arises close to the transition, i.e. canonical and microcanonical ensembles give different results. In particular, in the microcanonical ensemble negative specific heat regimes and temperature jumps are observed. Moreover, having access to dynamics one can study non-equilibrium processes. Among them, the most striking is the emergence of coherent structures in the repulsive model, whose formation and dynamics can be studied either by using the tools of statistical mechanics or as a manifestation of the solutions of an associated Vlasov equation. The chaotic character of the HMF model has been also analyzed in terms of its Lyapunov spectrum.

1 Introduction

Long-range interactions appear in the domains of gravity [1,2] and of plasma physics [3] and make the statistical treatment extremely complex. Additional features are present in such systems at short distances: the gravitational potential is singular at the origin and screening phenomena mask the Coulomb singularity in a plasma. This justifies the introduction of simplified toy models that retain only the long-range properties of the force, allowing a detailed description of the statistical and dynamical behaviors associated to this feature. In this context a special role is played by mean-field models, i.e. models where all particles interact with the same strength. This constitutes a dramatic reduction of complexity, since in such models the spatial coordinates have no role, since each particle is equivalent. However, there are several preliminary indications that behaviors found in mean-field models extend to cases where the two-body potential decays at large distances with a power smaller than space dimension [4–6].

The Blume-Emery-Griffiths (BEG) mean-field model is discussed in this book [5] and represents an excellent benchmark to discuss relations between canonical and microcanonical ensembles. Indeed, this model is exactly solvable in both ensembles and is, at the same time, sufficiently rich to display such interesting features as negative specific heat and temperature jumps in the microcanonical ensemble. Since these effects cannot be present in the canonical ensemble, this rigorously proves ensemble inequivalence. However, the BEG model has no dynamics and only the thermodynamical behavior can be investigated. Moreover, it is a spin model where variables take discrete values. It would therefore be amenable to introduce a model that displays all these interesting thermodynamical effects, but for which one would also dispose of an Hamiltonian dynamics with continuous variables, whose equilibrium states could be studied both in the canonical and in the microcanonical ensemble. Having access to dynamics, one can moreover study non equilibrium features and aspects of the microscopic behavior like sensitivity to initial conditions, expressed by the Lyapunov spectrum [7]. Such a model has been introduced in [8] and has been called the Hamiltonian Mean Field (HMF) model. In the simpler version, it represents a system of particles moving on a circle, all coupled by an equal strength attractive or repulsive cosine interaction. An extension of it to the case in which particles move on a 2D torus has been introduced in [9] and it has been quite recently realized that all such models are particular cases of a more general Hamiltonian [10].

The HMF model, that we introduce in Sect. 2, is exactly solvable in the canonical ensemble by a Hubbard-Stratonovich transformation. The solution in the microcanonical ensemble can be obtained only under certain hypotheses that we will discuss in Sect. 3, but detailed information on the behavior in the microcanonical ensemble can be obtained by direct molecular dynamics (MD) simulations. The model has first and second order phase transitions and tricritical points. Its rich phase diagram allows to test the presence of ensemble inequivalence near canonical first order phase transitions and, indeed, we find negative specific heat and temperature jumps in the microcanonical ensemble. Having access to dynamics, one can study metastability of out-of-equilibrium states. This is done in Sect. 4, where we analyze the emergence of a coherent structure in the repulsive HMF at low energy. Similar features are also discussed in another chapter of this book [11] for the attractive case near the second order phase transition. Section 5 is devoted to the study of the spectrum of Lyapunov exponents. The maximal exponent has a peak near the phase transition [9,12,13] and vanishes when increasing the number of particles with a universal scaling law in the whole high energy disordered phase. In a low energy range the Lyapunov spectrum has a thermodynamic limit distribution similar to the one observed for systems with short-range interaction [14].

2 The HMF Models

A generic two-body potential in a two dimensional square box of side 2π with periodic boundary conditions, a 2D torus, can be Fourier expanded as

$$V(x,y) = \sum_{\mathbf{k}=(k_x,k_y)} \exp(i\mathbf{k}\cdot\mathbf{r})\, V(k_x,k_y) \quad . \tag{1}$$

A sufficiently rich family of potential functions is obtained if we restrict to the first two momentum shells $|k| = 1$ and $|k| = \sqrt{2}$, we require that the potential is only invariant under discrete rotations by all multiples of $\pi/4$, and we assume that the Fourier coefficients on each shell are the same. This amounts to perform a truncation in the Fourier expansion of the potential (1), as done in studies of spherically symmetric gravitational systems in another chapter of this book [15]. We get

$$V(x,y) = a + b\cos x + b\cos y + c\cos x\cos y \ . \tag{2}$$

As the constant a is arbitrary and scaling b is equivalent to scale the energy, c remains the only free parameter. We consider N particles interacting through the two-body potential $V(x,y)$ and we adopt the Kac prescription [16] to scale the equal strength coupling among the particles by their number N. This scaling allows to perform safely the thermodynamic limit, since both the kinetic and the potential energy increase proportional to N[1]. By appropriately redefining the constants $a = 2\varepsilon + A, b = -\varepsilon, c = -A$ in formula (2) and using Kac prescription one gets the following potential energy

$$\begin{aligned} V_A = \frac{1}{2N}\sum_{i,j=1}^{N} & \varepsilon\left(1-\cos(x_i-x_j)\right) + \varepsilon\left(1-\cos(y_i-y_j)\right) \\ & + A\left(1-\cos(x_i-x_j)\cos(y_i-y_j)\right) \ , \end{aligned} \tag{3}$$

with $(x_i, y_i) \in]-\pi,\pi]\times]-\pi,\pi]$ representing the coordinates of i-th particle and $(p_{x,i}, p_{y,i})$ its conjugated momentum. The Hamiltonian of the HMF model is now the sum of this potential energy with the kinetic energy

$$K = \sum_{i=1}^{N}\left(\frac{p_{x,i}^2 + p_{y,i}^2}{2}\right) \ . \tag{4}$$

We get

$$H_{HMF} = K + V_A \ . \tag{5}$$

In the following, we will consider model (5) for $A = 0$ and both ε positive (attractive case) and negative (repulsive case). The $A \neq 0$ case will always have $A > 0$ and $\varepsilon = 1$.

[1] Kac prescription is, however, unphysical and it would be important to find a viable alternative.

3 Equilibrium Thermodynamics

In this section, we discuss the equilibrium thermodynamical results for model (3) in the canonical and microcanonical ensembles. Canonical results will be obtained analytically while, for the microcanonical ones, we will mostly rely on molecular dynamics (MD) simulations.

3.1 Canonical Ensemble for $A = 0$

For pedagogical reasons, we will initially limit our analysis to the case $A = 0$, for which the model reduces to two identical *uncoupled* systems: one describing the evolution of the $\{x_i, p_{x,i}\}$ variables and the other $\{y_i, p_{y,i}\}$. Therefore let us rewrite the Hamiltonian associated to one of these two sets of variables, named θ_i in the following. We obtain

$$H_0 = \sum_{i=1}^{N} \frac{p_i^2}{2} + \frac{\varepsilon}{2N} \sum_{i,j=1}^{N} [1 - \cos(\theta_i - \theta_j)] = K_0 + V_0 \tag{6}$$

where $\theta_i \in [-\pi; \pi[$ and p_i are the corresponding momenta. This model can be seen as representing particles moving on the unit circle, or as classical XY-rotors with infinite range couplings. For $\varepsilon > 0$, particles attract each other and rotors tend to align (ferromagnetic case), while for $\varepsilon < 0$, particles repel each other and spins tend to anti-align (antiferromagnetic case). At short distances, we can either think that particles cross each other or that they collide elastically since they have the same mass.

The physical meaning of this model is even clearer if one introduces the mean field vector

$$\mathbf{M} = M e^{i\phi} = \frac{1}{N} \sum_{i=1}^{N} \mathbf{m}_i \tag{7}$$

where $\mathbf{m}_i = (\cos\theta_i, \sin\theta_i)$. M and ϕ represent the modulus and the phase of the order parameter, which specifies the degree of clustering in the particle interpretation, while it is the *magnetization* for the XY rotors. Employing this quantity, the potential energy can be rewritten as a sum of single particle potentials v_i

$$V_0 = \frac{1}{2} \sum_{i=1}^{N} v_i \qquad \text{with} \qquad v_i = 1 - M\cos(\theta_i - \phi) \tag{8}$$

It should be noticed that the motion of each particle is coupled to all the others, since the mean-field variables M and ϕ are determined at each time t by the instantaneous positions of all particles.

The equilibrium results in the canonical ensemble can be obtained from the evaluation of the partition function

$$Z = \int d^N p_i d^N \theta_i \exp(-\beta H) \tag{9}$$

where $\beta = 1/(k_B T)$, with k_B the Boltzmann constant and T the temperature. The integration domain is extended to the whole phase space. Integrating over momenta, one gets:

$$Z = \left(\frac{2\pi}{\beta}\right)^{N/2} \int_{-\pi}^{\pi} d^N\theta_i \exp\left[\frac{-\beta\varepsilon N}{2}(1-\mathbf{M}^2)\right] \quad . \tag{10}$$

In order to evaluate this integral, we use the two dimensional Gaussian identity

$$\exp\left[\frac{\mu}{2}\mathbf{x^2}\right] = \frac{1}{\pi}\int_{-\infty}^{\infty}\int_{-\infty}^{\infty} d\mathbf{y} \exp\left[-\mathbf{y}^2 + \sqrt{2\mu}\mathbf{x}\cdot\mathbf{y}\right] \tag{11}$$

where $\mathbf{x}$ and $\mathbf{y}$ are two-dimensional vectors and μ is positive. We can therefore rewrite (10) as

$$Z = \left(\frac{2\pi}{\beta}\right)^{N/2} \exp\left[\frac{-\beta\varepsilon N}{2}\right] J \tag{12}$$

with

$$J = \frac{1}{\pi}\int_{-\pi}^{\pi} d^N\theta_i \int_{-\infty}^{\infty}\int_{-\infty}^{\infty} d\mathbf{y} \exp\left[-\mathbf{y}^2 + \sqrt{2\mu}\mathbf{M}\cdot\mathbf{y}\right] \tag{13}$$

and $\mu = \beta\varepsilon N$. We use now definition (7) and exchange the order of the integrals in (13), factorizing the integration over the coordinates of the particles. Introducing the rescaled variable $\mathbf{y} \to \mathbf{y}\sqrt{N/2\beta\varepsilon}$, one ends up with the following expression for J

$$J = \frac{N}{2\pi\beta\varepsilon}\int_{-\infty}^{\infty}\int_{-\infty}^{\infty} d\mathbf{y} \exp\left[-N\left(\frac{y^2}{2\beta\varepsilon} - \ln\left(2\pi I_0(y)\right)\right)\right] \tag{14}$$

where I_n is the modified Bessel function of order n and y is the modulus of $\mathbf{y}$. Finally, integral (14) can be evaluated by employing the saddle point technique in the mean-field limit (i.e. for $N \to \infty$). In this limit, the Helmholtz free energy per particle f reads as :

$$\beta f = -\lim_{N\to\infty}\frac{\ln Z}{N} = -\frac{1}{2}\ln\left(\frac{2\pi}{\beta}\right) + \frac{\varepsilon\beta}{2} + \max_y\left(\frac{y^2}{2\beta\varepsilon} - \ln(2\pi I_0(y))\right) \quad . \tag{15}$$

The maximum condition leads to the consistency equation

$$\frac{y}{\beta\varepsilon} = \frac{I_1(y)}{I_0(y)} \quad . \tag{16}$$

For $\varepsilon < 0$, there is a unique solution $\bar{y} = 0$, which means that the order parameter remains zero and there is no phase transition (see Figs. 1c,d). The particles are all the time homogeneously distributed on the circle and the rotors have zero magnetization. On the contrary, in the ferromagnetic case ($\varepsilon > 0$), the solution $\bar{y} = 0$ is unstable for $\beta \geq \beta_c = 2$. At $\beta = \beta_c$, two stable symmetric solutions appear through a pitchfork bifurcation and a discontinuity in the second derivative of the free energy is present, indicating a second order phase transition.

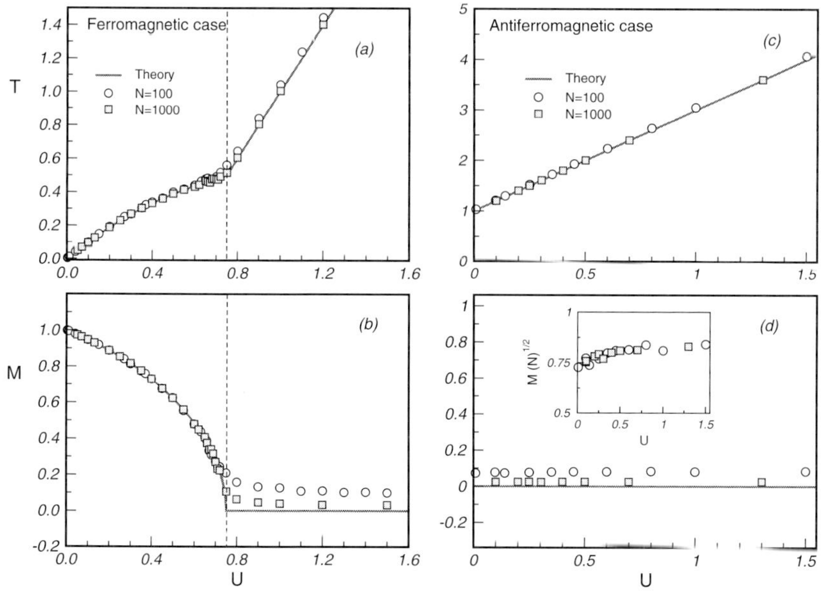

Fig. 1. Temperature and magnetization as a function of the energy per particle U for $A = 0$ and $|\varepsilon| = 1$ in the ferromagnetic (a),(b) and in the antiferromagnetic case (c),(d). Symbols refer to MD data for $N = 10^2$ and 10^3, while the solid lines refer to the canonical prediction obtained analytically in the mean-field limit. The vertical dashed line indicates the critical energy in the ferromagnetic model, located at $U_c = 0.75$, $\beta_c = 2$. The inset of panel (d) shows the rescaled magnetization $M \times \sqrt{N}$.

These results are confirmed by an analysis of the order parameter[2]

$$M = \frac{I_1(\bar{y})}{I_0(\bar{y})} \quad . \tag{17}$$

For $\varepsilon > 0$, the magnetization M vanishes continuously at β_c (see Fig. 1a,b), while it is always identical to zero in the antiferromagnetic case (see Fig. 1c,d). Since M measures the degree of clustering of the particles, we have for $\varepsilon > 0$ a transition from a clustered phase when $\beta > \beta_c$ to a homogeneous phase when $\beta < \beta_c$. We can obtain also the energy per particle

$$U = \frac{\partial(\beta f)}{\partial \beta} = \frac{1}{2\beta} + \frac{\varepsilon}{2}\left(1 - M^2\right) \tag{18}$$

which is reported for $|\varepsilon| = 1$ in Fig. 1. Panels (c) and (d) of Fig. 1 are limited to the range $U > 0$ because in the antiferromagnetic model a non-homogeneous state, a *bicluster*, can be generated for smaller energies. The emergence of this state modifies all thermodynamical and dynamical features as will be discussed in section 4.2.

[2] This is obtained by adding to the Hamiltonian an external field and taking the derivative of the free energy with respect to this field, evaluated at zero field.

The dynamics of each particle obeys the following pendulum equation of motion

$$\ddot{\theta}_i = -M \sin(\theta_i - \phi) \quad , \tag{19}$$

where M and ϕ have a non trivial time dependence, related to the motion of all the other particles in the system. Equation (19) has been very successfully used to describe several features of particle motion, like for instance trapping and untrapping mechanisms [8]. There are also numerical indications [8] and preliminary theoretical speculations [17] that taking the mean-field limit before the infinite time limit, the time-dependence would disappear and the modulus and the phase of the magnetization become constant. This implies, as we will discuss in Sect. 5, that chaotic motion would disappear. The inversion of these two limits is also discussed in the contribution by Tsallis et al [11].

3.2 Canonical Ensemble for $A \neq 0$

As soon as $A > 0$, the evolution along the two spatial directions is no more decoupled and the system cannot be described in terms of a single order parameter. Throughout all this section $\epsilon = 1$. A complete description of the phase diagram of the system requires now the introduction of two distinct order parameters:

$$\mathbf{M}_z = \left(\frac{\sum_i \cos z_i}{N}, \frac{\sum_i \sin z_i}{N} \right) = M_z \exp\left(i\phi_z\right) \tag{20}$$

where $z_i = x_i$ or y_i and;

$$\mathbf{P}_{x\pm y} = \left(\frac{\sum_i \cos(x_i \pm y_i)}{N}, \frac{\sum_i \sin(x_i \pm y_i)}{N} \right) = P_z \exp\left(i\psi_z\right) \quad . \tag{21}$$

It can be shown that on average $M_x \simeq M_y \simeq M$ and $P_{x+y} \simeq P_{x-y} \simeq P$: therefore, we are left with only two order parameters.

Following the approach of section 3.1, the canonical equilibrium properties can be derived analytically in the mean-field limit [9]. We obtain

$$\beta f = \frac{M^2 + P^2}{\beta} - \ln\left[\frac{G(M,P;A)}{\beta}\right] \tag{22}$$

with

$$G = \int_0^{2\pi} ds \; I_0\left(M + \sqrt{2A}P\cos s\right) \exp\left(M\cos s\right) \tag{23}$$

where s is an integration variable.

The energy per particle reads as

$$U = \frac{1}{\beta} + \frac{2 + A - 2M^2 - AP^2}{2} = K + V_A \quad . \tag{24}$$

Depending on the value of the coupling constant A, the single particle potential defined through $V_A = \frac{1}{2}\sum_i v_i$ changes its shape, inducing the different clustering

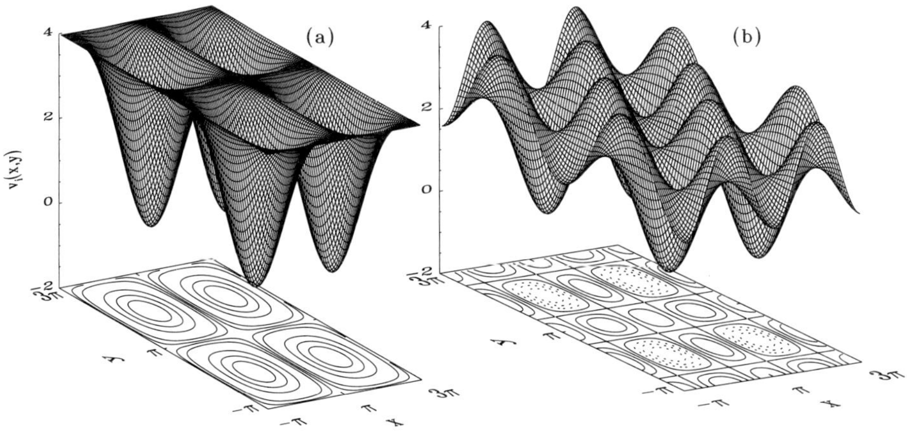

Fig. 2. Single particle potential $v_i(x, y)$ for $A = 1, \varepsilon = 1$ (a) and $A = 4, \varepsilon = 1$(b).

phenomena described below. For small values of A, v_i exhibits a single minimum per cell $(x, y) \in ([-\pi, \pi], [-\pi, \pi])$ (see Fig. 2a for $A = 1$), while for larger values of A four minima can coexist in a single cell (see Fig. 2b for $A = 4$). In the former case only one clustered equilibrium state can exist at low temperatures, while in the latter case, when all the four minima have the same depth a phase with two clusters can emerge, as described in the following.

Since the system is ruled by two different order parameters M and P, the phase diagram is more complicated than in the $A = 0$ case and we observe two distinct clustered phases. In the very low temperature regime, the system is in the clustered phase CP_1: the particles have all the same location in a single point-like cluster and $M \approx P \approx 1$. In the very large temperature range, the system is in a homogeneous phase (HP) with particles uniformly distributed, $M \approx P = O(1/\sqrt{N})$. For $A > A_2 \sim 3.5$, an intermediate two-clusters phase CP_2 appears. In this phase, due to the symmetric location of the two clusters in a cell, $M \sim O(1/\sqrt{N})$ while $P \sim O(1)$ [10]. We can gain good insights on the transitions by considering the line T_M (resp. T_P) where M (resp. P) vanishes and the phase CP_1 (resp. CP_2) looses its stability (see Fig. 3 for more details).

We can therefore identify the following four different scenarios depending on the value of A.

(I) When $0 \lesssim A \lesssim A_1 - 2/5$, one observes a continuous transition from the phase CP_1 to HP. The critical line is located at $T_M = 1/2$ ($U_M = 3/2 + A$) and a canonical tricritical point, located at $A_1 = 2/5$, separates the 1st order from the second order phase transition regions[3]

[3] The tricritical point has been identified by finding first the value of $\bar{P}(M, A, \beta)$ that minimizes $f(M, P; A, \beta)$ and then substituting the solution $\bar{P}$ in f. This reduces f to a function of M only. Now, the standard procedure described also for the BEG model [5] can be applied. This consists in finding the value of β and A where both the M^2 and the M^4 coefficient of the development of the free energy in powers of M vanishes

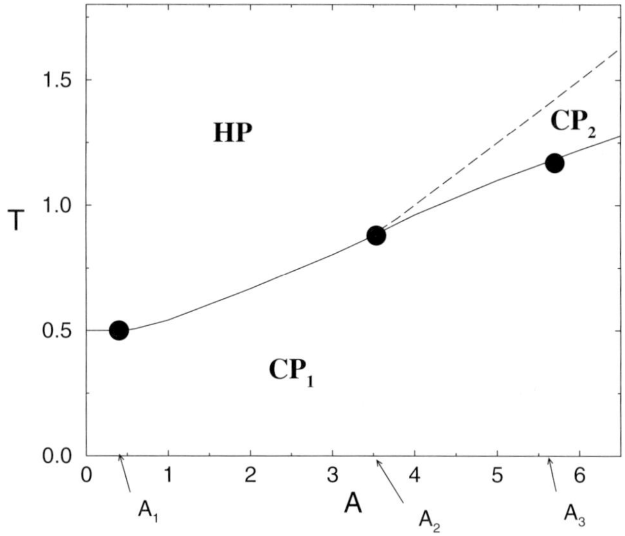

Fig. 3. Canonical phase diagram of model (3) reporting the transition temperatures versus the coupling parameter A. The solid (resp. dashed) lines indicate the T_M (resp. T_P) lines. The dots the points where the nature of the transitions change. A_1, A_2 and A_3 are the threshold coupling constants that determine the transition scenario $I \to IV$.

(II) When $A_1 < A < A_2 \approx 3.5$, the transition between CP_1 and HP is first order with a finite energy jump (latent heat). Inside this range of values of A one also finds microcanonical discontinuous transitions (temperature jumps) as we will see in Sect. 3.2.

(III) When $A_2 < A < A_3 \approx 5.7$, the third phase begins to play a role and two successive transitions are observed: first CP_1 disappears at T_M via a first order transition that gives rise to CP_2; then this two-clusters phase gives rise to the HP phase via a continuous transition. The critical line associated to this transition is $T_P = A/4$ ($U_P = 3A/4 + 1$).

(IV) When $A > A_3$, the transition connecting the two clustered phases, CP_1 and CP_2, becomes second order.

3.3 Microcanonical Ensemble

As we have anticipated, our microcanonical results have been mostly obtained via MD simulations, since we cannot estimate easily the microcanonical entropy S analytically for the HMF model[4]. However, we show below how far we can get, starting from the knowledge of the canonical free energy, using Legendre transform or inverse Laplace transform techniques. The following derivation is indeed valid in general, it does not refer to any specific microscopic model.

[4] It is possible using large deviation technique [18].

The relation that links the partition function $Z(\beta, N)$ to the microcanonical phase-space density at energy $E = U \cdot N$

$$\omega(E, N) = \int d^N p_i d^N \theta_i \, \delta(E - H) \tag{25}$$

is given by

$$Z(\beta, N) = \int_0^\infty dE \, \omega(E, N) \, \mathrm{e}^{-\beta E} \, , \tag{26}$$

where the lower limit of the integral ($E = 0$) corresponds to the energy of the ground state of the model. Expression (26) can be readily rewritten as

$$Z(\beta, N) = N \int_0^\infty dU \exp\left[N(-\beta U + \frac{1}{N}\ln(\omega(E, N))\right] \, , \tag{27}$$

which is evaluated by employing the saddle-point technique in the mean-field limit. Employing the definition of entropy per particle in the thermodynamic limit

$$S(U) = \lim_{N\to\infty}\left[\frac{1}{N}\ln\omega(U, N)\right] \quad , \tag{28}$$

one can obtain the Legendre transform that relates the free energy to the entropy:

$$-\beta F(\beta) = \max_U[-\beta U + S(U)] \qquad \text{with} \qquad \beta = \frac{\partial S}{\partial U} \quad . \tag{29}$$

Since a direct analytical evaluation of the entropy of the HMF model in the microcanonical ensemble is not possible, we are rather interested in obtaining the entropy from the free energy. This can be done only if the entropy S is a concave function of the energy. Then, one can invert (29), getting

$$S(U) = \min_{\beta>0}[\beta(U - F(\beta))] \qquad \text{with} \qquad U = \frac{\partial(\beta F)}{\partial \beta} \quad . \tag{30}$$

However, the assumption that S is concave is not true for systems with long-range interactions near a canonical first order transition, where a "convex intruder" of S appears [5,15], which gives rise to a negative specific heat regime in the microcanonical ensemble[5]. For such cases, we have to rely on MD simulations or microcanonical Monte-Carlo simulations [20].

An alternative approach to the calculation of the entropy consists in expressing the Dirac δ function in (25) by a Laplace transform. One obtains

$$\omega(E, N) = \frac{1}{2i\pi}\int_{-i\infty}^{+i\infty} d\beta \, e^{\beta E} \, Z(\beta) \tag{31}$$

where one notices that β is imaginary. As the partition function can be estimated for our model, we would just have to analytically continue to complex values of β.

[5] A convex intruder is present also for short range interactions in *finite* systems, but the entropy regains its concave character in the thermodynamic limit [15].

By performing a rotation to the real axis of the integration contour, one could then evaluate ω by saddle point techniques. However, this rotation requires the assumption that no singularity is present out of the real axis: this is not in general guaranteed. It has been checked numerically for the $A = 0$ model [19], allowing to obtain then an explicit expression for ω and confirming ensemble equivalence for this case.

However, when $A > A_1$, a first order canonical transition occurs, which implies the presence of a convex intruder and makes the evaluation of $S(U)$ through (30) impossible . For these cases, a canonical description is unable to capture all the features associated with the phase transition. For the microcanonical entropy, we have to rely here on MD simulations.

The MD simulations have been performed adopting extremely accurate symplectic integration schemes [21], with relative energy conservations during the runs of order $\sim 10^{-6}$. It is important to mention that the CPU time required by our integration schemes, due to the mean-field nature of the model, increases linearly with the number of particles.

Whenever we observe canonically continuous transitions (i.e. for $A \leq A_1$), the MD results coincide with those obtained analytically in the canonical ensemble, as shown in Fig. 1 for $A = 0$. The curve $T(U)$ is thus well reproduced from MD data, apart from finite N effects. It has been however observed that starting from "water-bags" initial conditions metastable states can occur in the proximity of the transition [11].

When $A > A_1$, discrepancies between the results obtained in the two ensembles are observable in Fig. 4. For example, MD results in the case $A = 1$, reported in Fig. 4a, differ clearly from canonical ones around the transition, exhibiting a regime characterized by a negative specific heat. This feature is common to many models with long-range [22,23] or power-law decaying interactions [4] as well as to finite systems with short-range forces [20]. However, only recently a characterization of all possible microcanonical transitions associated to canonically first order ones has been initiated [5,10].

For A slightly above A_1, the transition is microcanonically continuous, i.e. there is no discontinuity in the $T-U$ relation (this regime presumably extends up to $A \sim 1.2$). Before the transition, one observes a negative specific heat regime (see Fig. 4a). In addition, as already observed for the Blume-Emery-Griffiths model [5], microcanonically *discontinuous* transitions can be observed in the "convex intruder" region. This means that, at the transition energy, *temperature jumps* exist in the thermodynamic limit. A complete physical understanding of this phenomenon, which has also been found in gravitational systems [2], has not been reached. For $A > 1.2$, i.e. above the "microcanonical tricritical point", our model displays temperature jumps. This situations is shown in Fig. 5 for $A = 2$.

For $A > A_2$, we have again a continuous transition connecting the two clustered phases CP_1 and CP_2. This is the first angular point in the $T - U$ relation at $U \sim 3.65$ in Fig. 4b. The second angular point at $U \sim 4$ is the continuous transition, connecting CP_2 to HP. The transition at lower energy associated to the vanishing of M is continuous in the microcanonical ensemble with a negative

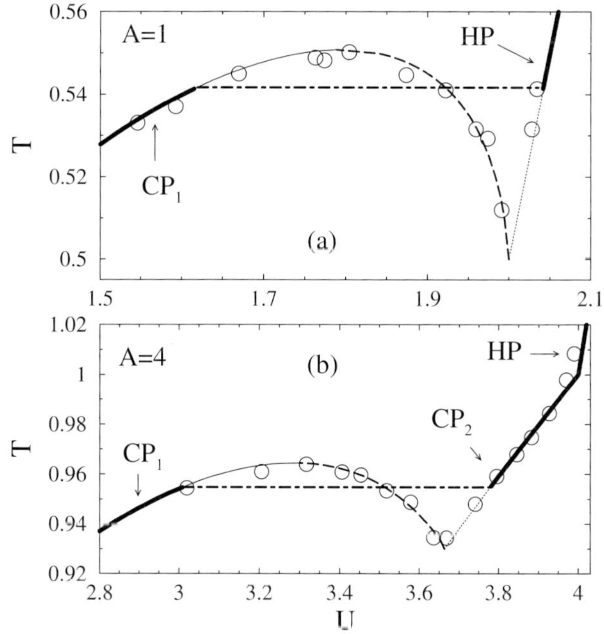

Fig. 4. Temperature-energy relation in the coexistence region for A=1 (a) and A=4 (b). Lines indicate canonical analytical results, while circles correspond to microcanonical MD simulations. Solid thick lines are equilibrium results, solid thin lines metastable states and dashed thin lines unstable states. The dash-dotted line is the Maxwell construction. Figure (a) refers to a first order transition from CP_1 to HP, (b) to discontinuous transition connecting the two clustered phases. In (b) the second order transition from CP_2 to HP associated to the vanishing of P is also shown. The MD results refer to model (3) with $N = 5000$ averaged over a time $t = 10^6$.

specific heat, while discontinuous in the canonical ensemble; the dash-dotted line indicates the transition temperature in the canonical ensemble, derived using the Maxwell construction). The second transition associated to the vanishing of P is continuous in both ensembles.

4 Dynamical Properties I: Out-of-Equilibrium States

4.1 Metastable States

Around the critical energy, relaxation to equilibrium depends in a very sensitive way on the initial conditions adopted. When one starts with out-of-equilibrium initial conditions in the ferromagnetic case, one finds quasi-stationary (i.e. long lived) nonequilibrium states. An example is represented by the so-called "water bag" initial condition: all the particles are clustered in a single point and the momenta are distributed according to a flat distribution of finite width centered around zero. These states have a lifetime which increases with the number of

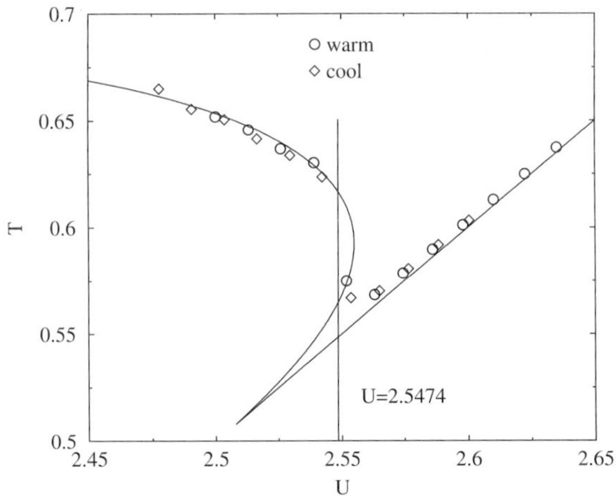

Fig. 5. Evolution of the temperature T versus the energy U in the case $A = 2$. The symbols refer to MD results obtained by successively cooling or warming a certain initial configuration. Each simulation has been performed at constant total energy and refers to a system of $N = 4,000$ particles integrated for a time $t = 10^6$. The solid lines are computed in the canonical ensemble and include also unstable and metastable cases. The solid vertical line indicating the transition energy has been estimated via a Maxwell construction performed in the microcanonical ensemble (for details see [5]).

particles N, and are therefore stationary in the continuum limit. In correspondence of these metastable states, anomalous diffusion and Lévy walks [24], long living correlations in $\mu-$space [25] and zero Lyapunov exponents [26] have been found. In addition, these states are far from the equilibrium caloric curve around the critical energy, showing a region of negative specific heat and a continuation of the high temperature phase (linear T vs U relation) into the low temperature one. It is very intriguing that these out-of-equilibrium quasi-stationary states indicate a caloric curve very similar to the one found in the region where one gets a canonical first order phase transitions, but a continuous microcanonical one, as discussed in section 3.3. In the latter case, however, the corresponding states are stationary also at finite N. The coexistence of different states in the continuum limit near the critical region is a purely microcanonical effect, and arises after the inversion of the $t \to \infty$ limit with the $N \to \infty$ one [25,26,11].

Similarly, the antiferromagnetic HMF ($A = 0$, $\varepsilon = -1$), where the particles interact through *repulsive* forces presents unexpected dynamical properties in the out-of-equilibrium thermodynamics. On the first sight, the thermodynamics of this model seems to be less interesting since no phase transition occurs as discussed above. However, thermodynamical predictions are again in some cases in complete disagreement with dynamical results leading in particular to a striking localization of energy. This aspect, as we will show below, is of course, again, closely related to the long-range character of the interaction and to the fact that such a dynamics is chaotic and self-consistent. We mean by this that all particles give a contribution to the field acting on each of them. One calls this phenomenon, *self-consistent chaos* [27]. In addition to the toy model that we consider here, we do think that similar emergence of structures, but even more importantly, similar dynamical stabilization of out-equilibrium states could be encountered in other long-range systems, as we briefly describe at the end of this section.

4.2 The Dynamical Emergence of the Bicluster in the Antiferromagnetic Case

In the antiferromagnetic case, (6) with $\varepsilon = -1$, the intriguing properties appear in the region of very small energies. To be more specific, if an initial state with particles evenly distributed on the circle (i.e. close to the ground state predicted by microcanonical or canonical thermodynamics) and with vanishingly small momenta is prepared, this initial condition can lead to the formation of unstable states. This process, discovered by chance, is now characterized in full detail [8,28–30].

As shown by Fig. 6, the density of particles is initially homogeneous. However a localization of particles do appear at a given time, in two different points, symmetrically located with respect to the center of the circle. This localized state, that we call *bicluster*, is however unstable (as shown again by Fig. 6), since both clusters are giving rise to two smaller localized groups of particles: this is the reason for the appearance of the first *chevron*. However, also this state

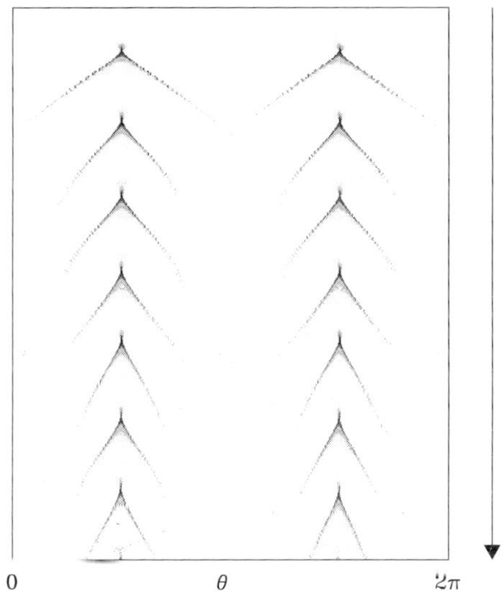

Fig. 6. Short-time evolution of the particle density in grey scale: the darker the grey, the higher the density. Starting from an initial condition with all the particles evenly distributed on the circle, one observes a very rapid concentration of particles, followed by the quasi periodic appearance of *chevrons*, that shrink as time increases.

is unstable, so that the first chevron disappears to give rise again to a localization of energy in two points. This state enhances the formation of a chevron with a smaller width and this phenomenon repeats until the width of the chevron is so small that one does not distinguish anymore its destabilization. Asymptotically one gets a density distribution displaying two sharp peaks located at distance π on the circle, a dynamically stable bicluster.

As we have shown in [28], the emergence of the bicluster is the signature of shock waves present in the associated hydrodynamical equations. Indeed, we found a strikingly good description of the dynamics of the particles by a non linear analysis of the associated Vlasov equation, which is mathematically justified [31] in the infinite N limit. The physical explanation of this problem can be summarized as follows. Once the Hamiltonian has been mapped to the Vlasov equation, it is possible to introduce a density $\rho(\theta, t)$ and a velocity field $v(\theta, t)$. Neglecting the dispersion in momentum and relying on usual non linear dynamics hierarchy of time-scales, one ends up with dynamical equations at different orders in a multiscale analysis. The first order corresponds to the linear dynamics and defines the plasma frequency of order one. However, a second timescale appears that is related to the previous one by the relationship $\tau = \sqrt{U}\, t$, where U is the energy per particle. When one considers initial conditions with a very small energy density, the two time scales are very different and clearly distinguishable by considering particle trajectories: a typical trajectory corresponds indeed to a very fast motion with a very small amplitude, superimposed to a slow motion with a large amplitude.

This suggests to average over the fast oscillations and leads to the spatially forced Burgers equation

$$\frac{\partial u}{\partial \tau} + u\frac{\partial u}{\partial \theta} = -\frac{1}{2}\sin 2\theta \qquad , \tag{32}$$

once the average velocity $u(\theta, \tau) = \langle v(\theta, t, \tau)\rangle_t$ is introduced. Due to the absence of dissipative or diffusive terms, (32) supports shock waves and these can be related to the emergence of the bicluster. By applying the methods of characteristics to solve (32), one obtains

$$\frac{d^2\theta}{d\tau^2} + \frac{1}{2}\sin 2\theta = 0\ . \tag{33}$$

which is a pendulum-like equation.

Figure 7 shows the trajectories, derived from (33), for particles that are initially evenly distributed on the circle. One clearly sees that two shock waves appear and lead to an increase of the number of particles around two particular sites, which depend on the initial conditions: these two sites correspond to the nucleation sites of the bicluster. Because of the absence of a diffusive term, the shock wave starts a spiral motion that explains the destabilization of the first bicluster and also the existence of the two arms per cluster, i.e. the chevron.

This dynamical analysis allows an even more precise description, since the methods of characteristics show that the trajectories correspond to the motion

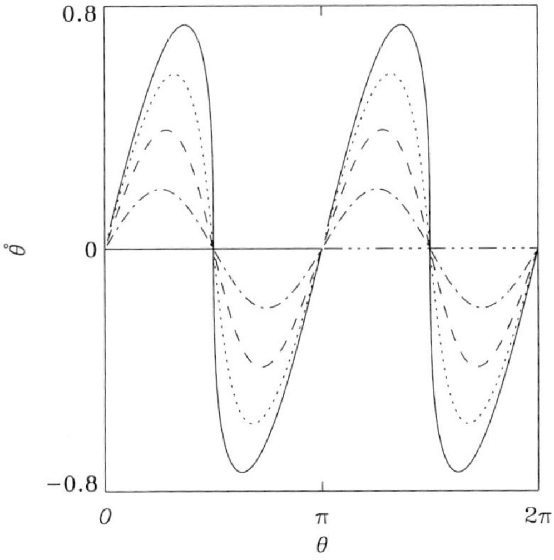

Fig. 7. Five successive snapshots of the velocity profiles $u(\theta, t)$ are shown including the initial state when all particles are uniformly distributed in space with a small velocity dispersion.

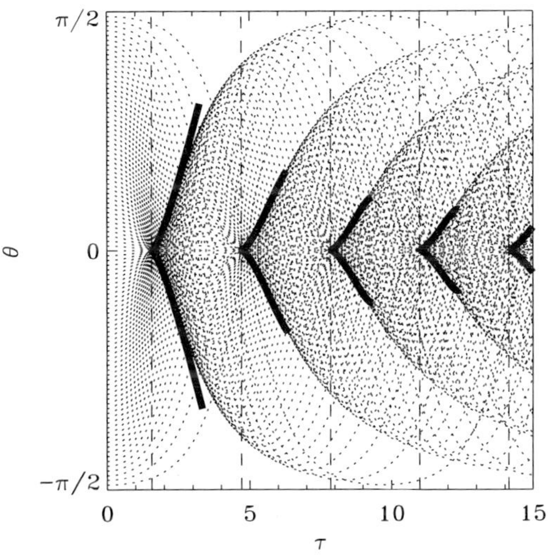

Fig. 8. Superposition of the caustics (thick full lines) over the characteristics (dotted lines) of particles that are initially evenly distributed between $-\pi/2$ and $\pi/2$.

of particles in the double well periodic potential $V(\theta, \tau) = -(\cos 2\theta)/4$. This potential is of mean field origin, since it represents the effect of all interacting particles. Therefore, the particles will have an oscillatory motion in one of the two wells. One understands thus that particles starting close to the minimum will collapse at the same time, whereas a particle starting farther will have a larger oscillation period. This is what is shown in Fig. 8 where trajectories are presented for different starting positions. One sees that the period of recurrence of the chevrons corresponds to half the period of oscillations of the particles close to the minimum; this fact is a direct consequence of the isochronism of the approximate harmonic potential close to the minimum. The description could even go one step further by computing the caustics, corresponding to the envelops of the characteristics. They are shown in Fig. 8 and testify the striking agreement between this description and the real trajectories: the chevrons of Fig. 6 correspond to the caustics.

4.3 Thermodynamical Predictions Versus Dynamical Stabilization

If the above description is shown to be particularly accurate, it does not explain why this state is thermodynamically preferred over others. Indeed, as shown before, thermodynamics predicts that the only equilibrium state is homogeneous. This result has been discussed in Sect. 3.1, where we have proven that magnetization is zero at all energies for the antiferromagnetic model and this has been also confirmed by Monte Carlo simulations [29]. However, since MD simulations are performed at constant energy, it is important to derive analytically the most probable state in the microcanonical ensemble. Since this model does not present ensemble inequivalence, we can obtain the microcanonical results by employing the inverse Laplace transform (31) of the canonical partition function. The microcanonical solution confirms that the maximal entropy state is homogeneous on the circle. It is therefore essential to see why the bicluster state, predicted to be thermodynamically unstable is instead *dynamically stable*.

The underlying reason rely on the existence of the two very different timescales and the idea is again to average over the very fast one. Instead of using the classical asymptotic expansion on the equation of motions, it is much more appropriate to develop an adiabatic approximation which leads to an effective Hamiltonian that describes very well the long time dynamics. Doing statistical mechanics of this averaged problem, one predicts the presence of the bicluter.

The theory that we have developed relies on an application of adiabatic theory, which in the case of the HMF model is rather elaborate and needs lengthy calculations [28] that we will not present here. An alternative, but less powerful, method to derive similar results has been used in [30]. On the contrary, we would like to present a qualitative explanations of this phenomenon, using a nice (but even too simple !) analogy.

This stabilization of unstable states can be described using the analogy with the inverted pendulum, where the vertical unstable equilibrium position can be made stable by the application of a small oscillating force. One considers a rigid rod free to rotate in a vertical plane and whose point-of-support is vibrated vertically as shown by Fig. 9a. If the support oscillates vertically above a certain frequency, one discovers the remarkable property that the vertical position with the center of mass above its support point is *stable* (Fig. 9b). This problem, discussed initially by Kapitza [32], has strong similarities with the present problem and allows a very simplified presentation of the averaging technique we have used.

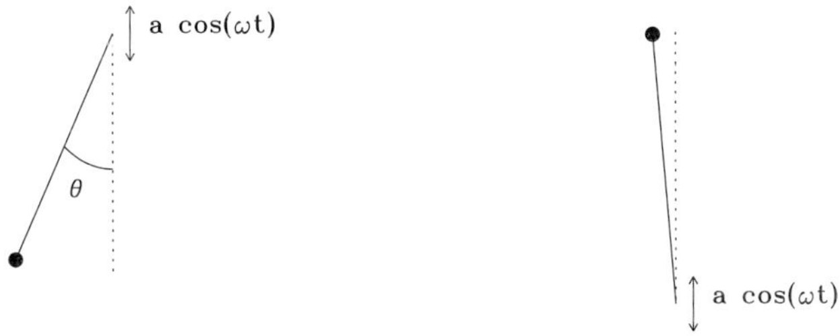

Fig. 9. Schematic picture of the inverted pendulum.

The equation of motion of the vibrating pendulum is

$$\frac{d^2\theta}{dt^2} + \left(\frac{\omega_0^2}{\omega^2} - a\cos t\right)\sin\theta = 0\ , \tag{34}$$

where ω_0 is the proper linear frequency of the pendulum, ω the driving frequency of the support and a the amplitude of excitation. Introducing a small parameter $\varepsilon = \omega_0/\omega$, one sees that (34) derives from the Lagrangian

$$\mathcal{L} = \frac{1}{2}\left(\frac{d\theta}{dt}\right)^2 + \left(\varepsilon^2 - a\cos t\right)\cos\theta \quad . \tag{35}$$

Here the two frequencies ω and ω_0 define two different time scales, in close analogy with the HMF model. Using the small parameter to renormalize the amplitude of the excitation as $A = a/\varepsilon$, and choosing the ansatz $\theta = \theta_0(\tau) + \varepsilon\,\theta_1(t,\tau)$ where $\tau = \varepsilon t$, the Lagrangian equations for the function θ_1, leads to the solution $\theta_1 = -A\cos t\sin\theta_0(\tau)$. This result not only simplifies the above ansatz, but more importantly suggests to average the Lagrangian on the fast variable t to obtain an effective Lagrangian $\mathcal{L}_{eff} = \langle\mathcal{L}\rangle_t = \frac{1}{2}\left(\frac{d\theta_0}{d\tau}\right)^2 - V_{eff}$, where the averaged potential is found to be

$$V_{eff} = -\cos\theta_0 - \frac{A^2}{8}\cos 2\theta_0 + \mathrm{Cste} \quad . \tag{36}$$

It is now straightforward to show that the inverted position would be stable if $A^2 > 2$, i.e. if $a\omega > \sqrt{2}\omega_0$. As the excitation amplitude a is usually small, this condition emphasizes that the two time scales should be clearly different, for the inverted position to be stable.

The procedure for the HMF model is analogous, but of course it implies a series of tedious calculations. Since we would like to limit here to a pedagogical presentation, we will skip such details that can be found in [28]. It is however important to emphasize that the potential energy in the HMF model is self-consistently determined and depend on the position of all particles. The magic and the beauty is that, even if this is the potential energy of N particles, it is possible to compute the statistical mechanics of the new effective Hamiltonian, derived directly from the effective Lagrangian via the Legendre transform. The main result is that the out-of-equilibrium state, (i.e. the bicluster shown in Fig. 6) corresponds to a statistical equilibrium of the effective mean-field dynamics. No external drive is present in this case, as for the inverted pendulum, but the time dependence of the mean field plays the role of the external drive.

The HMF model represents presumably the simplest N-body system where out-of-equilibrium dynamically stabilized states can be observed and explained in detail. However, we believe that several systems with long-range interactions should exhibit behaviours similar to the ones we have observed here. Moreover, this model represents a paradigmatic example for other systems exhibiting non-linear interactions of rapid oscillations and a slower global motion. One of this is the piston problem [33]: averaging techniques have been applied to the fast motion of gas particles in a piston which itself has a slow motion [34]. Examples can also be found in applied physics as for instance wave-particles interaction in plasma physics [35], or the interaction of fast inertia gravity waves with the vortical motion for the rotating Shallow Water model [36].

5 Dynamical Properties II: Lyapunov Exponents

In this section, we discuss the chaotic features of the microscopic dynamics of the HMF model. We mainly concentrate on the $A = 0$ case, presenting in detail the behavior of the Lyapunov exponents and the Kolmogorov-Sinai entropy both for ferromagnetic and antiferromagnetic interactions. We also briefly discuss some peculiar mechanisms of chaos in the $A \neq 0$ case. The original motivation for the study of the chaotic properties of the HMF was to investigate the relation between phase transitions, which are macroscopic phenomena, and microscopic dynamics (see [37] for a review) with the purpose of finding dynamical signatures of phase transitions [38]. Moreover, we wanted to check the scaling properties with the number of particles of the Lyapunov spectrum [14] in the presence of long-range interactions.

5.1 The $A = 0$ Case

In the $A = 0$ case (see formulae (6)-(19)), the Hamiltonian equations of motion are

$$\dot{\theta}_i = p_i \tag{37}$$

$$\dot{p}_i = -M \sin(\theta_i - \phi) \quad . \tag{38}$$

The Largest Lyapunov Exponent (LLE) is defined as the limit

$$\lambda_1 = \lim_{t \to \infty} \frac{1}{t} \ln \frac{d(t)}{d(0)} \tag{39}$$

where $d(t) = \sqrt{\sum_{i=1}^{N} (\delta\theta_i)^2 + (\delta p_i)^2}$ is the Euclidean norm of the infinitesimal disturbance at time t. Therefore, in order to obtain the time evolution of $d(t)$, one has to integrate also the linearized equations of motion along the reference orbit

$$\frac{d}{dt}\delta\theta_i = \delta p_i \ , \ \frac{d}{dt}\delta p_i \ = -\sum_j \frac{\partial^2 V}{\partial\theta_i\partial\theta_j}\delta q_j \ , \tag{40}$$

where the diagonal and off-diagonal terms of the Hessian are

$$\frac{\partial^2 V}{\partial\theta_i^2} = M\cos(\theta_i - \Phi) - \frac{1}{N} \tag{41}$$

$$\frac{\partial^2 V}{\partial\theta_i\partial\theta_j} = -\frac{1}{N}\cos(\theta_i - \theta_j) \quad , \ i \neq j \ . \tag{42}$$

To calculate the largest Lyapunov exponent we have used the standard method by Benettin et al [39]. In Fig. (10), we report the results obtained for four different sizes of the system (ranging from $N = 100$ to $N = 20,000$).

In panel (a), we plot the largest Lyapunov exponent as a function of U. As expected, λ_1 vanishes in the limit of very small and very large energies, where the system is quasi-integrable. Indeed, the Hamiltonian reduces to weakly coupled harmonic oscillators in the former case or to free rotators in the latter. For $U < 0.2$, λ_1 is small and has no N-dependence. Then it changes abruptly and a region of "strong chaos" begins. It was observed [8] that between $U = 0.2$ and $U = 0.3$, a different dynamical regime sets in and particles start to evaporate from the main cluster, in analogy with what was reported in other models [38,40]. In the region of strong chaoticity, we observe a pronounced peak already for $N = 100$ [41]. The peak persists and becomes broader for $N = 20,000$. The location of the peak is slightly below the critical energy and depends weakly on N.

In panel (b), we report the standard deviation of the kinetic energy per particle Σ computed from

$$\Sigma = \frac{\sigma_K}{\sqrt{N}} = \frac{\sqrt{\langle K^2\rangle - \langle K\rangle^2}}{\sqrt{N}} \quad , \tag{43}$$

where $\langle\bullet\rangle$ indicates the time average. The theoretical prediction for Σ, which is also reported in Fig. 10, is [42,43]

$$\Sigma = \frac{T}{\sqrt{2}}\sqrt{1 - \left[1 - 2M\left(\frac{dM}{dT}\right)\right]^{-1}} \quad , \tag{44}$$

where $M(T)$ is computed in the canonical ensemble. Finite size effects are also present for the kinetic energy fluctuations, especially for $U > U_c$, but in general there is a good agreement with the theoretical formula, although the experimental points in Fig. 10b lies systematically below it. The figure emphasizes that the behavior of the Lyapunov exponent is strikingly correlated with Σ: in correspondence to the peak in the LLE, we observe also a sharp maximum of the kinetic energy fluctuations. The relation between the chaotic properties

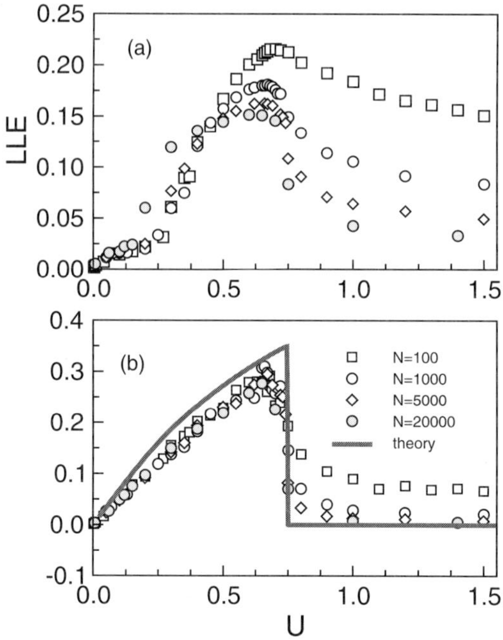

Fig. 10. Largest Lyapunov exponent LLE and kinetic energy fluctuations $\Sigma = \sigma_K/\sqrt{N}$ as a function of U in the $A = 0$ ferromagnetic case for different N sizes. The theoretical curve is shown as a full line, see text.

and the thermodynamics of the system, namely the presence of a critical point, can be made even more quantitative. An analytical formula, relating (in the $A = 0$ model) the LLE to the second order phase transition undergone by the system, has been obtained [42] by means of the geometrical approach developed in Refs. [44,37]. Using a reformulation of Hamiltonian dynamics in the language of Riemannian geometry, they have found a general analytical expression for the LLE of a Hamiltonian many-body system in terms of two quantities: the average Ω_0 and the variance σ_Ω of the Ricci curvature $\kappa_R = \Delta V/N = 1/N \sum_{i=1}^N \frac{\partial^2 V}{\partial q_i^2}$, where V is the potential energy and q_i are the coordinates of the system. Since in the particular case of the HMF model, we have

$$\frac{1}{N}\sum_{i=1}^{N}\frac{\partial^2 V}{\partial \theta_i^2} = M^2 - \frac{1}{N} = \frac{2K}{N} + 1 - 2U - \frac{1}{N}\ , \qquad (45)$$

the two quantities Ω_0 and σ_Ω can be expressed in terms of average values and fluctuations either of M^2 or of the kinetic energy K

$$\begin{aligned} \Omega_0 &= \langle M^2 \rangle - \frac{1}{N} = \frac{2}{N}\langle K \rangle + (1-2U) - \frac{1}{N} \\ \sigma_\Omega^2 &= N\sigma_{M^2}^2 = \frac{4}{N}\sigma_K^2\ . \end{aligned} \qquad (46)$$

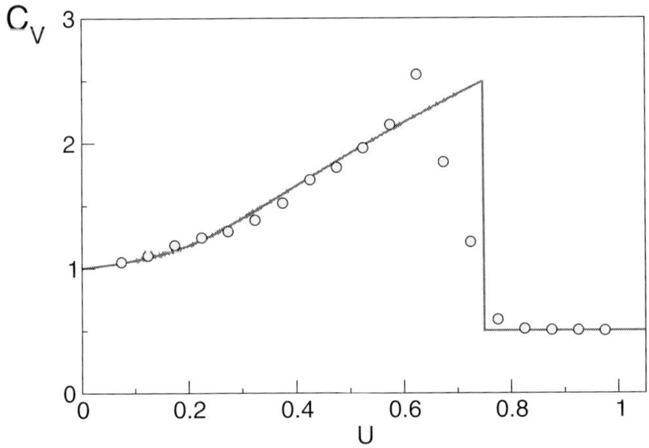

Fig. 11. Specific heat as a function of U in the $A = 0$ ferromagnetic case. The numerical simulation at equilibrium for a system with $N = 500$ is compared with the expected theoretical result (48).

The formula obtained by Firpo [42] relates the LLE, a characteristic dynamical quantity, to thermodynamical quantities like $\langle M^2 \rangle$ and σ_{M^2}, or $\langle K \rangle$ and σ_K, which characterize the macroscopic phase transition. For moderately small values of U, an approximation of the formula gives

$$\lambda \propto \frac{\sigma_K}{\sqrt{N}} = \Sigma \ . \tag{47}$$

This is in agreement with the proportionality between LLE and fluctuations of the kinetic energy found numerically in Fig. 10. This implies also a connection between the LLE and the specific heat, another quantity which is directly related to the kinetic energy fluctuation. In fact the specific heat can be obtained from Σ by means of the Lebowitz-Percus-Verlet formula [45]

$$C_V = \frac{1}{2}\left[1 - 2\left(\frac{\Sigma}{T}\right)^2\right]^{-1} \ . \tag{48}$$

In Fig. 11, we report the numerical results for the specific heat as a function of U for a system made of $N = 500$ particles, and we compare them with the theoretical estimate.

In the HMF model and for a rather moderate size of the systems, it is possible to calculate not only the LLE but all the Lyapunov exponents, and from them the Kolmogorov-Sinai entropy. We give first a succinct definition of the spectrum of Lyapunov exponents (for more details see [7]). Once the 2N-dimensional tangent vector $\mathbf{z} = (\delta\theta_1, \cdots, \delta\theta_N, \delta p_1, \cdots, \delta p_N)$ is defined, with its dynamics given by Eqs. (40), one can formally integrate the motion in tangent space up to time t, since the equations are linear,

$$\mathbf{z}(t) = J^t \mathbf{z}(0) \ , \tag{49}$$

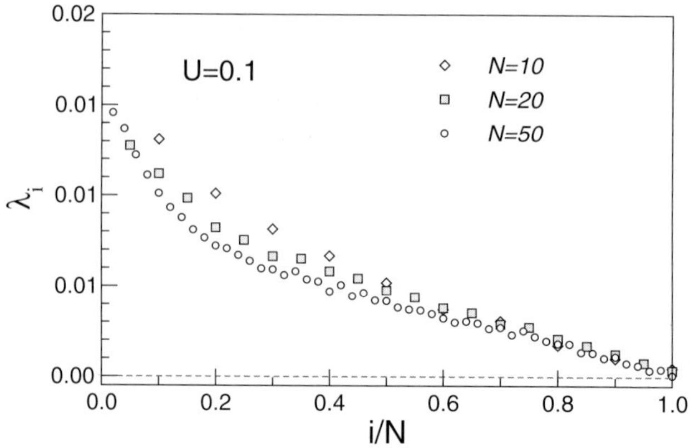

Fig. 12. Scaling of the positive part of the spectrum of Lyapunov exponents in the A=0 ferromagnetic case for $U = 0.1$.

where J^t is a $2N \times 2N$ matrix that depends on time through the orbit $\theta_i(t), p_i(t)$. The first k exponents of the spectrum $\lambda_1, \ldots, \lambda_k$, which are ordered from the maximal to the minimal, are then given by

$$(\lambda_1 + \cdots + \lambda_k) = \lim_{t\to\infty} \frac{1}{2t} \ln \mathrm{Tr} J_k^t (J_k^t)^* \quad , \tag{50}$$

where J_k^t is the matrix ($(J_k^t)^*$ its transpose) induced by J^t that acts on the exterior product of k vectors in the tangent space $\mathbf{z}_1 \wedge \cdots \wedge \mathbf{z}_k$. The spectrum extends up to $k = 2N$ and in our Hamiltonian system obeys the pairing rule

$$\lambda_i = -\lambda_{2N+1-i} \quad \text{for} \quad 1 \le i \le 2N \; . \tag{51}$$

The numerical evaluation of the spectrum of the Lyapunov exponents is a heavy computational task, in particular for the necessity to perform Gram-Schmidt orthonormalizations of the Lyapunov eigenvectors in order to maintain them mutually orthogonal during the time evolution. We have been able to compute the complete Lyapunov spectrum for system sizes up to $N = 100$. In Fig. 12, we report the positive part of the spectrum for different system sizes and an energy $U = 0.1$ inside the weakly chaotic region. The negative part of the spectrum is symmetric due to the pairing rule (51). The limit distribution $\lambda(x)$, suggested for short range interactions,

$$\lambda(x) = \lim_{N\to\infty} \lambda_{xN}(N) \; , \tag{52}$$

that is obtained by plotting λ_i vs. i/N and letting N going to infinity, is found also here for the N values that we have been able to explore. At higher energies, this scaling is not valid and a size-dependence is present [43].

The Kolmogorov-Sinai (K-S) entropy is, according to Pesin's formula [7], the sum of the positive Lyapunov exponents. In Fig. 13, we plot the entropy density S_{KS}/N as a function of U for different systems sizes.

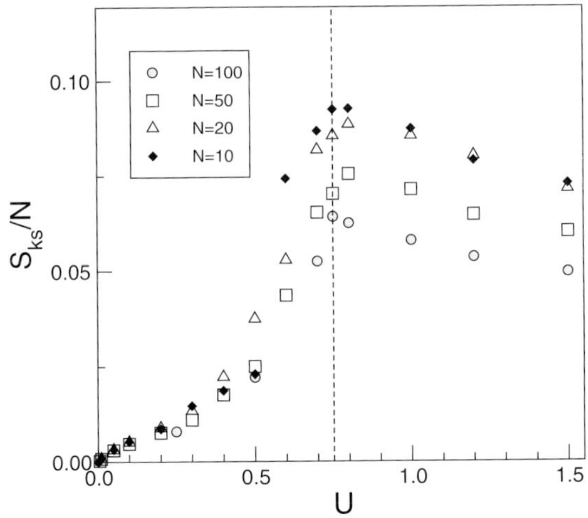

Fig. 13. S_{KS}/N as a function of U in the $A = 0$ ferromagnetic case.Numerical calculations for different systems sizes ranging from $N = 10$ to $N = 100$ are shown. The dashed line indicates the critical energy.

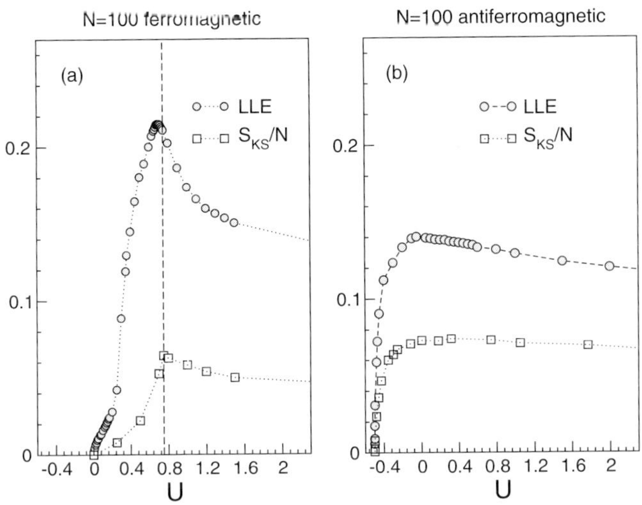

Fig. 14. LLE and S_{KS}/N as a function of U in the $A = 0$ ferromagnetic and antiferromagnetic cases for N=100.

As for the LLE, S_{KS}/N shows a peak near the critical energy, a fast convergence to a limiting value as N increases in the small energy limit, and a slow convergence to zero for $U \geq U_c$. A comparison of the ferromagnetic and antiferromagnetic cases is reported in Fig. 14. Here, for N=100, we plot as a function of U the LLE and the Kolmogorov-Sinai entropy per particle S_{KS}/N. In both the ferromagnetic and antiferromagnetic cases, the system is integrable in the limits of small and large energies. The main difference between the fer-

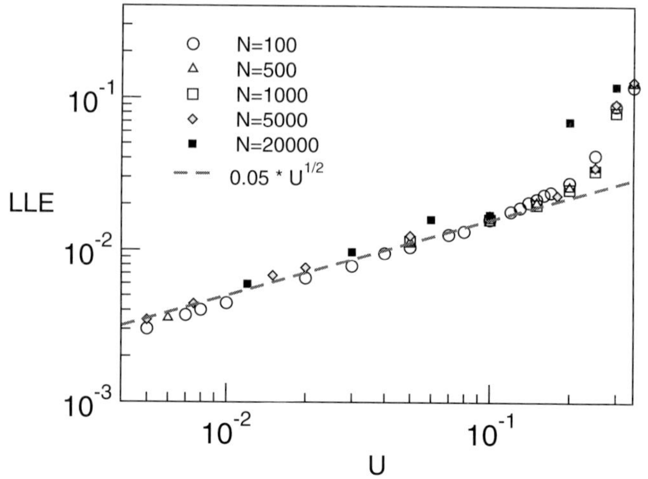

Fig. 15. Scaling properties of the LLE at low energies in the $A = 0$ ferromagnetic case. No N-dependence is observed for $U < 0.2$. The dashed line indicates a power-law $U^{1/2}$.

romagnetic and the antiferromagnetic model appears at intermediate energies. In fact, although both cases are chaotic (LLE and S_{KS}/N are positive), in the ferromagnetic system one observes a well defined peak just below the critical energy, because the dynamics feels the presence of the phase transition. On the other hand, a smoother curve is observed in the antiferromagnetic case. In the low energy regime, it is possible to work out [12] a simple estimate $\lambda_1 \propto \sqrt{U}$, which is fully confirmed for the ferromagnetic case in Fig. 15 for different system sizes in the range $N = 100, 20000$. The same scaling law is also valid in the antiferromagnetic case [46] and for the $A = 1$ [9] HMF model.

At variance with the N-independent behavior observed at small energy, strong finite size effects are present above the critical energy in the ferromagnetic case and for all energies for the antiferromagnetic case. In Fig. 16a, we show that the LLE is positive and N-independent below the transition (see the values $U = 0.4, 0.5$), while it goes to zero with N above. We also report in the same figure a calculation of the LLE using a random distribution of particle positions θ_i on the circle in (40) for the tangent vector. The agreement between the deterministic estimate and this random matrix calculation is very good. The LLE scales as $N^{-\frac{1}{3}}$, as indicated by the fit reported in the figure. This agreement can be explained by means of an analytical result obtained for the LLE of product of random matrices [47]. If the elements of the symplectic random matrix have zero mean, the LLE scales with the power 2/3 of the perturbation. In our case, the latter condition is satisfied and the perturbation is the magnetization M. Since M scales as $N^{-\frac{1}{2}}$, we get the right scaling of λ_1 with N. This proves that the system is integrable for $U \geq U_c$ as $N \to \infty$. This result is also confirmed by the analytical calculations of [42] and, more recently, of [51].

In the antiferromagnetic case, the LLE goes to zero with system size as $N^{-\frac{1}{3}}$ for all values of U.

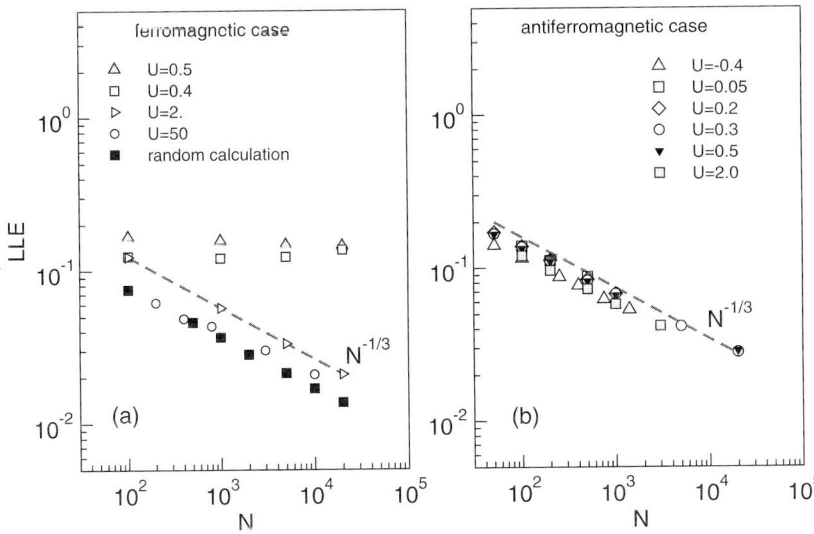

Fig. 16. Scaling of the LLE vs N for the $A = 0$ ferromagnetic and antiferromagnetic cases at various energies, see text. A power-law decreasing as $N^{-1/3}$ of the LLE is observed for overcritical energies in the ferromagnetic case and for all energies in the antiferromagnetic one. See text for further details.

Interesting scaling laws have also been found for the Kolmogorov-Sinai entropy in the ferromagnetic case: at small energies $S_{KS}/N \propto U^{3/4}$ with no size dependence, and $S_{KS}/N \propto N^{-1/5}$ for overcritical energy densities. The latter behavior has been found also in other models [48]. Concluding this section we would like to stress that the finite value of chaotic measures close to the critical point is strongly related to kinetic energy fluctuations and can be considered as a *microscopic dynamical indication* of the macroscopic equilibrium phase transition. This connection has been found also in other models and seems to be quite general [37,38,49,50]

The behavior of the HMF model as a function of the range of the interaction [13,4,6] and the dynamical features before equilibration [25,26] is discussed in a separate chapter of this volume in connection with Tsallis nonextensive thermodynamics [11].

5.2 Mechanisms of Chaos in the $A \neq 0$ Case

For $A = 0$, the origin of chaos is related to the non time-dependence of (19), since it is obvious that if the phase ϕ and the magnetization M would become constant the dynamics of the system will reduce to that of an integrable system. There are indeed preliminary indications [17] that in the mean-field limit $N \to \infty$, M and ϕ will become constant and $\lambda \to 0$. It should be noticed that this is true if the mean-field limit is taken before the limit $t \to \infty$ in the definition of the maximal Lyapunov exponent, and numerical indications were reported in [26]. When $A > 0$ we expect a quite different situation: indeed, even assuming that in

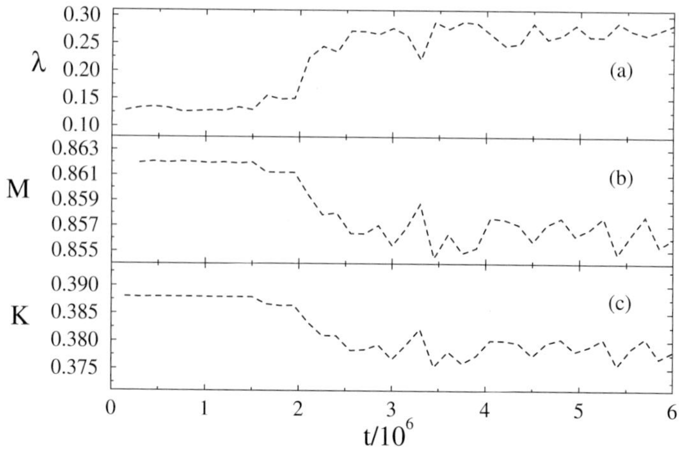

Fig. 17. Time evolution of the Largest Lyapunov Exponent (LLE) λ_1, of the magnetization M, and of the kinetic energy K are shown for a clustered initial condition for the model with $A = 1$ and $\varepsilon = 1$ at $U = 0.87$ and with $N = 200$.

the mean-field limit M and P and their respective phases will become constant, the dynamics will eventually take place in a 4-dimensional phase space and chaos can in principle be observed.

As already shown in [9], for $A = 1$ and $\varepsilon = 1$, two different mechanisms of chaos are present in the system for $U < U_c$: one acting on the particles trapped in the potential and another one felt by the particles moving in proximity of the separatrix. This second mechanism is well known and is related to the presence of a chaotic layer situated around the separatrix. The origin of the first mechanism is less clear, but presumably related to the erratic motion of the minimum of the potential well, i.e. to the time-dependent character of the equations ruling the dynamics of the single particle. Indications in this direction can be found by performing the following numerical experiment. Let us prepare a system with $N = 200$ and $U = 0.87$ (the critical energy is in this case $U_c \sim 2$) with a Maxwellian velocity distribution and with all particles in a single cluster.

For an integration time $t < 2 \times 10^6$, the Lyapunov exponent has a value $\lambda \simeq 0.13$. But when at time $t \sim 2 \times 10^6$, one particle escapes from the cluster, its value almost doubles (see Fig. 17). The escaping of the particle from the cluster is associated to a decrease of the magnetization M and of the kinetic energy K. This last effect is related to the negative specific heat regime: the potential energy V_A is minimal when all the particles are trapped, if one escapes then V_A increases and due to the energy conservation K decreases. As a matter of fact, we can identify a "strong" chaos felt from the particles approaching the separatrix and a "weak" chaos associated to the orbits trapped in the potential well. We believe that the latter mechanism of chaotization should disappear (in analogy with the $A = 0$ case) when the mean-field limit is taken before the $t \to \infty$ limit. Therefore we expect that for $N \to \infty$ the only source of chaotic behaviour should be related to the chaotic sea located around the separatrix. As already noticed in [9], the degree of chaotization of a given system depends strongly on the initial condition (in particular in the mean-field limit). In the latter limit, for initial condition prepared in a clustered configuration, we expect that $\lambda = 0$, until one particle will escape from the cluster.

6 Conclusions

We have discussed the dynamical properties of the Hamiltonian Mean Field model in connection with its thermodynamics. This apparently simple class of models has revealed a very rich and interesting variety of behaviours. Inequivalence of ensembles, negative specific heat, metastable dynamical states and chaotic dynamics are only some among them. During the past years these models have been of great help in understanding the connection between dynamics and thermodynamics when long-range interactions are present. Such kind of investigation is of extreme importance for self-gravitating systems and plasmas, but also for phase transitions in finite systems, such as atomic clusters or nuclei, and for the foundation of statistical mechanics. Several progresses have been done during these years. This contribution, although not exhaustive, is an effort to summarize some of the main results achieved so far. We believe that the problems which are still not understood will be hopefully clarified in the near future within a general theoretical framework. We list three important open questions that we believe can be reasonably addressed: the full characterization of the out-of-equilibrium states close to phase transitions; the clarification of the scaling laws of the maximal Lyapunov exponent and of the Lyapunov spectrum; the study of the single-particle diffusive motion [24] in the various non-equilibrium and equilibrium regimes.

Acknowledgements

We would like to warmly thank our collaborators Mickael Antoni, Julien Barré, Freddy Bouchet, Marie-Christine Firpo, François Leyvraz and Constantino Tsallis for fruitful interactions. One of us (A.T.) would also thank Prof. Ing. P. Miraglino for giving him the opportunity to complete this paper. This work has been partially supported by the EU contract No. HPRN-CT-1999-00163 (LOCNET network), the French Ministère de la Recherche grant ACI jeune chercheur-2001 N° 21-311. This work is also part of the contract COFIN00 on *Chaos and localization in classical and quantum mechanics*.

References

1. T. Padhmanaban, *Statistical Mechanics of gravitating systems in static and cosmological backgrounds*, in "Dynamics and Thermodynamics of Systems with Long-Range Interactions", T. Dauxois, S. Ruffo, E. Arimondo, M. Wilkens Eds., Lecture Notes in Physics Vol. 602, Springer (2002), (in this volume)

2. P.-H. Chavanis, *Statistical mechanics of two-dimensional vortices and stellar systems*, in "Dynamics and Thermodynamics of Systems with Long-Range Interactions", T. Dauxois, S. Ruffo, E. Arimondo, M. Wilkens Eds., Lecture Notes in Physics Vol. 602, Springer (2002), (in this volume)
3. Elskens, *Kinetic theory for plasmas and wave-particle hamiltonian dynamics*, in "Dynamics and Thermodynamics of Systems with Long-Range Interactions", T. Dauxois, S. Ruffo, E. Arimondo, M. Wilkens Eds., Lecture Notes in Physics Vol. 602, Springer (2002), (in this volume)
4. F. Tamarit and C. Anteneodo, Physical Review Letters **84**, 208 (2000)
5. J. Barré, D. Mukamel, S. Ruffo, *Ensemble inequivalence in mean-field models of magnetism*, in "Dynamics and Thermodynamics of Systems with Long-Range Interactions", T. Dauxois, S. Ruffo, E. Arimondo, M. Wilkens Eds., Lecture Notes in Physics Vol. 602, Springer (2002), (in this volume)
6. A. Campa, A. Giansanti, D. Moroni, Physical Review E **62**, 303 (2000) and Chaos Solitons and Fractals **13**, 407 (2002)
7. J.-P. Eckmann, D. Ruelle, Review of Modern Physics **57**, 615 (1985)
8. S. Ruffo, in *Transport and Plasma Physics*, edited by S. Benkadda, Y. Elskens and F. Doveil (World Scientific, Singapore, 1994), pp. 114-119; M. Antoni, S. Ruffo, Physical Review E **52**, 2361 (1995)
9. M. Antoni and A. Torcini, Physical Review E **57**, R6233 (1998); A. Torcini and M. Antoni, Physical Review E **59**, 2746 (1999)
10. M. Antoni, S. Ruffo, A. Torcini: Physical Review E, **66**, 025103 (2002)
11. C. Tsallis, A. Rapisarda, V. Latora and F. Baldovin *Nonextensivity: from low-dimensional maps to Hamiltonian systems*, in "Dynamics and Thermodynamics of Systems with Long-Range Interactions", T. Dauxois, S. Ruffo, E. Arimondo, M. Wilkens Eds., Lecture Notes in Physics Vol. 602, Springer (2002), (in this volume)
12. V. Latora, A. Rapisarda and S. Ruffo, Physical Review Letters **80**, 692 (1998)
13. C. Anteneodo and C. Tsallis, Physical Review Letters **80**, 5313 (1998)
14. R. Livi, A. Politi, S. Ruffo, Journal of Physics A **19**, 2033 (1986)
15. D.H. Gross, *Thermo-Statistics or Topology of the Microcanonical Entropy Surface*, in "Dynamics and Thermodynamics of Systems with Long-Range Interactions", T. Dauxois, S. Ruffo, E. Arimondo, M. Wilkens Eds., Lecture Notes in Physics Vol. 602, Springer (2002), (in this volume)
16. M. Kac, G. Uhlenbeck and P.C. Hemmer, Journal Mathematical Physics (N.Y.) **4**, 216 (1963); **4**, 229 (1963); **5**, 60 (1964)
17. S. Tanase-Nicola and J. Kurchan, Private Communication (2002)
18. J. Barré and F. Bouchet, unpublished (2002)
19. J. Barré, PHD thesis, ENS Lyon, unpublished (2002)
20. D.H.E. Gross, *Microcanonical thermodynamics, Phase transitions in "small" systems* (World Scientific, Singapore, 2000)
21. H. Yoshida, Physics Letters A **150**, 262 (1990); R. I. McLachlan and P. Atela, Nonlinearity **5**, 541 (1992)
22. P. Hertel and W. Thirring, Annals of Physics **63**, 520 (1971)
23. A. Compagner, C. Bruin, A. Roelse, Physical Review A **39**, 5989 (1989); H.A. Posch, H. Narnhofer, W. Thirring, Physical Review A **42**, 1880 (1990)
24. V. Latora, A. Rapisarda and S. Ruffo, Physical Review Letters **83**, 2104 (1999)
25. V. Latora, A. Rapisarda and C. Tsallis, Physical Review E **64**, 056124-1 (2001)
26. V. Latora, A. Rapisarda and C. Tsallis, Physica A **305**, 129 (2002)
27. D. Del Castillo-Negrete, *Dynamics and self-consistent chaos in a mean field Hamiltonian model*, in "Dynamics and Thermodynamics of Systems with Long-Range

Interactions", T. Dauxois, S. Ruffo, E. Arimondo, M. Wilkens Eds., Lecture Notes in Physics Vol. 602, Springer (2002), (in this volume)
28. J. Barré, F. Bouchet, T. Dauxois, S. Ruffo, *Birth and long-time stabilization of out-of-equilibrium coherent structures*, European Physical Journal B, to appear (2002) and Physical Review Letters **89**, 110601 (2002)
29. T. Dauxois, P. Holdsworth, S. Ruffo, European Physical Journal B **16**, 659 (2000)
30. M.C. Firpo, F. Leyvraz and S. Ruffo, Journal of Physics A, **35**, 4413 (2002)
31. W. Braun, K. Hepp, Communications in Mathematical Physics **56**, 101 (1977)
32. P. L. Kapitza, in *Collected Papers of P. L. Kapitza*, edited by D. Ter Harr (Pergamon, London, 1965), pp 714, 726
33. J. L. Lebowitz, J. Piasecki, Ya. Sinai "Scaling dynamics of a massive piston in an ideal gas" , in Hard ball systems and the Lorentz gas, 217-227, Encycl. Math. Sci., 101, Springer, Berlin (2000)
34. Ya. Sinai, Theoretical and Mathematical Physics (in Russian), **121**, 110 (1999)
35. M. Antoni, Y. Elskens, D. F. Escande, Physics of Plasmas **5**, 841 (1998); M-C. Firpo, *Etude dynamique et statistique de l'interaction onde-particule*, PhD Thesis, Université de Marseille (1999)
36. P. F. Embid, A. J. Majda, Com. in Partial Differential Equations **21**, 619 (1996)
37. L. Casetti, M. Pettini, E.G.D. Cohen, Physics Reports **337**, 237 (2000)
38. A. Bonasera, V. Latora and A. Rapisarda, Physical Review Letters **75**, 3434 (1995)
39. G. Benettin, L. Galgani and J.M. Strelcyn, Physical Review A **14**, 2338 (1976)
40. S.K. Nayak, R.Ramaswamy and C. Chakravarty, Physical Review E **51**, 3376 (1995); V. Mehra, R. Ramaswamy, Physical Review E **56**, 2508 (1997)
41. Y.Y. Yamaguchi, Progress of Theoretical Physics **95**, 717 (1996)
42. M-C. Firpo, Physical Review E **57**, 6599 (1998)
43. V. Latora, A. Rapisarda and S. Ruffo, Physica D **131**, 38 (1999); Physica A **280**, 81 (2000)
44. L. Casetti, C. Clementi and M. Pettini, Physical Review E **54**, 5969 (1996); L. Caiani, L. Casetti, C. Clementi and M. Pettini, Physical Review Letters **79**, 4361, (1997)
45. J.L. Lebowitz, J.K. Percus and L. Verlet, Physical Review A **153**, 250 (1967)
46. V. Latora, A. Rapisarda, S. Ruffo, Progress of Theoretical Physics Supplement **139**, 204 (2000)
47. G. Parisi and A. Vulpiani, Journal of Physics A **19**, L425 (1986)
48. T. Tsuchiya and N. Gouda, Physical Review E **61**, 948 (2000)
49. L. Caiani et al: Physical Review E **57**, 3886 (1998)
50. J. Barré and T. Dauxois, Europhysics Letters **55**, 164 (2001)
51. C. Anteneodo and R. Vallejos, Physical Review E **65**, 016210 (2002)

Index